Trophy Hunting

Nikolaj Bichel · Adam Hart

Trophy Hunting

Springer

Nikolaj Bichel
Wildlife Conservation Research Unit
WildCRU at Oxford University
Tubney, UK

Adam Hart
University of Gloucestershire
Cheltenham, UK

ISBN 978-981-19-9975-8   ISBN 978-981-19-9976-5 (eBook)
https://doi.org/10.1007/978-981-19-9976-5

© The Editor(s) (if applicable) and The Author(s), under exclusive license to Springer Nature Singapore Pte Ltd. 2023, corrected publication 2023

This work is subject to copyright. All rights are solely and exclusively licensed by the Publisher, whether the whole or part of the material is concerned, specifically the rights of translation, reprinting, reuse of illustrations, recitation, broadcasting, reproduction on microfilms or in any other physical way, and transmission or information storage and retrieval, electronic adaptation, computer software, or by similar or dissimilar methodology now known or hereafter developed.

The use of general descriptive names, registered names, trademarks, service marks, etc. in this publication does not imply, even in the absence of a specific statement, that such names are exempt from the relevant protective laws and regulations and therefore free for general use.

The publisher, the authors, and the editors are safe to assume that the advice and information in this book are believed to be true and accurate at the date of publication. Neither the publisher nor the authors or the editors give a warranty, expressed or implied, with respect to the material contained herein or for any errors or omissions that may have been made. The publisher remains neutral with regard to jurisdictional claims in published maps and institutional affiliations.

Cover credit: 'Horn measurements and weights of the great game of the world: being a record for the use of sportsmen and naturalists' by Rowland Ward. London, The Jungle, 1892. https://doi.org/10.5962/bhl.title.15203. Accessed at: https://www.biodiversitylibrary.org/bibliography/15203

This Springer imprint is published by the registered company Springer Nature Singapore Pte Ltd.
The registered company address is: 152 Beach Road, #21-01/04 Gateway East, Singapore 189721, Singapore

The Smallest of Trophies

I have been a fisherman all my life. I was born into it—to give you some idea, my mother's engagement present from my father was a Hardy trout rod instead of a ring. In fact, on its inaugural outing with my newly engaged parents, my mother fell into the river. My father, sensing possible catastrophe, leapt into the swirling water... and retrieved the rod! He proffers the defense that he hung my mother on a handy tree by her belt loop before diving in. History will judge, but nearly 50 years of marriage tells us she at least must have forgiven him.

The Atlantic Salmon lives in the sea and swims up the rivers to spawn. We fish for them at this stage of the journey—a job made hilariously hard by the fact that they do not feed while in the river. The salmon arrives as a stunning bar of silver and gradually turns into a flabby, colorful warrior by the time it spawns. But here's the thing; when at sea, salmon are attacked by sea lice. This has always been the case but was exacerbated by salmon farms. The lice look like tiny, flat tadpoles and they attach themselves to the salmon, usually around either the head or vent.

Sea lice die in fresh water, but their strangest quirk is that they have long tails that drop off in sections as they die, and this happens in a fairly predictable timescale. This means that you can tell how long a salmon has been in the river by examining the tail length of its sea lice. If the sea lice have long tails, it is a very fresh-run salmon. The measure is a matter of hours rather than days. In the past, most salmon would be killed but these days the emphasis is very heavily on preserving the salmon stocks—a job where, maybe counterintuitively, salmon fishermen have been right at the forefront of conservation.

Many people keep a personal record of the fish they've caught. For some, it's a meticulous record of dates, times, weather conditions, type of lure and so on (look out for references to "Garden Fly" which has long been a way of saying "worm" without future readers sucking their teeth in disapproval). Others will keep more of a scrapbook-style record with drawings, newspaper articles, photos of long dead companions—canine or human, as well as, in our case, sea lice.

I remember clearly from my childhood that buzz when a salmon was caught. The party sent me for the net as I was the fastest, or at least the keenest runner then. We would search the bank for a stone of suitable size to administer the coup de gras.

Someone would say "there's a priest in the car" (it's a type of cosh, a hammer-like object to kill the fish) but hear the thud of the stone before it could be retrieved. Then invariably another thud just to be sure. The excitement of returning to the car, laying the silver fish out on the bass to be admired by all as they trudged by. Then, on seeing the silver glint, bounced grinning back to base. The bass was a rush bag that was used to pack fish, as in those days they were rushed from the river to the train station and then on to the cities. It's said the sale of fish could pay for your holiday, although not in my lifetime.

The evening ritual was to fill in the fishing book, reflect on the day and, should a fish have been exceptionally fresh, sellotape into the book any sea lice that survived the journey from fish, via my pocket, the car, the kitchen table, and a viewing by any non-fishing and often confused house guests. I think our sellotaped records probably go back 80 years, depending on how much my grandmother put in her book. She was a note keeper though.

Best of all, these days the salmon even survive to tell the tale.

—*A. W., Scotland*

Contents

1	**Introduction**	1
	1.1 A Lion Dies	1
	1.2 Trophy Hunting as a Field of Research	4
	1.3 Defining Trophy Hunting	7
	1.4 Going Hunting	9
	References	14
2	**A History of Hunting and Hunting Perceptions**	19
	2.1 Prehistory	19
	2.2 Ancient Egypt and China	24
	2.3 Ancient Greece	26
	2.4 Ancient Rome	29
	2.5 Pre-Columbian America	31
	2.6 From the Middle Ages to the Nineteenth Century in Europe	33
	2.7 Colonial Hunting in India and Africa	41
	2.8 Theodore Roosevelt and North American Conservation	50
	2.9 The Transcendentalists	61
	2.10 Modern Animal Rights and Ecofeminism	72
	2.11 Recent Trends and Perceptions of Hunting	83
	References	87
3	**Hunting Trophies**	95
	3.1 Taxidermy	99
	3.2 Fish and Birds	107
	3.3 Skulls, Antlers, Skins, and Other Trophies	110
	3.4 Trophies and Record Books	113
	3.4.1 Rowland Ward	114
	3.4.2 Boone and Crockett	117
	3.4.3 International Council for Game and Wildlife Conservation	123
	3.4.4 Pope and Young	127
	3.4.5 Safari Club International	131

		3.4.6	Erongo Verzeichnis for African Game Animals	140
	3.5	A Critical Look at Trophy Records		141
		3.5.1	Do Record Books Honour Animals or Hunters?	142
		3.5.2	The Consequences of Record Books	145
	3.6	The Meanings of Hunting Trophies		148
		3.6.1	Dead Nature	148
		3.6.2	Achievement, Tribute, and Gratitude	156
		3.6.3	Trophies as Interior Decoration	159
	3.7	The Trophy Photo		160
	References			163
4	**Hunters and Antihunters**			**169**
	4.1	Hunter Demographics		170
	4.2	Hunting, Privilege, and Social Schisms		175
		4.2.1	Gender	176
		4.2.2	Race and Wealth	179
		4.2.3	Guns and God	180
	4.3	Hunting Motivations		184
		4.3.1	The Naturalistic Hunter—Nature	188
		4.3.2	The Utilitarian Hunter—Meat	191
		4.3.3	The Dominionistic Hunter—Sport	193
		4.3.4	Motivation Demographics and Prevalences	194
	4.4	Antihunters		197
	4.5	More on Hunting Motivations		200
		4.5.1	New Nature Hunting	200
		4.5.2	Rituals and (Male) Identity	203
		4.5.3	Getting Away	206
		4.5.4	The Conservation Justification	208
		4.5.5	Costly Signalling	210
		4.5.6	Collecting and Collections	212
		4.5.7	Record-Book Hunting	216
	References			216
5	**Trophy Hunting and Conservation**			**221**
	5.1	Hunting- and Trophy Trade Regulations		221
		5.1.1	CITES	222
		5.1.2	USA	226
		5.1.3	Europe	230
	5.2	Hunting and Conservation in Africa		232
		5.2.1	Lion Conservation	236
		5.2.2	Rural Livelihoods	241
		5.2.3	Eco-Imperialism and Living with Lions	248
	5.3	Hunting and Conservation in Eurasia and the Americas		252
		5.3.1	An Overabundance of Deer	257
		5.3.2	Problems of Overabundance	258

		5.3.3	The Role of Hunting in Deer Management and the Problems with Fair Chase	264
	5.4	Trophies and Genetics		270
		5.4.1	Trophy Deer Management and Deer Farming	270
		5.4.2	Evolution in Reverse	272
	5.5	Population Management Alternatives		274
	5.6	Photo Safari and Photographic Hunting		277
	5.7	Poaching		280
	References			285
6	**Fair Chase and Sport Hunting**			297
	6.1	Technological Handicaps		299
	6.2	Behavioural Handicaps		302
	6.3	Fair Chase and Animal Welfare		303
	6.4	Canned Hunting		310
	References			319
7	**Anthropomorphism and "Viralability"**			323
	7.1	#SaveMarius		323
		7.1.1	Individuality	325
		7.1.2	Empathy	328
		7.1.3	Innocence and Defencelessness	334
	7.2	Anthropomorphism, Empathy, Conservation, and Trophy Hunting		336
	References			339
8	**The Morality of Trophy Hunting**			341
	8.1	The Vices of Trophy Hunters		342
		8.1.1	Trophy Hunter Antipathy on Twitter	347
	8.2	Weighing Character Against Consequences		351
	References			354
9	**Trophy Hunting Now**			357
	9.1	Covid-19		357
	9.2	Post-Truth		359
	References			370
Correction to: Trophy Hunting and Conservation				C1

Abbreviations

ART	Age-Related Measuring System
B&C	The Boone and Crockett Club
B&C Records	Boone and Crockett Records of North American Big Game
BVC	Bubye Valley Conservancy
CAMPFIRE	Communal Areas Management Programme for Indigenous Resources
CBNRM	Community-Based Natural Resource Management
CIC	International Council for Game and Wildlife Conservation (Conseil International de la Chasse et de la Conservation du Gibier)
CITES	Convention on International Trade in Endangered Species of Wild Fauna and Flora
CoP	Conference of the Parties
DVC	Deer-Vehicle Collision
ESA	Endangered Species Act
EU	European Union
EV	Erongo Verzeichnis
EVE	Environmental Virtue Ethics
FACE	Federation of Associations for Hunting and Conservation of the EU
FAO	Food and Agriculture Organization of the United Nations
IUCN	International Union for Conservation of Nature
LEMIS	Law Enforcement Management Information System
NGO	Non-Governmental Organization
NRA	National Rifle Association
P&Y	Pope and Young
P&Y Records	Bowhunting Big Game Records of North America
PA	Protected Area
PH	Professional Hunter
PHASA	Professional Hunters' Association of South Africa
QDM	Quality Deer Management
RDC	Rural District Council
RW	Rowland Ward

RW Records	Rowland Ward's Records of Big Game
SAPA	South African Predator Association
SCI	Safari Club International
SCI Records	SCI Record Book of Big Game Animals
SCIF	Safari Club International Foundation
SVC	Savé Valley Conservancy
TCM	Traditional Chinese Medicine
TDM	Trophy Deer Management
UN	United Nations
UNEP	United Nations Environment Programme
UNESCO	United Nations Educational, Scientific, and Cultural Organization
USFWS	U.S. Fish & Wildlife Service
WildCRU	Wildlife Conservation Research Unit (Oxford University)
WNS	Ward's Natural Science

Chapter 1
Introduction

1.1 A Lion Dies

On the evening of the first of July, 2015, on Antoinette farm in the Gwaai Conservancy just outside Hwange National Park in Zimbabwe, a twelve-year-old male lion (*Panthera leo*) approached the carcass of a rotting bull elephant. The elephant had been killed elsewhere by a hunter who—days earlier—had removed the ivory, hide, and choice pieces of meat. What remained had been towed to Antoinette farm and was now being used as lion bait. As the lion began tearing into the elephant, he was struck and wounded by an arrow from a compound bow and disappeared into the bush. About 11 h later the hunting party caught up with the lion and it was killed with a second arrow. According to data from its tracking collar, it had only moved about 350 m away from the bait. Early news reports stated that the lion had been lured out of the Hwange National Park, but the tracking collar proved these to be false. The first news information about a 40-h gap between the wounding and the kill and about the final kill being made with a rifle were also false.

The tracking collar belonged to the Wildlife Conservation Research Unit (WildCRU) at Oxford University. WildCRU had been tracking this lion since 2009 as part of a study of lions in Hwange National Park that began in 1999. They had given the lion the ID code MAGM1, but for ease of reference in conversation they also gave him a nickname. That nickname was Cecil (Macdonald et al. 2016; Loveridge 2018: 7–9, 199–208).

WildCRU personnel lost contact with Cecil's GPS collar on July 4. On hearing that a lion had been shot in the Gwaai Conservancy, and being aware that there was no lion quota there for 2015, WildCRU raised the alarm to the Zimbabwe Parks and Wildlife Management Authority on July 7. Andrew Loveridge, one of the WildCRU researchers closest to Cecil, provides in *Lion Hearted: The Life and Death of Cecil & the Future of Africa's Iconic Cats* 2018) the best available account of the investigation that followed and the incriminating evidence of malicious intent that could be extrapolated from the movements and sudden disappearance of Cecil's tracking collar.

Cecil was a photo safari favourite, and news of his death was first spread by the tour guides who knew him well and were angered by his death. Information and speculation began circulating on forums and in local news between July 23 and 26. From July 26 to 28, mainstream media in Europe and the US picked up the Cecil story. Initial reports claimed the hunter was Spanish, but on July 27 the hunter who killed Cecil was revealed to be Walter Palmer, an American dentist from Minnesota who had allegedly paid $50,000 for the hunt. Palmer was (and still is) a passionate compound bow hunter, as evidenced by his many trophy photos circulating online. It also came to light that Palmer had been sentenced to one year of probation and required to paid around $3000 in fines in 2006 for killing a black bear forty miles from where his hunting permit was valid. He had tried to cover this infraction up by offering each of his guides $20,000 to lie about the incident.

On July 28, Palmer closed his dental practice, by then under siege by angry protesters. He wrote the following in a letter to his patients:

> In early July, I was in Zimbabwe on a bow hunting trip for big game. I hired several professional guides and they secured all proper permits. To my knowledge, everything about this trip was legal and properly handled and conducted. I had no idea that the lion I took was a known, local favorite, was collared and part of a study until the end of the hunt. I relied on the expertise of my local professional guides to ensure a legal hunt. I have not been contacted by authorities in Zimbabwe or in the U.S. about this situation, but will assist them in any inquiries they may have. Again, I deeply regret that my pursuit of an activity I love and practice responsibly and legally resulted in the taking of this lion. (Alexander et al. 2015)

Many celebrities condemned Palmer and his hunt on social media as the story emerged, but it did not truly go viral until late night talk show host Jimmy Kimmel's opening monologue on *Jimmy Kimmel Live!* on July 28. Here, Kimmel delivered a five-minute exposition of the events surrounding Cecil's death (some details of which turned out to be untrue), and a choked-up tirade against Walter Palmer. Kimmel's monologue ended with the website address www.wildcru.org on the screen, and with him encouraging his viewers to donate money to WildCRU to "show the world not all Americans are like this Jackhole" (Kimmel 2017), referencing a photo on-screen of Walter Palmer.

The WildCRU website received about 4.5 million visitors over the next 24 h, crashing the Oxford University host server. By August 4, WildCRU had received more than $750,000 in donations (Loveridge 2018: 221–223).

On July 29, online interest in Cecil's death peaked. Macdonald et al. (2016) show that searches on just that one day with the parameters "Cecil AND lion" yielded more than 11,000 hits in editorial media and almost 90,000 hits on social media across 125 languages. Between July and September, the story amassed 695,000 media hits (Macdonald et al. 2016: 4–7; Loveridge 2018: 227). Also on July 29, Theo Bronkhorst—the Professional Hunter who accompanied Walter Palmer—was charged with poaching (illegal hunting), as the Zimbabwe Parks and Wildlife Management Authority had not issued a lion quota in 2015 for the Gwaai Conservancy. The owner of Antoinette Farm where Cecil was killed, Honest Ndlovu, had accompanied Palmer and Bronkhorst and was charged with allowing an illegal hunt on his land. At this point Walter Palmer and his family had gone into hiding because of the death threats they had received. Online image searches for Walter Palmer produced many of his previous trophy photos, featuring dead bears, boars, lions, leopards, rhinos, and other species alongside a smiling Palmer, holding his compound bow.

On August 1, the Zimbabwe Parks and Wildlife Management Authority banned the hunting of lions, leopards, and elephants in the area surrounding Hwange National Park. This ban was suspended on August 10. Between August 1 and 4, Delta Airlines, United Airlines, and American Airlines all banned the transportation of lion, leopard (*Panthera pardus*), elephant (*Loxodonta africana*), and rhino (white rhino *Ceratotherium simum* and black rhino *Diceros bicornis*) trophies on their flights.[1]

While editorial media and social media mentions peaked at the end of July 2015 and new stories and posts per day declined after that date, the story continued to spread across the world throughout August. The story's spread was fastest in North America and Western Europe and slightly slower in Central- and South America. In Africa, the story spread slowly and not as much, but interest was more sustained (Macdonald et al. 2016: 7–8).

The Cecil story became something of a watershed moment in the media depiction of trophy hunting, at least in the UK media. In a study of trophy hunting coverage in the most popular UK media outlets between 2010 and 2020, Yeomans et al. (2022) found that 90% of all coverage occurred after the death of Cecil. The coverage, particularly after Cecil, tended towards being anti-trophy hunting in sentiment, with a decline in articles covering nuance and complexity following 2015. Most of the media coverage focussed on southern Africa, with that focus tightening after Cecil. The top five species mentioned were reflective of perceived charisma rather than likelihood of being hunted for a trophy, with lion taking the top spot followed by elephant (presumably African savannah elephant although the species name was rarely mentioned), rhino (presumably one of the two African species), leopard and bear (again, rarely at species level).

Keith Somerville suggests that the media reaction to Cecil was a turning point in anti-trophy hunting communication (Somerville 2019). In what he describes as "the wake of the media frenzy," debate over trophy hunting and movement towards trophy import bans was increased, and Cecil became "one of the cards that is perpetually stacked in opposition to trophy hunting." The Cecil killing clearly has great propaganda value for those who oppose trophy hunting, but Somerville goes further, categorising anti-trophy hunting campaign communications in propagandic terms. His case study, the pamphlet "Trophy Hunting and Conservation: An assessment of evidence regarding impacts and benefits of sport hunting on wildlife and habitat conservation" was distributed at a meeting in 2019 between the UK's Secretary of State for Environment, Michael Gove, and sustainable use supporters. The pamphlet is illustrative of many of the features we have observed in the post-Cecil trophy hunting debate, including cherry picking of facts, the use of misinformation, distraction, misuse of data and the mischaracterization and misidentification of experts. As

---

[1] Lions, leopards, rhinos, and elephants are known as the "Big Five" in trophy hunting circles, with cape buffalo (*Syncerus caffer*) being the fifth.

Somerville summarises: "The pamphlet is such a perfect example of card-stacking, cherry-picking and other manipulations that I will use it in the course on propaganda methods that I teach" (Somerville 2019).

Cecil the lion still features commonly in trophy hunting narratives even, at the time of writing, more than seven years after the event. A familiar rallying cry for those seeking to stop trophy hunting is "Justice for Cecil," perhaps driven in part by the fact that there was no real legal fallout for those involved in his death. After previously demanding that he be extradited, Zimbabwean authorities announced in October 2015 that Walter Palmer would not be prosecuted for a crime, as his paperwork was correct. As an accompanied hunting client, it was the professional hunter and landowner who had broken the law. Palmer is welcome to return to Zimbabwe as a tourist, but not to hunt. The charges against Theo Bronkhorst were dropped in November of 2016. However, as Loveridge reasons (2018: 226), he seems to have taken the brunt of the fall, having lost his business, hunting license, and Zimbabwe Professional Hunters and Guides Association membership. The case against Honest Ndlovu appears to have gone nowhere.

1.2 Trophy Hunting as a Field of Research

The killing of Cecil resulted in what some of Africa's wildlife researchers have called "an unprecedented global public media reaction" (Lindsey et al. 2016: 296), "the largest reaction in the history of wildlife conservation" (Macdonald et al. 2016: 2), and "a storm that would shake the foundations of the international trophy-hunting industry" (Loveridge 2018: 9).

Since 2015, France, Netherlands, and Australia have banned the import of trophies from lions and other species. Britain is, at the time of writing, considering a ban on all trophy imports while some countries now require better documentation of responsible and sustainable management before allowing trophy imports. South Africa, Africa's primary hunting destination, has recently promised to dismantle its captive lion-breeding industry and more than 40 airlines are refusing to transport hunting trophies (Carpenter and Konisky 2017). The US, which imports more than ten times as many trophies as the rest of the world combined, has banned the import of trophies from captive-bred lions (from South Africa) and changed the lion's status to "threatened" under the Endangered Species Act (ESA).

This is just some of the evidence of an intensification of global trophy hunting antipathy since Cecil was killed. The ongoing fight against trophy hunting is today spearheaded in the popular media—especially in Britain—by Eduardo Gonçalves' Campaign to Ban Trophy Hunting and its celebrity supporters—most prominently Ricky Gervais and Peter Egan. Since 2020, an even bigger challenge for the trophy hunting industry, as well as all other tourism in Africa and elsewhere, has of course been Covid-19, as here summarized by BBC News:

1.2 Trophy Hunting as a Field of Research

> A wildlife catastrophe is unfolding in Africa, according to park rangers and conservation experts. They say the closure of safari tourism, due to the coronavirus pandemic, is decimating the industry, and leading to an increase in poaching. The African tourism industry is worth almost $30 billion a year and employs almost four million people. Experts and rangers on the ground say they are seeing a surge in poaching as thousands of unemployed people dependent on the industry turn to wild animals for food. (Marshall 2020)

Despite all these challenges, trophy hunting is still practiced in a number of African countries, including South Africa, Namibia, Mozambique, Botswana, Tanzania, Zimbabwe and Zambia. Elephants, lions, rhinos, giraffes (*Giraffa*), and a few other especially charismatic species receive occasional international attention, but not as much as in the years immediately following 2015. Most African game species, including impala (*Aepyceros melampus*), greater kudu (*Tragelaphus strepsiceros*), common warthog (*Phacochoerus africanus*) and plains zebra (*Equus quagga*) are hunted without much public attention being paid to them. The lions near Hwange are also still subject to trophy hunting; Cecil's son Xanda, also wearing a tracking collar and monitored by WildCRU, was killed by a hunter in 2017, but this received much less public attention than Cecil's death.

What we have written so far has all portrayed trophy hunting as something that just takes place in Africa. This was a natural place to begin, as this is how most people have come to know of trophy hunting. Photos taken under an African sun of white hunters posing with dead African wildlife are the images that most likely come to mind when you hear of trophy hunting, but the activity and its definition is more nuanced than the Great White Hunter of Africa, Hemingwayesque stereotype. The hunting of deer, and the collection of their impressive antlers, takes place on a grand scale in Europe and the United States. Thousands of deer are killed there for every lion killed in Africa, yet deer are overabundant more often than threatened, and most people consider food a primary motivation for deer hunting. Overall, despite also having elements of "trophy hunting," deer hunting is far less controversial and deer hunting faces very different ethical and practical dilemmas than trophy hunting in Africa.

Trophy hunting can rile people up, and offend them, like very few other topics. The lack of understanding between hunting's proponents and opponents is more pronounced than ever and attempts at reasonable debate often fail. Further compounding this problem, antihunters prefer discussing individual animals, their emotions, suffering, and rights, and the act of killing or the moral character of hunters, while hunters and conservationists prefer discussing species, habitats, populations, and the consequences of killing. This does not make for meaningful conservations, much less compromises.

The gulf between hunters and antihunters is not the only challenge to negotiate. There is also a wide gulf between our western perception of African wildlife and the perceptions of those who live alongside these animals. In the West, lions and elephants are highly romanticized and anthropomorphized symbols of unspoilt, wild, and beautiful nature. To some Africans, who unbeknownst to many Westerners have to live with such species, elephants are also the pests that barge into their farms, trample their produce, and kill their friends and family. Lions are the predators that eat their livestock and attack, kill, and eat people. Lions and elephants, while loved and iconized in the West, are two of the most hated animal species among pastoralists and arable farmers in large parts of Africa (Packer 2015: 18).

The story of Cecil and the publicity it generated, the diversity of stakeholders, and all the other facets of the trophy hunting debate make it a complex but also opportune and interesting field of research. Hunting is not a recent field of study, as will be demonstrated and referenced throughout the chapters of this book. Sport hunting as training for war, a symbol of royal power over the land and its fauna, an obligatory pastime of nobility, and a masculine rite of passage has interested scholars since Ancient Greece at least (Plato circa 400 BC; Xenophon circa 400 BC; Anderson 1985; Barringer 2001; Kalof 2007; Allsen 2013). Sport hunting's role in wildlife conservation has received lots of scientific and autobiographical attention since the end of the nineteenth century (Mahoney 2009; IUCN 2009; Adams and Hamilton 2011; Hewitt 2015). Subsistence hunting in indigenous cultures, the cultural aspects of modern hunting, and the philosophy and ethics of hunting are decently covered fields of research (Loftin 1984; Causey 1989, 1996; Vitali 1990; Kellert 1996; Bronner 2004, 2008; Dizard 1994; 2003; Dickson 2009; Cahoone 2009). Most aspects of hunting have been or are being studied.

While hunting overall has a long and rich research history, trophy hunting research is more fragmented. Wildlife scientists have been very industrious in studying the conservation impacts (both positive and negative) of trophy hunting and especially lion conservation (Lindsey et al. 2006, 2007, 2012, 2016, 2017, 2020; Bauer et al. 2015, 2020; Creel et al. 2016; Dickman and Hinks 2019; Booth et al. 2020); other groups of wildlife scientists study the effect of trophy hunting on the genetics of wildlife populations (Coltman et al. 2003; Festa-Bianchet et al. 2009, 2014; Demarais and Strickland 2011; Jacobson et al. 2011; Douhard et al. 2016; Festa-Bianchet and Mysterud 2018); hunting trophies have been studied as souvenirs and collector's items (Marvin 2010, 2011; Poliquin 2012); and the ethics of trophy hunting and psychology of trophy hunters have received some attention as well (Gunn 2001; Lovelock 2015; Macdonald 2016; Batavia et al. 2018, 2020; Beattie 2020; Ghasemi 2020; Morris 2020). Overall though, there has been a concerning lack of interdisciplinarity in trophy hunting research and a lack of attempts to look at European and US trophy hunting of deer, boar, bear and other species, and African hunting tourism as different facets of the same debate. An approach that seeks to get to the bottom of trophy hunting needs to be interdisciplinary, because we cannot hope to understand trophy hunting unless we understand the trophy hunter, and we cannot hope to understand the trophy hunter unless we understand the hunting trophy. And we cannot fully understand the hunter or the trophy without understanding the history of hunting.

Trophy hunting is a divisive and complex subject with many different but also overlapping arenas of research. We need to zoom out and look at all the pieces of the puzzle. Achieving that perspective is the purpose of this book.

1.3 Defining Trophy Hunting

We have already written "trophy hunting" many times, but what do these words mean? What is trophy hunting?

A few people have asked if this book would be about people who compete in sports for trophies or men with trophy wives. To these questions, we must answer no. The key word is *hunting,* and it is hunting only in the literal, original sense of the word, and not in the sense in which one may be hunting for a date or a missing coin for a collection (though this kind of collecting is a highly relevant subject too, as we will show in Chapter 4.5.6). It is the kind of hunting that involves killing animals and, crucially, keeping a part of them as a souvenir.

Arriving at a good and operational definition of trophy hunting is challenging. One concern is that a good definition should take into consideration the common layman perception of trophy hunting—the trophy hunting stereotype—which is that of Americans and Europeans who pay large sums of money to hunt charismatic African megafauna. This stereotype has historically and colloquially been referred to as the "Great White Hunter."

Batavia et al. referred to this kind of stereotype in an article about the morality of trophy hunting, writing that

> by "trophy hunting" we refer specifically to the practice of Western (e.g., North American or European) individuals paying to hunt large mammals such as elephants (*Loxodonta Africana*) or Lions (*Panthera leo*). Only in this particular context do we consider what it means, and whether it is appropriate, for hunters to claim some part of an animal's body as a trophy. (Batavia et al. 2018: 2)

This definition is problematic for the purposes of this book, because the problem with relying *only* on the stereotypical layman understanding of trophy hunting is that the reference is, contrary to what the quoted paragraph says, not very specific, nor the context very particular. What, for example, do the authors mean by large mammals such as elephants or lions? The white-tailed deer (*Odocoileus virginianus*) is a large mammal. More than five million Americans hunt whitetail deer in the US every year, compared to only some thousand Americans in total, hunting in Africa. The typical whitetail trophy with the highest score in the *Boone and Crockett Records of North American Big Game*, a CIC gold medal red deer (*Cervus elaphus*) or a markhor (*Capra falconeri*) are arguably some of the most prestigious hunting trophies in the world, and the markhor is among the most expensive too. Batavia et al. seem unconcerned with trophy hunting for species like these. Do the authors then—by providing examples of two African species—refer only to hunting in Africa? In this case, Americans and Europeans paying to hunt for example trophy sheep in Siberia or Iran are also excluded from the definition of trophy hunting, but on what grounds? The authors do not discuss their definition further, and their article is entirely about African trophy hunting. Batavia et al.'s singling out of North American and European hunters could also be challenged. While this is the stereotype and where most hunting tourists still come from, other countries have seen strong growth in the numbers of hunting tourists, but none more so than China. CITES trade data show that while

China imported just a handful of hunting trophies from CITES species a year in the early 2000s, this had increased to hundreds or even thousands of imports a year just before COVID. The hunting industry is now increasingly turning its attention to the market in China, where legal domestic hunting is virtually unavailable, where gun control is extremely strict, and where the rich are continuously in search of novel experiences and ways to display their wealth. The first ever hunting trade show took place in 2019 in Shanghai and attracted more than 35,000 visitors to 80 stands that mainly represented safari hunting companies operating in Africa.

Lindsey et al. define trophy hunting as follows:

> '[T]rophy hunting' (also known as 'safari' or 'sport' hunting) is used to describe hunting by paying tourists, typically with the objective of selecting individuals with exceptional physical attributes (e.g., large horns, tusks, body size or skull length) and usually in the company of a professional hunting guide. (Lindsey et al. 2007: 456)

The references here to "safari" and "tourism," the mentioning of a hunting guide, not mentioning antlers as an exceptional physical attribute,[2] and the title of the article (*Economic and conservation significance of the trophy hunting industry in sub-Saharan Africa*) make it clear that these authors also consider trophy hunting only in an African context. This is confirmed by the rest of the article. It is a shame though that Lindsey et al. leave us with the impression that this is all trophy hunting is, rather than an expression of the authors' (and the public's) particular interest in the hunting of African wildlife.

Peer-reviewed literature on the subject tends to refer to trophy hunting without defining it. Most of the trophy hunting literature concerns trophy hunting in Africa, so some authors might simply rely on the readers' preconceived notions of stereotypical trophy hunting to do the definition work for them. Again, this leaves the reader with the impression that the stereotype of the trophy hunter holds true; it confirms the notion that African trophy hunting defines the subject in its entirety.

There is not necessarily any fault with the abovementioned authors' narrow conceptualization of trophy hunting other than them not specifying that these conceptualizations are indeed narrow. Trophy hunting tourism in Africa certainly incites and holds the general public's attention much more than domestic North American or European deer hunting or trophy hunting tourism for sheep in Central Asia. This is evidenced by how often news about trophy hunting in Africa is posted on social media and the extent to which it goes viral. The concerns about trophy hunting in Africa are generally more serious and urgent than any concerns about trophy hunting domestically in USA and Europe, because the species involved are often endangered or at least perceived to be so. They are also often highly charismatic, which is what leads to their viral potential in the first place. More on that in Chapter 7.

A definition aimed at facilitating a more comprehensive understanding of the subject of trophy hunting needs to be broad. It needs to move past the Great White Hunter stereotype and include domestic trophy hunting in Europe and the US as well as trophy hunting tourism outside Africa.

---

[2] Antlered species are typically hunted in Europe and the US and much less often in Africa. See Chapters 3.1 and 3.2 for more information about antlers, horns, and other types of trophies.

Here, the description of trophy hunting used by the International Union for Conservation of Nature (IUCN) is more useful:

> Trophy hunting generally involves the payment of a fee by a foreign or local hunter for a hunting experience, usually guided, for one or more individuals of a particular species with specific desired characteristics (such as large size or antlers). The trophy is usually retained by the hunter and taken home. Meat of hunted animals is usually used for food by local communities or the hunter. It may be a distinct activity or overlap with recreational or meat hunting. Many deer hunters, for example, may desire a trophy but also hunt for food or for the experience. (IUCN 2016: 4)

This definition suits our purposes better. For this book, when we say "trophy hunting" we will be referring to any hunting of a nonhuman animal where part of the animal is kept for personal display purposes and souvenir.

The downside to defining trophy hunting as any hunting that involves the taking of a trophy is that it includes many hunters who do not think of themselves as trophy hunters. Indeed, even hearing the term "trophy hunting" rather than "big game hunting" or simply "hunting" is upsetting to many hunters, and we have received criticism from hunters or hunting advocates on a regular basis for not trying to change the terminology. Many hunters have hunting trophies on their walls at home, but most of them would not call what they do trophy hunting. Nor would they call themselves trophy hunters. There are two good reasons why they would not. The first and most obvious reason is the stigma. "Trophy hunter" is not a neutral term. It has become loaded with negative connotations because it is used by a public with a largely negative perception of it, hence the Great White Hunter stereotype. Very few who hunt and happen to keep trophies (such as a deer skull and antlers) would call themselves trophy hunters, just as they would not say that they "murder" animals—a rhetoric commonly employed by antihunters. The second reason is that to most hunters, as will be expanded upon in Chapter 4, the trophy is just one aspect of hunting and one possible motivation for it. Being labelled a trophy hunter is misleading if the meat and the experience of participation in nature are more important hunting motivations than keeping a trophy as a memento.

However, it is exactly the trophy aspect of hunting—minor as it may be to many hunters—that interests us in this book. So, for this book's purposes, all hunters who keep trophies are considered trophy hunters. Our intention is not to stigmatize these hunters by lumping them in with the negatively perceived Great White Hunter stereotype. Our aim, by showing that there are many reasons for keeping trophies and many other hunting motivations at play, is to move towards a more nuanced understanding of the activity.

1.4 Going Hunting

Most of this book is concerned with the history, philosophy, and impacts of trophy hunting as well as characterizations of hunters and the meanings of trophies. One thing that it is not, nor is it intended to be, is a trophy hunting guidebook. It should still be useful, however, to illustrate how a trophy hunt may proceed in the modern world. This will give a flavour of what the activity involves and introduce some important terminology. To keep things simple, we will think about how a hypothetical hunter based in the USA (where, as we will see in later chapters, many overseas hunters originate) might go about hunting a trophy impala (*Aepyceros melampus*) in South Africa. The impala is a very commonly hunted species of antelope in South Africa, which is the most popular African hunting destination. In general, the process described below holds true for many other destinations and species but do bear in mind that trophy hunting is a complex activity that takes place across the globe. Inevitably, some details will differ for different regions and species. In Chapter 5, the situation in some other popular hunting destinations is discussed in more detail.

To get a feel for how such a hunt might proceed, we are going to imagine a hunter and follow their journey. The hunter of popular imagination, especially a hunter who travels the world to practice his hobby, is probably a white man, in his 50s, with plenty of money and a love of guns. He is also quite likely to be American. While this stereotype is, as we will see, not entirely unfair it does not accurately describe all hunters. Many different types of people hunt. To pull away from the popular stereotype for a moment let's imagine our hunter shoots a few whitetail deer every year, mostly for the freezer, but he likes to keep the antlers of any particularly good buck as a memento, a trophy if you like, of his hunting experiences. He may have grown up hunting whitetail deer and the occasional elk (*Cervus canadensis*), but he has long hankered to hunt under the big skies of Africa. Historical accounts by elephant hunters like Frederick Courtney Selous and WDM ("Karamojo") Bell, and more contemporary accounts by hunters like Peter Capstick, have whetted his appetite for "big game hunting," but it is the graceful, elegant impala with its slender lyre-shaped horns that has him captivated. Putting aside a little every month for the past eight years has provided him with enough money to make his dream of hunting an impala, maybe even one with record-breaking horns, a reality. This thumbnail sketch is the simplified biography of a real hunter that AH spoke with in South Africa. The hunter in question was a blue-collar worker from Montana.

With the dream in place, where does our imagined hunter start to look for opportunities to hunt in Africa? On the internet of course. Googling "hunting an impala" takes him straight to a company with a well-constructed website that offers hunting packages "to suit every wallet." Based in South Africa, the company is what it is called an "outfitter," a person or company that can arrange everything needed for a hunting trip. With a website crammed full of information on hunting in Africa, as well as photos of amazing locations and the various species that a client can hunt, our hunter is spoilt for choice. After much deliberation he opts for a seven-day "Bushveld Hunting Package." This package includes all-inclusive bush lodge accommodation, airport transfers, and the chance to hunt kudu, blue wildebeest (*Connochaetes taurinus*) and warthog as well as the dream species, impala. He can even hunt any other species from a long list he might come across, for an additional fee naturally. The overall cost excludes airfare and the taxidermy, packing and shipping of any trophies (like skins or horns), but otherwise $US8000 will make his African hunting dream come true. He books up that very evening. Flights are easily sorted out, and soon he is all set for the hunting trip of a lifetime.

A few days afterwards, at a time arranged via email and WhatsApp, the outfitter gets in touch from South Africa to find out more about what species our hunter wants, how he prefers to hunt (he wants to be on foot, African sun on his back and dust under his boots), what rifle he likes to shoot and so on. The outfitter talks his client through how to get trophies prepared, either by a local taxidermist or, if desired, the skins can be treated locally and then shipped to a taxidermist in the US. Apart from the skin and horns of an impala for his den, he isn't really sure what he wants, and there are such a wide range of options available. Not to worry, the outfitter tells him, there will be plenty of time to decide that later, although he might need more than the $US2000 he put aside for taxidermy if he gets some of the other species on the list. The outfitter also talks about Kobus, the Professional Hunter (the "PH") who will guide and direct the hunt, along with the trackers and skinners who will help on the hunt and in the vital first steps of preparing the hunting trophies.

After getting a feel for the client's basic needs, it is time to discuss firearms. Our hunter has a rifle chambered in .308 that he has used for more than 10 years to shoot deer. It is a large enough calibre for the animals he is going to hunt, but it is suggested that he might want to use a rifle available at the hunting lodge, chambered in .375. With a larger bullet and more "punch," this is a very commonly used calibre for African game and would even give the option to shoot a cape buffalo if our hunter's pockets are deep enough (which, he laments, they aren't). To keep things simple with airlines and customs, and to give him options, our hunter decides to pay a small fee to use the rifle in South Africa rather than bring his own. With the details sorted and pick up details arranged; he is all set.

Seven months later, much of the intervening time spent reading just about everything he can find on African hunting and watching hundreds of videos on YouTube, our hunter flies direct to OR Tambo International Airport in Johannesburg. After clearing through the airport, he is met in the Arrivals Hall by Piet, the general odd-job man at the hunting lodge and driver of the airport transfer. Very soon, our hunter is heading away from the city outskirts and is immersed in conversations with Piet about South Africa, hunting and wildlife. A couple of pleasant hours pass and the vehicle turns through some impressive gates into a beautiful bushveld property on the edge of a river. A large lunch buffet is already laid out on the shaded veranda of the main lodge and a well-stocked fridge has a good selection of cold beers and wines. On the counter are three bottles of excellent South African pinot noir and a selection of imported spirits. Deciding to save the alcohol for later, our hunter pours a long glass of freshly squeezed orange juice and sits down for a late lunch with the PH, Kobus, a softly-spoken man to whom our hunter immediately warms.

What happens now is pretty much the same wherever client-based hunting takes place. Walking in the bush with someone carrying a loaded rifle requires a degree of trust, as does putting your safety around wild animals in the hands of someone you barely know. That mutual trust needs to be built early. Luckily, Kobus and our hunter get along, and by the time lunch has ended, both men are happy that the week should work out nicely. The next step is for the PH to see whether the client can shoot.

Driving out in an open-backed pickup truck with safari seats mounted in the back (a vehicle known across South Africa as a bakkie, pronounced "bucky") is a very different experience from the earlier airport transfer. Bouncing along on dirt roads, rutted and baked hard by the sun, suddenly feels a lot more like Africa to our hunter. Scattering a sounder of warthogs as he leaves camp, his adventure begins.

The range visit is a rite of passage in any hunting trip, and different locations have different setups. The range our hunter finds himself about to shoot on is very basic compared to the range he is used to at home. An old iron framed bench with a wooden top makes a solid enough shooting rest though, and our hunter takes a seat at it on a battered but comfortable camp chair that has appeared from back of the vehicle. A rifle is carefully taken from the back seat and set up on the bench. At the other end of the range, 100 m away, is a massive cable drum resting on its circular edge, a huge round face sporting some pieces of paper stapled roughly to the wood. Settling behind the rifle, our hunter takes careful aim at a target drawn on to one of the pieces of paper. Squeezing the trigger, as he done thousands of times before, he releases the bullet downrange with a deafening roar. Our hunter quickly realises the .375 is, as Kobus wryly observes, "a lot more gun than your .308, eh?" Like many hunters though, ours is an experienced shooter and he quickly adapts to the increased recoil of the rifle he will be using. He takes some shots from the bench and from the shooting sticks Kobus produces from the back of the vehicle. These sticks form a tripod that a hunter can use in the field to steady the rifle and help to ensure a clean shot is made. Kobus is soon happy that his client can shoot well, and our hunter is happy that the rifle he will use is "zeroed." He wants to stalk his quarry and get to 50 m or closer, rather than shoot from any great distance. Even so, he needs to make sure that where he is aiming is an inch or two below the point where his bullets hit the paper target mounted on the drum. "Zeroing high" like this means that he will be able to aim and shoot at anything out to 200 m or so without worrying about adjusting his point of aim to allow for the fall of the bullet under gravity. That is good, because one thing he doesn't want to do is miss, or worse still, injure an animal without killing it. The thought of that has kept him awake more than once in the seven-month wait before coming to Africa. Not only does he not want to wound an animal, but he also knows that an animal wounded is an animal bought.

By this point there is only an hour-and-a-half of daylight left, so he heads back to camp for sundowners, via a game drive that surpasses his wildest imagination! Jackals run ahead of the vehicle in the famous African "golden hour" as he ticks off wildebeest, zebra, kudu, impala, steenbok (so small!), hartebeest, and eland from his "to see" list. Back at camp, one cold beer leads to another, and campfire talk turns to hunting, rifles, meat, American beer (and its shortcomings) and conservation. Our hunter goes to bed as happy as he can ever remember.

The next day our hunter's alarm wakes him at 6am and he is met on the veranda by hot coffee and a continental breakfast with fresh fruit. Refreshed and ready, he, Kobus and Wes, a very experienced tracker, head out in a bakkie to the far side of the property with the goal of hunting an impala. It takes the best part of thirty minutes to reach the drop off point, and with the early morning sun starting to make an impression on the cold dawn air, our hunter starts his first hunting foray. Carrying the unfamiliar rifle, chambered with a lipstick-sized .375 cartridge and three more waiting in the magazine, he checks and rechecks that the safety is on. Kobus gives him an encouraging smile, the PH noting his client's attention to gun safety and "muzzle awareness." Not all hunters are so careful and Kobus well remembers the time a client shot a hole in the side of a water tank, carrying a rifle the client claimed was unloaded. Holding the set of hunting sticks, Kobus leads the way off the track into the bush. The bakkie pulls away and is quickly lost from view.

Several hours pass as the hunting party walk slowly through the bush, occasionally cutting a dirt road covered in animal prints. There are plenty of animals around but being on foot and moving through the bush is a very different prospect from the game drive the evening before. Stopping regularly for a drink, our hunter starts to appreciate the fact that he broke his boots in well before he set foot in the bush. Three hours of tense bush walking carrying a 4.5 kg rifle, trying not to make a sound and watching out for thorns and snakes, is not everyone's idea of fun, but our hunter is still smiling. He is, after all, living his dream.

Suddenly, Wes raises his hand. The party has just rounded a thorn bush and thanks to a helpful headwind wandered within 120 m of a male impala. Standing alone, the old ram is magnificent. With huge, symmetrical horns, he is exactly what our hunter had dreamt of back home. The hunt is on.

Although the impala is close enough to shoot comfortably, Kobus notices that his client's adrenalin is pumping. Experience tells the PH that his client's shooting in the field might not match his excellent earlier performance at the range. The best shot from a bench can miss a barn from the inside once their blood is pumping. Given the wind is in their faces, and several thorny thickets can hide the approach, Kobus decides to stalk closer. Slowly working his way towards the impala, moving from one piece of cover to the next, our hunter and Kobus close the distance to just 60 m. Easing out from behind the thorn bush they are hunkered behind, Kobus slowly stands, puts up the shooting sticks and nods. His heart pounding so hard he can feel it in his head, our hunter raises the rifle and rests the wooden fore-end carefully in the space created by the three legs of the tripod. Every movement, every sound, feels amplified. The impala still hasn't seen them. Looking down the scope mounted on the rifle, the antelope comes into view, standing perfectly broadside and oblivious to the presence of humans. Standing still, broadside and close, the impala offers a simple shot, and our hunter knows it. That knowledge only adds to the pressure he is feeling. Desperately trying to control his breathing, he looks through the scope, following the rear of the front leg up to a point about a third of the way up the body—a classic "boiler room" shot that should result in a very quick death. Gently he squeezes the trigger. Afterwards, he doesn't remember anything about the sound of the shot or the recoil of the rifle.

What he does remember is looking up and not seeing the impala. Anywhere. He thinks he missed, fancying that he saw a flash of movement as the impala ran at the noise of the shot. But he didn't miss. Kobus pats him hard on the shoulder and whispers loudly, "good shot man, good shot." Our hunter is shaking. "Is he down?" he asks, also whispering. "Ya, he's down. He's done. You got him. Great shot!" Kobus sees his client is shaking, a response he has seen many times before. He also knows that the client worked the bolt of the rifle after the shot, chambering another round. He now has a shaking client, and a loaded rifle with the safety off. He gently tells his client to make the rifle safe, which the hunter does with trembling fingers under Kobus's watchful eye. Satisfied that the impala is dead and will not need a follow-up shot the party walks down to where the animal fell dead, about eight metres forward of where he was hit. He is as magnificent up close as he looked from a distance, and Wes sets the ram up for a trophy photo. Meanwhile, Kobus radios the camp and a bakkie is soon dispatched to collect the hunting party and the dead impala.

Kobus and our hunter, who has stopped shaking and feels a sense of euphoria mixed with a strong sense of relief, return to the lodge for a celebratory lunch. Wes accompanies the impala to the skinning shed where his experienced hands and a sharp knife carefully remove the skin that will be prepared and eventually shipped to the US. The carcass is taken to a refrigerated larder where it will hang for a day or so to firm up before being butchered for camp meat. The fillets will form the centre of a braai on the final day of the hunter's trip, while the next client will enjoy an impala roast on their arrival and, later, a lunchtime lasagne made from impala mince.

The rest of the trip sees our hunter add a warthog, a kudu, and a wildebeest to his tally. Coming to a deal with the local taxidermist to prepare them, he can just about afford to get the four skins and the impala horns back home. On the drive back to the airport he starts to plan his next trip—to hunt a cape buffalo with Kobus.

References

Adams, Kip P., and R. Joseph Hamilton. 2011. Management History. In *Biology and Management of White-Tailed Deer*, ed. David G. Hewitt, 355–378. Boca Raton: CRC Press.
Alexander, Harriet, Peta Thornycroft, and Aislinn Laing. 2015. Cecil the Lion's Killer Revealed as American Dentist. *The Telegraph*, July 28. http://www.telegraph.co.uk/news/worldnews/africaandindianocean/zimbabwe/11767119/Cecil-the-lions-killer-revealed-as-American-dentist.html (accessed 29 August 2017).
Allsen, Thomas T. 2013. *The Royal Hunt in Eurasian History*. Singapore: Institute of Southeast Asian Studies (ISEAS) Publishing.
Anderson, J.K. 1985. *Hunting in the Ancient World*. London: University of California Press.
Barringer, Judith M. 2001. *The Hunt in Ancient Greece*. Baltimore: Johns Hopkins University Press.
Batavia, Chelsea, Michael Paul Nelson, Chris T. Darimont, Paul C. Paquet, William J. Ripple, and Arian D. Wallach. 2018. The Elephant (Head) in the Room: A Critical Look at Trophy Hunting. *Conservation Letters* 12 (1): 1–6.
Batavia, Chelsea, Michael Paul Nelson, and Arian D. Wallach. 2020. The Moral Residue of Conservation. *Conservation Biology* 34: 1114–1121.
Bauer, Hans, Guillaume Chapron, Kristin Nowell, Philipp Henschel, Paul Funston, Luke T.B. Hunter, David W. Macdonald, and Craig Packer. 2015. Lion (*Panthera leo*) Populations Are Declining Rapidly Across Africa, Except in Intensively Managed Areas. *Proceedings of the National Academy of Sciences of the United States of America (PNAS)* 112: 14894–14899.
Bauer, Hans, Amy Dickman, Guillaume Chapron, Alayne Oriol-Cotterill, Samantha K. Nicholson, Claudio Sillero-Zubiri, Luke Hunter, Peter Lindsey, and David W. Macdonald. 2020. Threat Analysis for More Effective Lion Conservation. *Oryx* 56: 108–115.
Beattie, Geoffrey. 2020. *Trophy Hunting: A Psychological Perspective*. London and New York: Routledge.

References

Booth, Vernon R., Jones Masonde, Chuma Simukonda, and David H.M. Cumming. 2020. Managing Hunting Quotas of African Lions (*Panthera leo*): A Case Study from Zambia. *Journal for Nature Conservation* 55: 1–8.

Bronner, Simon. 2004. "This Is Why We Hunt": Social-Psychological Meanings of the Traditions and Rituals of Deer Camp. *Western Folklore* 63/1–2, Hunting and Fishing: 11–50.

Bronner, Simon. 2008. *Killing Tradition—Inside Hunting and Animal Rights Controversies*. Lexington: The University Press of Kentucky.

Cahoone, Lawrence. 2009. Hunting as a Moral Good. *Environmental Values* 18: 67–89.

Carpenter, Stefan, and David M. Konisky. 2017. The killing of Cecil the Lion as an Impetus for Policy Change. *Oryx* 53 (4): 698–706.

Causey, Ann S. 1989. On the Morality of Hunting. *Environmental Ethics* 11 (4): 327–343.

Causey, Ann S. 1996. Is Hunting Ethical. In *A Hunter's Heart—Honest Essays on Blood Sport*, ed. David Petersen, 1997 (first 1996), 80–89. New York: Henry Holt and Company.

Coltman, David W., Paul O'Donoghue, Jon T. Jorgenson, John T. Hogg, Curtis Strobeck, and Marco Festa-Bianchet. 2003. Undesirable Evolutionary Consequences of Trophy Hunting. *Nature* 426 (6967): 655–658.

Creel, Scott, Jassiel M'Soka, Egil Dröge, Eli Rosenblatt, Matthew S. Becker, Wigganson Matandiko, and Twakundine Simpamba. 2016. Assessing the Sustainability of African Lion Trophy Hunting, with Recommendations for Policy. *Ecological Applications* 26 (7): 2347–2357.

Demarais, Steve, and Bronson K. Strickland. 2011. Antlers. In *Biology and Management of White-Tailed Deer*, ed. David G. Hewitt, 107–145. Boca Raton: CRC Press.

Dickman, Amy, and Amy Hinks. 2019. *State of the Lion: Fragility of a Flagship Species—Report Prepared for the Lion Footprint Forum*, May 2019 (Unpublished Report Based on a Paper in Preparation).

Dickson, Barney. 2009. The Ethics of Recreational Hunting. In *Recreational Hunting, Conservation and Rural Livelihoods*, ed. Barney Dickson, Jon Hutton, and William M. Adams, 59–72. Oxford: Wiley-Blackwell.

Dizard, Jan E. 1994. *Going Wild—Hunting, Animal Rights, and the Contested Meaning of Nature*. Amherst: University of Massachusetts Press.

Dizard, Jan E. 2003. *Mortal Stakes—Hunting and Hunters in Contemporary America*. Amherst and Boston: University of Massachusetts Press.

Douhard, Mathieu, Marco Festa-Bianchet, Fanie Pelletier, Jean-Michel. Gaillard, and Christophe Bonenfant. 2016. Changes in Horn Size of Stone's Sheep Over Four Decades Correlate with Trophy Hunting Pressure. *Ecological Applications* 26 (1): 309–321.

Festa-Bianchet, Marco, and Ray Lee. 2009. Guns, Sheep, and Genes: When and Why Trophy Hunting May Be a Selective Pressure. In *Recreational Hunting, Conservation and Rural Livelihoods*, ed. Barney Dickson, Jon Hutton, and William M. Adams. Oxford: Wiley-Blackwell.

Festa-Bianchet, Marco, and Atle Mysterud. 2018. Hunting and Evolution: Theory, Evidence, and Unknowns. *Journal of Mammalogy* 99: 1281–1292.

Festa-Bianchet, Marco, Fanie Pelletier, Jon T. Jorgenson, Chiarastella Feder, and Anne Hubbs. 2014. Decrease in Horn Size and Increase in Age of Trophy Sheep in Alberta Over 37 Years. *The Journal of Wildlife Management* 78 (1): 133–141.

Ghasemi, Benjamin. 2020. Trophy Hunting and Conservation: Do the Major Ethical Theories Converge in Opposition to Trophy Hunting? *People and Nature* 3: 77–87.

Gunn, Alastair S. 2001. Environmental Ethics and Trophy Hunting. *Ethics and the Environment* 6 (1): 68–95.

Hewitt, David G. 2015. Hunters and the Conservation and Management of White-Tailed Deer (*Odocoileus virgianus*). *International Journal of Environmental Studies* 72 (5): 839–849.

IUCN. 2009. Big Game Hunting in West Africa—What Is Its Contribution to Conservation?" *Papaco Studies* 2. https://www.iucn.org/content/big-game-hunting-west-africa-what-its-contribution-conservation (accessed 18 June 2020).

IUCN. 2016. IUCN Briefing Paper: Informing Decisions on Trophy Hunting. https://www.iucn.org/sites/dev/files/iucn_sept_briefing_paper_-_informingdecisionstrophyhunting.pdf (accessed 6 January 2020).

Jacobson, Harry A., Charles A. DeYoung, Randy W. DeYoung, Timothy E. Fulbright, and David G. Hewitt. 2011. Management on Private Property. In *Biology and Management of White-Tailed Deer*, ed. David G. Hewitt, 453–479. Boca Raton: CRC Press.

Kalof, Linda. 2007. *Looking at Animals in Human History*. London: Reaktion Books.

Kellert, Stephen Robert. 1996. *The Value of Life—Biological Diversity and Human Society*. Washington: Island Press/Shearwater Books.

Kimmel, Jimmy. 2017. Opening Monologue. Jimmy Kimmel Live!, 28 August. https://www.youtube.com/watch?v=saHGvxFAhE0 (accessed 1 September 2017).

Lindsey, P., R. Alexander, G. Balme, N. Midland, and J. Craig. 2012. Possible Relationships Between the South African Captive-Bred Lion Hunting Industry and the Hunting and Conservation of Lions Elsewhere in Africa. *African Journal of Wildlife Research* 42 (1): 11–22.

Lindsey, Peter A., R. Alexander, L.G. Frank, A. Mathieson, and S.S. Romañach. 2006. Potential of Trophy Hunting to Create Incentives for Wildlife Conservation in Africa Where Alternative Wildlife-Based Land Uses May Not be Viable. *Animal Conservation* 9 (3): 283–291.

Lindsey, Peter, James Allan, Peadar Brehony, Amy Dickman, Ashley Robinson, Colleen Begg, Hasita Bhammar, Lisa Blanken, Thomas Breuer, Kathleen Fitzgerald, Michael Flyman, Patience Gandiwa, Nicia Giva, Dickson Kaelo, Simon Nampindo, Nyambe Nyambe, Kurt Steiner, Andrew Parker, Dilys Roe, Paul Thomson, Morgan Trimble, Alexandre Caron, and Peter Tyrrell. 2020. Conserving Africa's Wildlife and Wildlands Through the COVID-19 Crisis and Beyond. *Nature Ecology & Evolution* 4: 1300–1310.

Lindsey, Peter A., Guy A. Balme, Paul J. Funston, Philipp H. Henschel, and Luke T. B. Hunter. 2016. Life after Cecil: Channelling Global Outrage into Funding for Conservation in Africa. *Conservation Letters* 9 (4): 296–301.

Lindsey, P.A., L.S. Petracca, P.J. Funston, H. Bauer, A. Dickman, K. Everatt, M. Flyman, P. Henschel, A.E. Hinks, S. Kasiki, A. Loveridge, D.W. Macdonald, R. Mandisodza, W. Mgoola, S.M. Miller, S. Nazerali, L. Siege, K. Uiseb, and L.T.B. Hunter. 2017. The Performance of African Protected Areas for Lions and Their Prey. *Biological Conservation* 209: 173–149.

Lindsey, Peter A., P.A. Roulet, and S.S. Romañach. 2007. Economic and Conservation Significance of the Trophy Hunting Industry in Sub-Saharan Africa. *Biological Conservation* 134: 455–469.

Loftin, Robert W. 1984. The Morality of Hunting. *Environmental Ethics* 6 (3): 241–250.

Lovelock, Brent. 2015. Troubled-Shooting: The Ethics of Helicopter-Assisted Guided Trophy Hunting by Tourists for Tahr. In *Animals and Tourism—Understanding Diverse Relationships*, ed. Kevin Markwell, 91–105. Channel View Publications.

Loveridge, Andrew J. 2018. *Lion Hearted: The Life and Death of Cecil & the Future of Africa's Iconic Cats*. New York: Regan Arts.

Macdonald, David W. 2016. Report on Lion Conservation with Particular Respect to the Issue of Trophy Hunting. Wildlife Conservation Research Unit (WildCRU), University of Oxford.

Macdonald, David W., Kim S. Jacobsen, Dawn Burnham, Paul J. Johnson, and Andrew J. Loveridge. 2016. Cecil: A Moment or a Movement? Analysis of Media Coverage of the Death of a Lion, *Panthera leo*. *Animals* 6 (26): 1–13.

Mahoney, Shane. 2009. Recreational Hunting and Sustainable Wildlife Use in North America. In *Recreational Hunting, Conservation and Rural Livelihoods*, ed. Barney Dickson, Jon Hutton, and William M. Adams, 266–271. Oxford: Wiley-Blackwell.

Marshall, Claire, 2020. How the Covid-19 Pandemic Is Threatening Africa's Wildlife. BBC News, May 7. https://www.bbc.com/news/av/world-africa-52564615/how-the-covid-19-pandemic-is-threatening-africa-s-wildlife (accessed 28 May 2020).

Marvin, Garry, 2010. Living with Dead Animals?—Trophies as Souvenirs of the Hunt. In *Hunting Philosophy for Everyone: In Search of the Wild Life*, ed. Nathan Kowalsky, 107–117. Wiley-Blackwell.

References

Marvin, Garry. 2011. Enlivened Through Memory—Hunters and Hunting Trophies. In *The Afterlives of Animals—A Museum Menagerie*, ed. Samuel J.M.M. Alberti, 202–217. Charlottesville and London: University of Virginia Press.

Morris, S.P. 2020. A Moral Defense of Trophy Hunting and Why It Fails. *Sport, Ethics and Philosophy* 15: 386–399.

Packer, Craig. 2015. *Lions in the Balance—Man-Eaters, Manes, and Men with Guns*. Chicago and London: The University of Chicago Press.

Poliquin, Rachel. 2012. *The Breathless Zoo—Taxidermy and the Cultures of Longing*. Pennsylvania State University Press.

Somerville, Keith. 2019. Propaganda in the Trophy Hunting Debate—Card-Stacking, Cherry-Picking, Name-Calling & KISS. Conservation Frontlines Editorial, July—Volume I-3. https://www.conservationfrontlines.org/2019/07/propaganda-in-the-trophy-hunting-debate-card-stacking-cherry-picking-name-calling-kiss/ (accessed 19 October 2022)

Yeomans, N., E. Droge, D. Hare, and A. G. Hart. 2022. Ten years of coverage of trophy hunting in UK newspapers. *Frontiers in Conservation Science*. https://doi.org/10.3389/fcosc.2022.1061295.

Vitali, Theodore. 1990. Sport Hunting: Moral or Immoral? *Environmental Ethics* 12: 69–82.

Chapter 2
A History of Hunting and Hunting Perceptions

To understand trophy hunting, we must first understand hunting, which requires us to take a deep dive into its history and place in human life and society. This chapter will explore how hunting developed, and how its role and attitudes towards it have changed, from our early days in the Pleistocene until today. Given the topic of this book, we will naturally pay special attention to events, developments, and persons that can be related to hunting ethics and especially trophy hunting. A comprehensive history of hunting is not the objective of this book, so this chapter will be exhaustive neither geographically nor temporally. Many parts of the world will not be discussed if developments there were similar to developments elsewhere that are already covered. In addition, subsistence hunting cultures are still to be found across the modern world in a wide variety of locations and geographical contexts, and hunting, in all its guises, continues to play a crucial role. The sociological, nutritional, and ecological roles of subsistence hunting are topics that have been, and continue to be, well studied. While it is certainly the case that subsistence hunters across the world may retain trophies, this is only marginally related to the trophy hunting that we explore in this book. Consequently, subsistence hunting will be discussed only in a pre-agricultural context. For a broader and deeper overview of the topic, we recommend the book *Hunting Wildlife in the Tropics and Sub-tropics* by Julia Fa, Stephan Funk and Robert Nasi (2022), and the chapter *Subsistence Hunting* by Raymond Anthony and Gary Varner in the *Routledge Handbook of Animal Ethics* (Anthony and Varner 2019).

2.1 Prehistory

The prehistory of hunting is firstly a story of the Pleistocene, the period often called the Ice Age that started around 2.5 million years ago and lasted until around 12,000 years ago. Secondly, it is a story of the Pliocene; the period immediately prior to the Pleistocene that started perhaps five million years ago. It is hard to be precise, as the wooden weapons and tools of the Pliocene are far less well-preserved for archaeologists than the stone tools characteristic of Pleistocene. In terms of time, less than one percent of our history of hunting is a story of the Holocene—the recent 12,000 or so years, following the end of the last ice age. This period of human history is characterized by the advent of agriculture, animal husbandry, the development of extensive settlements and the shift towards a sedentary "modern" lifestyle. In fact, our history of hunting is not even mostly "our" story, as anatomically modern humans, our species *Homo sapiens*, have only been around for 300,000 years or so. Before then, the history of hunting was a story of such relatives and ancestors as *Australopithecus*, *H. habilis*, *H. erectus*, and *H. neanderthalensis*.

How much these early hominids hunted and relied on hunting for survival is a long-standing point of scholarly dispute, as is the influence that hunting has had on our evolution. There are two principal schools of thought: one claims that today we are humans primarily because we were forced to take up hunting (and cannibalism) and survive on what we killed, while the other claims that bar the occasional hunting of smaller animals, we were more prey than predator during the Pleistocene, surviving mostly on a vegetarian diet combined with some scavenging.

In 1924 the Australian anatomist and palaeontologist Raymond Dart discovered the 2.8-million-year-old skull of a small child in South Africa, a discovery that become known as the "Taung Child" named for the location in which it was discovered. It was the first discovery of the genus *Australopithecus*, which Dart named *A. africanus*. It looked much like the skull of an ape, but with more human characteristics than previous discoveries. Most interestingly, it was found in what had not been forest, the natural habitat of apes, for millions of years (Dart 1925). In the first few years following this discovery, Dart reasoned that the Taung Child was an ancestor of ours who had abandoned the vegetarian diet, which could not be sustained outside the forests, and was forced to scavenge and hunt small animals. He also proposed that the child had developed bipedalism and a larger brain to escape and outsmart the predators of the open plains, thus setting the species on the road towards humanity. As more fossils were found from the mid-1920s until the mid-1950s, Dart's theory gradually changed. These new *A. africanus* fossils were found amidst the remains of large prey species, which Dart surmised must have been killed by *A. africanus*, using animal bones as weapons. Dart's theory went from describing an ancestor who was mostly a prey to describing an ancestor who was among the top predators on Africa's plains and a vicious cannibal. In the shadow of the second world war, this was an appropriate explanation for modern humans' proclivity for war, violence, and hunting (Cartmill 1996: 1–10; Hart and Sussmann 2005: 25–32). As an interesting postscript, more recent work by the palaeontologist Lee Berger, examining damage to the Taung child's skull, concluded that the child had fallen prey to an eagle. So, whether this ancestor was a hunter can be disputed, but the evidence certainly points to them having been hunted (Berger and McGraw 2007).

From the mid-1950s until the late-1970s the idea of the killer-ape *Australopithecus* took hold. The book *Man the Hunter* was published in 1968 and became highly influential in determining how human history was framed, and the role of hunting (and understanding the hunter) in interpreting our past. Now largely discredited, the book referenced a hallmark text by Washburn and Lancaster, whose conclusions echoed those of Dart:

> Human hunting is made possible by tools, but it is far more than a technique or even a variety of techniques. It is a way of life, and the success of this adaptation (in its total social, technical, and psychological dimensions) has dominated the course of human evolution for hundreds of thousands of years. In a very real sense our intellect, interests, emotions, and basic social life—all are evolutionary products of the success of the hunting adaptation. (Washburn and Lancaster 1968: 293)

The idea, which centralises the role of hunting, and by extension (male) hunters took hold, and in 1976, famous Hollywood screenwriter-turned-anthropologist Robert Ardrey further popularized the idea of hunting as having shaped our evolution by giving it a catchy name. The *hunting hypothesis* argued that our killing lifestyle went hand in hand with evolution for several million years:

> If among all the members of our primate family the human being is unique, even in our noblest aspirations, it is because we alone through untold millions of years were continuously dependent on killing to survive. (Ardrey 1976: 11)

Roughly summarized, Ardrey's claims were that bipedalism developed because we needed to chase down prey, that we learned to create and grasp tools and weapons because we were weak and had no other ways to kill effectively, and that our large brains and our abilities to communicate effectively developed because taking down large prey requires planning and cooperation. Our sexual division of labour came about because infants needed nursing, and because other foods needed to be foraged and scavenged while the men hunted. Ardrey stated that "[b]irds do not fly because they have wings; they have wings because they fly" (Ardrey 1976: 120). He used this rationale to explain our hunting as well: We do not hunt because we are human; we are human because we hunt.

The hunting hypothesis, as a general concept and as the more developed assertions of Ardrey and others, was widely accepted until the late 1970s at least, despite criticisms of its archaeological foundation being advanced from the late 1960s. Brain demonstrated, for instance, that leopards were the likelier cause of death for many of the hominids that Dart had thought killed each other (Brain 1970: 1116–1118). Increasing evidence that *Homo*—appearing far later than *Australopithecus*—was more scavenger than hunter also posed problems for the hypothesis and the romanticised vision of early humans as dominant hunters.

The hunting hypothesis has not been completely abandoned, but it has faced continued criticism and competition. More recent and conflicting archaeological findings and better interpretations of previous findings, as just exemplified by Brain's leopard wounds and the evidence of scavenging *Homo* are one reason, but another reason is that we now know much more about the behaviour of other primates. The extent to which chimpanzees hunt cooperatively, eat meat, and even use weapons (Morris and Goodall 1977; Stanford 2003: 127–131) showed that the type of hunting summarised by Ardrey was never as uniquely human as he and others claimed. Also, while being a predator could plausibly explain some key aspects of human evolution, so too could being prey. In *Man the Hunted* Hart and Sussman demonstrate this, and further claim that australopithecine dentition does not point towards carnivory (Hart and Sussman 2005: 23). The gender bias inherent in the hunting hypothesis has also long been noted. Slocum argued, back in 1971, that since the hunting hypothesis emphasizes only the role of men as hunters, it effectively ignores any influence that women might have had on human evolution (Phelps 2007: 1–3).

Finally, opposition to the hunting hypothesis may come not solely from conflicting archaeological evidence or shaky theoretical foundation but also from, as Cartmill here intimates, a socio-political discomfort:

> Most of the brickbats hurled at the hunting hypothesis of human origins came from the direction of the left. Marxist critics disliked the Dart-Ardrey-Lorenz picture of human nature because it blamed many of our social ills on human ancestry rather than on our economic systems. Pacifists disliked the notion that an instinctive, ineradicable joy in killing is one of the causes of war. And feminists looked with understandable suspicion on the assumption that "the male specialized as hunter, the female as domestic" from the very start of human evolution. Since the early 1970s, anthropologists (who tend to be left-wing, pacifist, and feminist) have accordingly tended to dismiss the hunting model as a myth that serves reactionary interests. (Cartmill 1996: 18)

While the oldest stone tools have been dated to about 3.3 million years ago (Harmand et al. 2015), widespread use of stone tools is generally taken to stem from about 2.5 million years ago, at which point archaeological evidence of them is more common. Yet even with physical evidence to hand, it is still hard to say whether stone hand-axes were used for killing, scavenging, digging, or chopping, but evidence has been found that they were used for sharpening spears, which is a possible indicator of hunting (Stanford 2003: 145).

Homo erectus was the first *Homo* species to spread from Africa to Europe, doing so around a million years ago. These early humans were using hand-axes (i.e. rocks sharpened to an edge and a point) as their multipurpose tool although refinement to more of a chopping shape happened as they spread to Asia (Smith 1999: 385). Regardless of how instrumental hunting has been in turning our ape-like ancestors into *H. sapiens*, there is little doubt that by the time we reached that stage (around 300,000 years ago [Hublin et al. 2017]), hunting was of greater importance for our survival than it had been previously. The earliest spear points, assumed to have been used for hunting, are about 500,000 years old (Wilkins et al. 2012), but they are common from around 300–250,000 years ago (Smith 1999). This matches them with *H. heidelbergensis, H. erectus* and presumably the earliest of *H. neanderthalensis* and *H. sapiens* (Langdon 2005: 31).

By 20,000 years ago, the tools and methods of hunting had become much more advanced, as here described by Stanford:

> European *Homo sapiens* ate the big mammals: bison, reindeer, red deer, and wild horse. *Homo sapiens* could have killed these potentially dangerous animals only through cooperation and much planning. These hunters, armed with spears, darts, bows and arrows, harpoons, and other projectiles, learned to ambush prey by chasing them into narrow canyons and to follow migratory herds in their annual wanderings. Hunting became strategic and well coordinated. And these modern walkers had the necessary locomotor efficiency and cognitive powers to follow, find, and stalk their prey. (Stanford 2003: 168)

2.1 Prehistory

But the new tools and techniques soon proved too effective. While the main driver seems to have been climate change (Stewart et al. 2021), wasteful hunting practices, using fire to slaughter large herds of animals and wolves as hunting companions—portrayed in late cave paintings (Isaacson 2001: 24)—probably contributed to the extinction of many species of large mammals towards the end of the Pleistocene (Kalof 2007; Dembitzer et al. 2022).

After the appearance of anatomically modern *H. sapiens*, there were several developments occurring between 100,000 and 10,000 years ago that have significance to hunting, either directly or indirectly. The domestication of wolves was important, as the role of hunting with dogs in the first primary literature about hunting from classical antiquity demonstrates. Technological developments were also important. Archaeologists have found evidence of bow and arrow use from 64,000 years ago in South Africa (Lombard and Phillipson 2010); cave paintings of prey animals and hunting parties are common from 40,000 to 12,000 years ago (Aubert et al. 2014); ceramics and pottery – important precursors to agriculture—were developed 26,000 to 10,000 years ago; and livestock was first domesticated 20,000 to 10,000 years ago (Kalof 2007: 6–10).

Notably relevant to trophy hunting is evidence that the teeth of foxes, stags, wolves, bears, and lions were engraved and perforated for use as pendants and jewellery 55,000 years ago. These were often found in human burial sites and dwellings. Elephant and mammoth ivory, antlers, and bones or skulls from horses, bears, rhinos, mammoths, elephants and practically any other large animal were engraved, painted, or sculpted into statuettes and seem to have been common forms of Upper Palaeolithic décor. Seashells and fish bones turned into ornaments and decorations were also common. Abstract patterns, lines, dots, and circles were commonly engraved on teeth and bones (Bahn and Vertut 1997: 25–26; 86–87; 92–98). There is of course no way to tell whether a bone or tooth turned into a trophy-like decoration was obtained through hunting or scavenging, but it seems reasonable to assume that both categories were represented. Recent research published in *nature* shows how archaeologists discovered dozens of what they characterized as hunting trophies from large ungulates, cleaned and treated much as your would today, in a neanderthal cave in Spain (Baquedano et al. 2023). The earliest discovered antler trophy, hanging on the wall of a prehistoric house, as a hunter would do it today, is 6500 years old and was uncovered in 2012 in Bapska, Croatia. It was a 12-point set of antlers (Archaeology News Network 2012).

The possible significance of cave paintings should not be ignored either in a trophy hunting context. Randall Eaton suggested in his article *The Evolution of Trophy Hunting* that cave paintings could be a form of hunting trophies:

> The oldest paintings and sculptings of Paleolithic hunters may have represented trophies. The predominance of large herbivores and dangerous carnivores, the accuracy of depiction as to species, sex and age, and the emphasis on the head region all suggest trophyism [the hunting and display of trophies], as do conventions such as the turning of antlers and horns sideways (to dramatize the trophy value of an animal), the clumping together and superimposition of animal paintings (as if representing individual collections), and the association with handprints, which could have been a hunter's means of identifying trophies. Moreover,

species painted are organized in cave chambers according to trophy values. (Eaton 1978: 115)

The trophy element here is interesting, but even more so, perhaps, is the record book element. With accurate drawings of animals and handprints, possibly for identification, palaeolithic cave walls suddenly appear remarkably similar to pages in modern trophy record books with details of animals recorded alongside the names of the hunters who killed them. We return to such trophy scoring systems and record books in Chapter 3.4.

2.2 Ancient Egypt and China

At 5000 to 1000 BC, agriculture, riding, and animal husbandry in general spread quickly across the world. Tools and weapons of wood, stone, tooth, and bone were supplemented and then replaced by copper and gold, and then later with bronze, and eventually iron. Our story of hunting continues with archaeological evidence that is supplemented by the first primary literature about hunting. We can begin our exploration of hunting in the ancient world with the situation in Egypt, circa 2400 BC, summarized here by Patrick Houlihan:

> For the well-to-do, the desert hunt frequently took place in well-stocked confined areas, the animals being corralled together. The earliest prototype for this spirited theme seems to be the large wall scene in the Fifth Dynasty mortuary temple of Sahure at Abusir. The monarch is portrayed standing outside the walls of an extensive, fenced-off enclosure, forming a kind of hunting park or reserve,[1] into which a great concourse of wild desert animals has been driven by attendants.[2] He fires a rain of arrows at the mêlée of fleeing wild-life, slaying many, while his dogs are in full pursuit of others. In this and other depictions we see the frantic movements of great swarms of hunted wild animals that had often been driven by beaters or lured by food and water into the vast enclosures fenced in by tall, strong, spanning nets and occasionally surrounded by deep ditches. Here the hunt was carefully managed, virtually guaranteeing kings and courtiers a lavish yield of prize trophies. (Houlihan 1996: 42–43)

These few lines indicate several important hunting developments. First, agriculture and animal husbandry were developing in Egypt already by 5500 BC, so by 2400 BC hunting for nourishment had likely become far less important (Houlihan 1996: 41). Houlihan's description captures a hunt that is purely for fun, sport, and the display of power, wealth and dominance over nature. Second, hunting has become a recreational activity for the rich and privileged, and especially royalty. This theme, of hunting as a recreation for the rich and privileged, is one that will repeatedly surface,

---

[1] According to Allsen (2013: 36), this kind of "park" was more like a relatively small and temporary enclosure, into which the animals were driven immediately prior to the killing. As such, it was a far different setting than the later hunting parks in China, to be described shortly.

[2] Egyptian deserts with enough animals to corral for a slaughter like this may sound strange to us today, and rightly so. Due to climate change, desertification, overhunting, agriculture, and pest control, Egypt was already in 500 BC a land of very little wildlife, and today there is even less than that (Houlihan 1996: 44–45).

as we progress through to the Middle Ages and the present day. Third, this is some of the earliest historical evidence of "canned hunting," a term that is applied to the killing of animals inside an enclosure. Canned hunting is a very controversial topic in trophy hunting today, and it is something we will get back to in Chapter 6.4.

Hunting's association with royalty and displays of power was by no means particular to Egypt. As Thomas Allsen states in *The Royal Hunt in Eurasian History*, "the vast majority of the royal houses and aristocracies of Eurasia made some use of the chase in the pursuit and maintenance of their social and political power" (Allsen 2013: 14).

The royal lion hunts by kings of the Assyrian empire—most notably Ashurnasirpal II who reigned 883–859 BC and Ashurbanipal (669–631 BC)—are also thoroughly documented on wall reliefs uncovered in what is today Iraq, near Mosul. Dick (2006) argued that Assyrian kings identified with the kingly lions and—by hunting and killing them—became "creatures of nature" like them, ruling over the common people, who in turn represented domesticated animals, as a shepherd rules over his flock. The defeated lion, he writes, "is never mutilated but is treated with respect, unlike the fate of the king's human foes" (Dick 2006: 245). This point is supported by Bonatz (2004), who writes that when Ashurbanipal had defeated his enemy Teumman, ruler of Elam 664–653 BC, he had his head cut off and delivered to him. He then mutilated it, spat on it, and hung it in a tree in his garden to look upon while dining with his wife, all documented on the wall reliefs. The difference in treatment of lion heads and human heads here demonstrated how the Assyrian king stood above other kings but treated lions as his equals. The lion hunt figures so prominently on wall reliefs that it's considered central to the maintenance of Assyrian rule. We will return to the subject of human headhunting and how it influences interpretations of hunting trophies in Chapter 3.6 (Albenda 1975; Bonatz 2004; Dick 2006).

In ancient China, similarly, "[k]ings and barons killed leopards and deer, under carefully controlled conditions, to demonstrate their power and glory" (Schafer 1968: 319). Starting from at least 675 BC, China created large animal parks for royal hunting, for separating wild animals from farmers' crops, and for demonstrating China's control over nature. The famous Chinese philosopher Mencius (372–289 BC) was an advocate for humans' dominion over nature and a need to eliminate (by killing or sequestering) animals that would otherwise compete with people for land and edible wildlife. The parks had different purposes, but they also had different restrictions. Some were the domain of the privileged classes and just for hunting, while others were purely for the emperor and served various ceremonial functions. Some parks were even open to the public, who would be allowed to hunt hares or other small animals (Schafer 1968: 319–324).

The animal parks were numerous and well-documented around the third century BC. Emperor Qin Shi Huang, founder of the Qin dynasty, created a hunting preserve—*Shang Lin* or "Supreme Forest"—which was in later Han dynasty times given back to the people for agriculture and hunting, but other grandiose and exclusive supreme forests were created in its stead. The largest Han dynasty park was encircled by a wall of over 200 kms. This was no mere walled-in section of wild nature. It was carefully controlled, littered with gardens and villas for the emperor's enjoyment, and containing many exotic plants and animals. Dangerous animals like tigers (*Panthera tigris*) seem to have been kept in cages inside the park (Schafer 1968: 326–332; Allsen 2013: 41–46).

In the Chinese Supreme Forest, the distinction between hunting and combat sometimes seemed to disappear:

> The ground of the Supreme Forest was prepared for the great winter hunt by the royal foresters. They burned clear a large open space, and cut away brambles. Beaters, hunters and athletes readied themselves for the onslaughts of wild beasts and forest demons with spells and periapts. When the royal party arrived, the birds and beasts were driven into the cleared areas, and the slaughter began: […] Frenzied animals lunged and clawed at each other—tigers fought with buffaloes. Part of the display was no hunt at all. There were combats between man and beast, wrestlers against wild creatures, tests of strength. Strong men even grappled with lions, leopards, bears, rhinoceroses, yaks and elephants—and some stranger beasts. Some of these animals at least were killed. The scene was not dissimilar from the exhibitions of the emperor Commodus in the Roman arena. (Schafer 1968: 332)

Ancient Greece also had animal parks, some of which were shrines and sacred groves to honour the gods. Hunting in these sacred groves was strictly forbidden, and Greek mythology has many examples of people being punished for doing so. Other animal parks were effectively vast private gardens (Hughes 2014: 90–96). Animal parks, hunting parks, paradise parks, game parks, and sacred groves as they were varyingly called were found throughout the Assyrian and the vast Persian and Mongolian empires, in India and in Western Europe, where they would later become known as medieval deer parks (Allsen 2013: 34–51).

The scale and scope of animal parks in China would have required considerable management and husbandry to maintain, but the first known examples of game management for conservation and hunting purposes were, according to Aldo Leopold, found in the Mongol Empire during the 1259–1294 reign of Kublai Khan. Kublai Khan instigated what we would now recognise as hunting seasons, decreeing that between March and October, it was forbidden to kill any kind of hares, deer, or large birds. In addition to hunting seasons, Kublai Khan created hunting preserves with shelters, game crops, and winter feeding essentially identical to modern day practices associated with pheasant stocking or deer park management (Leopold 1986: 6–7).

2.3 Ancient Greece

The first descriptions of hunting in ancient Greek literature are in Homer's Iliad and Odyssey from around 800 BC. Both poems contain literal and metaphorical references to hunting. Judith Barringer noted that "Homeric epic repeatedly draws analogues between hunting and battle, using hunting metaphors, particularly lion hunting, to describe battle situations between Achaeans and Trojans" (Barringer 2001: 43). On only two occasions in the Odyssey is food the purpose of hunting; a goat is hunted for food on the island of the Cyclops by Odysseus and his men, and

2.3 Ancient Greece

a stag is hunted for food by Odysseus alone. Otherwise, though the meat is usually eaten, the purposes of hunting in the Odyssey and Iliad are fun, sport, revenge, self-defence, training for war, and initiation into adulthood (Alepidou 2017: 11, 18–21).

Greek mythology is full of hunting fiction (see e.g. Anderson 1985: 1–17; Cartmill 1996: 28–36; Barringer 2001), but one of the first to write non-fiction accounts of hunting was Xenophon, a Greek military leader, philosopher, and student of Socrates, living around 400 BC. The first chapter of Xenophon's *Cynegeticus*[3] (usually translated to "*On Hunting*") concerns Artemis, the Goddess of the hunt, her twin brother Apollo, and the importance of hunting to heroes of ancient Greece. The rest of it is a very detailed and descriptive handbook for hunting different quarry with dogs.[4] One may learn from *Cynegeticus* which breeds of dogs are best for which kinds of hunting, how to raise hunting dogs and handle them in the field, and even what the proper names for hunting dogs are. The most common quarry in Greece was hare, so hare hunting is described at great length. Second most common was deer, followed by boar, the pursuit of which could be very dangerous. The book also describes how various nets should be made and handled, how to build and use a leghold trap for deer hunting, and how the changing of the seasons influences the hunt.

Only Chapters 2 and 12 of *Cynegeticus* contain any extensive general (non-technical) observations on hunting. The following passage is the beginning of Chapter 2. Xenophon here hints at hunting's importance to the transition from boyhood to manhood, and at the social status associated with hunting:

> The first efforts of a youth emerging from boyhood should be directed to the institution of the chase, after which he should come to the rest of education, provided he have the means and with an eye to the same; if his means be ample, in a style worthy of the profit to be derived; or, if they be scant, let him at any rate contribute enthusiasm, in nothing falling short of the power he possesses. (Xenophon circa 400 BC 1897: Ch. 2)

The status aspect is further emphasized, as Xenophon continues as such:

> The net-keeper should be a man with a real passion for the work, and in tongue a Hellene, about twenty years of age, of wiry build, agile at once and strong, with pluck enough to overcome the toils imposed on him, and to take pleasure in the work. (Xenophon circa 400 BC 1897: Ch. 2)

Since the netkeeper is a slave, the clear assumption is that the hunter will be a member of the slave-owning classes. The "breed" of the slave is here described in rather the same way as Xenophon describes the breed of the dogs in Chapters 3 and 4, only in less detail. As noted by Anderson, the netkeeper in this account is just another piece of property and treated as such (Anderson 1985: 37).

---

[3] It has been debated whether Xenophon was really the writer of Cynegeticus (see Gray 1985).

[4] This "how-to" genre of hunting literature has only increased in popularity. In a search for hunting books on Amazon, the popular majority almost all have titles such as "Hunting Big Mule Deer: How to Take the Best Buck of Your Life," "The Ultimate Guide to Black Bear Hunting," and "Hunting for Food: Guide to Harvesting, Field Dressing and Cooking Wild Game."

After several chapters of practical guidance and advice, Xenophon returns to general hunting philosophy in Chapter 12, emphasizing now the importance of hunting as education and preparation for war:

> With regard to methods of procedure in the hunting-field, enough has been said. But there are many benefits which the enthusiastic sportsman may expect to derive from this pursuit. I speak of the health which will thereby accrue to the physical frame, the quickening of the eye and ear, the defiance of old age, and last, but not least, the warlike training which it ensures. To begin with, when some day he has to tramp along rough ways under arms, the heavy infantry soldier will not faint or flag — he will stand the toil from being long accustomed to the same experiences in capturing wild beasts. In the next place, men so trained will be capable of sleeping on hard couches, and prove brave guardians of the posts assigned them. In the actual encounter with the enemy, they will know at once how to attack and to carry out the word of command as it passes along the lines, because it was just so in the old hunting days that they captured the wild game. If posted in the van of battle, they will not desert their ranks, because endurance is engrained in them. In the rout of the enemy their footsteps will not falter nor fail: straight as an arrow they will follow the flying foe, on every kind of ground, through long habituation. Or if their own army encounter a reverse on wooded and precipitous ground beset with difficulties, these will be the men to save themselves with honour and to extricate their friends; since long acquaintance with the business of the chase has widened their intelligence. (Xenophon circa 400 BC 1897: Ch. 12)

The connection between hunting and warfare is echoed throughout classical Greek literature, and so is the relationship between hunting, respectability, and sportsman-like behaviour. Plato, Xenophon's fellow student under Socrates, wrote the following in Book Seven of *Laws*:

> My friends, let us pray that you are never seized by any desire or love for hunting at sea, or angling or catching water-creatures at all, or taking the lazy way out using lobster-pots, which do your work for you whether you are asleep or awake. [...] Again, the desire to catch birds – calling for low and slavish cunning – may that idea not occur to any of our young either. That just leaves our athletes with the pursuit and hunting of land animals. Here, what they call night hunting, the sport of lazy men taking it in turns to sleep, is not to be recommended, any more than that hunting with breaks for rest when people rely on nets and traps, rather than the triumph of an energetic soul, to overcome the fierce strength of wild animals. That only leaves, for everybody, the best kind – the pursuit of four-footed animals using horses and dogs and their own bodily effort, when they hunt animals and get the better of them, in every case, by their own running, their own blows, and their own missiles; these are the ones who hunt with their own hands, and who care about the courage which comes from the gods. (Plato circa 400 BC 2016: Book 7, p. 823)

Trapping, fishing, creeling (using pots and other devices for trapping fish and crustaceans), and similar activities that are safe, easy, and done purely to obtain food are clearly looked down upon as activities of lesser men. "Athletes" hunt to train themselves for war and leadership. Only lowly working men were motivated to hunt by sustenance.

There were however plenty of these professional hunters who did small-scale hunting to supply the local markets with meat, fur, feathers, and leather. Greece also imported large quantities of ivory from Africa and India (Hughes 2014: 93–94). Though important, these hunting activities were without associated status. The respectable hunting activities of Greek sportsmen were physically taxing, not motivated by food, and required slaves and horses or dogs. Respectable hunting was an activity for the young athletes of Athens who could vote, had the means to own slaves, horses, and dogs, and who did not have to hunt for food. This was a class that could "live by the labor of others and consequently spent its time engaged in warfare and [...] horsemanship, athletics, hunting and philosophy" (Barringer 2001: 44). These Greek citizens and their horses could be called upon to go to war, so the connection between hunting and warfare was strong for a reason. Hunting and warfare also became thoroughly intertwined in Greek art, as noted by Barringer:

> By the end of the sixth century, the two types of imagery, hunt and battle, begin to fuse as martial elements occasionally spring up in hunting scenes: hunters wear helmets and greaves and carry shields; swords become primary weapons; spears play a secondary role; even chlamydes[5] are poised on a bent or outstretched arm, recalling images of shields; and in some instances, such armor manqué[6] is visually paired with a shield. In these cases, a visual metaphor for the hunt as battle is achieved, and the ideas of hunt and battle become one—that is, hunting is warfare. (Barringer 2001: 42–43)

That hunting needed to be physical training for combat influenced the ethics of the hunt. Executing animals in an enclosure (which would in modern times be termed canned hunting) as Egyptian royalty did would not be challenging and would therefore be useless as a training exercise. Respectable hunting required physical exertion, rough living, patience, keen observation, and similar qualities; the point of it would be lost if it were easy. These qualities tend now to be associated with "fair chase," the subject of Chapter 6, which is perceived as more ethical hunting.

Lastly, in relation to Xenophon's *Cynegeticus*, the very short Chapter 11 is also of some interest. It concerns the killing of "lions, leopards, lynxes, panthers, bears and all other such game" (Xenophon circa 400 BC: Ch. 11) in foreign countries using poisoned bait and pitfall traps. These activities were probably not considered hunting by the Greeks. As suggested by Anderson (1985: 56), this was just practical extermination of dangerous predators either in self-defence or to protect livestock. We make note of this here because using such methods, usually illegally (poaching), to protect humans and livestock is an issue in present day wildlife management debates that have connections with trophy hunting, especially in Africa. We will return to this issue in Chapter 5.

2.4 Ancient Rome

From Athens, we skip ahead a few hundred years to Rome, where canned hunting and the slaughtering of animals for amusement and public spectacle had become normalised as part of cultural life. From 200 BC to 300 AD, it was common to see humans and animals killed in the Roman amphitheatres. Lions were an important part of the spectacles, but they had to be captured in Africa and then transported to

---

[5] "A short mantle or cloak worn by men in ancient Greece" (Oxford English Dictionary).

[6] "Defective, spoilt; missing, lacking" (Oxford English Dictionary).

Rome. This was the case for much of the wildlife used for entertainment. The reasons for such spectacles were probably numerous, as Kalof summarizes:

> Roman spectacles of slaughter have long captured the attention of scholars, generating numerous theories to account for the public killing of exotic animals. It has been argued that animals were publicly slaughtered in Rome to establish a sense of control over the strange and the spectacular, to display exotic animals as the novel booty of foreign exploits, to legitimately kill less powerful beings, to prove wealth by destroying exotic and expensive animals, to fulfil a desire to watch animals by combining the pleasure of killing and torture with the pleasure of watching the swift and spirited activity of wild beasts, and to promote the spread of agriculture by reducing the number of crop-destroying animals. (Kalof 2007: 27)

Hunting, as understood by the Greeks, was less common in Rome. Few of the rich Romans who owned large tracts of land in Italy around 150 BC seemed to enjoy hunting. It was a hobby valued mostly by those who had a special admiration for Greek culture and lifestyle, and who surrounded themselves with Greeks. Some found the endeavour silly and pointless, when you could just have meat brought or watch animals being killed in the arena. Others thought the activity too frivolous; important people should have more important tasks to undertake (Anderson 1985: 83–90). A few scholars even questioned the morality of hunting (or rather the slaughtering of animals for entertainment that passed for hunting at the time); Plutarch implied, like Immanuel Kant much later, that brutality towards animals breeds brutality towards men, while Pliny the Elder mourned the wickedness of humans when compared to animals (Cartmill 1996: 40–45).

The Roman emperor Trajan (Emperor from 98–117 AD) and his successor Hadrian (117–138 AD) were both hunting enthusiasts. Trajan was praised by Pliny the Younger for his style of hunting, so opposite to that of Trajan's predecessor Domitian (81–96 AD). Domitian hunted in the ways of Egyptian royalty, where he—overweight and unable to ride—could kill trapped animals with a bow and without having to move. Of Trajan's hunting, on the other hand, Pliny wrote the following:

> What form of relaxation have you, save to range the forest glades, rouse the wild beasts from their lairs, ascend the huge ridges of the mountains, and set your feet upon the beetling crags? No hand of man assists you; no human footprint guides you; and in the midst of these occupations you visit the sacred groves and present yourself before the divine powers. Once this was the training of the young and their delight. These arts were the education of future generals—to compete against fugitive beasts in swiftness, against the bold in strength, against the cunning in skill. It was considered no mean honor, in time of peace, to keep off from the fields the invasions of wild beasts and, as it were, to deliver from siege the works of the countrymen. A false claim used to be laid to this glory by emperors who could not rightly assert it. Their usurpation took the form of having wild beasts tamed and broken by imprisonment and thereafter (what else?) turned loose for their own disport. Then, with feigned skill, they finished them off. But Trajan gives as much effort to the chase as to the kill, and the greatest labor, that of finding the game, brings him the greatest pleasure. (Pliny, Panegyricus 81.1-3, quoted in Anderson 1985: 101–102)

Pliny praises the Greek practice of training young men into future generals through the tough and challenging hunt, and he denounces hunting practices that require no toil or skill. Pliny was probably referring to Domitian, but the description would also fit the later Commodus (emperor from 180–196 AD). Commodus was known for his vicious slaughter of both animals and men in the Circus. Pliny's last sentence may be the most important, as it suggests a very advanced and unusual hunting ethic for its time; it suggests that the success of a hunt should rely on the joy of the chase rather than the joy of killing.

Last on the Greek and Roman agenda is Arrian, who was a Roman noble and friend of Hadrian's. He studied philosophy in Athens and become known as the "young Xenophon." Like Xenophon, Arrian also wrote a book about hunting that he called *Cynegeticus*, but Arrian's version is often translated as *Arrian on Coursing*. In this book, Arrian summarizes and dismisses large parts of the original *Cynegeticus* as outdated, with the knowledge of different breeds of dogs and horses and the skill in breeding them having advanced quite substantially. Agriculture had also progressed significantly, so Arrian's hunting is more purely for sport than Xenophon's. The meat—while secondary to the other benefits of the hunt—was still important in Xenophon's time, hence reference to traps and trapping techniques. Overall though, Arrian's *Cynegeticus* is much like Xenophon's, presenting in a similar guidebook fashion the most recent knowledge on dog breeds, horse breeds, and hunting techniques and equipment (Anderson 1985: 106–121).

2.5 Pre-Columbian America

Most literature about ancient hunting culture, much of which has been covered already, focuses on developments in Eurasia. Exploring and interpreting the history of hunting culture in America is hampered by the fact that Native American culture is mainly oral, so there is little primary literature to work from. While a philosophy of hunting surely developed among the original pre-Columbian Native, or Indigenous Americans (Paleo-Indians and later Archaic Indians[7]), little has been written down. What we know of this age, we know almost solely from archaeological artefacts.

Humans arrived and settled in the Americas probably between 30,000 and 15,000 years ago, most likely migrating there from Siberia across the Bering Strait, which formed a land-bridge (Beringia) during parts of the Pleistocene. Concrete evidence of human settlement dates to 15,500 years ago, but some argue for the existence of earlier cultures.[8] The earliest coherent culture that archaeologists are certain of is the Clovis culture that existed from around 9500 to 8900 BC. Clovis culture spread—with tremendous speed—across North- and South America, likely in

---

[7] To the best of our knowledge, the terms "Indian" and "American Indian" are—we suppose, contrary to popular belief—generally not perceived of as insulting by many Indians, if not used disrespectfully or in an insulting context. In recognition of there being no clearly unambiguous and agreed-upon general term, American Indians use a variety of names for themselves, including both of the above, and "Native Americans" (Thomas 2000: 17). Many Indians do however take issue with the description of their ancestors as "prehistoric," as it implies that they did not have a history before being given one by white American scholars (Thomas 2000: vii–viii).

[8] See for example Meltzer and Dillehay (1999), Meggers (2010: 7–15), and Pringle (2011).

pursuit of migrating mammoths and other large animals. The Clovis were hunters and gatherers who killed large mammal species like mammoths and bison, and smaller species like deer and rabbits with stone-tipped spears. As in many other places around the world, however, the end of the ice age, the fast spread of human populations, and a mass extinction of megafauna coincided. Whether this was because of climate change, overhunting or a combination is a point of some archaeological disagreement although overhunting seems to be winning ground in recent times (Dembitzer et al. 2022). Regardless of the reasons for their demise, horses, camels, mastodons, giant sloths, some subspecies of bison, the short-faced bear, the American lion, the dire wolf, the American cheetah, and mammoths all went extinct around the time of the late Clovis culture (Fagan 1995: 67–86; Thomas 2000: 19–29; Frison 2004: 38–52; Stewart et al. 2021).

From around 9000 BC, numerous other cultures replaced the Clovis, a major one of which was the Folsom. With mammoths extinct, the most important meat on the Great Plains had become the bison (*Bison bison*). Bison were still hunted on foot with stone-tipped spears like those used by Clovis,[9] but a strategy of stampeding large herds over the edge of a cliff was also often used. Places were this could be done are called bison jumps, and more than a hundred have been identified in North America. Some Plains Indians remained bison hunters until after Columbus, some settled in areas where fishing was the main source of protein, while others were mostly foragers and hunters of smaller game (Fagan 1995: 101–107; Calloway 1999: 14–15).

Between 7000 and 5000 BC a technique of grinding and polishing stones developed and was used to make more advanced tools, alongside the conventional method of flaking[10] for making projectile points. This period also records the first evidence of domesticated dogs (Meggers 2010: 18–21). Starting from around 5600 BC, the cultures are labelled Archaic Indian instead of Paleo-Indian, based on rather sudden changes in the designs of projectile points, notably notches in the sides for much stronger fastening to spear shafts. Aside from refined stonework, the hunting and gathering life of the Archaic Plains Indians of 1000 BC was presumably much like it had been for the Paleo-Indians since 10,000 BC. Bison were still the most important source of food for most cultures, and bison jumps were still a common way of killing them (Fagan 1995: 130–138; Thomas 2000: 52–60). Some tools and items were made from copper by 2000–1000 BC, but this does not signify metallurgy or a copper age as such, as the Indians had no knowledge of smelting. They simply hammered crude copper into different shapes (Meggers 2010: 23).

The earliest signs of cultivation of wild plants seem to be from around 7000 BC in South America and 5000 BC in North America, but it was a slow beginning. For a few thousand years, these early farming steps did not have any considerable impact. By about 4000 BC, farmed corn had become an important source of food in

---

[9] Hunting from horseback was a post-Columbian development. In general, as Calloway (1999: 14) writes, "[t]he way of life that popular stereotypes depict as typical of all Indians at all times never existed in most of North America and was not even typical of the Great Plains until the eighteenth and nineteenth centuries." Even the bow and arrow was not widely used until around 1000 AD (Calloway 1999: 15).

[10] Hitting tiny flakes off a rock until it is shaped to an edge and a point.

parts of South America, but agriculture did not drastically change the lives of North American Indians until around 500 AD. At this time bison hunting was also at its height and the bow and arrow was first used for hunting.

Along with the cultivation of corn and beans by American Indians on the eastern Great Plains, the making and use of pottery also became widespread in the first millennium AD. A more sedentary lifestyle naturally followed for the tribes that relied on corn farming. Some settlements were much larger than one might imagine from popular media representations, which tend to portray settlements as small camps:

> The Mississippian town of Cahokia was a thriving urban market center. Founded around A.D. 700 close to the confluence of the Missouri, Mississippi, and Illinois rivers and occupied for about seven hundred years, Cahokia had a population of between 10,000 and 30,000 at its height, about the population of medieval London. (Calloway 1999: 22)

With permanent settlements followed more burial mounds, which had not previously been part of the culture.[11] Testament to the continued importance of hunting even for farmers, bison skulls have often been found in these burial mounds (Fagan 1995: 134–139, 145–150; Calloway 1999: 16–24; Meggers 2010: 29–34).

With only archaeological evidence to analyse, it is hard to speculate about ancient American Indian hunting ethics. We know that bison jumping killed very large numbers of bison at a time, but they did not seem to go to waste. Indians would keep meat frozen or turn it into pemmican—a durable and high-energy mixture of fat, meat, and berries (Fagan 1995: 130, 133). There are no signs that bison were overhunted until after the Europeans introduced rifles and horseback riding in the fifteenth century. Besides, until around 500–1000 AD and maybe even later, North American archaic Indians still needed to hunt bison to survive. In contrast, Agriculture and animal husbandry had developed much earlier in Greece, Rome, and Egypt, so hunting there was not essential for survival.

2.6 From the Middle Ages to the Nineteenth Century in Europe

While the relationship between social standing and certain kinds of hunting had been evident for a few thousand years already, it would become more pronounced during the early Middle Ages in Europe. From around the seventh century already, hunting rights were being appropriated by European monarchs and most or all forms of hunting and trapping increasingly became illegal for anyone but the ruling elites. Likewise, the ownership of hunting dogs was restricted to the "hunting classes." Nobles were practically forced to participate in hunting or risk social ostracism, and boys of noble birth were taught to hunt or hawk[12] from the age of seven (Edward

---

[11] Although the first burial mounds are from around 3000 BC, it seems it was not a widespread tradition until much later (Calloway 1999: 21).

[12] Hawking, with a history in Egypt dating back to antiquity or even earlier and still a popular hobby in the Middle East today, was an extremely common form of noble hunting in Europe all

circa 1413: 125; Kalof 2007: 55; Almond 2011: 13–15). Hunting was a demonstration of power and influence not only because participation was exclusive, but also because the activity occupied and involved large parts of the countryside. A hunting monarch demonstrated and affirmed his power over the people through his physical occupation of the spaces in which they lived (Knoll 2004: 10; Poliquin 2012: 157–158). Deer hunting trophies were common in medieval castles from around the fifteenth century onwards. Trophies could be antlers only, or half-skull mounts with antlers, a style known in modern day trophy mounting as a European mount or half-skull mount. Such trophies also served the purpose of demonstrating the nobility's rightfully inherited control over the natural landscape and the peasantry:

> Generations of hunting trophies decorating hunting halls were read as evidence of generational status: rows and rows of antlers demonstrated that the current owner was nobly born. In a sense, collections of antlers functioned symbolically like portraits of ancestors. Both portraits and trophies endowed descendants with the weight and prestige of an ancestral lineage. Both fulfilled the demands of witness, nostalgia, and memory. Authority and authenticity lingered in both objects. In a sense, both were storytellers for future generations, who could say, Here is evidence of my status, and my father's status, and my grandfather's before him. (Poliquin 2012: 158–159)

Aside from hawking, the English nobility—according to Middle English hunting literature—had three preferred ways of hunting. The first and most noble of hunts was the par force hunt: the mounted stag hunt with hounds.[13] The parforce hunt would begin with a huntsman tracking down a suitable stag and returning to the hunting party with its general location. A tracking hound was then used to locate it again for the entire party and to flush it out, after which the stag was pursued by the noble or nobles on horseback, accompanied by hounds. The hounds were replaced with fresh ones as they tired. Eventually the stag was so exhausted that a nobleman or honoured guest could simply walk up to it and stab it. The second form of deer hunt was the drive. In a drive, servants would walk around in a wooded area and flush deer and other wildlife towards the waiting bowhunters. This is a very common way of hunting today as well, but with rifles or shotguns instead of bows, as bowhunting is usually more of a solitary endeavour. The third kind was the boar (*Sus scrofa*) hunt, which was essentially the same as the parforce hunt for deer. Boar are a far more aggressive and dangerous quarry though and would often turn around and attack hounds and hunters. The boar was usually dispatched with long spears (Rooney 1993: 3–5).

Just like the Greeks, English nobles had very particular ideas about honourable hunting. Here Rooney contrasts English hunting to French hunting:

---

through the Middle Ages and Renaissance, and foxhunting has been around since the late Middle Ages. However, despite both being historically very significant and—albeit to a lesser extent—still practiced today, we will not give them much attention here. We focus on trophy hunting, and while some foxhunters keep trophies, these are not the reason foxhunting is widely criticized. Hawking is less controversial than foxhunting or trophy hunting, and here neither do trophies play an important role for the hunters or their critics, though it is technically possible to hunt prey as large as roe deer with a golden eagle. For the reader especially interested in the history of hawking, there is probably no better book than John Cummins' *The Art of Medieval Hunting: The Hound and The Hawk* from 2003.

[13] A stag is an adult male red deer. It was also called a hart.

2.6 From the Middle Ages to the Nineteenth Century in Europe

There is no doubt that Englishmen also speared otters and rid their land of pests, but they didn't write about it. These were not noble activities, and certainly did not qualify as hunting. The hunt involved riding on horseback to a pack of hounds pursuing one of the noble quarry: the hart (male red deer), wild boar, or hare. The hind (female red deer), buck (fallow deer) and roedeer could scrape through as worthy victims, and in theory the wolf was a noble quarry, though the hunt for the wolf rarely features in English literature. (Rooney 1993: 3)

Hunting propriety became a discipline in itself for the English nobility. Since every noble hunted, individual nobles had to find other ways of standing out. While the athlete hunters of Greece sought to demonstrate their strength, endurance, courage, and perseverance, the British nobles in the Middle Ages competed in gentility; they sought to surpass each other in knowledge of the proper rituals and vocabulary of the hunt. This is splendidly explained and illustrated by Cartmill:

The fantastic and amusing terms that Elizabethan huntsmen applied to groupings of different species—a *sloth* of bears, an *exaltation* of larks, a *trip* of goats, a *richesse* of martens—are familiar today to lovers of quaint words. Every other aspect of the aristocratic hunt also afforded a different lexicon for each species of game. For example, a male fallow deer during its first six years of life was known successively as a *fawn*, *pricket*, *sorel*, *sore*, *buck of the first head*, and *buck*, while the corresponding terms for a red deer were *calf*, *brocket*, *spade*, *staggart*, *stag*, and *hart*; and some authorities held that the word *hart* should be reserved for a stag slain by a prince. To drive a red deer out of hiding was to *unharbour* it—but one was said to *rouse* a fallow deer, *find* a roebuck, *rear* a boar, *start* a hare, *bolt* a rabbit, *unkennel* a fox, *dig* a badger, *raise* a wolf, *bay* a marten, and *vent* an otter. Even the signals given to the dog pack were supposed to vary with the prey species. It was proper to say "to him, to him" when unleashing dogs to flush a red deer out of hiding; but if the quarry was a boar, the huntsman was expected to use the plural number ("to them")—a grammatical nicety presumably wasted on hunting dogs, even in Elizabethan England. The unhappy hunter who used the wrong pronoun to the dogs, called a brocket a pricket, or said "raise" when he meant "rouse" seldom escaped with mere derision. A public spanking with the flat of a hunting knife was the customary penalty for many such misuses of ritual language—for example, uttering the forbidden word "hedgehog" during a deer hunt. (Cartmill 1996: 64–65, emphasis in original)

Knowledge of hunting terminology became a way for nobles to set themselves apart in conversation from nonhunting commoners and from other nobles who did not keep up with the ever-changing decorum.[14]

---

[14] Hunting vernacular remains confusing today—and probably in many languages. Let us illustrate the complexity with some comparisons between Danish and British/American hunting terminology: In English, the word *deer* generally refers to the whole Cervidae family, including males and females of all ages. The Danish word for *deer* is *hjort*, yet *hjort* also shares etymology with the old English word *hart*, referring most commonly to a male red deer after its fifth year, which is also called a *stag*. The English word *deer* is etymologically related to the Danish word *dyr*, which just means *animal*, yet it can in a hunting context also refer to just Cervidae, like the word deer. When used by the general public, *hjort* refers to all male and female Cervidae of all ages, like the word *deer*; yet when used by hunters and wildlife managers it refers instead to males of red deer (krondyr), fallow deer (dådyr—*Dama dama*), and sika deer (sika—*Cervus nippon*) after their first year. Hunters and wildlife managers refer to all Cervidae as *hjortevildt*, which translates directly to something like *wild deer* or *deer wildlife*. The word *vildt* alone—beside being the adjective *wild*—corresponds to the English word *game*, meaning any huntable wildlife. Male deer in general are called *bucks* in American English. Buck is almost identical to the word *buk* in Danish, yet the Danish word *buk*

Flamboyant outfits, often entirely impractical for hunting, were another way for nobles to distinguish themselves. Hunters who were more dedicated to the actual chase, however, learned from the poaching peasantry that camouflage and freedom of movement yielded better chances of success. It gradually became commonplace for nobles to dress in comfortable greens for hunting success, and to dress flamboyantly when needing to impress. Some nobles would even dress in their reds and pinks but have camouflaged huntsman servants do the actual hunting on their behalf (Almond 2011: 48–51; 97–99).

Just as Plato's *Laws* showed us in ancient Greece, there was a clear distinction in the English aristocracy between hunting for sport and hunting for food. With hunting being the privilege of the ruling class—and their being more concerned with looking good than hunting efficiently—it tended to take a lot of nobles a long time to acquire very little meat. This was especially the case when an entire hunting party and their servants would spend a whole day exhausting a single stag. Yet venison was, especially during winter, an important part of the noble diet.[15] For food collection—as opposed to what the nobles considered hunting—deer would be driven into fixed nets and killed in a way that was otherwise considered a hunting method of the peasantry (Kalof 2007: 52–53; Almond 2011: 18–20).

The artificial breeding and trapping of deer and pheasants in enclosed parks was increasingly important in the fifteenth century onwards (Leopold 1986: 11). One of the reasons was that wild deer were getting scarcer due to overhunting. Another was that outright war on wildlife served as a proxy for war between English nobles:

> In addition to its traditional recreational role for the nobility, hunting also provided opportunities for military exercise and training, and the vigorous display of military skills and, during the relatively peaceful reign of James I, illegal hunting served as a substitute for war. Feuds between the English gentry were acted out by poaching game from the rival's deer parks, and illegal hunting took on a ritualized form of traditional land warfare, with the gentry dressed in full war attire, including chain mail and helmets. Manning writes that these subtle forms of rebellion were similar to the use of plays, masques and entertainment to air political grievances and engage in controversy, and poaching was considered preferable to the more overt violence of late medieval England. When applied to the practice of poaching

---

refers to the male roe deer (rådyr—*Capreolus capreolus*) and to male goats (and a male goat in English is also a buck) and not to any of the *hjorte*. While the male roe deer is not a *hjort*, roe deer are still somehow included under the umbrella of *hjortevildt*. A juvenile roe deer is a *lamb* (rålam) while juveniles of red deer, fallow deer, and sika deer are *calves* (kronkalv, dåkalv, sikakalv). The females of all deer in English are called *does*, which is etymologically related to the Danish *då*, although då refers only to the female fallow deer. The female roe deer is just *rå*, while the females of red deer and sika deer are *kronhind* and *sikahind*, adopting the old English word *hind* for a female red deer of more than two years.

[15] Meat in general had great cultural and nutritional importance. The King James Bible, translated in England in the early fifteenth century, serves as an example of how the words food and meat were considered synonymous: "29 And God said, Behold, I have given you every herb bearing seed, which is upon the face of all the earth, and every tree, in the which is the fruit of a tree yielding seed; to you it shall be for *meat*. 30 And to every beast of the earth, and to every fowl of the air, and to every thing that creepeth upon the earth, wherein there is life, I have given every green herb for *meat*: and it was so" (Genesis 1: 29–30, emphasis added).

as warfare, Manning notes that the military term 'havoc' means to spoil and pillage, for example, to completely destroy the deer in a hunting park. (Kalof 2007: 83–85)

Here is evidence once again of hunting's close historical association with warfare. Keith Thomas further notes that "hunting was notoriously a military exercise and a training ground for cavalry. It taught men and horses to endure hardships, to cross difficult terrain and to become experts in battle tactics" (Thomas 1983: 183). This subject is further explored in Daniel Beaver's *Hunting and the Politics of Violence Before the English Civil War* (2008).

With the excessive amount of hunting ritual and terminology followed eventually a tremendous amount of hunting literature. Most of it was first written in French or Latin—another way to separate nobility from the peasantry—and then translated to languages that commoners might understand. The French fashion of royal hunting had become the European standard in the Late Middle Ages and Renaissance (Cartmill 1996: 66) and these books were written for noble hunters by noble hunters. The most famous of these books was finished in 1389 by Count Gaston de Foix, also known as Gaston Fébus (1331–1391). *Livre de Chasse* was lauded, even more than 500 years later, as "indisputably the most famous hunting book of all times" (Baillie-Grohmann 1904: xii) and has been described as "probably the most influential hunting text" (Almond 2011: 30) and "one of the most influential texts of its era" (Klemettilä 2015: 3). Hunting was such an important activity for the gentry that the most notable literature about hunting would naturally be among the most notable literature overall. *Livre de Chasse* was translated into English and expanded between 1406 and 1413 by Edward of Norwich, the second Duke of York, as *The Master of Game*. The manuscripts were adapted to modern language for a modern audience in 1904 by passionate big game hunter William Baillie-Grohmann and his wife Florence Baillie-Grohmann. A foreword was included by a friend of William Baillie-Grohmann and a man we shall return to later, the 26th US President, Theodore Roosevelt (Roosevelt 1904b).

Like Arrian's *Cynegeticus* before it, and Xenophon's *Cynegeticus* before that, *The Master of Game* is principally a manual for hunting with hounds, and it follows the same formula as its predecessors. Aside from a brief description of the book's chapters, the prologue is mainly—viewed with modern eyes—a rather comical proof of why the hunter is a good and godly person who will not go to Hell. The principal argument is that idle hands are the devil's playthings, and hunting takes up so much time and effort that the hunter is too busy and exhausted to stray from a righteous path:

> When a man is idle and reckless without work, and be not occupied in doing some thing, he abides in his bed or in his chamber, a thing which draweth men to imaginations of fleshly lust and pleasure. (Edward circa 1413: 5)

One gathers from such comments that, again like both versions of *Cynegeticus*, *The Master of Game* was written with the privileged class in mind, which had only hobbies to occupy their time. Chapters 2–11 describe different quarry. Separate chapters are dedicated to hare, different deer, wild boar, wolf, fox, badger, wild cat, and otter. Chapters 12–18 are about greyhounds, alauntes, spaniels, and mastiffs and their manners, habits, sicknesses, and corruptions. Chapters 19–27 include all manner of advice for how to hunt and how to conduct oneself on a hunt.

It would be a mistake, as Richard Almond's *Medieval Hunting* demonstrates, to conclude from the aristocratic hunting literature and the illegality of commoners' hunting, that commoners did not kill and eat wildlife. The common peasants did not read or write books about hunting, but they did poach, even if the need for secrecy demanded that they used different methods than the nobles. *Livre de chasse* is among the few aristocratic hunting books to also describe the trap- and snare hunting methods used by peasants.

Commoners hunted everything that could be eaten, and almost everything *could* be eaten. Wolves and foxes were not, but were hunted with impunity regardless, as they were threats to livestock, competition for hunted species, and had valuable pelts. Venison was the best meat that a commoner could hope for, and it was sought after not only for the quality and quantity of the meat, but for the fact that it was also the best of the noble dining table. Venison gave peasantry the chance to feel like nobility for a short while, if only they dared to risk poaching deer.

The commoners' reliance on, and fascination with, hunting is evident in their folk heroes. Robin Hood was a hero and a rebel because he stole from the rich and gave to the poor, but also because he and his men poached the King's deer. Poaching was a no less important part of his rebellion than redistributing the wealth of the nobles. The deer symbolized nobility, and deer belonged to them exclusively; poaching and eating deer was both a pretension of nobility and a symbolic rebellion against it.

After being admired all through the Middle Ages and early Renaissance, criticism of hunting began to surface in the sixteenth and seventeenth centuries. In the early 1500s Desiderius Erasmus ridiculed it and all the pretence surrounding it in *The Praise of Folly*, and Thomas More noted the depravity of it in *Utopia*. He considered it a murderous, cruel, and primitive activity. Later and more significantly, Michel de Montaigne questioned whether humans were above animals in any morally significant way. This was a great affront to Christianity, which grants man dominion over animals. Shakespeare often used the hunt as a metaphor for both murder and rape, where it had hitherto more often been used as a romantically and erotically charged metaphor for playful chase and courtship. In art as well, the morbid sides of hunting became more commonplace. The antihunting sentiment became evident even as hunting authors increasingly found it necessary to add disclaimers or justifications in their hunting manuals. Cartmill relates this development to modern antihunting sentiments:

> Sixteenth-century antipathy to hunting thus seems to be a real phenomenon, not just a literary convention. This antipathy was something new; there is nothing much like it in the art and literature of the Middle Ages. A medieval master of the hunt would have been astonished if somebody had come up to him and accused him of being a murderer. In the five centuries that have gone by since Thomas More started talking that way, this sort of rhetoric has grown commonplace, and hunters today have learned to expect it. (Cartmill 1996: 84)

It should be noted, however, that the antipathy to hunting in the 1500s probably had more to do with the lower classes rebelling because of widespread starvation than it did with them sharing Montaigne's moral persuasion. In fact, the philosophy of the early seventeenth century went in a direction opposite to that of Montaigne. (Cartmill 1996: 77–91; Kalof 2007: 83–87)

This was a time of great progress in hard sciences like mathematics, astronomy, physics, and engineering, and this affected philosophers as well. Thomas Hobbes considered animals unable to engage in social contracts and therefore outside moral concern, and René Descartes considered animals' lack of speech "evidence not only that beasts have less reason than man, but that they have no reason at all" (Descartes 1637: 46). He also argued that

> they have no minds at all, and that it is nature which acts in them, according to the disposition of their organs, as we see that a clock, which is composed of nothing but wheels and springs, can count the hours and measure time more accurately than we can with all the care at our command. (Descartes 1637: 47)

If animals were mindless machines, then people were free to hunt, vivisect, and torture for the fun or education of it, which they did, according to Keith Thomas:

> The seventeenth century was an age when country gentlemen would entertain their visitors by putting their dogs to chase tame ducks or by throwing a goose or chicken into a pike-infested pond to watch its struggles. At country fairs there were contests at biting off the heads of live chickens or sparrows. Even a highly cultivated figure like the economist Dudley North could casually remark that, when he and other young gentlemen were with the English traders at Smyrna, they, for diversion's sake, 'tied a dog they had no great respect for to a bush; and fell on him with their scimitars till they had hewed him to pieces to show what heroes they could be upon occasion'. (Thomas 1983: 148)

The winds changed again in the eighteenth century. The idea of ethical vegetarianism goes back at least to Pythagoras (circa 570–495 BC), but it was extremely rare. It began, however, to appear often, and be promoted in both discussions and literature in the late seventeenth and eighteenth century (Cartmill 1996: 92–107; Kalof 2007: 97–99). Keith Thomas notes that "growing concern about the treatment of animals […] was one of the most distinctive features of late-eighteenth-century English middle-class culture" (Thomas 1983: 144). By the mid- and late eighteenth century, the Cartesian idea that animals are unable to feel pain was widely disputed. Jean-Jacques Rousseau and later Jeremy Bentham were notable proponents of the moral consideration of animals based on sentience and capacity for suffering. Rousseau stated that

> an end can also be made to the ancient disputes regarding the participation of animals in the natural law. For it is clear that, lacking intelligence and liberty, they cannot recognize this law; but since they share to some extent in our nature by virtue of the sentient quality with which they are endowed, one will judge that they should also participate in natural right, and that man is subject to some sort of duties toward them. It seems, in effect, that if I am obliged not to do any harm to my fellow man, it is less because he is a rational being than because he is a sentient being: a quality that, since it is common to both animals and men, should at least give the former the right not to be needlessly mistreated by the latter. (Rousseau 1754: 35–36)

Bentham later agreed that

> [t]he day may come when the non-human part of the animal creation will acquire the rights that never could have been withheld from them except by the hand of tyranny. The French have already discovered that the blackness of the skin is no reason why a human being should be abandoned without redress to the whims of a tormentor. Perhaps it will some day be recognised that the number of legs, the hairiness of the skin, or the possession of a tail, are equally insufficient reasons for abandoning to the same fate a creature that can feel? What else could be used to draw the line? Is it the faculty of reason or the possession of language? But a full-grown horse or dog is incomparably more rational and conversable than an infant of a day, or a week, or even a month old. Even if that were not so, what difference would that make? The question is not Can they reason? or Can they talk? but Can they suffer? (Bentham 1789: 143–144)

One of the early calls for animal rights was made ironically but ended up being quite influential. Thomas Taylor's *A Vindication of the Rights of Brutes* (1792) was an anonymously written response to Mary Wollstonecraft's 1790 *A Vindication of the Rights of Woman*. It was written as a reductio ad absurdum; if women should have the same rights as men, then the same argument could be made for animals or even plants. While made entirely in jest, the arguments Taylor advances are very similar, or identical, to claims being made in earnest today; namely, that the abilities to reason and speak of animals and their capacity for emotions should afford them rights, rather than any human dispositions toward them (Anker 2004).

The care for animals, with notable exceptions such as Montaigne, had hitherto been rooted in sentience and capacity for suffering (Rousseau and Bentham) or concern for humans, i.e., the worry that someone who tortures animals is in danger of moving on to people (Thomas 1983: 150–153). Immanuel Kant represented the latter line of thought:

> If a man shoots his dog because the animal is no longer capable of service, he does not fail in his duty to the dog, for the dog cannot judge, but his act is inhuman and damages in himself that humanity which it is his duty to show towards mankind. If he is not to stifle his human feelings, he must practise kindness towards animals, for he who is cruel to animals becomes hard also in his dealings with men. We can judge the heart of a man by his treatment of animals. (Kant 1963: 240)

A Vindication of the Rights of Brutes broke with this line of thought and advocated (again, ironically) for the rights of animals based purely on their inherent. The notion that animals are purely for the use and benefit of man was hereafter slowly being supplemented by thinking that fit the bill of Montaigne, and Taylor's ironic text: God cares equally for man, plant, and beast and they are to be treated well because they, aside from feeling pain, are God's creations and have intrinsic worth. In 1824 this animal rights and welfare movement in Britain resulted in the founding of the Society (later to become the Royal Society) for the Prevention of Cruelty to Animals, the first animal welfare organisation in the world.

Hunting did not go free from criticism. The previously described par force hunting of the European nobles in game parks did not reach its climax until the eighteenth century, yet it was increasingly being criticized for its cruelty. While it was defensible to kill quickly for food, sport hunting, where the purpose of the hunt very obviously was the hunt itself and not the resulting meat, met with increasing levels of disapproval. Deer hunting with hounds was hard to defend, as these park-raised deer could be killed for their meat in much faster and less stressful ways. In the case of hare hunting, it was obvious that the scant amount of meat yielded by a single hare did not justify the time and effort spent, unless the enjoyment of the chase was the real motivation.

Hunting with hounds or hawks was still the noblest form of hunting, but from the seventeenth century onwards, bird hunting with guns was becoming gradually more popular. This practice was also condemned by the new animal welfare movement (Thomas 1983: 146, 155–163; Knoll 2004: 12). In the legislation that followed, however, it was not the English upper classes' forms of hunting that got outlawed, which is perhaps a testament to the aristocracy's political power. Keith Thomas notes that the Acts of Parliament that were passed were "against cruelty to horses and cattle (1822), against cruelty to dogs (1839 and 1854), and against baiting and cock-fighting (1835 and 1849)" (Thomas 1983: 149). And as he continues, "it was the sports with a strong proletarian following which were outlawed – cock-throwing, bull-baiting and cock-fighting – whereas the gentlemen's fox-hunting, fishing and shooting survived unscathed" (Thomas 1983: 185–186).

So far, the European way of relating to nature had been to assert dominance over it. Gardens were carefully trimmed and sculpted, and wild nature was avoided and even demonized. Rousseau was among the first and most influential to break with this tradition with his admiration and celebration of wild, untamed nature. Wild nature was in nineteenth century Europe considered a spiritual and healthy respite from civilized city life. This view of nature as spiritual and harmonic, and individual animals as precious and valuable, sometimes contrasted with the simultaneously developing view of nature as brutal and hierarchical, inspired by Darwinist thought and breakthroughs in biology and palaeontology.

2.7 Colonial Hunting in India and Africa

In contrast to hunting itself, relatively little has been written about the history of the hunting trophy. Its beginnings are largely a matter of interpretation. Tools, weapons, and clothing have been made from bones, horns, teeth, skins, and pelts since the Palaeolithic, but these were probably not really what we would consider trophies, as their purposes were that of practical utility and survival. We have mentioned that engraved teeth and bones and other decorated animal parts have been found from the Upper Palaeolithic period about 50,000 years ago, but we can only guess as to their significance. Being items not of practical utility, they have likely been decorations, religious totems, pendants for luck, as well as physical evidence of hunting prowess, i.e., hunting trophies. Antlers—much later—have decorated the walls of medieval castles in Europe. While they were surely also trophies of the hunt, their main purpose was, as mentioned previously, to serve as evidence of the legitimacy of noble dominance and rightful ownership of land.

An argument could be made that hunting trophies as we fully know and think of them today have their roots mainly in the systematic measuring and scoring of trophies from colonial hunting in India and Africa in 1892 when taxidermist Rowland Ward published the first trophy record book, *Horn Measurements and Weights of the Great Game of the World: Being a Record for the use of Sportsmen and Naturalists*, later *Rowland Ward's Records of Big Game*. We shall return to Rowland Ward and his taxidermic work in Chapter 3.1 and to this and other trophy record keeping systems in Chapter 3.4. For now, we shall cover the advent of colonial hunting in India and Africa and the hunting tourism and trophy collecting trends that it facilitated. Our modern focus is mostly on Africa, as hunting has been banned in India since the Wildlife Protection Act of 1972 but as we will see, colonial hunting in India played a significant part of the development of modern trophy hunting.

John MacKenzie writes the following about hunting trophies in his immensely important book *The Empire of Nature—Hunting, Conservation and British Imperialism* (1988):

> It is difficult to establish exactly when the passion for trophies began. They had no doubt been used for practical purposes and as decoration in an earlier period, but not in any systematic way, neither as mementos of specific hunting occasions nor in the pursuit of records. The eighteenth-century interior was surely innocent of them. The craze for them was a product of the Romantic period. Roualeyn Gordon Cumming, who was born in 1820, relates in his memoirs that he had adorned the bedroom of his parental home at Altyre in Morayshire with trophies of his youthful, and presumably local, hunting exploits before he went to Eton. (MacKenzie 1988: 28–29)

We shall return to Cumming shortly. MacKenzie continues later, now speaking in more general terms about nineteenth century hunting trophies from India and Africa:

> To a certain extent the fashion for bedecking interiors with such trophies was another nineteenth century attempt to reinvoke the Middle Ages. The great stairwells of the Gothic house, the appearance of specific male sanctums like the billiard room or the smoking room, the effort to recreate the medieval 'hall', all lent themselves to adornment with trophies. In an age when most middle-class and aristocratic families had some members serving in India, later in Africa, or some other part of the world, it was not long before the local stags were joined by the heads and horns of exotic animals. By the end of the century such trophies were the fruits of upper-class tourism. Fine skins, particularly of the great cats, appeared on floors and walls. Like all collecting interests, it stimulated a desire to possess representative collections from specific areas and a desire for rarities to fill gaps. Soon bourgeois stay-at-homes were aping the taste, and trophies began to appear in homes further down the social scale, if only as hat racks. This effort to bask in the reflected glories of the Hunt generated a considerable trade in trophies which was to stimulate legislative action in African colonies in the early twentieth century. (MacKenzie 1988: 29–30)

Many of the nineteenth century European hunters in Africa came from previous military stationing in India. They were as such already thoroughly rooted in shooting traditions, reflecting an obsession with sport hunting among British officers in India. English hunting vocabulary even adopted the Indian words for hunter (shikar) and hunting (shikari), as it also did with the Swahili word for journey or expedition (safari). The shooting culture in India came partly out of boredom, partly out of the social expectations of that class (nobility being expected to hunt in Medieval Europe carried over to the British being expected to hunt in India), partly out of needing to protect local communities from tigers and their crops from deer, antelopes, and elephants, and partly out of hunting, once again, being training for combat. And there was plenty of combat to be had in India and Africa; British military men might hardly distinguish between shooting big game and shooting rebellious locals. Hunting was also a pastime of royalty and nobility in India, and some hunting expeditions put even the royal stag hunts in Europe to shame. Kings would take to the field for months at a time, their hunts accompanied by tens of thousands of men and horses, hundreds of elephants, and even most of their armies' artillery (Johnson 1827; MacKenzie 1988: 168–180).

While there were many hunting traditions in India, no quarry stands out more than the tiger, and as the interest in Indian trophies intensified in the nineteenth century, the tiger skin was the ultimate trophy. Yet the reasons for hunting tigers were also political and symbolic. As was the case with rulers in Ancient China and Egypt, killing the most powerful animals demonstrated a ruler's power over nature, and the tiger was considered the most powerful animal in India.[16] By hunting tigers, the British also picked up on traditions firmly held by the Mughal rulers, thus proving the fitness of the British to rule:

> In hunting tigers during the nineteenth century, then, the British drew upon two powerful meanings, both associated with Indian kingship. On the one hand, they consciously sought to emulate the Mughals and other Indian rulers by killing tigers. On the other hand, in the course of destroying these animals, they were also symbolically staging the defeat of Tipu Sultan and other Indian rulers who dared to get in the way. (Sramek 2006: 661)

William Storey similarly writes that the colonial hunting of lions and tigers in Africa and India symbolized "the triumph of culture over nature and of the colonist over the colonized" (Storey 1991: 149).

While hunting tigers from the back of an elephant is perhaps considered the pinnacle of British hunting in India, virtually every sort of hunting could be had somewhere in the country: hunting with rifles, shotguns, muzzleloaders, nets, or spears (for pigsticking from horseback), on foot or from the backs of horses or elephants, and with dogs in any number of roles. Hunters could even use cheetahs (*Acinonyx jubatus*) in the place of dogs for killing smaller quarry if they so desired. Hunting quarry in India could be wildfowl, deer, tigers, elephants, lions, rhinos, bears, boars, goats, sheep, gaur (also known as the Indian bison), and much more, available on plains, in forests, in mountains, deserts, or any of other habitats across India's vast landscape (Johnson 1827; MacKenzie 1988: 168–180).

Many books from the early nineteenth century cover the sporting life of the British colonists in India. Two of the most famous ones, as thoroughly covered by MacKenzie, were Thomas Williamson's *Oriental Field Sports* from 1808 and Daniel

---

[16] Overall, tigers are and were considered the most dangerous wildlife in India, but local exceptions may apply, as indicated by this paragraph in one hunting anecdote: "The bears of these parts are the common Indian sloth bear (*U. labiatus*), and their coats are so open and coarse that they yield but poor trophies. They do such a lot of damage, however, and are so dangerous to the people who have to frequent these forests, that as a rule it is right never to spare one. In my experience the people fear a bear here much more than a tiger" (Markham 1895: 38). Sloth bears in present day India may also be a problem and their attacks, frequently focused on the face of their victims, can produce horrific injuries, and may be fatal.

Johnson's *Sketches of Indian Field Sports* from 1822. Since its first publication in 1886, another useful source of Indian hunting anecdotes, trophy descriptions, and many odd Indian natural history observations has been the *Journal of the Bombay Natural History Society*. Many of these stories of "sport," as it is invariably called, also feature detailed descriptions by the authors of the trophy dimensions of their kills, demonstrating how prevalent the interest in detailed trophy scoring was already. The following is a quote from A. M. Markham's *Some Indian Stalking and Shooting* where he evaluates his newly shot sambar deer (*Rusa unicolor*):

> The head proved not so fine as it had appeared in life (they hardly ever do I think), but it is considerably over the average for these parts. The horns are nearly symmetrical—left $38\frac{1}{2}''$, right $38''$[17]; the spread between the points of the outer (shorter) tines is $34''$, and that from outside to outside of the beam, at the point of widest spread, is $37\frac{3}{4}''$. The brow antlers are $15\frac{1}{2}''$; the circumference of the beam, from above the brow antlers all the way up to the spring of the tines, is $6''$, and at the latter point it is $8\frac{7}{8}''$. Below the brow antlers the circumference is $8''$, and that of the burr is $10''$. (Markham 1895: 36)

In another edition, we have English professor Harold Littledale in the article *Rough Notes of Travel and Sport in Kashmir and Little Thibet*, providing an account of his markhor (*Capra falconeri*) kills after about two months of hunting in Kashmir and Little Thibet in 1888:

> May 31st. — Having now got ten markhor (measuring 46, 43½, 40, 38½, 34, 28, 25, 24, 22½, and 18 inches), and the weather being very hot, I mean to take it easy and loaf back to Kashmir. I shall perhaps pick up some ibex on the way in Derrell or Loyen Harda. If not, I do not much care, as I shall come this side again, and spend a winter over the passes, and try for some specially big heads up beyond Gilgit or in Chilas. (Littledale 1889: 115)

The casualness with which Littledale here describes his ten markhor trophies is remarkable, considering the trophy fees for markhor today, which can exceed $US150,000 (Geo News 2021).

In the late 1800s and early 1900s, it was clear that over a century of both local and British hunting had taken its toll on India's wildlife. Rhinos had been hunted down to very low numbers, and buffalo and gaur had disappeared from most of their historical range. Indian elephants (*Elephas maximus indicus*) had been one of the favourite quarries of the British, hunted both for sport and to protect crops, but they were protected from most hunting in 1879. Lions had been hunted almost to extinction by the late nineteenth century but survived in the Saurashtra district of Kathiawar peninsula, largely within the Gir Forest, in Gujarat. They had been hunted there by Junagadh nobles, who also invited British officers and Indian royalty to hunt. The decline in lion numbers was halted by the Nawab whose royal hunting ground was home to most of the remaining lions (Somerville 2019). Tigers (*Panthera tigris*), relentlessly pursued by the British for over a century for their skins and to protect livestock, became endangered by the 1930s. In excess of 80,000 tigers, twice that number of leopards, and 200,000 wolves were killed in India between 1875 and

---

[17] Despite Markham's apparent disappointment at the size, 38 inches is a very large sambar trophy by today's standards in most of its global range.

2.7 Colonial Hunting in India and Africa

1925, and Asiatic cheetahs (*Acinonyx jubatus*) were hunted to local extinction by the 1960s (Velho et al. 2012). As was almost always the case, the response to dwindling wildlife populations was mainly to restrict hunting privileges to a smaller and smaller colonial elite, while labelling local hunting as poaching (MacKenzie 1988).

Not until the Wildlife Protection Act of 1972 did India implement effective conservation measures to halt the threat of unsustainable hunting. This act effectively banned all hunting of wildlife but for a few circumstances such as the prevention of crop damage, elimination of problem animals, or scientific research. Hunting remains of great cultural importance in rural indigenous communities, however, but it has all effectively become poaching since 1972, and is as such practiced more extensively in remote rural regions, further from effective law enforcement. As noted by Aiyadurai et al., the collection and display of skulls and other animal remains as trophies can be a part of indigenous Indian hunting cultures:

> [T]hose who have converted [to Christianity] stop displaying animal skulls in their huts but continue to hunt. A stuffed red panda *Ailurus fulgens*, trophies of sambar *Rusa unicolor*, skins of barking deer *Muntiacus muntjak* and Chinese pangolin *Manis crassicaudata* are some of the wildlife parts that we found in government staff quarters that had been received as gifts or occasionally bartered for by Army personnel. (Aiyadurai et al. 2010)

We next turn to Africa, where the first large scale hunting by Europeans followed from the settlement of the Cape area of South Africa by the Dutch in the middle of the seventeenth century. The Dutch arrived at the Cape in 1652 to an area that is said to have been teeming with wildlife, of which there would be no trace (except for the many locations named after the wildlife that used to inhabit the areas) only 150 years later. Wildlife was considered to be competition to livestock for water and forage, a cheap source of food, clothing, and trade items. Wildlife was dispensed with as fast as the Dutch were able to explore and settle down as traders and farmers (Boers). Ostrich (*Struthio camelus*) eggs and feathers were very important for trade, but it was ivory that was the most valuable trading good. When the British took over in Southern Africa at the end of the eighteenth century and added a new influx of colonists, they built on Dutch experiences with trade. The British also greatly stepped up the killing of wildlife and trading efforts with Europe over the next few decades. Ostrich and many other species were gradually exterminated from the region, except where ostriches were being farmed for their eggs and feathers. Ivory, hides, horns, skins, and ostrich feathers constituted over 75% of the colony's trade in 1831. In the 1830s and 1850s the farming of merino sheep occupied vast tracts of land on which no competition for grazing was tolerated; all grazing game was exterminated to make room for sheep, and the frontier for hunting was pushed continuously further inland (MacKenzie 1988: 86–92).

Hunting in Africa in the late eighteenth century was not yet really a sport, and there seemed to be less interest in hunting trophies as well. Things started to change in the first decades of the nineteenth century, however, as European naturalists began taking a scientific interest in Africa's flora and fauna and bringing back specimens for museums. MacKenzie highlights Andrew Smith, a medical officer at the Cape between 1821 and 1837, as notably important. Smith's expeditions and his *Illustrations of the Zoology of South Africa* in five volumes was instrumental in inspiring a new breed of hunter-scientists. One of these was William Cornwallis Harris, who "lamented the decline and disappearance of game in the Cape Colony, and suggested that African hunters, confronted with vast herds in the interior, were bent on extermination" (MacKenzie 1988: 95). Apparently not one for self-reflection, when Harris himself reached the hunting frontier on his expedition in the 1830s, where game was still abundant, he

> filled his wagons with so much ivory that he was 'reluctantly compelled to leave the ground strewed with that valuable commodity'. Soon afterwards he found it so easy to kill elands that he salted their tongues and briskets for his homeward journey, leaving the carcases to the vultures. [...] On his journey from Kuruman to Mzilikazi's capital he left so many quaggas and wildebeest wounded that local hunters could emerge to finish them off with their spears. Before the expedition was over his tally of big game passed the 400 mark. (MacKenzie 1988: 95–96)

Nineteenth century colonial hunting in Africa is well represented by such select individuals as Harris, because it was often their writings and illustrations and souvenirs of the hunt—brought back to Europe—that inspired others to go to Africa on similar expeditions. For Harris, it was mostly his paintings that garnered interest in Europe.

Angela Thompsell in *Hunting Africa* (2015: 12–13) attributes the first widespread European interest in sport hunting and hunting trophies to Roualeyn Gordon Cumming. Brian Herne in *White Hunters* (1999: 3–5) similarly mentions Cumming as one of the first legendary European hunters in Africa. Cumming, a Scot, was stationed in India before going back to Scotland and then off to Africa in 1843. His killing spree was even more impressive than Harris', and so was his lack of self-reflection. He writes in the introduction to his book, *Five Years of a Hunter's Life in South Africa*, later republished in a shorter version as *The Lion Hunter in South Africa*, that

> the strong love of sport and admiration of Nature in her wildest and most attractive forms became with me an all-absorbing feeling, and my greatest possible enjoyment was to pass whole days and many a summer night in solitude, where, undisturbed, I might contemplate the silent grandeur of the forest and the ever-varying beauty of the scenes around. (Cumming 1850: vii)

This all sounds very romantic, but, as MacKenzie shows, it seems a little at odds with his behaviour in Africa as a hunter:

> By the late nineteenth century a hunting code had developed, and adherence to it was the mark of the gentleman sportsman. Cumming flouted every provision of that as yet unformulated code. He fired indiscriminately into herds; he shot females as often, or more often, than males; he indulged in long shots with little hope of success. He frequently shot at a water hole at night. [...] The meat was often left for scavengers, although it is true that anything up to 200 African camp followers accompanied him to take advantage of the remains he left in his wake. Often, however, there was more than people could eat. (MacKenzie 1988: 98)

Cumming at one time used around 57 shots to kill an elephant. As the firearms of the time were muzzleloaders, this must have taken quite a while, even if he used several rifles and had local gun bearers to reload them for him. Killing swiftly and painlessly was not something that concerned Cumming. He once paralyzed and wounded an elephant with bullets and then resolved to "contemplate" the elephant before killing it: "There I sat in my forest home, coolly sipping coffee, with one of the finest elephants in Africa awaiting my pleasure behind a neighbouring tree" (Cumming 1874: 15).

Whatever sentiments in favour of the consideration of animals were growing back in Europe at the time did not translate into any widespread condemnation of hunters like Cumming in Africa. His *Five Years* was a tremendous success, despite it being an unapologetic admission of all his horrible treatment of wildlife; he was celebrated as a hero on his return to Europe. Cumming was a proud trophy hunter before he even went to India, and the collections of trophies and souvenirs that he brought back from Africa filled nine wagons and formed the backbone of his collection and museum.

Other Europeans than Cumming became famous for hunting in Africa and returning home with huge collections of trophies over the next five decades. William Cotton Oswell, like Cumming, had been closely involved with David Livingstone's expeditions in Africa and returned with a collection of trophies large enough to establish his own museum in Kent. The second Lord Rothschild, Walter, opened a museum in Tring in England with African trophies of his own and had many prominent hunters collect for him. And Percy Powell-Cotton hunted in both Africa and Asia between 1887 and 1939 and turned his home in Kent into a museum of hunting taxidermy.

Until the mid-nineteenth century most colonial hunting took place in the southern part of the continent, but from 1850 to 1880, white hunters appeared in great numbers in Central Africa. One of these was the legendary British hunter and collector Frederick Courtney Selous (1851–1917), namesake of the Selous game reserve in southern Tanzania, one of Africa's largest wildlife and hunting reserves. Nineteen-year-old Selous left for Africa in 1870, wanting to be an elephant hunter and ivory trader. He travelled in 1872 to what is today Zimbabwe and was the most famous and successful ivory hunter in Africa by the age of 25. Part of Selous' obituary in the *Catalogue of the Selous Collection of Big Game in the British Museum (Natural History)* describes the hunting years of his career in Africa as such:

> Until 1881 Selous devoted himself mainly to elephant hunting, save for one holiday to England in 1875. In the former year he returned home for the second time, and shortly afterwards published 'A Hunter's Wanderings in Africa,' a book which ran through five editions, and took rank immediately among the classical works on African hunting. His keen interest in topography led to a succession of contributions to the Journal of the Royal Geographical Society, which in due time awarded him its gold medal. By the end of 1881 he was back at the Cape, and for the next six years wandered about the Matabele and other territory to the Zambesi, devoting himself mainly to procuring specimens of the African fauna for museums and private collections at home. Many of the finest specimens to be seen in the Natural History Museum were procured by Selous during these years. (Regan 1921: iv)

And *Nature* wrote this in 1919:

> [The] Selous collection of big game trophies, which has been presented to the Natural History Museum by Mrs. Selous, is, without doubt, the finest ever brought together as the product of one man's gun. It consists of some five hundred specimens shot by the late Capt. F. C. Selous, D.S.O.,[18] during a period of nearly forty years, some of the trophies dating from his earliest days' as a hunter. The greater part of the collection is African, but there are many specimens from Canada, Newfoundland, the southern Carpathians, and Asia Minor. (Nature 1919: 274)

After hunting and writing in North America for some years at the end of the century, Selous returned to Africa and accompanied Theodore Roosevelt on his 1909–1910 expedition for the Smithsonian. Captain Frederick Courtney Selous was killed in the first world war in East Africa in 1917 at the age of 65 by a German sniper (Regan 1921: v–vii).

Selous occasionally acted as one of the first "White Hunters" in the sense where it refers to hunting guides, better known in the industry as professional hunters (PH) (see Chapter 1.4). Professional hunters in Africa help hunting tourists—"clients"—track down desired and legal quarry, make sure it is safe to shoot, advise on where and how to shoot, etc. They are also armed and ready to shoot if the client only wounds the animal or if the party is charged by an animal. Such tourism, and therefore such a profession, took off in the late nineteenth and early twentieth century. Examples of other early professional hunters are Bill Judd, Alan Black, Arthur Cecil Hoey, and R. J. Cunninghame. The latter acted as the hunting guide for Carl Ethan Akeley (see Chapter 3.1) and for Selous and Roosevelt in 1909–1910 (Herne 1999: 4–14).

Professional hunters and their clients as well as professional ivory hunters relied on up to hundreds of local people to act as porters, servants, scouts, cooks, and other roles. They also relied on the goodwill of tribal leaders to access hunting grounds. Everything was generally bought with meat. The advanced weaponry of white hunters could dispatch large animals like elephant, rhino, hippopotamus, and buffalo much more effectively and efficiently than local people were capable of, so the communities were generally happy to trade their services for large quantities of meat (MacKenzie 1988: 121–146).

The rapid destruction of wildlife was obvious already in the 1800s, and British hunting legislation was introduced in South Africa in 1822—a first step towards officially replacing the customary rules of hunting that had guided behaviour before European colonization. Booth and Chardonnet note to this development that

> [t]he introduction of hunting regulations, modelled in European laws and systems, is largely responsible for the disappearance of customary hunting rules, especially the application of the colonial concept of *res nullius* to game. By depriving rural communities of the sovereign right to use 'their' traditional resources, this legal status rid hunters of responsibility towards something that no longer belonged to them. The nationalization of land and wildlife resources contradicted the notion of an ancestral right to community use. And those who traditionally enjoyed the prestigious status of hunter within their communities thus became common reprehensible poachers. (Booth and Chardonnet 2015: 5–6, emphasis in original)

July to November would be closed season, a hunting license was needed for the open season, and the killing of certain game (elephants, hippos, and bonteboks)

---

[18] Distinguished Service Order: A British military decoration.

would need a special license from the governor. This legislation expanded in the 1880s to 1900, as such species—adopting the hunting traditions and vernacular of medieval Europe—were designated "royal game."

Some blamed the colonialists for the decimation of Africa's wildlife, but others would blame local people; Selous for example claimed that 997 out of a thousand elephants were killed by them (MacKenzie 1988: 206). Legislators also seemed to be of the latter persuasion, as local communities were broadly excluded from hunting through a combination of not allowing them to own firearms and banning all other means of hunting:

> In most British colonies Africans were excluded from hunting. The process was begun by stringent gun laws, introduced in Kenya and Southern Rhodesia [today Zimbabwe], for example, in the late 1890s. In some colonies African licenses could be taken out, but since most African techniques were banned few actually did so. As the game legislation was developed the contrast between humane and cruel killing was often explicitly introduced, as in the 1928 Game Amendment Ordinance in Kenya. Humane killing could be effected only by modern firearms; pitfalls, snares, traps, nets and drives were prohibited. In Kenya the sale of snares, gins and traps was forbidden in 1937, while in Southern Rhodesia the possession of snares was made illegal in 1938. Sometimes it was specifically enacted that no 'native' could hunt except as an assistant to a European, in which case he could not bear firearms. In Kenya a 'resident' for the purposes of a resident's license was defined in such a way as to exclude Africans. (MacKenzie 1988: 209)

The destruction of game was considered a necessary part of the European settlement of Africa, but in the late nineteenth century, the colonial administrations tried to halt this destruction by ensuring that locals were excluded from hunting and that the value of game was converted "from a direct economic resource in ivory, meat, hides and skins into an indirect one, a means of raising revenue from 'sport' and tourism" (MacKenzie 1988: 201).

So, while hunting tourism—trophy hunting—had for a long time been glamourized in adventure literature by European hunter-authors and by the first ranking of trophies in *Rowland Ward's Records of Big Game*, it took hold primarily as a direct consequence of political decisions in early twentieth century colonial administrations. This development also led to the establishment of game parks for hunting and national parks for nonconsumptive tourism where even the colonial elite would later be banned from hunting (MacKenzie 1988: 200–224).

It was however not until after the First World War that the establishment of the hunting industry as we know it today really took off. Kenya pioneered wildlife conservation through hunting with the East African Professional Hunters Association in 1934. This lasted until hunting was banned there in 1977. This association was instrumental in developing wildlife tourism and conservation practices. In 1966 came the Tanzanian Professional Hunters Association, which is still in operation today. The system of restricting hunting to hunting blocks, subject to license fees and hunting permits, was likewise implemented in Kenya first (Booth and Chardonnet 2015).

2.8 Theodore Roosevelt and North American Conservation

The nineteenth century in America was mostly a continuation of the European colonists' previous two centuries of religious warfare against America's wilderness and indigenous population. But attitudes to wildlife were slowly starting to change (Payne 1996: 9–28). The first closed seasons, hunting bans, and bans on certain destructive hunting methods were instituted in the late eighteenth century. In the late nineteenth century, warden systems, protection of non-game birds, hunting license requirements, and bag limits also began to appear. Arkansas was the first state to ban market hunting (the commercial hunting of species for meat) in 1875, and by 1880 all states had at least rudimentary game laws (Leopold 1986: 12–14).

An ideal of the "hunter" developed in America as some came to admire the European-inspired romantic idea of the one-with-nature hunter, who ventures into uncharted, wild nature, cherishes the simple life in the wild, and connects spiritually with his quarry; an idea that is heavily criticized today for its racist and colonialist portrayal of the home of many indigenous peoples as "wilderness" and "pristine nature." There was little frontier left in Europe however, which made the character all the more interesting. This hunter archetype came to represent frontiersman legends from nineteenth century American folklore. Examples of such folk heroes include Grizzly Adams, Daniel Boone, Davy Crockett, Jim Bridger, and Kit Carson. Most famous of all was perhaps the fictional Nathaniel "Natty" Bumppo, aka Hawkeye, aka La Longue Carabine, aka many other aliases, from James Fenimore Cooper's *Leatherstocking* novels. The best known of Cooper's novels is *The Last of the Mohicans*, more famous yet because of the 1992 movie that starred Daniel Day-Lewis as Hawkeye.

Toward the end of the nineteenth century, when European imperialism had spread to most of the world and North America was settled from east to west, another archetype of hunter appeared, thanks in no small amount to European hunting in Africa:

> The Great White Hunter is exactly the opposite [of the Hawkeye hunter]: an upper-class white man wearing a conspicuously "civilized" costume, who leads an army of servile natives on a foray into the bush to kill for thrills, glory, and trophies. For the Romantic hunter, the Man in the Buckskin Suit, the hunt is an act of loving communion with nature. For the imperial and Darwinian hunter, the Man in the Pith Helmet, it is an assertion of his competitive superiority over the natives and other local fauna. (Cartmill 1996: 137)

While this perfectly describes the negative modern trophy hunter stereotype, it also rather aptly describes the 26th president of USA (1901–1909), Theodore Roosevelt. Roosevelt is a central figure in hunting history and the understanding of trophy hunting, so his life will be covered in some depth here. Firstly, he is one of the best historical examples of a hunter who sincerely loved wild nature and natural science, and lived for conservation, but who also just really liked to shoot at animals with guns for the fun of it. It is a combination of characteristics that appears to both puzzle and anger many opponents of hunting (and many hunters), so it is an important type of hunter to examine.

2.8 Theodore Roosevelt and North American Conservation

Secondly, Roosevelt is one of a few key individuals responsible for putting conservation on the political agenda in the United States of America and making the public aware of it in the late 19th and early twentieth century. He was personally involved with this shift as the founding father of the Boone and Crockett Club, an organisation whose importance for the North American model of wildlife conservation and modern trophy scoring methodology cannot be overstated.

Theodore Roosevelt, Teedie to his parents and Teddy to his later admirers (a nickname he himself greatly disliked), had a privileged upbringing. His father, Theodore Roosevelt, Sr., was a rich glass merchant, involved in the founding of both the Society for the Prevention of Cruelty to Animals and the American Museum of Natural History. His uncle, Robert Barnhill Roosevelt, authored several books about America's declining wildlife. Both were influential in shaping Theodore's life as a naturalist and conservationist. Theodore was very curious about the natural world from a very young age. He wrote his first essay about ants at the age of seven,[19] and he started his own natural history museum at the age of nine. He wrote diaries about his animal observations and later made attempts at scientific essays about them. On the occasions where his parents were travelling without him, his letters to them were mostly requests for specimens for his museum and excited reports of the goings-on of nature as it surrounded him (Cutright 1985: 1–15).

What set Theodore apart from other children who might have been interested in the outdoors is that even as a child, his interest was always deeply scientific. He hated anthropomorphized talking animals and any portrayals of animals that were unscientific or dumbed down, as they commonly were (and are) in children's books about wildlife. One of his favourite books as a child (though he would surely have required the aid of his family to grasp any of it) was Darwin's *The Origin of Species*, which he often carried around as an adult (Brinkley 2010: Ch. 1, I–II; Ch. 2, IV). Evolution fascinated him, and he was delighted that humans are the biological relatives of apes. As Brinkley writes, "Roosevelt swallowed natural selection hook, line, and sinker. For the rest of his life, in fact, he used evolutionary theory as his guiding light; it illuminated his views on everything from politics to geography to fatherhood" (Brinkley 2010: Ch. 2, IV).

Theodore disdained cruelty to animals. His diaries portray a boy who was deeply troubled whenever he saw animals caused unnecessary pain, and the Roosevelts were instrumental in the 1866 founding of the American Society for the Prevention of Cruelty to Animals by lending their support to Henry Bergh. Theodore took great pride in what his family did for the humane movement of the 1860s and 70s, and he followed a code for sportsmanlike hunting (sometimes in theory more so than in practice) and collecting from a very young age (Brinkley 2010: Ch. 1, III; Ch. 2, I–III).

---

[19] Cutright writes that the title of this lost essay was *The Foregoing Ant*, while Brinkley names it *The Foraging Ant*. It could be that young Roosevelt meant to write about the foraging ant but misspelled it. This is not unlikely, considering that he also wrote about both "soilder ants" and "solider ants." Roosevelt seemed to have a penchant for typos, misspellings, and made-up words (Cutright 1985: 6, 11–13: Brinkley 2010: Ch. 1).

While Theodore felt terrible about the torture or senseless killing of animals, he was never opposed to the killing of animals for food or science, and he no doubt—and not wholly incorrectly—considered his own pursuits very scientific. At around age twelve, in his continued pursuit of faunal scientific knowledge, Theodore picked up an interest in taxidermy, and he began learning the craft from esteemed taxidermist John Graham Bell, who used to travel with the legendary ornithologist John James Audubon, one of Theodore's favourite authors. Studying under Bell, Theodore built on his keen interest in ornithology (he could already name most birds local to the New York area by either flight or song) by learning how to prepare birdskins,[20] as they were (and still are) prepared for museums.

With Theodore's interest and training in taxidermy came an impressive amount of killing, especially after his father presented him, at the age of 14, with his first shotgun. His boyhood journals document his observations and collection of specimens in detail. He shot mostly birds, but any wild animal that crossed his path was in danger of ending up in his museum (Cutright 1985; Brinkley 2010: Ch. 2). As noted by the ornithologist Elliott Coues, however, Roosevelt was simply doing what any good ornithologist must do to learn: "The true ornithologist goes out to study birds alive and destroys some of them simply because that is the only way of learning their structure and technical characters" (Coues 1890: 15).

As Theodore was about to enter Harvard University at age seventeen, he was still determined to be a man of science, but only if he could conduct his research in the field. He did not have much respect for naturalists who only observed nature through a microscope. Theodore was very engaged in the Nuttall Ornithological Club and Harvard Natural History Society and kept busy with lots of other memberships and sports. He spent much of his time in the Adirondack Mountains and Maine Woods when his studies allowed it. His determination to become a man of natural science changed in his senior year and he (heavily influenced by his first wife-to-be Alice Lee) decided to pursue a career in law instead. While studying in Columbia Law School, he rose to prominence in the Republican Party, and was in 1881 (aged 23) elected to the New York State Assembly. As he no longer had enough time to curate his childhood museum, Roosevelt donated a large part of his collection, including 622 birdskins, to the Smithsonian. This acquainted him with famous ornithologist Spencer Fullerton Baird, who was museum secretary at the time (Cutright 1985: 97–136).

Between 1884 and 1892, Theodore split his time between cattle ranching and wilderness hunting in far-away Dakota and pursuing his political career in New York City. His seclusion in Dakota, while initiated in 1883, was prolonged as a reaction to the tragic, unrelated deaths of both his wife and his mother on February 14, 1884. He did the bulk of his life's big game hunting in these years, and wrote *Hunting Trips of a Ranchman* in 1885, *Ranch Life and the Hunting Trail* in 1888, and *The Wilderness Hunter* in 1899, three of his most famous hunting treatises (Cutright 1985: 144–161; Brinkley 2010: Ch. 6). The following lines about the destruction of

---

[20] A birdskin is the skin, head, feet, and feathers of a bird, removed from the body, stuffed, and sewn back together. It is not mounted in any kind of pose, as it is filled with soft stuffing only.

the American buffalo—also known as the bison—by hunters are from *Hunting Trips of a Ranchman*:

> Gone forever are the mighty herds of the lordly buffalo, a few solitary individuals and small bands are still to be found scattered here and there in the wilder parts of the plains; and though most of these will be very soon destroyed, others will for some years fight off their doom and lead a precarious existence either in remote and almost desert portions of the country near the Mexican frontier, or else in the wildest and most inaccessible fastnesses of the Rocky Mountains; but the great herds, that for the first three quarters of this century formed the distinguishing and characteristic feature of the Western plains, have vanished forever. [...]
>
> Less than a score of years ago the great herds, containing many millions of individuals, ranged over a vast expanse of country [...]
>
> But since that time their destruction has gone on with appalling rapidity and thoroughness; and the main factors in bringing it about have been the railroads, which carried hordes of hunters into the land and gave them means to transport their spoils to market. [...]
>
> The bulk of this slaughter was done in the dozen years from 1872 to 1883; never before in all history were so many large wild animals of one species slain in so short a space of time.
>
> The extermination of the buffalo has been a veritable tragedy of the animal world. (Roosevelt 1885: 258–260)

All three books—but *The Wilderness Hunter* more so than the others—are testament to Roosevelt being both a gifted writer and one of the foremost experts on North American big game of his time. Brinkley describes him as "one of the most astute wildlife observers our country has ever produced" (Brinkley 2010: Ch. 1).

Roosevelt's life converged with the life of George Bird Grinnell in 1887, and this led to the founding of the Boone and Crockett Club soon after. Yet despite Roosevelt sometimes being credited with the founding of the American conservation movement, it is more accurate to think of the late nineteenth century conservation efforts as a culmination of processes that began a century earlier. John F. Reiger's *American Sportsmen and the Origins of Conservation* (1975) demonstrates that the origins of North American conservation ought to be attributed to a widespread concern for wildlife among hunters and anglers in the 1870s. This too, however, was merely a renaissance in notions of sportsmanlike hunting and angling (Reiger 2001: 1–4). Sixth US president John Quincy Adams wrote the following in his diary in 1787:

> Rain'd in the fore part of the day, but cleared up in the afternoon. I went with my gun down upon the marshes, but had no sport. Game laws are said to be directly opposed to the liberties of the subject; I am well persuaded that they may be carried too far, and that they really are in most parts of Europe. But it is equally certain that where there are none, there is never any game; so that the difference between the country where laws of this kind exist and that where they are unknown must be that in the former very few individuals will enjoy the privilege of hunting and eating venison, and in the latter this privilege will be enjoy'd by nobody. (Adams 1903: 23)

This passage neatly ties together feudal hunting in Europe and hunting in the US. While it was obvious early on—as indicated in the quote above—that a lack of hunting regulations inevitably led to a lack of game, hunting laws and privileges that "smacked of the same Old-World tyranny America had just thrown off" (Reiger

2001: 5) were so contrary to the American ideals of freedom and liberty that they were a hard sell in the New World. This could be said for any kind of government enforced conservation, which unavoidably must trample some amount of personal freedom. In the late nineteenth and early twentieth century, American sportsmen recognized both that something had to be done if there was to be any hunting and wildlife in the future, and that appeals to authority were bound to fail. Instead, a "code of the sportsman" was developed, which appealed not to authority but to a sense of propriety and decency in the individual hunter or angler.

This was a code that, when followed, turned the mere hunter or angler into a sportsman:

> Both in books and magazine articles, early sportsmen continually pushed for the adoption of an Old World-derived code of conduct in the field. They emphasized that sportsmen should be "gentlemen," suggesting that they were members of the upper classes who should have nothing in common with the lower-class "market" (commercial), or "pot" (meat), hunter and fisherman. The first group presumably killed without restraint, because the more game they took, the more money they made; the second group supposedly killed simply to fill stomachs, without any appreciation of the aesthetics of the sport and the rules for giving game a fair chance of escape. Sportsmen, too, ate what they bagged, believing that wildlife should never be wasted, but they demanded that it must be taken in an appropriate way, meaning that the experience and skill of the sportsman should be challenged before being able to make a kill. (Reiger 2001: 7)

It is remarkable how similar this passage is to the passage from Plato's *Laws*, quoted earlier. As Plato celebrated the athlete, so did this group of nineteenth century hunters and anglers celebrate the sportsman and gentleman, and all showed equal disdain for meat- and market hunters. It is important to note, however, that Plato objected to the methods of meat- and market hunters because their hunting was not proper training for war and leadership, while the nineteenth century sportsmen worried about the lack of concern for conservation that would follow from a material motivation.[21] The sportsman, on the other hand, does not hunt to sell and will not go hungry without the meat. His motivation is his own enjoyment, which, along with his esteem in the eyes of fellow sportsmen, is lessened if he behaves unsportingly.

A few hunting manuals, promoting the sportsman's code, started to appear around 1780, and the genre gained in popularity hereafter, with more titles on the market from the 1820s. *The American Shooter's Manual* from 1827 by Jesse Y. Kester is one of the early examples. Kester prefaced his book with these words:

> Living in a country destitute of game laws, and almost without any legal restrictions, in regard to its destruction, every man, whose leisure, or circumstances will permit, may become a shooter. But of the multitude who shoot, few indeed, will be found entitled to the appellation of sportsmen ; they are very generally game killers, and nothing more. (Kester 1827: x)

---

[21] This is not to say that hunting was no longer considered training for war in the twentieth century. Theodore Roosevelt, for example, wrote that "[i]n the Civil War the soldiers who came from the prairie and the backwoods and the rugged farms where stumps still dotted the clearings, and who had learned to ride in their infancy, to shoot as soon as they could handle a rifle, and to camp out whenever they got the chance, were better fitted for military work than any set of mere school or college athletes could possibly be (Roosevelt 1906: 150).

To turn America's game killers into sportsmen (and common folk into gentlemen), Kester sought to educate on such subjects as the proper training of and hunting with various species of dogs, proper shooting (at birds in flight; never sitting) without wounding, gun care and safety, and the various species of game birds (partridge, wood cock, snipe, pheasant, rail, grouse, duck, plover, and wild turkey) (Kester 1827: 144–200; Reiger 2001: 9–11). The elitism in Kester's writing is inescapable: the sportsman is one who cares for his quarry of today and his hunting of next year, but he is also just a better class of citizen because of the great knowledge he possesses. This separation of classes based on knowledge of hunting of course recalls feudal Europe again. The difference is that in the 1830s and 40s in the US, the important knowledge was of zoology, and such relevant subjects as dog training and gun handling, rather than just familiarity with the everchanging rituals and language of the hunt.

An influential writer of hunting and fishing literature in the 1840s and 50s was Henry William Herbert, writing under his better-known pen name of Frank Forester. He wrote in a period of protracted war between sportsmen—fighting for game laws, habitat protection and their "code"—and what Kester called "game killers"—fighting against the tyranny of the elite for American freedom and oftentimes their livelihood. In 1848 Herbert published his famous work, *Frank Forester's Field Sports of the United States and British Provinces of North America* in two volumes, in which he writes in the preface to the first, that

> it appeared to me that such a work was needed, at this juncture, and that this publication might possibly tend, in some small degree, to avert the impending doom which seems to have gone forth from the democracy of the land against game of all sorts. (Herbert 1868: ix)

He continues in the introduction, recollecting and repeating words that he wrote in a previous publication:

> "There is perhaps", I remarked, "no country in the world which presents, to the sportsman, so long a catalogue of the choicest game, whether of fur, fin, or feather, as the United States of North America ; there is none, probably, which counts more numerous, or more ardent, devotees ; there is none, certainly, in which the wide-spread passion for the chase can be indulged, under so few restrictions, and at an expense so trifling.
>
> "Yet all this, notwithstanding, it is to be regretted greatly that there is no country in which the nomenclature of these *feræ naturæ*, these roving denizens of wood, wold and water, is so confused and unscientific ; none in which their habits are so little known, and their seasons so little regarded ; none, in which the gentle craft of Venerie is so often degraded into mere pot-hunting ; and none, in which, as natural consequence, the game that swarmed of yore in all the fields and forests, in all the lakes, rivers, bays, and creeks of its vast territory, are in such peril of becoming speedily extinct. (Herbert 1868: 11–12)

Herbert does not lay the blame for the perilous situation on any one party alone. He blames the ignorance and the selfish, short-sighted practices of farmers, hunters, poachers, and city dwellers alike. Herbert spoke in favour of legal measures backed by consequences for lawbreaking; he argued that there was a "very singular, if not incomprehensible, characteristic of the people of the United States, to disregard and violate all laws, even laws of their own making" (Herbert 1868: 18–19). As to why the rural population of lower classes may do so, he speculated that they are of a perception that these laws are passed by and for the rich and privileged in the city (Herbert 1868: 19), thus connecting squarely the lawbreaking mentality with European hunting history.

Herbert's message did not stand alone as sportsmen of the time were characterized—despite his insistence that conservation was for everyone and unrelated to European elitism—by affluence and influence. The sports clubs that they established counted many prominent names among their members,[22] and the books, magazines, and periodicals they published gained traction up through the 1850s. They advocated bans on market hunting and fishing, closed seasons for game, the sportsman's code, restraint in killing, permanent protection of nongame wildlife, and wilderness preservation. This progress came to a halt and a setback with the civil war in 1861 however (Reiger 2001: 40–44).

By 1870 the railroad had made almost every part of the land easily accessible. Combined with increasingly cheaper firearms, ammunition, and fishing gear, wildlife faced greater pressure than ever. Conflicts between sportsmen and commercial hunters and fishermen escalated and frequently ended in physical confrontations. A more civilized battle was fought by sportsmen with the written word, and four very important magazines were soon to be published: *American Sportsman* from 1871, *Forest and Stream* from 1873, *Field and Stream* from 1874, and *The American Angler* from 1881 (Cutright 1985; Reiger 2001: 46–49).

Forest and Stream became the most important of the four, due in no small part to editor Charles Hallock, who, according to Reiger, was ahead of his time on conservation issues:

> Less than ten years after the Civil War, he was calling for protection of watersheds and scientific management of forests; establishment of uniform game laws dictated by geography, habitat, and migration patterns, rather than judicial accident; creation of a science and industry of fish culture that would develop new strains of game fish and restock depleted waters; abatement of water pollution; and experimentation into methods for domesticating and "farming" fur-bearing animals. (Reiger 2001: 49–50)

His business partner and fellow Yale alumnus George Bird Grinnell took over *Forest and Stream* in 1880 and was owner and editor-in-chief until 1911. He also founded the National Audubon Society in 1886. Hallock and Grinnell's editorial work at *Forest and Stream* prompted American sportsmen to organize themselves into clubs at an unprecedented rate. In 1878, five years after the first issue, hundreds of sportsmen's clubs had been formed. Grinnell continued throughout his career to inspire sportsmen and criticize the American ideals that placed progress and consumption above all else (Reiger 2001: 49–51). He was also adamant that lack of hunting legislation was not the major concern; lack of enforcement was (Reiger 2001: 93).

At this junction, we meet up with Theodore Roosevelt again. Grinnell and Roosevelt became friends after Roosevelt wrote *Hunting Trips of a Ranchman* in 1885, about which Grinnell had some critical comments. Their friendship and

---

[22] See Reiger (2001: 62) for a long, yet far from exhaustive list of famous people of the eighteenth–nineteenth centuries who hunted or fished.

shared concerns led to the founding of the Boone and Crockett Club (B&C) in 1887, named after famous pioneers and hunters Daniel Boone and Davy Crockett. Roosevelt was president of B&C from 1888 to 1893, and his efforts and political influence here, combined and coordinated with Grinnell's efforts as editor of *Forest and Stream* and the efforts of many other prominent B&C members, made a significant difference for forest and wildlife preservation (Cutright 1985: 165–187). The Yellowstone Park Protection Act of 1894, for example, allowed army personnel to combat widespread poaching problems in the country's first national park. In David Petersen's words, "B&C was significantly, if not single-handedly, responsible for rescuing the American bison, the wapiti, and other then-endangered species from looming oblivion" (Petersen 2010: 47). The crowning achievement and "victory" of the sportsmen movement over commercial interests, if one can single out anything, may be the passing of the Lacey act in 1900, named after congressman John F. Lacey (fellow sportsman and B&C-member from 1902). The Lacey act proved of great import in protecting wildlife from commercial exploitation. It regulated wildlife trade, punished offences, and combatted invasive species. Thomas Dunlap explained about the progress in conservation laws around the period where B&C's influence was greatest, that

> [p]rior to the turn of the century individual states had regulated the killing of a few species, but only to assure a continued supply of meat. They had done so by a few laws that had rarely been enforced—and then only by the local sheriff. After 1880 several states passed new wildlife laws: they regulated the killing of birds and mammals; they limited bags, seasons, and methods; and they set up game commissions and appointed game wardens. Their purpose, too, was new: to assure a supply of game for sport hunting. (Dunlap 1988: 51)

Roosevelt's political career progressed rapidly in the last few years of the nineteenth century. He was first elected governor of New York State in 1898, ending his career in the navy, and then elected vice-president in 1900. When president William McKinley was shot in September 1901, Theodore Roosevelt, at 42, became the youngest president in American history.

Wasting no time, Roosevelt sent, in his 1901 State of the Union speech (and in many other speeches and writings to come during his presidency), a clear message about the priority of forests:

> The fundamental idea of forestry is the perpetuation of forests by use. Forest protection is not an end of itself; it is a means to increase and sustain the resources of our country and the industries which depend upon them. The preservation of our forests is an imperative business necessity. [...]
>
> The forest reserves will inevitably be of still greater use in the future than in the past. Additions should be made to them whenever practicable, and their usefulness should be increased by a thoroughly business-like management. [...]
>
> Certain of the forest reserves should also be made preserves for the wild forest creatures. All of the reserves should be better protected from fires. Many of them need special protection because of the great injury done by live stock [sic], above all by sheep. The increase in deer, elk, and other animals in the Yellowstone Park shows what may be expected when other mountain forests are properly protected by law and properly guarded. Some of these areas have been so denuded of surface vegetation by overgrazing that the ground breeding birds,

including grouse and quail, and many mammals, including deer, have been exterminated or driven away. [...]

In cases where natural conditions have been restored for a few years, vegetation has again carpeted the ground, birds and deer are coming back, and hundreds of persons, especially from the immediate neighborhood, come each summer to enjoy the privilege of camping. Some at least of the forest reserves should afford perpetual protection to the native fauna and flora, safe havens of refuge to our rapidly diminishing wild animals of the larger kinds, and free camping grounds for the ever-increasing numbers of men and women who have learned to find rest, health, and recreation in the splendid forests and flower-clad meadows of our mountains. The forest reserves should be set apart forever for the use and benefit of our people as a whole and not sacrificed to the shortsighted greed of a few. (Roosevelt 1901: 11–12)

Roosevelt held the presidency from 1901 to 1909. Aided by Grinnell and now also forest management expert, chief forester, and fellow B&C member, Gifford Pinchot (1865–1946), Roosevelt managed to use the authority given to the president with the 1891 Forest Reserve Act to create from public lands (and protect from the lumber industry) close to 150 million acres of forest reserve. His predecessors had created 50 million in total (Cutright 1985: 215–222). Roosevelt also established the US Forest Service, created the first 51 wildlife refuges and 18 national monuments, and added five national parks to the existing five (Cutright 1985: 220–235). Yet his greatest accomplishment may be that he taught the American public the importance of conservation and the appreciation of unspoiled nature. But, lest we forget, Roosevelt was also a trophy hunter and a colonialist. He assumed these roles once again on the greatest hunting trip of his life in Africa in 1909 to 1910.

Only three weeks after relinquishing the presidency to William Howard Taft (who, despite Roosevelt backing his candidacy, would later fire Pinchot and undo much of Roosevelt's conservation work), Roosevelt set off to Africa. He was persuaded to do so by Carl Ethan Akeley (see Chapter 3.1) during a White House dinner (Jeffers 2003: 131). Roosevelt left for Africa, accompanied by a handful of naturalists and his son, Kermit Roosevelt. The expedition was co-sponsored[23] by the Smithsonian National Museum of Natural History, and its stated purpose was to collect specimens for the museum. Roosevelt's *African Game Trails* (1910) is his account of this trip.

Roosevelt conformed to the Great White Hunter stereotype of the late nineteenth century to the letter on this expedition. This was not young Theodore roughing it in the wilderness, living off the land with a local to lead the way; this was ex-president Roosevelt of the United States of America on an expedition for the Smithsonian, leading an indigenous company of 15 soldiers and 260 porters, gunbearers, and other helpers. The job of all these helpers was to carry, among other items, tents, hundreds of traps for small creatures, guns, and ammunition, four tons of salt for curing skins, and Roosevelt's personal and famous "pig-skin library," a collection of Roosevelt's favourite books, bound in pig skin leather (Roosevelt 1910: 17; 521–522; Jeffers 2003: 202).

---

[23] Roosevelt himself covered some of the expedition costs, and Andrew Carnegie donated funds to cover the larger part (once at first, and then again when the expedition ran out of money) (Jeffers 2003: 137).

Further adding to the negative stereotype was Roosevelt's blatantly racist and imperialist views, expressed in passages like this:

> The dark-skinned races that live in the land vary widely. Some are warlike, cattle-owning nomads; some till the soil and live in thatched huts shaped like beehives; some are fisherfolk; some are ape-like, naked savages, who dwell in the woods and prey on creatures not much wilder or lower than themselves. (Roosevelt 1910: viii)

Like this:

> At least part of the high inland region of British East Africa can be made one kind of "white man's country," and to achieve this white men should work heartily together, doing scrupulous justice to the natives, but remembering that progress and development in this particular kind of new land depend exclusively upon the masterful leadership of the whites. (Roosevelt 1910: 7)

And like this:

> It is pleasant to be made to realize in vivid fashion the progress the American negro has made by comparing him with the negro who dwells in Africa untouched, or but lightly touched, by white influence. (Roosevelt 1910: 9)

Roosevelt's (and those of others, like John Muir, to be covered shortly) negative perceptions and disregard for indigenous people laid the foundation for many problems with neo-colonialism in conservation, where predominantly white-led international NGOs with fortress conservation approaches have been responsible for forceful relocations and in some cases even assault, rape and murder of indigenous people who get in the way of "pristine" nature. Modern criticism of national parks, as national parks are often based on displacement of people, and of racism and colonialism in conservation is portrayed well in Mbaria and Ogada's *The Big Conservation Lie*. This history implicates historical figures covered in this chapter like Grinnel and Pinchot in addition to Muir and Roosevelt (Mbaria and Ogada 2016).

Theodore and Kermit personally shot 512 animals on the Africa expedition, including 17 lions (respectively 9 and 8), 11 elephants (8 and 3), 20 rhinos (13 and 7), 29 zebras (20 and 9), 9 giraffes (7 and 2), and many more ungulates and birds (Roosevelt 1910: 457–459). Roosevelt noted at the end of his list of kills—seemingly feeling that he needed to defend the tally—that

> Kermit and I kept about a dozen trophies for ourselves; otherwise we shot nothing that was not used either as a museum specimen or for meat—usually for both purposes. We were in hunting-grounds practically as good as any that have ever existed, but we did not kill a tenth nor a hundredth part of what we might have killed had we been willing. The mere size of the bag indicates little as to a man's prowess as a hunter, and almost nothing as to the interest or value of his achievement. (Roosevelt 1910: 459)

The Roosevelt expedition ended up collecting 23,151 specimens of plants and animals for the Smithsonian, which altogether took the museum eight years to catalogue. While the animals shot by Theodore and Kermit were a numerically small part of this, they must have more than made up for it in proportion of the total weight. The tally seems excessive by any standards, and the number of animals shot did not go unremarked at the time. The Lieutenant-Governor of the East African Protectorate,

Sir Frederick Jackson, himself a keen hunter, was critical of the Roosevelt expedition, denouncing it for exceeding "all reasonable limits of hunting" (Somerville 2019).

There was, as is hopefully plain now, a curious duality and internal conflict in Theodore Roosevelt, as we suppose there is in many trophy hunters, and which makes their positions and passions so hard to explain or even understand. Roosevelt loved wild nature; of this there can be no doubt. But he was also, throughout his life, a competitive sportsman. He enjoyed shooting at animals for the simple fun and sport of it, and he enjoyed the competitive element of hunting trophies.

While on one hand deeply saddened by the looming extinction of the bison, Roosevelt was in equal measure eager to shoot one himself, before the species disappeared. He finally managed to do so on a hunting trip in the Dakota Badlands in 1884 (Brinkley 2010: Ch. 6, I-II). Daniel Payne recounts a similar incident concerning the last elks of the Badlands near Roosevelt's ranch:

> In *Hunting Trips of a Ranchman*, he notes how the elk had nearly vanished from the area near his ranch, attributing this in part to "their occasional fits of stupid panic, during whose continuance hunters can now and then work greater slaughter in a herd" (208). He was aware that this slaughter meant the elk were "seemingly doomed to total destruction at no distant date" (209), a fate that "can be looked upon only with unmixed regret by every sportsman and lover of nature" (210). However, Roosevelt seems to be mourning the end of elk-hunting more than the end of the elk; when he hears that a few stragglers have been sighted near his ranch, he sets out and kills "probably the last of his race that will ever be found in our neighbourhood," with very little evident regret (226). (Payne 1996: 111)

Also, while working tirelessly to protect non-game birds from being killed for fun or for use in the fashion industry (colourful feathers were all the rage in fashion), he personally—over the course of his lifetime—killed hundreds if not thousands of them for what he deemed were "scientific purposes." Surely, Roosevelt was a man of science, but almost just as surely, shooting birds was just a great bit of fun, and he probably killed a lot more than was strictly necessary for his scientific endeavours.

Roosevelt held very anthropocentric views of nature, and his conservation efforts were largely—when not explicitly for the benefit of future hunters—for the sake of future generations of humanity. As Payne notes, Roosevelt did not ponder the role of humans in relation to nature in any profound way (Payne 1996: 109–112). He did not really scrutinize—as we shall shortly see that Thoreau did—his own desire to hunt and kill. Roosevelt's writings are rife with the preaching of the ideals of fair chase and sportsmanship, invariably followed by his personal hunting anecdotes that show a worrying lack of anything approaching either.

Roosevelt's political accomplishments for wildlife and conservation greatly outdid and overshadow his contributions to environmental philosophy and hunting ethics. The transcendentalists, on the other hand, tend to be the opposite.

2.9 The Transcendentalists

The American origins of conservation and concrete political steps in its favour may, as Reiger argued, be found among early nineteenth century sportsmen and, later, individuals like Roosevelt, Grinnell, and Pinchot. The modern origins of preservation may be found among those who became known—disparagingly at first—as the transcendentalists.

The transcendentalist appreciation of wild nature is—while still embracing ecological science—spiritual, emotional, aesthetic, intuitive, and poetic, and thus very recognizably influenced by European thought:

> The contribution of European ideas, in natural science toward the discipline of ecology (e.g., Linnaeus, Malthus, Darwin), and in social science and literature toward a modern philosophy and aesthetics of nature (e.g., Diderot, Rousseau, Goethe, and Wordsworth), created much of the intellectual foundation necessary to the rise of a growing party of nature in America. This new view of nature, an essential element in the Romantic Movement, later became a part of American Transcendentalism. (LaFreniere 1990: 41–42)

Jean-Jacques Rousseau especially, LaFreniere argues, is key to understanding the roots of transcendentalism. First, because of Rousseau's direct influence on transcendentalist writers and second, because of the impact that especially *La Nouvelle Héloïse*, *Emile*, and *Lettres sur la Botanique* had on the American public's perceptions of nature. Eastern philosophy and religion served as important literary backgrounds as well (Brooks 2014: xiv).

Paul Brooks wrote perhaps the finest account of the transcendentalist environmental history and perception in *Speaking for Nature—The Literary Naturalists, from Transcendentalism to the Birth of the American Environmental Movement* (1980). We will limit coverage of transcendentalism to the four authors that are most relevant in a hunting context: Henry David Thoreau, John Burroughs, John Muir, and Aldo Leopold. We will also note the influence of William Temple Hornaday, though he is not considered a transcendentalist.

Thoreau has always been used more by antihunters than hunters, yet his stance on hunting was nothing if not complicated and ambivalent. The two key texts that we may use to interpret his attitude are *Higher Laws* in *Walden* from 1854 and *Chesuncook* in *The Maine Woods* from 1864, but Altherr (1984) shows that Thoreau's journals are also important source materials. Let us examine *Walden* first.

Thoreau was quite explicit about the duality in him, regarding the killing of animals:

> I found in myself, and still find, an instinct toward a higher, or, as it is named, spiritual life, as do most men, and another toward a primitive rank and savage one, and I reverence them both. (Thoreau 1882: 226)

While being both drawn to and reviled by hunting, he did see it as an essential part of human development, both for individuals and for humanity. This paragraph constitutes some of his strongest arguments in favour of hunting:

When some of my friends have asked me anxiously about their boys, whether they should let them hunt, I have answered, yes, — remembering that it was one of the best parts of my education, — make them hunters, though sportsmen only at first, if possible, mighty hunters at last, so that they shall not find game large enough for them in this or any vegetable wilderness, — hunters as well as fishers of men. [...] There is a period in the history of the individual, as of the race, when the hunters are the "best men," as the Algonquins called them. We cannot but pity the boy who has never fired a gun; he is no more humane, while his education has been sadly neglected.

Yet he proceeds in the very next line with powerful antihunting rhetoric:

This was my answer with respect to those youths who were bent on this pursuit, trusting that they would soon outgrow it. No humane being, past the thoughtless age of boyhood, will wantonly murder any creature, which holds its life by the same tenure that he does. (Thoreau 1882: 228–229)

He writes this as someone who has hunted and fished himself as an adult, but adds further that "I have found repeatedly, of late years, that I cannot fish without falling a little in self-respect" (Thoreau 1882: 230). The following lines conclude the parts of *Walden* that are quoted most often by antihunters and vegans:

Whatever my own practice may be, I have no doubt that it is a part of the destiny of the human race, in its gradual improvement, to leave off eating animals, as surely as the savage tribes have left off eating each other when they came in contact with the more civilized. (Thoreau 1882: 232)

In *Chesuncook* from *The Maine Woods*, Thoreau reports on a hunting trip for moose that he joined in 1853, one year before he published *Walden*. The concluding remarks about how Thoreau had had enough of moose-hunting after witnessing one being killed—is important:

I think that I could spend a year in the woods, fishing and hunting, just enough to sustain myself, with satisfaction. This would be next to living like a philosopher on the fruits of the earth which you had raised, which also attracts me. But this hunting of the moose merely for the satisfaction of killing him, not even for the sake of his hide, without making any extraordinary exertion or running any risk yourself, is too much like going out by night to some wood-side pasture and shooting your neighbour's horses. (Thoreau 1864: 122)

He continues shortly after:

For one that comes with a pencil to sketch or sing, a thousand come with an axe or rifle. What a coarse and imperfect use Indians and hunters make of nature! (Thoreau 1864: 123)

Despite this apparent dislike of hunters (and native Americans), Altherr notes that Thoreau's journals are full of his observations of the tallies of his hunting acquaintances, in whom he took a keen interest. He rarely expressed negative views of these hunters or their kills and gladly ate what kills they brought him, but he felt like a murderer if he killed animals himself (Altherr 1984: 348–349). For some years in his diaries there are passages displaying pure admiration of hunters, their craft, and their authentic closeness with nature (Altherr 1984: 353–354). Other journal entries express delight at the sound of guns and the thrill of the chase (Altherr 1984: 351–352).

2.9 The Transcendentalists

What might we conclude about Thoreau's stance on hunting then? He saw hunting as an essential step in human development, both for the boy growing up and for the human race. He thought sportsmen knew nature better than anyone else, and that "[h]e who is only a traveller [in nature] learns things at second-hand and by the halves, and is poor authority" (Thoreau 1882: 227). After all, Thoreau had often hunted and fished himself. He even stated that "perhaps the hunter is the greatest friend of the animals hunted, not excepting the Humane Society" (Thoreau 1882: 227) when pondering the increased scarcity of game that also concerned Roosevelt and Grinnell. He derided farmers and city folk for not hunting, and even Emerson for sitting inside and writing about nature while the hunter was outside, experiencing it (Altherr 1984: 354).

Clearly, though, he also prided himself on having passed through the stage of hunting and fishing (with occasional relapses), as he hoped all of humanity eventually would. He intermittently looked down upon hunting as primitive and crude. This was especially the case when hunting was conducted unsportingly, i.e. without eating the meat or using the hide, without physical effort, or when it caused unnecessary suffering. Yet rarely—even when the moose was killed in *Chesuncook*, which he considered a tragedy—did he forego his share of the meat: "We had moose-meat fried for supper. It tasted like tender beef, with perhaps more flavor, sometimes like veal" and "[a]fter breakfasting on moose-meat, we returned down Pine Stream" (Thoreau 1864: 120, 125).

To interpret Thoreau's stance on hunting as anything but often self-contradictory would, it seems, be very generous. As Altherr says: "Hunters appeared to him alternately as grand actors on the stage of nature or as petty killers skulking through the shadows" (Altherr 1984: 359). So perhaps the point that we should take away from Thoreau is that strongly conflicting emotions are exactly what a hunter ought to feel? That it is not *right* to kill an animal if one is not both excited and saddened at the same time? That it is acceptable to embrace one's savage side only if one is also aware and remorseful of such savagery? Such ambiguous feelings are indeed a part of hunting for many hunters, as we shall discuss further in Chapter 3.6. Stoneham interprets the hunting ethic of Thoreau throughout *Chesuncook* and *The Allegash and East Branch* (both in *The Maine Woods*) as follows:

> A hunter establishes right relations with the natural world when he knows his prey, understands its biotic community, and can coexist in that community with grace. When he chooses to responsibly hunt an animal, he must limit his hunt to the animal that he can responsibly harvest, and then he or she must use all of its meat and its hide - and waste nothing. When a hunter does that, Thoreau intimates, then he or she is acting rightly in the world. (Stoneham 2003: 9)

This, as we shall see, resembles the environmental ethic advocated by Aldo Leopold, but for the sake of chronology, let us first turn to John Burroughs and John Muir.

Perhaps self-contradictory attitudes toward hunting run in the family of transcendentalists, for we see signs of it in the works of John Burroughs as well. Here is what he wrote in his 1905 account of a camping trip to Yellowstone in 1903 with Roosevelt, *Camping and Tramping with Roosevelt*:

Some of our newspapers reported that the President intended to hunt in the Park. A woman in Vermont wrote me, to protest against the hunting, and hoped I would teach the President to love the animals as much as I did, — as if he did not love them much more, because his love is founded upon knowledge, and because they have been a part of his life. She did not know that I was then cherishing the secret hope that I might be allowed to shoot a cougar or bobcat; but this fun did not come to me.

And he continues shortly after:

I have never been disturbed by the President's hunting trips. It is to such men as he that the big game legitimately belongs, — men who regard it from the point of view of the naturalist as well as from that of the sportsman, who are interested in its preservation, and who share with the world the delight they experience in the chase. Such a hunter as Roosevelt is as far removed from the game-butcher as day is from night. (Burroughs 1907: 6–7)

Yet here is what he had written twelve years earlier in *The Way of Sportsmen* from the book *Riverby* in a short but scathing anecdote about a hunter's encounter with a deer:

But a man in the woods with a gun in his hand is no longer a man — he is a brute. The devil is in the gun to make brutes of us all. (Burroughs 1895: 279)

He shares in the same chapter another anecdote of a moose hunt, during which the hunters tracked a moose through the snow for several days until it finally tired:

On the morning of the sixth day he had made up his mind to travel no farther, but to face his enemies and have it out with them. As he heard them approach, he rose up from his couch of snow, mane erect, his look fierce and determined. Poor creature, he did not know how unequal the contest was. How I wish he could at that moment have had a Winchester rifle, too, and had known how to use it. There would have been fair play then. With such weapons as God had given him he had determined to meet the foe, and if they had had only such weapons as God had given them, he would have been safe. But they had weapons which the devil had given them, and their deadly bullets soon cut him down, and now probably his noble antlers decorate the hall of his murderer. (Burroughs 1895: 280–281)

We here learn not only of Burroughs' attitude towards guns—that they are the work of the devil—but of his opinion of trophies as well.

So how does one fit these attacks on hunters who killed a moose in a fair chase that required tremendous time and effort with Burrough's unconditional praise for Theodore Roosevelt? Burroughs even shares his experience of Roosevelt showing him his "bearskins of all sorts and sizes on the floors, panther and lynx skins on the chairs, and elk heads and deer heads on the walls" (Burroughs 1907: 98), but wrote nothing to indicate any discomfort with Roosevelt's hunting trophies. Perhaps we should attribute to Burroughs the same rationale as to Thoreau; that one ought to feel ambivalent about taking life? Maybe hunting to Burroughs was simultaneously a primitive activity and one of the most genuine and pure ways in which one can connect with and participate in nature. Another possible explanation, however, is that Burroughs was simply less inclined to direct confrontation and political-environmental activism than fellow transcendentalist John Muir, a criticism that has also been directed at Thoreau (see Payne 1996: 49–50). Roosevelt may also just have

2.9 The Transcendentalists

been too likeable as a character and too intimidating as a potential adversary. This would certainly fit his reputation.

John Muir's stance on hunting is more straightforward than those of Thoreau and Burroughs. The following is part of a letter from Muir to Henry Fairfield Osborn, who was president of the American Museum of Natural History for 25 years:

> I have often written on the subject [of hunting], but mostly with non-effect. The murder business and sport by saint and sinner alike has been pushed ruthlessly merrily on, until at last protective measures are being called for, partly I suppose because the pleasure of killing is in danger of being lost from there being little or nothing left to kill, and partly, let us hope from a dim glimmering recognition of the rights of animals and their kinship to ourselves. (Muir 1904)

In *A Thousand Mile Walk to the Gulf*, published based on his journals in 1916, two years after his death, Muir wrote the following:

> October 22. This morning I was easily prevailed upon by the captain and an ex-judge, who was rusticating here, to join in a deer hunt. Had a delightful ramble in the long grass and flowery barrens. Started one deer but did not draw a single shot. The captain, the judge, and myself stood at different stations where the deer was expected to pass, while a brother of the captain entered the woods to arouse the game from cover. The one deer that he started took a direction different from any which this particular old buck had ever been known to take in times past, and in so doing was cordially cursed as being the "d---dest deer that ever ran unshot." To me it appeared as "d---dest" work to slaughter God's cattle for sport. "They were made for us," say these self-approving preachers; "for our food, our recreation, or other uses not yet discovered." As truthfully we might say on behalf of a bear, when he deals successfully with an unfortunate hunter, "Men and other bipeds were made for bears, and thanks be to God for claws and teeth so long." Let a Christian hunter go to the Lord's woods and kill his well-kept beasts, or wild Indians, and it is well; but let an enterprising specimen of these proper, predestined victims go to houses and fields and kill the most worthless person of the vertical godlike killers, oh! that is horribly unorthodox, and on the part of the Indians atrocious murder! Well, I have precious little sympathy for the selfish propriety of civilized man, and if a war of races should occur between the wild beasts and Lord Man, I would be tempted to sympathize with the bears. (Muir 1916: 121–122)

Muir's opinion of hunting and hunters should need no further clarification after these passages. It does bear mentioning, however, that Muir seemed to have a good time camping with President Roosevelt in Yosemite National Park for three days in 1903 (perhaps because Roosevelt—just as when he camped with Burroughs—did not bring a gun) (Muir 1903; Roosevelt 1915: 27–28). And just like Thoreau, Muir expressed no qualms about eating hunters' venison for several meals when he was their guest (Muir 1916: 113, 121).

There is perhaps no better representation in history of the clash between the anthropocentric conservation view and the nature-centric preservation view than the clash between John Muir and Roosevelt's chief forester Gifford Pinchot. Pinchot was a forest management expert and the first of his kind in USA. He had received his education in Europe, which was—forestry science-wise—more advanced than the US. Pinchot and Muir became friends in the 1890s. It was always clear, however, that Pinchot's market economic idea of protecting forests only to the extent that the highest yield of timber was ensured did not harmonize with Muir's vision of forests protected for their beauty and spiritual value, and for the sake of the trees themselves and the creatures living in and amongst them. Their friendship petered out in 1897, when Pinchot publicly declared that there was nothing wrong with letting sheep ("hoofed locusts" to Muir) graze in forest reserves (Wilkins 1995: 190–204).

Pinchot's market-economic approach to forest conservation had a great influence on Roosevelt. Pinchot is also credited with the statement in 1903 that

> [t]he object of our forest policy […] is not to preserve the forests because they're beautiful or wild or the habitat of wild animals; it is to ensure a steady supply of timber for human prosperity. Every other consideration is secondary. (Wilkins 1995: 202)

Roosevelt repeated this position in the same year in a slightly softened but otherwise identical manner:

> First and foremost, you can never afford to forget for one moment what is the object of our forest policy. That object is not to preserve the forests because they are beautiful, though that is good in itself; nor because they are refuges for the wild creatures of the wilderness, though that, too, is good in itself; but the primary object of our forest policy, as of the land policy of the United States, is the making of prosperous homes. (Roosevelt 1904a: 8)

The conflict between Muir and Pinchot was a public one, taking place in books and magazines. Each of them tried to win over an increasingly urbanized public that was still interested in economic growth but also nurtured a growing sense of sentimentality about the wilderness—as one does when one no longer lives in it:

> It was to this middle and upper class audience Muir and Pinchot pitched their respective crusades: Muir's to preserve the American wilderness as a sanctuary for spiritual renewal, a great garden free from machines in perpetuity; Pinchot's to conserve resources once thought limitless for the continued prosperity of the American nation and the continued growth of American industry. (Smith 1998: 759)

Aside from being a poet and naturalist, Muir—unlike Thoreau and Burroughs—was a political activist. Among his many accomplishments he was the first president of The Sierra Club, founded in 1892. John Muir has in recent years, like Roosevelt, been heavily criticized for his racist attitudes toward, and complete disregard for, Native Americans and their homes and lifestyle in his preservationist approach to nature protection, disparagingly referred to today as "fortress conservation." Yellowstone was a product of this approach and the eviction of Native Americans from the park was a part of it. Muir's interpretation of wilderness as "pristine" if it was untouched by white men and of this pristine characteristic being valuable in itself left little room for acceptance of the fact that it was often not at all untouched by the people who made it their home. He wrote the following now rather infamous passage in *Mountains of California* about a group of Native Americans that he met in the passes:

> Occasionally a good countenance may be seen among the Mono Indians, but these, the first specimens I had seen, were mostly ugly, and some of them altogether hideous. The dirt on their faces was fairly stratified, and seemed so ancient and so undisturbed it might almost possess a geological significance. The older faces were, moreover, strangely blurred and divided into sections by furrows that looked like the cleavage-joints of rocks, suggesting exposure on the mountains in a cast-away condition for ages. Somehow they seemed to have no right place in the landscape, and I was glad to see them fading out of sight down the pass. (Muir 1894: Chapter 5)

2.9 The Transcendentalists

We conclude this overview of the transcendentalists—and suitably so, as he is in many ways a compromise between Muir and Pinchot—with Aldo Leopold. His best-known work, *A Sand County Almanac and Sketches Here and There* (1949), was accepted for publication just a few months before he died of a heart attack at the age of 61 while trying to put out a fire on a neighbour's property. It is a work of art in all aspects and one of the seminal texts in modern environmental philosophy.

Aldo Leopold was a forester like Pinchot, and both worked for the Forest Service. Leopold was educated at the Yale School of Forestry, which had been founded by Pinchot's family and flourished during Theodore Roosevelt's presidency. Leopold can be considered a compromise between Muir and Pinchot because his *Land Ethic* combined the utilitarian rationale of mostly timber-based conservation with the transcendentalist's holistic, aesthetic, and spiritual appreciation of nature. The seminal sentence in Leopold's *Land Ethic* states that "[a] thing is right when it tends to preserve the integrity, stability, and beauty of the biotic community. It is wrong when it tends otherwise" (Leopold 1968: 224–225). For general interpretations of and expansions upon Leopold and his Land Ethic, we refer the reader to the extensive writings of J. Baird Callicott (1989 and others), as we shall focus on Leopold's thoughts on hunting.

As Pinchot might be regarded as the father of forest management in North America, so might Leopold be regarded as the father of the game management profession (today called wildlife management). *Game Management* was the title of his 1933 book and his field of professorship at the University of Wisconsin. Aldo Leopold began hunting as a teenager, accompanying his father. He wrote in the *Red Legs Kicking* essay in *Sketches* about his early days of hunting that "[l]ike most aspiring hunters, I was given, at an early age, a single-barreled shotgun and permission to hunt rabbits" and proceeds to recount the stories of his first kills of ducks and partridges (1968: 120–121). Leopold was taught the ideas of sportsmanship in hunting from a young age. Aged 17, he went big game hunting in Yosemite with his father. Much like the young Roosevelt, Leopold spent most of his time outdoors (Meine 1987: 22–24).

A newly graduated forester in 1909, Leopold led a timber reconnaissance team in the wild countryside of the Apache National Forest in Arizona. He here had the experience that made him write the following famous passage about a wolf-encounter many years later:

> In those days we had never heard of passing up a chance to kill a wolf. In a second we were pumping lead into the pack, but with more excitement than accuracy: how to aim a steep downhill shot is always confusing. When our rifles were empty, the old wolf was down, and a pup was dragging a leg into impassable slide-rocks.
>
> We reached the old wolf in time to watch a fierce green fire dying in her eyes. I realized then, and have known ever since, that there was something new to me in those eyes—something known only to her and to the mountain. I was young then, and full of trigger-itch; I thought that because fewer wolves meant more deer, that no wolves would mean hunters'

paradise. But after seeing the green fire die, I sensed that neither the wolf nor the mountain agreed with such a view. (Leopold 1968: 130)[24]

In 1919 while employed in the Forest Service, Leopold published the article *Wild Lifers vs. Game Farmers: A Plea for Democracy in Sport*. The scarcity of game due to habitat loss and overhunting had led to the question of whether to increase restrictions on public hunting or to breed hunting game for release into the wild. Leopold argued for hunting restrictions, stating that game farmers merely care about producing something to shoot. The "wild lifer" wishes to maintain wild populations for the benefit of every person who enjoys the outdoors as well as for the wildlife as an end in itself (Leopold 1919).

In the years 1924–1928 Leopold stepped up his effort in writing articles about game management and in favour of wilderness preservation (Meine 1987: 27–33). *Wilderness as a Form of Land Use* provides some of Leopold's strongest arguments for wilderness preservation, and he echoes ideas about sport hunting in this article that should appear familiar by now:

> Hunting for sport in its highest form is an improvement on hunting for food in that there has been added, to the test of skill, an ethical code which the hunter formulates for himself and must often execute without the moral support of bystanders. (Leopold 1925: 137)

Leopold's attitudes toward traditional Pinchot-inspired conservation models and making wilderness accessible to everyone with roads had become increasingly critical, and eventually led to these famous words from *Marshland Elegy* in *Sketches*:

> To build a road is so much simpler than to think of what the country really needs. A roadless marsh is seemingly as worthless to the alphabetical conservationist[25] as an undrained one was to the empire-builders. Solitude, the one natural resource still undowered of alphabets, is so far recognized as valuable only by ornithologists and cranes.
>
> Thus always does history, whether of marsh or market place, end in paradox. The ultimate value in these marshes is wildness, and the crane is wildness incarnate. But all conservation of wildness is self-defeating, for to cherish we must see and fondle, and when enough have seen and fondled, there is no wilderness left to cherish. (Leopold 1968: 101)

---

[24] These are beautiful lines, but they are not entirely truthful. As Meine reasons, "it would be years, and many an antiwolf campaign later, before he would come to regret the incident. Pioneers did not pause to bemoan the loss of a 'varmint.'" (Meine 1987: 27). Leopold wrote a short piece in 1915—*The Varmint Question*—in which he calls for extensive predator control (Leopold 1915), and in 1920—11 years after the wolf-incident—he wrote that "[i]t is going to take patience and money to catch the last wolf or lion in New Mexico, but the last one must be caught before the job can be called successful" (Leopold 1920 in Brooks 2014: 257). Aldo Leopold's view on predators changed dramatically in the face of problems with excess deer before he was ready to write *Sketches*, but the view he held in 1920—that predators are varmints and the sworn enemies of sportsmen, better exterminated as soon as possible—was common among naturalists at that time.

[25] "Alphabetical conservationists" here and "alphabetical conservation" elsewhere in Leopold's writings are references to the abbreviations of the many agencies established in 1933 under the presidency of Franklin D. Roosevelt as a part of New Deal. These were commonly referred to as alphabet agencies, and the one that Leopold mainly alludes to here is the Civilian Conservation Corps (CCC). Leopold, however, also employs the nickname to convey an inability of by-the-book, box-ticking conservationists to—unlike cranes and ornithologists—notice or appreciate intangible values of nature such as solitude and wildness.

2.9 The Transcendentalists

Leopold left the Forest Service in 1928 for the Sporting Arms and Ammunitions Manufacturers' Institute (Meine 1987: 33). Being the leading expert in a field that barely existed and was not very respected at the time was not a lucrative business though, and times were tough for Leopold's family until he found permanent employment in 1933 as professor of game management at the University of Wisconsin and published *Game Management*. He became a beloved teacher there, and his field gained in popularity and turned into today's wildlife management discipline (Meine 1987: 33). Leopold conducted national game management research and wrote game policy that set the direction for US wildlife policy for 40 years.

Writing as an ecologist and game management expert, Leopold adopted a language that was sometimes far from the colourfully poetic and typically transcendentalist language for which we also know him. The very first sentence of Chapter 1 of *Game Management* states that "[g]ame management is the art of making land produce sustained annual crops of wild game for recreational use." It continues shortly after: "Like the other agricultural arts, game management produces a crop by controlling the environmental factors which hold down the natural increase, or productivity, of the seed stock" (Leopold 1986: 3). The ethical implications of his choice of words are important. As game management in its early years needed to be accepted as a science, it had to adopt a scientific language and approach. The way to do this was to adopt the language of agricultural science; deer are crops, and crops are harvested, so deer are harvested.

The language that hunters use or should use for their activity is still a subject of debate and controversy today. On one hand, some hunters wish to portray hunting as just another form of farming: carefully managed to be sustainable and no different in its ethical implications than agriculture. On the other hand, some hunters are deeply disturbed by this language, feeling that it is disrespectful and demeaning to talk about the deer they just killed as if it were a cabbage being picked from a field. Some hunters also feel that owning up to killing is more honest than calling it harvesting.

William Temple Hornaday (1854–1937), lastly, was famous as both taxidermist and writer. He would often agree with Roosevelt on the goals of saving forests and wildlife, but he would just as often disagree with him on how to attain these goals. Hornaday shared some of John Muir's later aversions to much of what Gifford Pinchot stood for, as he would favour preservation, i.e. the complete protection of forests from human interference, over the anthropocentric conservation (Cutright 1985: 202, 216).

One of Hornaday's great books is *Our Vanishing Wild Life* from 1913. He—like Roosevelt—was deeply saddened by America's dwindling wildlife populations and had—like Roosevelt—also written about the extermination of the bison (Hornaday 1889). Unlike Roosevelt, Hornaday became critical of hunting[26]:

---

[26] Hornaday seems, like Thoreau, to have suffered from some internal conflicts in relation to hunting. In 1940, many years after writing *Our Vanishing Wild Life*, Hornaday called the appointment of Aldo Leopold as Professor of Game Management at the University of Wisconsin "a helpful gesture in the struggle to save American game and sport from finally going over the precipice." Reiger further explains that Hornaday has often been both overemphasized and misinterpreted by modern historians, especially concerning hunting: "Because of a penchant for self-flattery, William

I have been a sportsman myself; but times have changed, and we must change also. When game was plentiful, I believed that it was right for men and boys to kill a limited amount of it for sport and for the table. But the old basis has been swept away by an Army of Destruction that now is almost beyond all control. We must awake, and arouse to the new situation, face it like men, and adjust our minds to the new conditions. The three million gunners of to-day must no longer expect or demand the same generous hunting privileges that were right for hunters fifty years ago, when game was fifty times as plentiful as it is now and there was only one killer for every fifty now in the field.

The fatalistic idea that bag-limit laws can save the game is to-day *the curse of all our game birds, mammals and fishes!* It is a fraud, a delusion and a snare. That miserable fetich has been worshipped much too long. Our game is being exterminated, everywhere, by blind insistence upon "open seasons," and solemn reliance upon "legal bag-limits." If a majority of the people of America feel that so long as there is any game alive there must be an annual two months or four months open season for its slaughter, then assuredly we soon will have a gameless continent. (Hornaday 1913: x, emphasis in original)

It is worth taking a closer look at *Our Vanishing Wild Life* because it presents many historically and contemporarily relevant concerns about hunting. One of these is the distinction Hornaday makes between "The Gentlemen Sportsmen" and "The Gunners." Hornaday bore no ill will towards the man he called the Gentleman Sportsman, and in fact attributes "the best and the most of the game-protective laws now in force in the United States and Canada" to these hunters, who he considered "the very bone and sinew of wild life preservation" (Hornaday 1913: 56, 54). He described the Gentlemen Sportsmen as deeply nature-loving men who do not care about shooting as much as the law allows (the bag limit), but rather stop when they have "enough" and who had now in 1913 given up hunting altogether and picked up a camera instead. He adds that one may find the Gentlemen Sportsmen in the Boone and Crockett Club—of which he himself was still an associate member in 1913 (Grinnell 1913: 509)—and many others like it. He reiterated the point, however, that any Gentleman Sportsman who still hunts is "a soldier in the Army of Destruction" who does not "stand on the summit with the men who now protect the game *and do not shoot at all!*" (Hornaday 1913: 56, emphasis in original). This Gentleman Sportsman passage seems like a personal message to Theodore Roosevelt and his friends, first acknowledging all that they had done for conservation, but then pleading with them to cease hunting altogether.

Hornaday does not have the same respect for "The Gunner" as for the sportsman. The gunner is someone who stands firmly on his right to shoot all legal game and

---

Temple Hornaday has been accepted by literal-minded historians as one of the greatest wildlife conservationists. As the first Director of the New York Zoological Park and a crusader for the bison, Hornaday deserves a place in the history of conservation—but not the one he chose for himself. Far more important conservationists like George Bird Grinnell and Charles Sheldon believed that Hornaday actually did more harm than good by splitting the ranks of conservationists through his bitter attacks on anyone who did not agree precisely with "Mr. Hornaday," as he often called himself. In recent years antihunting historians have quoted him for his assaults on hunters, failing to understand that he was against those hunters who, in his opinion, abused the sport, but he certainly was not against all hunting. Hornaday believed that only he and a small minority of hunters who agreed with his often erroneous notions of wildlife management were the true adherents of the code of the sportsman" (Reiger 2001: 253).

as much as the law allows him, regardless of its scarcity or even utility. Hornaday claims that the gunner outnumbers the true gentleman sportsman by 500 to 1. The view that whatever is not forbidden is permitted, according to Hornaday, is "the spirit of Extermination, clothed in the robes of law and justice" (Hornaday 1913: 56).

Hornaday blamed market hunters even more though, reasoning that they have little reason to even abide by the law:

> The true sportsman hunts during a very few days only each year. The market gunners shoot early and late, six days a week, month after month. When game is abundant, the price is low, and a great quantity must be killed in order to make it pay well. When game is scarce, the market prices are high, and the shooter makes the utmost exertions to find the last of the game in order to secure the "big money". (Hornaday 1913: 65)

Hornaday's criticism was not all directed at individual hunters though. He also blamed the conditions under which they thrived, and which passively condoned their behaviour. Much too generous bag-limits and a gun-manufacturing industry that was too good at selling too effective guns were partly to blame. Like Aldo Leopold, Hornaday thus criticised the role that commercial interests and technological development played in hunting (see also Chapter 6.1). Though he added that many clubs for sportsmen—including B&C—condemned the kinds of weapons that were the main concern.

Hornaday did not disapprove of hunting in principle or of gentlemanly hunters; he simply argued that hunting had become irresponsible because of the current state of America's wildlife. Yet he did write passages to indicate the same appreciation of nature that we saw in Thoreau's writings:

> Really, it is to me very strange that gunners never care to save game birds on account of their beauty. One living bob white on a fence is better than a score in a bloody game-bag. A live squirrel in a tree is poetry in motion; but on the table a squirrel is a rodent that tastes as a rat smells. Beside the ocean a flock of sandpipers is needed to complete the beautiful picture; but on the table a sandpiper is beneath contempt. A live deer trotting over a green meadow, waving a triangular white flag, is a sight to thrill any human ganglion; but a deer lying dead, — unless it has an exceptionally fine head, — is only so much butcher's meat. (Hornaday 1913: 204)

Interestingly, "unless it has an exceptionally fine head" does hint at a personal appreciation for fine hunting trophies that seems at odds with the rest of the passage. In truth, the vast majority of Hornaday's ire was directed at bird hunters and not big game hunters. This had to do with the ease with which one could kill hundreds of birds compared to the difficulty of killing more than one deer. He also thought the character of bird hunters and big game hunters was fundamentally different: "It is my impression that of the men who shoot, it is only among the big game hunters that we find much genuine admiration for game animals, or any feeling remotely resembling regard for it" (Hornaday 1913: 203–204).

Toward the end of *Our Vanishing Wild Life*, Hornaday gave his definition of the sportsman:

> A Sportsman is a man who loves Nature, and who in the enjoyment of the outdoor life and exploration takes a reasonable toll of Nature's wild animals, but not for commercial profit,

and only so long as his hunting does not promote the extermination of species (Hornaday 1913: 382).

Whether Roosevelt was a sportsman according to this definition is debatable. He certainly loved nature and the rugged outdoor life and took a toll of nature's wild animals. One may question, though, if the toll was always reasonable, and whether his shooting of a bison for its trophy before it went extinct did not promote exactly its extermination.

2.10 Modern Animal Rights and Ecofeminism

Chapter 2.6 indicated some roots of the animal rights movement with Montaigne in the sixteenth century, and Rousseau, Kant, Bentham, and Taylor in the seventeenth and eighteenth century. We also touched on the early nineteenth century shift to a concern for animals for the sake of the animals themselves (biocentric) rather than for the moral character of the human who would mistreat them (anthropocentric). Thoreau and Muir's attitudes to hunting also come close to what we might associate with animal rights today. However, most concern for individual animals discussed so far has been in relation to animal cruelty and not in relation to the killing of animals that takes place in animal husbandry or hunting. Criticism of hunting has—as with Hornaday for example—been based on largely anthropocentric concerns for sustainability and conservation.

We return to the subject of animal rights by way of Henry Stephens Salt, who wrote the book *Animals' Rights Considered in Relation to Social Progress* in 1892. Salt represents an animal rights paradigm that is critical of all killing of all animals in addition to their mistreatment, and promotes veganism or at least vegetarianism. Salt—himself a vegetarian on ethical grounds—was not the first of his kind, but he has become one of the most influential. Peter Singer's words in the preface to the 1980 edition of *Animals' Rights* provide a good argument for highlighting Salt:

> Though not the first, *Animals' Rights* is, I believe, the best of the eighteenth- and nineteenth-century works on the rights of animals. Every time I re-read Salt's book—and I have now read it several times—I marvel at how he anticipates almost every point discussed in the contemporary debate over animal rights. Defenders of animals, myself included, have been able to add relatively little to the essential case Salt outlined in 1892; but we can console ourselves with the fact that our opponents have been able to come up with few objections that Salt has not already dealt with. (Singer 1980: viii)

In Chapter 1 of his book, Salt makes a general case for rights. Yet what is important to Salt is not to prove that rights exist, so much as it is to show that whatever humans have (which may be rights or may be something akin to what we think of as rights; he refers to a "distinctive individuality"), there can be no good reason to deny that animals have the same:

2.10 Modern Animal Rights and Ecofeminism

> Animals, as well as men, though, of course, to a far less extent than men, are possessed of a distinctive individuality, and, therefore, are in justice entitled to live their lives with a due measure of that "restricted freedom" to which Herbert Spencer alludes.[27] (Salt 1980: 9)

The chapter of *Animals' Rights* that is particularly relevant to the subject at hand is Chapter 5 on "Sport, or Amateur Butchery." Salt makes it clear here that he tolerates neither the hunter nor his pastime:

> Now, on the very face of it, this amateur butchery is, in one sense, the most wanton and indefensible of all possible violations of the principle of animals' rights. If animals—or men for that matter—have of necessity to be killed, let them be killed accordingly; but to seek one's own amusement out of the death-pangs of other beings, this is saddening stupidity indeed! (Salt 1980: 68)

Emphasizing here and elsewhere the immorality of deriving pleasure from killing animals, Salt is leaning into a virtue ethics criticism as well. We will discuss this further in Chapter 8. Salt proceeds by arguing that sportsmen must be wholly incapable of understanding or sympathizing with the suffering they cause, and that their inability to comprehend the wrongness of their actions also serves as their only possible excuse: "and being, in the great majority of instances, a man of slow perception, he naturally finds it much easier to follow the hounds than to follow an argument" (Salt 1980: 68).

For the sportsmen's contribution to conservation, Salt had equally little respect. He wrote about the British conservation of foxes that, as it serves only the purpose of ensuring a population of foxes to kill, it is hardly something for which the fox should be grateful. He does not comment on sportsmen's contributions to conservation in North America. To the hunters' argument that hunting is a way of connecting with nature, he responds:

> As for the nonsense sometimes talked about the beneficial effects of those field-sports which bring men into contact with the sublimities of nature, I will only repeat what I have elsewhere said on this subject, that "the dynamiters who cross the ocean to blow up an English town might on this principle justify the object of their journey by the assertion that the sea-voyage brought them in contact with the exalting and ennobling influence of the Atlantic". (Salt 1980: 71)

The subject of animal rights did not receive much scholarly attention for many years following the initial publishing of Salt's book. Nor did his book receive all that much attention until after he died. The great resurgence of the animal rights debate happened in the seventies,[28] and this was largely because of Peter Singer and Tom Regan. It should be noted here that Peter Singer is a utilitarian and as such does not technically believe in individual rights at all but in maximizing overall well-being. However, he is often in academic literature and among activists and NGOs considered a pillar of the animal rights movement, and he has been so intertwined with it that he

---

[27] "Every man is free to do that which he wills, provided he infringes not the equal freedom of any other man" (Spencer 1892: 46).

[28] Although in 1942, Disney released the motion picture cartoon Bambi. Bambi was a major boost to the animal rights movement and to antihunting sentiments specifically.

has often even used the language of rights himself. As such, we cover Singer under the heading of animal rights. He fights for similar abolitionist goals as proponents of animal rights after all; he just has different arguments for doing so.

While Singer and Regan are central to the animal liberation and animal rights movements, the subject of hunting never occupied much of their attention. Peter Singer's magnum opus is *Animal Liberation* from 1975, and Tom Regan's equivalent is *The Case for Animal Rights* from 1983. Peter Singer did not broach the subject of hunting in *Animal Liberation*, so we must look elsewhere for his pertinent views. One relevant instance was during a talk that Singer held at Williams College in 2009, where a member of the audience asked a question about how to manage deer populations when their natural predators have been exterminated, and whether hunting would then be an acceptable form of population control. Singer acknowledges that overpopulation of deer is a problem in many places and proceeds as such:

> Certainly let me say this, it is far more ethical – if you're a good shot that is – it's far more ethical to go out and shoot a deer than it is to go to the supermarket and pick up a ham or a chicken or something like that. I think a lot of the people have it sort of the wrong way around because they see the image of the hunter, and they don't see the image of how the chicken or the pork was produced. I would rather see other methods [than hunting] of controlling deer. [...] We control [our own] population not by randomly shooting people but by encouraging people to use fertility control. I think we could do that for animals. We need a bit of science to tell us how to do it. That would be ideal, but admittedly we don't have it at the moment, and there can be cases where animals need to be… their numbers need to be controlled, and if we don't have better techniques, the best technique is to get someone who is really an expert marksman to shoot them with a single bullet so that they don't know what's going on. The problem with hunting of course is a lot of people are not so expert. You get wounded animals and suffering animals, so I'm not actually going to defend hunting, but I can recognize that there are cases where killing might not be the greatest evil that you could do. (Singer 2009: 1:05:40, transcription by NB)

Another comment that Peter Singer has made about hunting was during a discussion with Nigel Warburton for *Ethics Bites* in 2008 about the ethics of using animals. Singer has just prior to this passage emphasized that factory farming is the worst example of what we do to animals to eat meat, and that free-range farming would be preferable:

> Nigel: And presumably if you go down to eating game, animals that have lived in the wild that have been killed very quickly with gunshot perhaps - that presumably is more morally acceptable than eating free range meat which has to be taken to be slaughtered.
>
> Peter: I think that's generally true. It will perhaps depend on particulars, how good a shot you are, how reliably you can put a bullet through the brain of a moving animal and kill it instantly[29] rather than have it wounded and escaping. But if you are a good shot it's better to go hunting for your animals than go down to a supermarket for it.

---

[29] We cannot be sure if Singer misspoke here or if he knows very little about hunting, but hunters generally aim for the chest (heart and lungs) of their prey. Hunters rarely aim for the brain, and especially not if the animal is moving. The exception to this would be if the animal (e.g., a wild boar or an elephant) is running directly towards the hunter to attack, as the flank would then not be exposed. The brain is a very small target and game animals tend to move their head quickly and often while keeping their chest relatively stationary. Aiming for the brain would lead to many misses and woundings, even if all hunters were expert marksmen. Culling operations may differ

2.10 Modern Animal Rights and Ecofeminism

> Nigel: See, what I like about what you're saying is that it's so reasoned. Everything is consistent; you're looking for an argument that makes distinctions based not on how you feel about killing or animal suffering, but about what the implications are rationally. Most people I've met who are vegetarians have a complete sense of revulsion about eating meat, or harming animals in any way, that they might not be able to justify rationally. Perhaps they're not moral vegetarians?
>
> Peter: Some of them may be aesthetic vegetarians; that is they are repulsed by the idea of killing an animal. And I agree that there is something that does repel me about it. Even if I convince myself by the argument that we were just having, that if you have deer that are reproducing and there are no predators and they will die in winter from lack of food and they will die a slower death and therefore it's ok to shoot them. Even if I were to convince myself of that, I still wouldn't like the idea of watching that beautiful animal suddenly drop dead and be reduced to a carcass and a piece of meat. But speaking as a moral philosopher I do have to really consider whether this is just a 'yuk' response that we might have to other things that actually are quite defensible or justifiable or whether there's a serious ethical reason for not doing it. (Singer 2008)

What we may gather from Singer's statements in these two interviews are the following three points.

First, while Singer does not go as far as condoning hunting, he is more forgiving of hunting than of factory farming, provided that the hunter is able to kill quickly. This is based on the rationale that a wild animal has lived a much less torturous and stressful life than conventional livestock.

Second, since the alternatives to hunting as population control in the absence of apex predators are not practically/fiscally viable, hunting—provided, once again, that the hunter is a good marksman—may be an acceptable solution.

Third, Singer sees a distinction between moral vegetarianism and aesthetic vegetarianism and—using himself as an example—says it is possible to be a vegetarian and disapprove of hunting on the basis of an emotional and instinctual 'yuk' response, rather than moral reasoning, and therefore independently of hunting's moral defensibility.

This 'yuk' point will be relevant again for Chapter 8. For now, we will move on to the views of Tom Regan. In the article *The Moral Basis of Vegetarianism* from 1975, Regan accepted that hunting is preferable to factory farming, but then emphasized that the great moral importance lies in killing and not in causing pain:

> For the animals slaughtered on even the most otherwise idyllic farms, as well as those shot in the wild, are just as much killed, and just as much dead, as the animals slaughtered under the most ruthless of conditions. (Regan 1975: 210)

So, while he granted that causing less pain is inherently better than causing more pain, neither way of killing is acceptable. Surprisingly though, Regan did towards the end of the article allow for some leeway for deer hunting. The reasoning seems

---

though and head or neck shots may be taken. Most hunters would consider this irresponsible, given the high chance of missing a lethal shot, and no one would ever aim for the head of a deer in flight. Hunting ethics outlined by groups overseeing hunting, like South Africa's Professional Hunter's Association of South Africa and the British Association for Shooting and Conservation, also advise against head shots.

utilitarian in nature and at odds with his usually uncompromising animal rights stance:

> This is not to say that practices that involve taking the lives of animals cannot possibly be justified. [...] What we would have to show in the present case, I think, in order seriously to consider approving of such a practice, is (1) that such a practice would prevent, reduce or eliminate a much greater amount of evil, including the evil that attaches to the taking of the life of a being who has as much claim as any other to an equal natural right to life; (2) that, realistically speaking, there is no other way to bring about these consequences; and (3) that we have very good reason to believe that these consequences will, in fact, obtain. Now, perhaps there are some cases in which these conditions are satisfied. For example, perhaps they are satisfied in the case of the eskimo's killing of animals *and in the case of having a restricted hunting season for such animals as deer*. (Regan 1975: 211–212, emphasis added)

This concession, we can surmise, is probably due to this article being a defence of vegetarianism more than of animal rights per se. In *The Case for Animal Rights*, eight years later, Regan was more specific about his stance on hunting:

> Standard justifications of the "sport" of hunting—that those who engage in it get exercise, take pleasure in communion with nature, enjoy the camaraderie of their friends, or take satisfaction in a shot well aimed—are lame, given the rights view. All these pleasures are obtainable by engaging in activities that do not result in killing any animal (walking through the woods with friends and a camera substitutes nicely), and the aggregate of the pleasures hunters derive from hunting could only override the rights of these animals if we viewed them as mere receptacles, which, on the rights view, they are not. (Regan 1983: 353–354)

To the argument that wildlife management through hunting reduces the herd to fit the range, thereby saving a larger number of animals from starvation and disease, Regan's response is threefold. First, not all hunters are sharpshooters, so there is great risk of wounding. A slow death from a gunshot wound is not guaranteed to be any less excruciating than starving to death. Second, the "maximum sustainable yield" idea in wildlife management suffers from simply aiming to have more animals to kill in the future, rather than fewer. This increases the aggregate amount of death and suffering. Third, Regan is a deontologist, which means that actions are right or wrong based not on their consequences but on whether they violate the rights of others. Thus, Regan denies on principle that wildlife management through hunting could be morally defensible, even if it led to less total death and suffering. Wildlife management is distinctly utilitarian, and no positive outcome could justify hunting:

> Policies that lessen the total amount of harm at the cost of violating the rights of individuals, whether these individuals are moral agents or patient, and, if the latter, human or animal, are wrong. Even if it were true, which it is not, that the philosophy of maximum sustainable yield would lead to a reduction in the total amount of death and suffering for undomesticated animals, it still would not follow that we should accept that philosophy. As it systematically ignores the rights of wild animals, so does it systematically violate them. (Regan 1983: 356)

It seems any leeway that Regan might have allowed deer hunting in 1975 is absent by 1983. Herein lies the major difference between Singer and 1983-Regan. Singer, a utilitarian and therefore concerned with aggregate amounts of pain and pleasure for humans and animals, could be forced to accept hunting under certain circumstances. Regan, a deontologist and as such concerned with the morality of the hunting activity

2.10 Modern Animal Rights and Ecofeminism

itself, regardless of the results that it may bring about, should be uncompromising in his condemnation of it. Regan is unconcerned with wildlife management also protecting forests from being overbrowsed by deer. The scope of Regan's animal rights does not extend to said animals' habitats.

Lastly, Regan responds to the argument that what hunters do is no different than what all predators do in nature. He does so by distinguishing between moral agents and moral patients. Humans are moral agents[30] with a duty to respect the rights of others. Animals are moral patients[31] with no comprehension of morality, so they can do no moral right or wrong (Regan 1983: 357–358).

As Singer is the pragmatist compared to Regan, so is Regan the pragmatist compared to Gary L. Francione. Francione is one of the most productive and well-known authors on animal rights of the last twenty years. He is unbending in his abolitionist stance, frequently criticizing other vegan- or animal rights advocates and organisations who accept merely animal welfare or "baby steps" towards abolitionism:

> [A] corollary of the rejection of animal use is that abolitionists do not support campaigns for the reform of animal use. That is, abolitionists do not support campaigns for more "humane" treatment of animals as a supposed incremental step on the road to eventual abolition. [...]
>
> [I]f animal exploitation cannot be justified morally, then we should not be promoting campaigns to (supposedly) make such exploitation more "humane." If animal use is morally wrong, promoting "humane" use ignores the injustice of the institution of animal use just as promoting "humane" slavery ignores the injustice of that institution. (Francione and Charlton 2017: 36)

Francione is very active on blogs and social media, and an update on Facebook from September 19, 2017 presents his view on hunting quite clearly:

> Hunting season has just started again. Where I live, the deer aren't even frightened of humans. They just stand there and look at you. There is a mom who has two twins who is around a lot. The fawns are always with her. Tonight, I saw some hunters loading a deer I think was she into the back of the truck. Her fawns will be without her tonight. They will be terrified and they will never again be with their mom. And for what? For palate pleasure and whatever thrill these guys got from shooting an arrow into a doe who wouldn't move if she saw them coming?
>
> Hunters are morally no different from those who go to the store and buy animal products. Those who eat meat or dairy or eggs also participate directly in the exploitation of animals. They are also responsible for the broken hearts of cows and calves who are separated and for all of the other tragedies that occur so that people can satisfy their palate pleasure or fashion sense.

---

[30] "Moral Agents are individuals who have a variety of sophisticated abilities, including in particular the ability to bring impartial moral principles to bear on the determination of what, all considered, morally ought to be done and, having made this determination, to freely choose or fail to choose to act as morality, as they conceive it, requires" (Regan 1983: 151).

[31] "In contrast to moral agents, *moral patients* lack the prerequisites that would enable them to control their own behavior in ways that would make them morally accountable for what they do. A moral patient lacks the ability to formulate, let alone bring to bear, moral principles in deliberating about which one among a number of possible acts it would be right or proper to perform. Moral patients, in a word, cannot do what is right, nor can they do what is wrong" (Regan 1983: 152).

> But there is a psychological difference with respect to anyone who does the killing themselves. I simply cannot understand the urge to kill. I cannot imagine looking at a doe and two fawns drinking at a creek and wanting to shoot an arrow into the doe while her fawns stand there and watch. And I am so very glad that I can't understand it. (Francione 2017)

Singer's utilitarian reasoning leads him to the conclusion that the carnivorous human who shoots game is more moral than the one who picks meat up in the supermarket. Francione reaches the conclusion that there is no moral difference. He indicates, however, that there is a psychological one. Francione thinks that those who kill are a special kind of depraved, compared to those who only eat meat and have others do the killing.

On his blog, *abolitionistapproach*, Francione somehow becomes the unlikely defender of infamous trophy hunters. On November 20, 2013, Francione wrote a post called "Some Thoughts on Melissa Bachman and the Lion"[32] about the media backlash that Bachman faced following her lion trophy photo, circulating online. His opinion is that there is no moral difference between Bachman killing this lion for her own unnecessary enjoyment and non-vegans consuming animal products for their equally unnecessary enjoyment:

> The best justification we have for imposing suffering and death on those billions of animals, many of whom have had lives far more hideous than the lion Bachman slaughtered, is that they taste good. [...] there is *no* coherent moral distinction between her and most of us. The fact that Bachman kills "charismatic species" and the rest of us just kill chickens, pigs, cows, and fish is *completely* irrelevant. (Francione 2013, emphasis in original)

Kendall Jones is another young, American celebrity trophy hunter who receives her share of death threats. Francione comes to her defence as well, writing in a post from July 2, 2014, headlined "Oh, No: Another Woman Does What Everyone Else Does" that there is no moral difference between her and anyone else who is not vegan (Francione 2014).

On July 29, 2015, just short of a month after Cecil was killed, Francione posted under "Adventures in Moral Schizophrenia: Cecil the Lion":

> If you are upset about the killing of Cecil the Lion and you are not a vegan, then you are suffering from moral schizophrenia. There is no difference between "Cecil" and all of the animals you eat who do not have names but who value their lives as much as the lion valued his. (Francione 2015)

In these quotations, Francione raises important points about species' charisma and individuality. He argues that killing lions is no more morally wrong than killing a less charismatic animal and that killing an animal without a name is just as wrong as killing a "Cecil." The importance of animal naming and animal charisma is not to be considered briefly or lightly, and we will reflect on these themes in Chapter 7.

Decker and Brown summarize the animal rights view of wildlife management through hunting as such:

---

[32] Melissa Bachman is an American hunter and host of a hunting show, Deadly Passion, who posed on her Facebook page with a lion she had killed on November 1, 2013.

2.10 Modern Animal Rights and Ecofeminism

> The fundamental problem with wildlife management, as animal rights advocates see it, is that it does not respect the right to life of individual animals in a population. The animal rightist does not believe animals should be killed for recreational purposes. Consequently, wildlife should not be managed so a harvestable surplus is available for hunting, nor should hunting be used to reduce populations when a surplus occurs "naturally." Basically, animal rightists see wildlife management as an activity producing a surplus of particular species (i.e., game) to perpetuate needless killing. (Decker and Brown 1987: 600)

Yet there are of course, as already discussed, numerous shades of animal rightists just as there are numerous shades of hunters. Jan E. Dizard's *Going Wild—Hunting, Animal Rights, and the Contested Meaning of Nature* from 1994 is an excellent case study of how an extremely wide range of interests and ethics—animal rights, animal welfare, conservation, preservation, hunting, logging, biodiversity, watershed protection and so on—clash over the question of whether to use hunting as a way of managing deer populations at the Quabbin reservoir in Massachusetts. Simon Bronner's *Killing Tradition—Inside Hunting and Animal Rights Controversies* from 2008 likewise offers insightful case study analyses of conflicts between animal rights and hunting, focusing on the differing interpretations of and attitudes toward certain cultural and traditional aspects of hunting.

A general tenet of the animal rights stance on hunting is related to the movement's close affiliation with the civil rights movement and the women's rights movement. Hunters' and wildlife managers' disregard for the rights of individual animals has traditionally been compared to historical disregards for the rights of non-white races and women. Peter Singer thus employs the concept of "speciesism" to demonstrate that, according to the animal rights view, the difference between considering some races more important than others and considering some species more important than others is arbitrary.

The women's rights movement's connection with the animal rights movement finds an avenue of expression through ecofeminism. With hunting historically being a male-dominated activity, it is a subject of interest to those who belong somewhere in the rather diverse ecofeminism category. Jon Littlefield, investigating masculinity in deer hunting subcultures, generalizes that

> [w]hile a broad range of views exist in the ecofeminist paradigm, they tend to center around an observation that men have historically used their dominion over women and nature to create patriarchal societies, and that such societies foster social structures and behavioral norms that serve to maintain male advantage over females and the natural world. (Littlefield 2010: 98)

Ecofeminists commonly consider hunting a way for men to express their masculinity and assert their dominance. The phallic interpretations of guns, shooting, and hunting trophies are obvious and commonly pointed out. These analyses—while perhaps trying to make men who hunt look foolish and primitive—do not necessarily condemn hunting as ethically wrong; yet they are closely tied to other interpretations of hunting that certainly do. Ecofeminists, and especially those who support vegetarian ecofeminism, draw parallels between men's exploitation of women and men's exploitation of (Mother) nature. In the case of hunting, the comparison between the rape of women and the exploitation (rape) of nature is also common.

One example of this comes from *Rape of the Wild* (1989). Collard and Contrucci are critical of Dart and Ardrey's Hunting Hypothesis and emphasize the importance of gathering for early human evolution and development. Hunters, in their view, contributed mostly to violence and patriarchy:

> Given the frequency with which men wage wars and commit violent crimes, proponents of the hunter theory of evolution would have been more credible if they had limited themselves to their observation that 'men enjoy hunting and killing'. For it is not *hunting* that formed our intellect, emotions, etc., but rather those hunters who, finding pleasure in the hunt, abandoned the ancient rituals of atonement that had accompanied the killing of animals for food. Thus, *some* hunters in *some* parts of the world developed a form of power based on the model of hunting. These hunter-kings spread their value systems through the violence of wars, destroying nature, killing animals, raping women and in general abusing those they enslaved. It is in this context that human intellect, interests, emotions, and basic social life evolved in patriarchy. (Collard and Contrucci 1989: 34, emphasis in original)

Women, on the other hand, are interpreted as prime movers of the peaceful aspects of culture that flourish in harmony with the natural world, rather than in opposition to it. The authors are critical of many of the theories and hypotheses, propagated by men, of large-scale hunting and killing, evidenced by archaeological findings. The authors argue that many such findings can be attributed just as well to animals fleeing from forest fires or other natural disasters and suggest that feminist theories do not find traction in a male-dominated scientific culture. Collard and Contrucci's objections to the wildlife management and conservation arguments approximate the animal rights arguments; hunters only care about conservation to the extent that they can keep on killing. But they also claim that hunters' love of nature is a lie, and that it is merely the feeling of exploiting nature that they love.

Another ecofeminist perspective on hunting is found in Kheel (1996). Kheel characterizes three archetypes of hunters, based on how they justify the activity: the happy hunter, the holist hunter, and the holy hunter.

The happy hunter is the sport hunter. He hunts for fun and excitement and feelings of rugged manliness, and his prey is properly characterized as "game" because that is what hunting is. This hunter believes that hunting is psychologically important, and the psychological benefits to the hunter are a main argument for it. The happy hunter cares about fair chase, because making hunting more challenging and giving the game a better chance at survival is the sporting attitude. If it is too easy, there is no sport in it (Kheel 1996: 33–35). Kheel mentions Theodore Roosevelt and Ernest Hemingway as examples of the happy hunter, and we do not find fault in that. Roosevelt cared deeply about nature and conservation and accomplished more in its name than any other American, but when one considers his own hunting adventures and his insistence that hunting provides important manly development for both the nation and the individual, it is fair to interpret him as a happy hunter first.

Kheel's second archetype is the holist hunter, whose key arguments for hunting are not the psychological needs of the happy hunter, but the ecological benefits of the hunting activity. The holist hunter typically adopts the wildlife management rhetoric of Aldo Leopold. Game is not thought of as "game" in a sporting sense but rather as a crop to be harvested and managed sustainably in an agricultural sense, and

2.10 Modern Animal Rights and Ecofeminism

the holist hunters see their part in managing game herds as essential to avoiding overpopulation, followed by overbrowsing and death from starvation or disease. Kheel makes counterarguments here that appear familiar by now:

> While hunters claim to be responding to nature's unfortunate excesses, the game management journals reveal another story. For example, according to an article in *the Journal of Wildlife Management*, "The primary management plan has been the one directed at increasing the productivity of the whitetail deer through habitat manipulation and harvest regulation ... to produce optimum sustained deer yields ... and hunter satisfaction" (34: p.92). In short, holist hunters are intent on "managing" animals so that sufficient numbers will remain for them to kill. (Kheel 1996: 35–36)

The holist hunter also sees himself as participating in nature in the role of (or in lieu of) the natural predator. To this, Kheel responds as such:

> Holist hunters, however, overlook the vast differences between human predation and natural predation. Whereas natural predators prey on the old, the weak, and the sick, human hunters typically select the biggest and healthiest animals to kill. As a consequence, hunters promote what Teale has called a kind of "evolution in reverse" (48: p. 161). Moreover, sport hunters overlook the extent to which their own actions have produced the problems that they claim to resolve. Sport hunters have pursued a deliberate policy of eliminating natural predators in numerous areas throughout the country, precisely so that they can claim the status of predators for themselves. (Kheel 1996: 36)

"Evolution in reverse" has become one of the most prominent arguments against hunting as wildlife management and especially against trophy hunting. We will discuss it further in Chapter 5.4.

Kheel's third category is the holy hunter:

> For the holy hunter, hunting is not a means of recreation, nor is it a form of work. For the holy hunter, hunting is a religious or spiritual experience. As James Swan has stated, for many it is their religion (47: p. 35). Holy hunters contrast their spiritual attitude of reverence and respect with the crass and superficial mentality of the typical sportsman or happy hunter. Although they too emphasize the notion of emotional self-restraint, they see it as a by-product of a transformed world view. Hunting is akin to a religious rite. (Kheel 1996: 36)

Holy hunters often compare themselves and their spiritual connection with their prey to that of native-American subsistence hunters who after a successful hunt might give thanks to nature and to the animal for "giving" itself to the hunter. Feelings associated with the kill are not the simple feelings of joy and excitement of the happy hunter, but rather ambivalent feelings that include also sorrow and guilt for having had to take a life to survive. Kheel points out that the problem with this philosophy is that modern hunters living in the modern world cannot just unproblematically adopt a subsistence philosophy; taking life is no longer technically necessary for survival, which Kheel argues makes the guilt that modern holy hunters feel seem like a charade (Kheel 1996: 36–37).

Referencing hunting's history as a male activity, rite of manhood, and training for war, Kheel tries to find themes shared between the three archetypes of hunters. She finds the analogy with sexual pursuit and men's pursuit of women instructive, specifically, again, the rape analogy:

> Rape is wrong because it is a violation of another living being. Significantly, the literature on rape argues that rapists are not motivated by the urge to fulfil a sexual drive, nor are they out of control. Similarly, hunting may be seen as a symbolic attempt to assert mastery and control over the natural world. (Kheel 1996: 39)

The ecofeminist comparisons between hunting and rape and assertions of power and the portrayals of women as altogether too gentle and nurturing to hunt raise the question of how to interpret the steadily increasing numbers of female hunters both in Europe and the US (more on hunter demographics in Chapter 4.1). Fitzgerald (2005) argues that the increase in women hunting could be partly attributed to recruitment strategies, meaning state agencies and hunting organisations trying to improve their image and compensate for dwindling overall membership numbers by attracting women. She also notes, emphasizing hunting's historical importance, that "the entrance of some women into sport hunting and the intense effort to recruit them may be partially due to the need to reaffirm the national identity in the face of increasing globalization" (Fitzgerald 2005: 98). Some women may be drawn to the conservationist rhetoric of hunting proponents, which Fitzgerald herself—for ecofeminist and animal rights reasons—does not quite accept (Fitzgerald 2005: 99–100).

Mary Zeiss Stange and her 1997 book, *Woman the Hunter*, represent a view that stands in opposition to traditional ecofeminist interpretations of hunting and hunters. She sees the woman hunter as the bravest and most genuine of the ecofeminists. Stange is a hunter herself, and some of her hunting anecdotes precede each chapter of the book. She raises in the introduction the important point that the woman who hunts

> [u]psets the equilibrium of the conventional interpretations on both sides. This no doubt helps to account for the virtual invisibility of women hunters, both in most popular literature (positive and negative) about hunting and in the various strains of feminist discourse. Regarding the former, the facts that some women in some cultures have always hunted, and that more women in Western cultures are hunting than perhaps ever before, are understandably unsettling to macho-traditionalists who want hunting to remain a rite of initiation into masculine culture and an occasion for "male bonding." [...]
>
> That women's hunting has not appeared within the purview of feminist analysis owes, I demonstrate in what follows, to the fact that American ecofeminism often unwittingly, but nonetheless deliberately, accepts and perpetuates traditional gender stereotypes about femaleness and femininity, and about women's identity with nature—their "innate" passivity and nonviolence—against men's inborn aggressiveness and alienation from the nonhuman world. (Stange 1997: 2)

One of Stange's chief messages is that the feminist portrayal of women as only nonviolent and nurturing is a historical oversight, ignoring plenty of examples of women hunting and female hunting cultures. Stange argues that traditional ecofeminist interpretations of hunting are a disservice to the feminist cause, because they maintain that women are victims in a them-versus-us narrative. Women should, instead of being victims, claim hunting and the role of aggressor as their rightful domain on equal terms. Those who are brave enough to do so often have to overcome stigma, peer pressure, and prejudice from both sides of the fence, but they end up stronger for it, and they break down exactly these barriers.

Finally, concerning female hunters, Debra Merskin (2010) wanted to find out if Kheel's categories of happy hunter, holistic hunter, and holy hunter would fit women who hunt as well as they fit men. She asked female hunters online about why they hunt and found that "[t]he explanations and justifications were remarkably the same as Kheel's three hunter types" (Merskin 2010: 229). Most had traits from all three types, just like the men. Merskin concluded that

> [a] goal of feminist theory is to break away from narrow either-or, this-or-that dichotomous gender roles. I argue, as does Mary Zeiss Stange, that an ecofeminist consideration of women and hunting must consider women as multi-dimensional. Unlike stereotypes of women that say we are only and always the emotional, passive, nurturing sex, Artemis was capable of both caring and killing. The discourse of women who hunt reminds us that aggression is a *human* characteristic. (Merskin 2010: 235–236, emphasis in original)

2.11 Recent Trends and Perceptions of Hunting

The general trend in the US (for reasons of available statistics, we focus mostly on US hunting in this part) before Covid-19 was a decline in hunters, and 2016 saw the lowest hunting participation for at least 25 years. Stephen Kellert wrote already in 1996 that the proportion of Americans who hunt has been declining for a quarter of a century while opposition to hunting has increased. Hunting participation is very reliant on rural family hunting traditions, and with the decline of rural living, hunter recruitment and retention has been declining (Kellert 1996: 69).

Duda et al. concur here:

> It takes a hunter to make a hunter. Almost all hunters are initiated into hunting before the age of 20, usually by a father or father figure, in a rural environment. […] The presence of other family members who hunt, amount of exposure to hunting, and the presence of the "hunting culture" are of utmost importance in hunting initiation. Rarely does hunting initiation occur outside these parameters. There are few other paths of initiation and no other paths of even relatively the same significance. (2010: 67)

Even if there were no other factors contributing to the decline of hunters, this would inevitably lead to fewer and fewer hunters. If almost no one begins hunting without having been introduced to it by (usually) a father, then for every father who stops hunting, does not have a child, or does not introduce his child to hunting, there will be one hunter fewer in the next generation.

Declining numbers of US hunters are driven by three demographic factors, which are "the trend toward urbanization, the aging of the American population, and the declining proportion of the U.S. population that is white/Caucasian" (Duda et al. 2010: 60). Since hunters are often, as we will show in Chapters 4.1 and 4.2, white, rural, and middle-aged, it follows that hunter recruitment would likely decline.

Ironically, along with hunting's urbanization problem—understood here as rural residents becoming urban residents and giving up hunting—comes a tendency towards de-urbanization, which only exacerbates the problem. The de-urbanization trend is that urban residents move to the suburbs or countryside. Yet while hunters often stop hunting when they move to the city, urban residents rarely begin hunting when they move out of it. These people tend to have grown up in an urban environment, have had no first-hand experiences with hunting, and may well not know any hunters who could introduce them to the activity. These new rural residents are often of an antihunting persuasion and not likely to allow any hunting on their new properties. The urbanization and de-urbanization trends put together create a situation where rural communities trade hunters for antihunters, lose viable hunting grounds, and experience increasing amounts of hunter-antihunter conflicts. The remaining hunters—despite their dwindling numbers—crowd together on fewer and smaller available hunting grounds, which, in turn, leads to more of them giving up hunting. Further, wildlife conservation becomes more challenging for state agencies because in addition to fewer hunters meaning less funding, some areas where game populations were previously kept in check by hunters are now not hunted at all, while other areas become overhunted.

Unsurprisingly, optimism is in short supply among hunters, tellingly exemplified here by Dizard:

> Since most hunters do not see a bright future for hunting, it is easier to shrug off their children's disinterest in the way Mac Braziel did when he said of his son's disinterest, "He won't be missing much anyway." Mac said this with a distinct look of resignation mixed with disappointment on his face. He knew that, in fact, his son would be missing something that Mac felt was important—hence the disappointment. Mac also knew that in all likelihood, even if his son did take up hunting, it would not be the sort of hunting that Mac has experienced—hence the resignation. (Dizard 2003: 180)

In addition to the demographic factors, shifting attitudes have also often been brought up as a reason for the decline in hunting participation. Kellert wrote that "opposition to hunting has dramatically increased during the twentieth century, particularly during the past fifty years" (Kellert 1996: 69), and Dizard (2003: 171–179) does an excellent job portraying the general pessimism of the hunters he interviewed about the future of hunting and the increasing amount of conflicts with antihunters and especially animal rights activists. It is likely, though, that contemporary hunting opposition seems stronger than it has before because those who are strongly opposed to hunting are even *more* strongly opposed today, more confrontational, and much more vocal on social media about their opposition. While hunter recruitment is still down, more recent studies indicate that the hunting *approval* trend might be reversing. Duda et al. (2010: 44) show that 73% approved and 22% disapproved of hunting nationwide in 1995, 75% approved and 17% disapproved in 2003, and 78% approved and 16% disapproved in 2006. Similar trends were seen in several state-wide studies (Duda et al. 2010: 48).[33] Statistics look even more convincing for some European countries. Sweden, for example, reported an 87% acceptance rate for hunting in 2014, which has been steadily increasing since 1971 where it was at 70%,

---

[33] It is important to note that the general American public often does not realize that hunting does not include poaching. When surveys describe hunting as "regulated hunting" or "legal hunting," respondents demonstrate higher approval ratings than when it is just called "hunting," meaning that some nonhunters take "hunting" to include both hunting and poaching. Hunters generally take for granted that "hunting" is legal, as illegal hunting is called poaching.

according to the Federation of Associations for Hunting and Conservation (FACE) (FACE 2014: 12).

Regardless of the support or opposition to hunting, the attitudes of antihunters seem to have only a minor influence on hunter retention, at least if the self-reporting of hunters is to be believed.[34] In one 2008 study, inactive hunters were asked 25 questions about how much certain factors had influenced them to not hunt in recent years. 40% indicated that "Amount of free time" had strongly influenced their decision, 35%"Family obligations," 34% "Work obligations," 33% "Loss of interest," and 17% "Not enough access." "Other people's negative opinions" and "Harassment by anti-hunters" was indicated as a strong influence by only 2% and 1% respectively—the least influential of all 25 factors measured. "Poor behavior of hunters" was at 11%, so hunters are seemingly much more bothered by other hunters' conduct than they are by the opinions or actions of antihunters (Duda et al. 2010: 66).

Fees from hunting licenses, taxes on fishing- and hunting equipment, tags, stamps, etc. are the primary sources of funding for state- and federal wildlife management agencies, and hunting is the standard way of controlling game populations. This means the decline of hunters poses serious funding and management problems:

> Decades of decline in the total number of licensed hunters in the United States has resulted in major biological, economic, and social consequences [...]. Wildlife agencies have endured reduced conservation funding, have experienced difficulty managing impacts of some species (e.g., some ungulates) and generally fear erosion of the North American Model of Conservation, which emphasizes a sustainable connection between people and public trust resources such as wildlife. (Larson et al. 2013: 1)

Consequently, Hunter Recruitment and Retention has become an important focus for wildlife managers' work and research. On the bright side, evidence shows that percentages of the general population with an interest in going hunting are considerably higher than percentages who hunt or have tried hunting, so there is a strong basis for recruitment success. Forty-four percent of youth, for example, indicated an interest in hunting in 2003 and 19% that they were *very* interested, even while youth are responsible for most of the decline in hunter recruitment over the past few decades (Duda et al. 2010: 209). Getting started without having easy access or anyone in the immediate family to take you is difficult, even if the interest is there, which is why mentoring shows signs of being one of the most promising instruments for Hunter Recruitment and Retention (Ryan and Shaw 2011).

Attitudes toward hunting depend heavily on hunting motivations. Duda et al. (2010: 39–42) compared several single-state studies from 1989 to 2006 of hunting approval rates and found that between a low 66% of respondents in one study and a high 96% in another study (average 83%) approved of hunting for meat, 62–87% (average 77%) approved of hunting for wildlife management/population control, 70–87% (average 78%) for pest control/protection of human property, 25–55% (average 40%) for sport/challenge/recreation, and 10–22% (average 16%) for a trophy. Around 80% on average disapproved of hunting for a trophy.

---

[34] It is worth considering that even in an anonymous survey, hunters may not like to admit it if they are bothered by the opinions of antihunters.

Byrd et al. (2017) report that 87% find hunting for food acceptable, 72% for wildlife management, and a surprisingly high 37% for a trophy.[35] According to Bird et al., trophy hunting acceptance depends heavily on respondent age, with young respondents much less likely to indicate acceptance. Other notable factors increasing approval percentages of trophy hunting by 10% or more were knowing someone who hunts, being male, and having visited a fair or livestock operation. Incidentally, knowing someone who hunts, according to all research, shows a stronger correlation with general hunting approval than any other characteristic for nonhunters.

Fifty-one percent of respondents in Byrd et al. think that hunting is an important rural tradition, and between a third to half agree that hunting helps manage wildlife populations (31%), reduce damage to agricultural crops (45%), and keep nature in balance (48%). Conversely, 39% think that hunting leads to species becoming threatened or endangered, 37% think that hunters often ignore safety rules, and about a third think that hunting for sport or recreation is cruel.

The most common reasons for disapproving of hunting is that it is morally wrong, which is given as a reason by 50 to 80% of antihunters, depending on the study, and—with similar proportions—that it causes pain to animals. The latter reason is based mostly on antihunter beliefs that wounding shots by careless hunters happen often, that the wounded animals are not tracked down, and that bullets are a painful way to die (Duda et al. 2010: 36).

Antihunters' widespread worry about prolonged pain indicates that opposition to hunting may in part be due to an assumption of poor marksmanship more so than an opposition in principle to the killing. Indeed, evidence suggests that the American public harbours more ill feelings toward hunters than they do toward hunting. A study from 1978 found that

> the top problems perceived by the public to be associated with hunting had little to do with hunting itself (at least not ethical hunting), but were associated with individual hunters and their poor behaviour (e.g., hunters fail to track wounded animals, hunters shoot animals that they are not allowed to shoot, hunters ignore safety regulations, hunters trespass, hunters shoot too close to highways, hunters don't know what they are shooting at). (Duda et al. 2010: 37)

Other studies, all summarized in Duda et al., found that 78% of respondents were bothered by disrespectful hunters and that even among hunters—in one Minnesota study—there were 27% who disapproved of hunting for recreation, with the commonly stated reason being the disrespectful conduct of other hunters. A national study from 1995 showed 64% agreeing with the statement that "a lot of hunters violate hunting laws or practice unsafe behavior while hunting" (Duda et al. 2010: 48), while only 23% disagreed. Fifty percent of respondents had the perception that many hunters drink alcohol while hunting and 74% of hunters said that hunters

---

[35] It might be relevant to note that Byrd et al. asked whether trophy hunting is "acceptable," while the studies summarized by Duda et al. tend to use the phrasing of "approval" or "support." It is quite likely that some respondents are willing to accept something, even if they do not support it or approve of it, which could explain the seemingly more positive attitude towards trophy hunting in Byrd et al.

violate laws knowingly. Target species of hunters are another important factor in determining support or opposition to hunting. Seventy-eight percent of Americans moderately or strongly approved of deer hunting in 2006 with wild turkey at 75%, small game at 71%, waterfowl at 69%, elk at 60%, black bear at 47%, and mountain lion at 42%, indicating clearly the lower approval of hunting for predators. This seems to be the case even when predator populations are problematically high and cause frequent human-animal conflicts (Duda et al. 2010: 43–44). Opposition to hunting can also be due to the opposition to firearms, which was the reason given by a third of nonhunting respondents in the abovementioned survey from Minnesota (Duda et al. 2010: 36). The opposition to hunters being stronger than the opposition to hunting is a central theme for the opposition to trophy hunting, and we shall return to this in Chapter 8.

References

Adams, John Quincy. 1903. *Life in a New England Town: 1787, 1788—Diary of John Quincy Adams*. Boston: Little, Brown, and Company.
Aiyadurai, Ambika, Navinder J. Singh and E. J. Milner-Gulland. 2010. Wildlife hunting by indigenous tribes: A case study from Arunachal Pradesh north-east India. *Oryx* 44 (4): 564–572. https://doi.org/10.1017/S0030605309990937.
Albenda, Pauline. 1975. Lions on Assyrian Wall Reliefs. *The Journal of the Ancient near Eastern Society of Columbia University* 6: 1–27.
Alepidou, Apostalia. 2017. From Hunter to Hero: Hunting narratives in Odyssey 19 and Iliad 9. Master's Thesis, University of Leiden. http://hdl.handle.net/1887/45712) (accessed 20 June 2017).
Allsen, Thomas T. 2013. *The Royal Hunt in Eurasian History*. Singapore: Institute of Southeast Asian Studies (ISEAS) Publishing.
Almond, Richard. 2011 (First 2003). *Medieval Hunting*. Gloucestershire: The History Press.
Altherr, Thomas L. 1984. "Chaplain to the Hunters": Henry David Thoreau's Ambivalence Toward Hunting. *American Literature* 56 (3): 345–361.
Anderson, J.K. 1985. *Hunting in the Ancient World*. London: University of California Press.
Anker, Peder. 2004. A Vindication of the Rights of Brutes. *Philosophy and Geography* 7 (2): 259–264.
Anthony, Raymond, and Gary Varner. 2019. Subsistence Hunting. In *The Routledge Handbook of Animal Ethics*, ed. Bob Fischer, 211–222. New York: Routledge.
Archaeology News Network. 2012. 6,500 Year Old Hunting Trophy Found In Eastern Croatia. Archaeology News Network, July 27. https://archaeologynewsnetwork.blogspot.com/2012/07/6500-year-old-hunting-trophy-found-in.html (accessed 21 January 2021).
Ardrey, Robert. 1976. *The Hunting Hypothesis*. Glasgow: William Collins Sons and Co Ltd.
Aubert, M., A. Brumm, M. Ramli, T. Sutikna, E.W. Saptomo, B. Hakim, M.J. Morwood, G.D. van den Bergh, L. Kinsley, and A. Dosseto. 2014. Pleistocene cave art from Sulawesi, Indonesia. *Nature* 514: 223–227.
Bahn, Paul G., and Jean Vertut. 1997. *Journey through the Ice Age*. London: Weidenfeld and Nicolson.
Baillie-Grohmann, William A. 1904. Introduction. In William A. Baillie-Grohman and F. N. Baillie-Grohman (eds.), 2005 in Edward of Norwich, circa 1413, *The Master of Game*, xi-xviii. Philadelphia: University of Pennsylvania Press.

Baquedano, Enrique, Juan L. Arsuaga, Alfredo Pérez-González, César Laplana, Belén Márquez, Rosa Huguet, Sandra Gómez-Soler, Lucía Villaescusa, M. Ángeles Galindo-Pellicena, Laura Rodríguez, Rebeca García-González, M.-Cruz Ortega, David M. Martín-Perea, Ana I. Ortega, Lucía Hernández-Vivanco, Gonzalo Ruiz-Liso, Juan Gómez-Hernanz, José I. Alonso-Martín, Ana Abrunhosa, Abel Moclán, Ana I. Casado, Marina Vegara-Riquelme, Ana Álvarez-Fernández, Ángel C. Domínguez-García, Diego J. Álvarez-Lao, Nuria García, Paloma Sevilla, Hugues-Alexandre Blain, Blanca Ruiz-Zapata, M. José Gil-García, Adrián Álvarez-Vena, Teresa Sanz, Rolf Quam, and Tom Higham. 2023. A symbolic Neanderthal accumulation of large herbivore crania. *Nature Human Behaviour*. https://doi.org/10.1038/s41562-022-01503-7.

Barringer, Judith M. 2001. *The Hunt in Ancient Greece*. Baltimore: Johns Hopkins University Press.

Beaver, Daniel C. 2008. *Hunting and the Politics of Violence before the English Civil War*. New York: Cambridge University Press.

Bentham, Jeremy. 1789. *An Introduction to The Principles of Morals and Legislation* in Jonathan Bennett, 2017 http://www.earlymoderntexts.com/assets/pdfs/bentham1780.pdf (accessed 10 October 2017).

Berger, L.R., and W.S. McGraw. 2007. Further evidence for eagle predation of, and feeding damage on, the Taung child. *South African Journal of Science* 103 (11–12): 496–503.

Bonatz, Dominik. 2004. Ashurbanipal's headhunt: An Anthropological Perspective. *Iraq* 66: 93–101.

Booth, Vernon and Philippe Chardonnet (eds.). 2015. *Guidelines for improving the administration of sustainable hunting in sub-Saharan Africa*. FAO: sub-Regional Office for Southern Africa. http://www.fao.org/3/a-bo583e.pdf) (accessed 6 March 2020).

Brain, C.K. 1970. New Finds at the Swartkrans Australopithecine Site. *Nature* 225: 1112–1119.

Brinkley, Douglas. 2010. *The Wilderness Warrior—Theodore Roosevelt and The Crusade for America*. New York: HarperCollins. https://erenow.com/biographies/the-wilderness-warrior-theodore-roosevelt-and-the-crusade-for-america/ (accessed 14 December 2017).

Brooks, Paul. 2014 (First 1980). *Speaking for Nature—The Literary Naturalists, from Transcendentalism to the Birth of the American Environmental Movement*. New York: Dover Publications Inc.

Burroughs, John. 1895. *Riverby*. Boston and New York: Houghton, Mifflin and Company.

Burroughs, John. 1907. *Camping and Tramping with Roosevelt*. Boston and New York: Houghton Mifflin Company.

Byrd, Elizabeth, John G. Lee, and Nicole J. Olynk Widmar. 2017. Perceptions of Hunting and Hunters by U.S. Respondents. *Animals* 7/83: 1–15.

Callicott, J. Baird. 1989. *Defense of the Land Ethic—Essays in Environmental Philosophy*. Albany: State University of New York Press.

Calloway, Colin G. 1999. *First Peoples—A Documentary Survey of American Indian History*. Boston and New York: Bedford/St. Martin's.

Cartmill, Matt. 1996 (First 1993). *A View to a Death in the Morning—Hunting and Nature Through History*. Cambridge: Harvard University Press.

Collard, Andrée with Joyce Contrucci. 1989. *Rape of the Wild—Man's Violence Against Animals and the Earth*. Bloomington and Indianapolis: Indiana University Press.

Coues, Elliott. 1890. *Handbook of Field and General Ornithology—A Manual of the Structure and Classification of Birds with Instructions for Collecting and Preserving Specimens*. London: Macmillan and Co.

Cumming, Roualeyn Gordon. 1850. *Five Years of a Hunter's Life in the far Interior of South Africa*, 2nd ed., Vol. 1/2. London: John Murray, Albemarle Street.

Cumming, Roualeyn Gordon. 1874 (First 1850). *Five Years of a Hunter's Life in the far Interior of South Africa*, Vol. 2/2. New York: Harper and Brothers.

Cummins, John. 2003. (First 1988). *The Art of Medieval Hunting: The Hound and the Hawk*. Edison: Castle Books.

Cutright, Paul Russell. 1985. *Theodore Roosevelt, The Making of a Conservationist*. Urbana and Chicago: University of Illinois Press.

References

Dart, Raymond. 1925. Australopithecus africanus The Man-Ape of South Africa. *Nature* 115: 195–199.

Decker, Daniel J., and Tommy L. Brown. 1987. How Animal Rightists view the "Wildlife Management-Hunting System." *Wildlife Society Bulletin* 15 (4): 599–602.

Dembitzer, Jacob, Ran Barkai, Miki Ben-Dor, and Shai Meiri. 2022. Levantine Overkill: 1.5 Million Years of Hunting Down the Body Size Distribution. *Quaternary Science Reviews* 276.

Descartes, René. 1637 (First 1965). Discourse on Method. In *Discourse on Method, Optics, Geometry, and Meteorology*, trans. Paul J. Olscamp, 2001. Cambridge: Hackett Publishing Company.

Dick, Michael B. 2006. The Neo-Assyrian Royal Lion Hunt and Yahweh's Answer to Job. *Journal of Biblical Literature* 125 (2): 243–270.

Dizard, Jan E. 2003. *Mortal Stakes—Hunting and Hunters in Contemporary America*. Amherst and Boston: University of Massachusetts Press.

Duda, Mark Damian, Martin Jones, and Andrea Criscione. 2010. *The Sportsman's Voice—Hunting and Fishing in America*. Responsive Management/Venture Publishing.

Dunlap, Thomas R. 1988. Sport Hunting and Conservation, 1880–1920. *Environmental Review* 12 (1): 51–60.

Eaton, Randall. 1978. The Evolution of Trophy Hunting. *Carnivores* 1 (1): 110–121.

Edward of Norwich, circa 1413 in William A. Baillie-Grohman and F. N. Baillie-Grohman (eds.). 2005. *The Master of Game*, 1–268. Philadelphia: University of Pennsylvania Press.

Fa, Julia E., Stephan M. Funk, and Robert Nasi. 2022. *Hunting Wildlife in the Tropics and Subtropics*. Cambridge: Cambridge University Press.

FACE. 2014. Sweden—A Country of Endless Hunting Opportunities. Federation of Associations for Hunting and Conservation of the EU, December 4. http://www.face.eu/sites/default/files/attachments/hunting_and_wildlife_management_i_sweden_2014-04-10_.pdf (accessed 24 October 2019).

Fagan, Brian M. 1995 (First 1991). *Ancient North America—The Archaeology of a Continent*, 2nd ed. London: Thames and Hudson.

Fitzgerald, Amy J. 2005. The Emergence of the Figure of "Woman-the-Hunter:" Equality or Complicity in Oppression. *Women's Studies Quarterly* 33 (1/2): 86–104.

Francione, Gary L., and Anna E. Charlton. 2017. Animal Rights. In *The Oxford Handbook of Animal Studies*, ed. Linda Kalof, 25–42. New York: Oxford University Press.

Francione, Gary L. 2017. Gary L. Francione: The Abolitionist Approach to Animal Rights. Facebook, September 9. https://www.facebook.com/abolitionistapproach/posts/1659307934088954 (accessed 4 May 2020).

Francione, Gary L. 2013. Some Thoughts on Melissa Bachman and the Lion. In *Animal Rights: The Abolitionist Approach*, November 20. http://www.abolitionistapproach.com/thoughts-melissa-bachman/#.U7QYbUD_naE (accessed 4 May 2020).

Francione, Gary L. 2014. Oh, No: Another Woman Does What Everyone Else Does. In *Animal Rights: The Abolitionist Approach*, July 2. https://www.abolitionistapproach.com/oh-another-woman-everyone-else/ (accessed 4 May 2020).

Francione, Gary L. 2015. Adventures in Moral Schizophrenia: Cecil the Lion. In *Animal Rights: The Abolitionist Approach*, July 29. https://www.abolitionistapproach.com/adventures-in-moral-schizophrenia-cecil-the-lion/ (accessed 4 May 2020).

Frison, George C. 2004. *Survival by Hunting: Prehistoric Human Predation and Animal Prey*. Berkeley and Los Angeles: University of California Press.

Geo.tv. 2021. Markhor Trophy Hunting Permit Auctioned at Record Price. Geo News, November 5. https://www.geo.tv/latest/380492-markhor-trophy-hunting-permit-auctioned-at-record-price (accessed 20 June 2022).

Gray, V.J. 1985. Xenophon's 'Cynegeticus.' *Hermes* 113 (2): 156–172.

Grinnell, George Bird. 1913. List of Members. In George Bird Grinnell (ed.), *Hunting at High Altitudes—The Book of the Boone and Crockett Club*, 505–511. New York: Harper and Brothers.

Harmand, Sonia, Jason E. Lewis, Craig S. Feibel, Christopher J. Lepre, Sandrine Prat, Arnaud Lenoble, Xavier Boës, Rhonda L. Quinn, Michel Brenet, Adrian Arroyo, Nicholas Taylor, Sophie Clément, Guillaume Daver, Jean-Philip Brugal, Louise Leakey, Richard A. Mortlock, James D. Wright, Sammy Lokorodi, Christopher Kirwa, Dennis V. Kent and Hélène Roche. 2015. 3.3-Million-Year-Old Stone Tools from Lomekwi 3, West Turkana, Kenya. *Nature* 521: 310–315.

Hart, Donna, and Robert W. Sussman. 2005. *Man the Hunted—Primates, Predators, and Human Evolution*. New York: Westview Press.

Herbert, William Henry. 1868 (First 1848). *Frank Forester's Field Sports of the United States and British Provinces of North America*. New York: W. A. Townsend and Adams.

Herne, Brian. 1999. *White Hunters—The Golden Age of African Safaris*. New York: Henry Holt and Company.

Hornaday, William T. 1889. *The Extermination of the American Bison*. Washington, DC: Government Publishing Office.

Hornaday, William T. 1913. *Our Vanishing Wild Life—Its Extermination and Preservation*. New York: Charles Scribner's Sons.

Houlihan, Patrick. 1996. *The Animal World of the Pharaos*. London: Thames and Hudson.

Hublin, Jean-Jacques, Abdelouahed Ben-Ncer, Shara E. Bailey, Sarah E. Freidline, Simon Neubauer, Matthew M. Skinner, Inga Bergmann, Adeline Le Cabec, Stefano Benazzi, Katerina Harvati and Philipp Gunz. 2017. New Fossils from Jebel Irhoud, Morocco and the Pan-African Origin of *Homo sapiens*. *Nature* 546: 289–292.

Hughes, Gugulethu. 2022. The Truth About Conservation: Privatization, Racism, and Control. ScoonTv, June 21. https://www.scoontv.com/the-truth-about-conservation-privatization-racism-and-control/ (accessed 22 June 2022).

Hughes, J. Donald. 2014 (First 1994). *Environmental Problems of the Greeks and Romans*. Baltimore: Johns Hopkins University Press.

Isaacson, Rupert. 2001. *The Wild Host—The History and Meaning of the Hunt*. London: Cassell and Co.

Jeffers, H. Paul. 2003. *Roosevelt the Explorer—Teddy Roosevelt's Amazing Adventures as a Naturalist, Conservationist, and Explorer*. New York: Taylor Trade Publishing.

Johnson, Daniel. 1827 (First 1822). *Sketches of Indian Field Sports*, 2nd ed. London: Robert Jennings, 2, Poultry.

Kalof, Linda. 2007. *Looking at Animals in Human History*. London: Reaktion Books.

Kant, Immanuel in Louis Infield (trans.) 1963 (First 1930). *Lectures on Ethics*. New York: Harper and Row. https://www.uta.edu/philosophy/faculty/burgess-jackson/Duties%20Towards%20Animals%20and%20Spirits%20(1930,%201963).pdf (accessed 7 September 2017).

Kellert, Stephen Robert. 1996. *The Value of Life—Biological Diversity and Human Society*. Washington, DC: Island Press/Shearwater Books.

Kester, Jesse Y. 1827. *The American Shooter's Manual: Comprising Such Plain and Simple Rules, as are Necessary to Introduce the Inexperienced Into a Full Knowledge of All that Relates to the Dog, and the Correct Use of a Gun; Also a Description of the Game of this Country*. Philadelphia: Carey, Lea and Carey.

Kheel, Marty. 1996. The Killing Game: An Ecofeminist Critique of Hunting. *Journal of the Philosophy of Sport* 23 (1): 30–44.

Klemettilä, Hannele. 2015. *Animals and Hunters in the Late Middle Ages—Evidence from the BnF MS fr. 616 of the Livre de Chasse by Gaston Fébus*. New York: Routledge.

Knoll, Martin. 2004. Hunting in the Eighteenth Century. An Environmental History Perspective. *Historical Social Research* 29/3 (109): 9–36.

LaFreniere, Gilbert F. 1990. Rousseau and the European Roots of Environmentalism. *Environmental History Review* 14 (4): 41–72.

Langdon, John C. 2005. *The Human Strategy—An Evolutionary Perspective on Human Anatomy*. New York: Oxford University Press.

Larson, Lincoln R., Daniel J. Decker, Richard C. Stedman, William F. Siemer, Meghan S. Baumer, and Jody W. Enck. 2013. *Hunter Recruitment and Retention in New York: A Framework for*

Research and Action. Human Dimensions Research Unit, Department of Natural Resources, Cornell University.

Leopold, Aldo. 1915. The Varmint Question. In *The River of the Mother of God—And Other Essays by Aldo Leopold*, ed. Susan L. Flader and J. Baird Callicott, 1991, 47–48. Wisconsin and London: The University of Wisconsin Press.

Leopold, Aldo. 1919. Wild Lifers vs. Game Farmers: A Plea for Democracy in Sport. In *The River of the Mother of God—And Other Essays by Aldo Leopold*, ed. Susan L. Flader and J. Baird Callicott, 1991, 62–67. Wisconsin and London: The University of Wisconsin Press.

Leopold, Aldo. 1925. Wilderness as a Form of Land Use. In *The River of the Mother of God—And Other Essays by Aldo Leopold*, Susan L. Flader and J. Baird Callicott, 1991, 134–142. Wisconsin and London: The University of Wisconsin Press.

Leopold, Aldo. 1968 (First 1949). *A Sand County Almanac, And Sketches Here and There*. New York: Oxford University Press.

Leopold, Aldo. 1986 (First 1933). *Game Management*. Madison and London: The University of Wisconsin Press.

Littledale, H. 1889. Rough Notes of Travel and Sport in Kashmir and Little Thibet. *The Journal of the Bombay Natural History Society* 4: 98–118.

Littlefield, Jon. 2010. Men on the Hunt: Ecofeminist insights into masculinity. *Marketing Theory* 10 (1): 97–117.

Lombard, Marlize, and Laurel Phillipson. 2010. Indications of Bow and Stone-Tipped Arrow Use 64 000 Years Ago in KwaZulu-Natal, South Africa. *Antiquity* 84 (325): 635–648.

MacKenzie, John M. 1988. *The Empire of Nature—Hunting, Conservation and British Imperialism*. Manchester and New York: Manchester University Press.

Markham, A.M. 1895. Some Indian Stalking and Shooting. *The Journal of the Bombay Natural History Society* 9: 34–41.

Mbaria, John, and Mordecai Ogada. 2016. *The Big Conservation Lie: The Untold Story of Wildlife Conservation in Africa*. Auburn: Lens&Pens Publishing LLC.

Meggers, Betty J. 2010 (First 1972). *Prehistoric America—An Ecological Perspective*, 3rd ed. New Brunswick: AldineTransaction.

Meine, Curt. 1987. Aldo Leopold's Early Years. In *Companion to A Sand County Almanac, Interpretive and Critical Essays*, J. Baird Callicott, 17–39. Madison: The University of Wisconsin Press.

Meltzer, David J., and Tom D. Dillehay. 1999. The Search for the Earliest Americans. *Archaeology* 52 (1): 60–61.

Merskin, Debra. 2010. The New Artemis? Women Who Hunt. In *Hunting Philosophy for Everyone: In Search of the Wild Life*, ed. Nathan Kowalsky, 225–238. Wiley-Blackwell.

Morris, Kathryn, and Jane Goodall. 1977. Competition for Meat Between Chimpanzees and Baboons of the Gombe National park. *Folia Primatologica* 28 (2): 109–121.

Muir, John. 1894. *The Mountains of California*. https://vault.sierraclub.org/john_muir_exhibit/writings/the_mountains_of_california/ (accessed 20 June 2022).

Muir, John. 1903. Letter from [John Muir] to [Theodore] Roosevelt, 1903 Mar 27. Online Archive of California. http://www.oac.cdlib.org/ark:/13030/kt7s203673/?layout=metadata&brand=oac4 (accessed 20 June 2022).

Muir, John. 1904. Letter from John Muir to [H. F.] Osborn, [1904 Jul 16]. *Calisphere* http://content.cdlib.org/ark:/13030/kt4s2034w1/?order=3&brand=calisphere (accessed 8 March 2018).

Muir, John. 1916. *A Thousand-Mile Walk to the Gulf*. Boston and New York: Houghton Mifflin Company.

Nature. 1919. The Selous Collection. *Nature* 5-6-1919, 2588/103: 274.

Payne, Daniel G. 1996. *Voices in the Wilderness—American Nature Writing and Environmental Politics*. Hannover and London: University Press of New England.

Petersen, David. 2010 (First 2000). *Heartsblood: Hunting, Spirituality, and Wildness in America*. Washington, DC and Covelo: Island Press/Shearwater Books.

Phelps, Norm. 2007. *The Longest Struggle—Animal Advocacy from Pythagoras to PETA*. New York: Lantern Books.

Plato, circa 400 BC in Malcolm Schofield (ed.) and Tom Griffith (trans.). 2016. *Plato—The Laws*. Cambridge: Cambridge University Press.

Poliquin, Rachel. 2012. *The Breathless Zoo—Taxidermy and the Cultures of Longing*. Pennsylvania State University Press.

Pringle, Heather. 2011. Texas Site Confirms Pre-Clovis Settlement of the Americas. *Science* 331: 1512.

Regan, C. Tate. 1921. Preface. In *Catalogue of the Selous Collection of Big Game in the British Museum (Natural History)*, ed. J. G. Dollman. London: Taylor and Francis.

Regan, Tom. 1975. The Moral Basis of Vegetarianism. *Canadian Journal of Philosophy* 5 (2): 181–214.

Regan, Tom. 1983. *The Case for Animal Rights*. London, Melbourne, and Henley: Routledge and Kegan Paul.

Reiger, John F. 2001 (First 1975). *American Sportsmen and the Origins of Conservation*, 3rd ed. Corvallis: Oregon State University Press.

Rooney, Anne. 1993. *Hunting in Middle English Literature*. Cambridge: The Boydell Press.

Roosevelt, Theodore. 1885. *Hunting Trips of a Ranchman*. New York: G.P. Putnam's Sons.

Roosevelt, Theodore. 1888. *Ranch life and the Hunting Trail*. New York: The Century Co.

Roosevelt, Theodore. 1901. State of the Union Message to the Senate and House of Representatives. http://www.theodore-roosevelt.com/images/research/speeches/sotu1.pdf (accessed 17 January 2018).

Roosevelt, Theodore. 1904a. Our Forest Policy. *The Plant World* 7/1: 8-11.

Roosevelt, Theodore. 1904b. Foreword. In *The Master of Game*, ed. William A. Baillie-Grohman and F. N. Baillie-Grohman, 2005, xix–xxix. Philadelphia: University of Pennsylvania Press.

Roosevelt, Theodore. 1906. *The Strenuous Life*. New York: Charles Scribner's Sons.

Roosevelt, Theodore. 1915. John Muir: An Appreciation. *Outlook* 109, January 16, 1915: 27–28: https://babel.hathitrust.org/cgi/pt?id=coo.31924066364351;view=1up;seq=41 (accessed 8 March 2018).

Rousseau, Jean-Jacques. 1754. Discourse on the Origin of Inequality. In *Jean-Jacques Rousseau—Basic Political Writings, Discourse on the Sciences and the Arts, Discourse on the Origin of Inequality, Discourse on Political Economy, On the Social Contract*, trans. and ed. Donald A. Cress, 1987, 25–109. Cambridge: Hackett Publishing Company.

Ryan, Elizabeth L., and Bret Shaw. 2011. Improving Hunter Recruitment and Retention. *Human Dimensions of Wildlife* 16: 311–317.

Salt, Henry Stephens. 1980 (First 1892). *Animals' Rights Considered in Relation to Social Progress*. London: Centaur Press Limited.

Schafer, Edward H. 1968. Hunting Parks and Animal Enclosures in Ancient China. *Journal of the Economic and Social History of the Orient* 11 (3): 318–343.

Singer, Peter. 1975. *Animal Liberation: A New Ethics for Our Treatment of Animals*. New York: Jonathan Cape.

Singer, Peter. 1980 (First 1892). Preface. In *Animals' Rights Considered in Relation to Social Progress*, ed. Henry Stephens Salt. London: Centaur Press Limited.

Singer, Peter. 2008. Peter Singer on Using Animals. Interview for Ethics Bites in association with Open University, OpenLearn, February 13. http://www.open.edu/openlearn/history-the-arts/culture/philosophy/human-use-animals (accessed 8 May 2018).

Singer, Peter. 2009. "Peter Singer: "The Ethics of What We Eat" talk at *Williams College*, December 15. https://www.youtube.com/watch?v=UHzwqf_JkrA (accessed 8 May 2018).

Smith, Andrew B. 1999. Archaeology and Evolution of Hunters and Gatherers. In *The Cambridge Encyclopedia of Hunters and Gatherers*, ed. Richard B. Lee and Richard Daly, 384–390. Cambridge: Cambridge University Press.

Smith, Michael B. 1998. The Value of a Tree: Public Debates of John Muir and Gifford Pinchot. *The Historian* 60 (4): 757–778.

References

Somerville, Keith. 2019. *Lions and Humans: Conflict, Conservation and Coexistence.* London: Routledge.

Spencer, Herbert. 1892 (First 1891). *Justice: Being Part IV of the Principles of Ethics.* New York: D. Appleton and Company.

Sramek, Joseph. 2006. "Face Him like a Briton": Tiger Hunting, Imperialism, and British Masculinity in Colonial India, 1800–1875. *Victorian Studies* 48 (4): 659–680.

Stanford, Craig. 2003. *Upright—The Evolutionary Key to Becoming Human.* New York: Houghton Mifflin Company.

Stange, Mary Zeiss. 1997. *Woman the Hunter.* Boston: Beacon Press.

Stewart, Mathew, W. Cristopher Carleton, and Huw S. Groucutt. 2021. Climate Change, not Human Population Growth, Correlates with Late Quaternary Megafauna Declines in North America. *Nature Communications* 12 (965): 1–15.

Stoneham, Michael. 2003. Thoreau's Hunting Ethos: Teaching Right Natural Relations in Modern America. *The Thoreau Society Bulletin* 282: 8–10.

Storey, William K. 1991. Big Cats and Imperialism: Lion and Tiger Hunting in Kenya and Northern India, 1898–1930. *Journal of World History* 2 (2): 135–173.

Taylor, Thomas. 1792. *A Vindication of the Rights of Brutes.* London: Edward Jeffrey, Pallmall.

Thomas, David Hurst. 2000. *Exploring Native North America.* New York: Oxford University Press.

Thomas, Keith. 1983. *Man and the Natural World—A History of the Modern Sensibility.* New York: Pantheon Books.

Thompsell, Angela. 2015. *Hunting Africa—British Sport, African Knowledge and the Nature of Empire.* New York: Palgrave Macmillan.

Thoreau, Henry David. 1864. *The Maine Woods.* Boston: Ticknor and Fields.

Thoreau, Henry David. 1882 (First 1854). *Walden'*, 19th ed. Boston: Houghton, Mifflin and Company.

Velho, Nandini, Krithi K. Karanth, and William F. Laurance. 2012. Hunting: A Serious and Understudied Threat in India, a Globally Significant Conservation Region. *Biological Conservation* 148: 210–215.

Washburn, Sherwood L., and Chet S. Lancaster. 1968. The Evolution of Hunting. In *Man the Hunter*, ed. Richard B. Lee and Irven DeVore, 293–303. Chicago: Aldine Publishing Press.

Wilkins, Jayne, Benjamin J. Schoville, Kyle S. Brown, and Michael Chazan. 2012. Evidence for Early Hafted Hunting Technology. *Science* 338: 942–946.

Wilkins, Thurman. 1995. *John Muir—Apostle of Nature.* Norman and London: University of Oklahoma Press.

Xenophon, circa 400 BC: *On Hunting—A Sportsman's Manual—Commonly Called Cynegeticus* in H. G. Dakyns (trans.) 1897 (web edition published by eBooks@Adelaide, 2014: https://ebooks.adelaide.edu.au/x/xenophon/x5hu/complete.html) (accessed 20 June 2017).

Chapter 3
Hunting Trophies

The origin of the word trophy is the Greek *tropaion*, which means "of defeat." If a Greek city state had routed an enemy in battle, they would erect a tropaion on the battlefield, which was usually a tree, adorned with the armour of the defeated enemy and inscribed with details of the battle. The tropaion was dedicated to one or more gods, so it would be sacrilege if the enemy were to destroy it. The original tropaion was a temporary installation but these simple and temporary trophies later turned into permanent and sometimes enormous monuments erected in cities. This became common practice in the Roman empire, where a *tropaeum* would be erected to commemorate victories and show that Rome was favoured by the gods. The first of such tropaeums was erected in 121 BC (Nielson 1983).

Tonio Hölscher (2006) describes the tropaion as a way of turning successful single events such as victories in battles into long-term political power and stability. This view is pertinent in relation to the hunting trophy, not because a hunting trophy is able to give its owner political power and stability,[1] but because its creation is a way of turning a short-term single-event success—i.e. the successful hunt—into an object of permanence that bespeaks characteristics of its owner and serves as a reminder of the event.

Aldo Leopold wrote that

> [t]he trophy, whether it be a bird's egg, a mess of trout, a basket of mushrooms, the photograph of a bear, the pressed specimen of a wild flower, or a note tucked into the cairn on a mountain peak, is a *certificate*. It attests that its owner has been somewhere and done something—that he has exercised skill, persistence, or discrimination in the age-old feat of overcoming, outwitting, or reducing-to-possession. (Leopold 1968: 169, emphasis in original)

---

[1] What could be considered a noteworthy exception to this is the traditional lion hunting for lion trophies (paws) that takes places in some local communities. Here the lion trophy, serving as proof of having killed a lion (with a spear, usually), can give the hunter status, power, respect, and even access to sex with women in the local community (Amy Dickman, personal communication).

This interpretation fits well with the etymology of trophy. "Of defeat" may be interpreted not just as having literally defeated an enemy, but figuratively as having overcome a challenge, or to simply have been successful in an endeavour that required some effort or skill. A trophy implies accomplishment. The head and antlers of a deer turned into a trophy may of course symbolize that the deer was defeated, but it may also demonstrate that the hunter succeeded in being very patient or quiet, overcoming challenging conditions, or making a very difficult shot.

Being a certificate of accomplishment is one of the two important characteristics of the hunting trophy. The other is that it is a *souvenir*.

Rachel Poliquin writes:

> Any object can become a souvenir. But some objects are purposely crafted as souvenirs. Photographs, locks of hair, bronzed baby shoes, trophy cups, tombstones, and hunting trophies all belong to a particular class of objects that have been created specifically to focus and retain memory. (Poliquin 2012: 150)

All trophies are souvenirs, but all souvenirs are not trophies. If you bought a trinket on a holiday, you would call it a souvenir; if you participated in a competition that required something of you (perhaps some skill or determination) and won the same trinket, it would be a souvenir *and* a trophy. The item is the same, but its status as a trophy was earned. The trophy is therefore often a more valuable possession than the souvenir, even if the objects are technically identical. The personal and sentimental value of a trophy usually far exceeds its monetary value.

A hunting trophy is a souvenir of the hunt, made to retain the memory of the hunt and the animal, and it is a certificate of accomplishment. Essential to both certificates of accomplishment and souvenirs, and therefore to hunting trophies, is that they are personal. Trophies only retain their original meaning and value for whoever earned them and for whom they retain memories. The hunting trophy ceases to be a hunting trophy if it passes to someone else, because it will cease both to be a souvenir and a certificate of accomplishment. Garry Marvin comes to a similar conclusion:

> Hunting trophies have "souvenir qualities" in that they are tokens of heightened moments, sometimes emblematic of the place where the animal was hunted, and are objects that are brought back from elsewhere to be displayed as mementoes of experience in that other place. Where they differ from other souvenirs, however, is that authentic trophies cannot be purchased; they must be created by the hunter. There is no such thing as a generic hunting trophy as there are generic souvenirs, and no hunter would display as a hunting trophy an animal that she/he had not personally hunted and killed; a trophy shot by someone else would have no significance for another hunter. (Marvin 2011: 210)

While hunters might display shed antlers[2] found in the forest because they—while perhaps not a trophy—will still be a souvenir, Marvin argues here that they will not display a set of antlers from a deer that was killed by another hunter.[3] If the trophy

---

[2] Antlered species like deer shed their antlers annually. See Chapter 3.3 for more on antlers.

[3] This is only partly true. There are definitely trophy hunters who will display hunting trophies as their own, which they have bought or which come from animals that someone else killed, just as someone might display and pretend to have won a medal that they never won. One motivation for doing so can be to get one's name in a trophy record book. Marvin's underlying meaning—it must

were only a souvenir, the idea of displaying another hunter's trophy would be less grievous, but because it is also a certificate of accomplishment, the inference is that one would be trying to take credit for someone else's accomplishment.

Clive Phillipps-Wolley said something similar:

> To the man who has won them fairly, the mighty antlers and fierce masks which hang in his hall or study are treasures beyond price. As to the men who *buy* such trophies, they are not of our guild, nor is it easy to comprehend them or their motives. (Phillipps-Wolley 1894: 413, emphasis in original)

And Karen Jones here references Roosevelt on the subject:

> For the gentleman sport hunter, the display of trophies without having killed the animal struck as unbridled heresy. Theodore Roosevelt explained the qualification: "Nothing adds more to a hall or room than fine antlers when they have been shot by the owner, but there is always an element of the absurd in a room furnished with trophies of the chase which the owner has acquired by purchase." […] The politics of ownership featured highly in the cultural meanings of the sporting trophy. The trophy could not be purchased, it had to be earned to be fully possessed. (Jones 2015: 244)

A hunting trophy bought as a joke or a kitschy decoration by a nonhunter, on the other hand, might be displayed as hunters would display their own trophies. This is because its meaning has changed; it has ceased to be a hunting trophy. We may all call it a hunting trophy because we would not know what else to call it, but the *meaning* of the item is no longer that of a hunting trophy. It is no longer a certificate of accomplishment. No longer in the original hunter's possession, it has become something else, such as an expression of a certain interior decoration style or sense of humour. What it has become is less important; it is only important that it is no longer a hunting trophy, even if we, for lack of a proper word, would still call it so. We find support here again in Marvin, who wrote that

> hunting trophies as material objects are primarily markers of what developed between the hunter and the hunted, the experiences of the hunter prior to the death of the animal. Their significance does not principally reside in what they are at present (although they do have such significance in terms of their taxidermied qualities) but rather in their power to evoke the past, a past that can be relived by the hunter and by only that hunter. (Marvin 2011: 203–204)

Similarly, when Theodore Roosevelt went to Africa to collect specimens for the Smithsonian, not all the animals he killed became hunting trophies, even if Rowland Ward, the taxidermist, treated all of them the same. The taxidermy that went to the Smithsonian became collections, specimens, and display pieces. The taxidermy that went to Roosevelt himself, although often made in the same way and from the same species, became hunting trophies. He had hunted the animals and to him they were souvenirs and certificates. Rachel Poliquin, with similar reasoning, differentiates between hunting trophies and other forms of taxidermy by arguing that a hunting trophy is infused with a personal narrative about a quest to kill an animal (2012: 149).

---

be assumed—is that an *honorable* hunter would never display a trophy from an animal they did not hunt. See Chapter 3.5.2 for further discussion of what trophy records lead some hunters to do.

It is important to note that for part of an animal killed in hunting to be a hunting trophy, it must be kept by the hunter mainly for non-utility purposes, just like the main purpose of a souvenir is usually non-utilitarian. Meat, while it may be kept in a freezer and eaten over a long time, is not a hunting trophy; its utility is to be eaten. Bones, horns, tusks, or antlers may be ground up and used for medicinal purposes in Traditional Chinese Medicine (TCM). Though their actual utility as medicine is often questionable at best, they are also (at least in this context) not hunting trophies, because their *intended* purpose is utilitarian. Whether they actually work is irrelevant. The poaching of rhinos in Africa for their horns is in popular media often confused and intermingled with trophy hunting. The poachers, however, do not keep these horns; they are not souvenirs. The only relevant motivation is the sales value of the horn, i.e. its utility. Also, rhinos that are killed for their horns are killed by poachers and not trophy hunters. Poaching is by definition illegal and trophy hunting is legal. Illegal trophy hunting would be poaching. We shall return to the discussion of poaching and trophy hunting in Chapter 5.7.

Other examples worthy of mention could be the elephant's foot that is turned into an umbrella stand or the bear turned into a drink holder. These should be considered hunting trophies. Though the elephant's foot has utility as an umbrella stand, its main purpose is that of souvenir and trophy. Any bucket can be an umbrella stand; one does not need an elephant's foot for that. The utility of this trophy clearly comes second to its function as souvenir and display item. The utility is merely an excuse for displaying the elephant's foot as a trophy. Yet these trophies, to emphasize our previous point, are still only trophies when they belong to the hunter who killed the animal. If they pass to someone else, they will lose their meaning as certificates of accomplishment.

If one finds a pair of shed antlers or some other part of a dead animal, and brings it home as a souvenir, this is also not what we consider a trophy, as no accomplishment or "earning" of the trophy was involved. It is however important to note that some record books do consider picked up trophies such as shed antlers eligible for admission, so in a technical record book sense, they can be hunting trophies.

Catching and collecting butterflies or other invertebrates is something that may be considered closely related to trophy hunting. It involves chasing and killing animals and displaying them as souvenirs and collector's items. It is, however, very far from what one would commonly consider hunting or trophy hunting, and today it has also become a rather marginal hobby. As such, insect collecting is outside the scope of this book. As fishing is usually considered an activity separate from hunting, we will not be considering fishing further in our analysis of trophy hunting, but we have included some information about fishing trophies in Chapter 3.2.

A trophy must also be dead. As Bostock (1993) explains in greater detail, history—from antiquity and up through the Middle Ages in Egypt, Greece, Rome, China, and Northern Europe—is full of examples of powerful rulers keeping exotic animals—like living trophies—to display their power, influence, wealth, or even divinity. Sometimes these were received as presents from foreign rulers, and sometimes they were hunted and captured. The practice is not widespread today and not what anyone understands by trophy hunting in the modern context, so we will not consider it further.

While perhaps not the case for hunters themselves, we figure that the quintessential "hunting trophy" to most nonhunters will be shoulder mount taxidermy, so we will in what follows begin our deep dive into the practical descriptions of hunting trophies by explaining what taxidermy is. Other kinds of hunting trophies will be covered afterwards.

3.1 Taxidermy

The Oxford English Dictionary defines taxidermy as "the art of preparing and preserving the skins of animals, and stuffing and mounting them so as to present the appearance, attitude, etc. of the living animal." This fits well with Patrick Morris' three-part working definition: "1. To preserve skin 2. To insert 'stuffing' 3. To pose the animal in lifelike form" (Morris 2010: 8).

The word "taxidermy" derives from the Greek words *taxis* (arrange) and *derma* (skin), and it refers to both the craft and the final product. The verbs to describe the process have historically been "stuffing," "mounting," or "modelling," and sometimes the questionable "taxidermize." "Stuffing" originated in the eighteenth century, when taxidermy was little more than shoving an animal skin full of cotton and rags—like producing furniture or a teddy bear. The results were often horrendous. Taxidermists prefer the words mounting or modelling today, as the days of stuffing are long gone (Asma 2001: 9–10; Morris 2010: 3–6; Jones 2015: 228–230).

There are many kinds of taxidermy. The most common big game hunting taxidermy is the shoulder mount, which is the head and neck of the hunted species, mounted onto a plaque. The shoulder mount is often favoured for antlered or horned species. Substantially rarer but more controversial examples are the head mount of a roaring lion or an elephant with its long, curving tusks. Rarer still are the full-body mounts, like a bear, rearing on its hind legs, or a leopard on a branch. Reserved for museums and the extremely wealthy is the kind of taxidermy that can be termed "habitat groups," defined by Pat Morris as "(a) more than one animal, (b) depicted engaged in some typical behavioural activity and (c) displayed among a reasonably accurate representation of its immediate habitat" (2010: 318). Finally, there is the diorama, which is even larger and represents not only a group of full mounts, but an entire natural scenery or habitat, including both flora and fauna.

As noted previously, taxidermy is also on rare occasions given a basic function, such as a bear turned into a drink platter holder, a giraffe turned into a chair, an elephant's foot turned into an umbrella stand, or the head of a rhinoceros turned into a drinks cabinet. These are all authentic examples, courtesy of Roland Ward Ltd., to which we will return shortly. A similar sort of joke item in taxidermy is the "jackalope," which is a mythical cross between a jackrabbit and an antelope. It is usually made by attaching deer antlers (not antelope horns, as the name seems to indicate) to a mounted jackrabbit (which is not a rabbit but a hare).

For hunting trophy taxidermy, the shoulder mount is by far the most common. A full mount taxidermic display costs much more to have made and requires a large area of open floor space to display, so it is only relevant for museums and very large houses. A shoulder mount only needs a solid wall.

The typical mounting process for mammal taxidermy is as follows:

First step is to skin the animal with as few incisions as possible in places that are easy to conceal. We might add here an explanation from John Wootters, famous hunting author and writer for Field and Stream, Sports Afield, and many other outdoor magazines, on how to prepare a freshly killed deer for a shoulder mount in the field:

> The first consideration in skinning is proper care of the cape, and the most common mistake is cutting it too short. The cape, or headskin, should include not only the entire neck but the brisket and shoulders as well, for the typical shoulder mount. The best way to deal with this most important part of the skin, unless you're an expert skinner and caper, is to make a cut all the way around the buck's shoulders, and then to "case" the skin forward to the head proper. This means simply skinning the neck forward without making any longitudinal cuts at all. Then sever the entire head where the spinal column joins the base of the skull, and lay the head with attached cape aside. (Wootters 1977: 226–227)

After removing the antlers or horns, the second step is to preserve the skin, either with a chemical or by tanning it. Arsenical soap used to be standard, but healthier alternatives exist today. The skin would then previously have been stretched, shaped, and stitched around the cleaned skull and (for a full mount) an artificial body with a balsa wood core in the desired posture. Since the 1950s it has been commonplace to produce manikins in the image of skull and body, which are lighter than bone and save time in skull cleaning. This involves sculpting a model of the animal in a clay-like substance and using it to make plaster moulds. Ready-made moulds are also available for skulls and bodies of a generic shape and size. Modelling paste is then added to the mould under the skin and manipulated through the skin for adjustments and better fitting. Finally, horns or antlers are slotted back into the moulds in their prepared spaces, and eyes made of glass are inserted (Moyer 1953: 48–67; Asma 2001: 30; Morris 2003: 111–112).

The historical roots of taxidermy are popularly traced back to Ancient Egyptian embalming practices. However, As Morris notes, Egyptian mummies do not really fit the definition of taxidermy. Mummies were not posed to look lifelike for display but merely preserved for religious and cultural reasons. They also were not "stuffed" in the taxidermic sense. Nevertheless, they were still efforts to preserve the husk of a dead human or animal just as taxidermy is, so the association is understandable. A thorough description of Egyptian mummification practices is available in Browne (1893). Palaeolithic practices of tanning skin could also be related to taxidermy's history, but the main purpose of preservation in these cases was to keep warm, and the animal skins were only stuffed with the living humans wearing them.

The taxidermic tradition as we know it today is much more recent, dating back only to the early or mid-sixteenth century in Europe, particularly the Netherlands. Morris surmises that the first taxidermic manuals are from 1555, 1622, and 1626,

3.1 Taxidermy

explaining how to model birdskins over an artificial body[4] (Moyer 1953: 3–7; Morris 2010: 8–10). Taxidermy in the sixteenth century was mostly a curiosity, and animals were often collected by the noble and the rich and displayed in their "Cabinets of Curiosities," containing whatever strange and exotic items they could get their hands on, including

> horns of the unicorn, farm animals joined at birth, monstrously deformed humans, griffon claws, dragons, hippo tusks and fossil fish; who could tell what was real and what was not? It all inflamed imagination and curiosity. (Morris 2010: 13)

In the seventeenth century the interest in taxidermy shifted towards the scientific. Collections of taxidermy were assembled and used for the purposes of teaching and studying. The Cabinets of Curiosities and their odd contents made evident the need for classification and order, and in the eighteenth century, Carl Linnaeus provided this order. The Linnaean taxonomy however also made it clear how many species were still missing from taxidermic collections. This inspired an interest in acquiring more, which lasted well into the twentieth century (Morris 2010: 10–13).

The earliest taxidermy still extant is from the mid-sixteenth century. It heavily features crocodiles, as their skin becomes extremely durable when dried. Reptiles in general make for durable mounts, and so does any animal with a carapace or shell, such as armadillos. Birds and mammals have proven fragile, succumbing to either insects if dry or fungal mould if damp, so very early examples of these have been lost. There has historically also been a tendency in taxidermy, even at museums, to discard old specimens when a newer one was obtained. The oldest extant mammal specimens are from the late seventeenth century (Morris 2010: 14–25).

Morris separates the customers of taxidermists into three groups. The first is collectors and museums, who often want specimens for scientific purposes or public display. The second group consists of domestic customers who wish to preserve a pet or buy a decorative household item without any intention of collecting taxidermy. The third is the hunters or fishermen who wish to preserve a trophy (Morris 2010: 229). We will focus on the final category.

Taxidermy is a complicated craft. Achieving a lifelike result that captures the right mood, expression, and natural behaviour of a species requires considerable experience and skill. Combined with the size and weight of some hunting trophies (the hide of an elephant can weigh more than half a ton), this makes it difficult for small, local taxidermists to handle the business brought to them by trophy hunters. To meet big game hunting demands for quality, quantity, and size, a few very large taxidermy companies started to appear in the latter half of the nineteenth century. Big game hunters travelling the world at the time were mostly British, so the big taxidermists were as well. Morris makes special note of four British companies: Van Ingen & Van Ingen (based in India, see Morris 2006), Edward Gerrard & Sons (see Morris 2004), Peter Spicer & Sons, and Rowland Ward Ltd. Rowland Ward was the most influential taxidermist of its time and is most relevant to trophy hunting today because of the Rowland Ward record book.

---

[4] We return to bird trophies in Chapter 3.2.

James Rowland Ward started his company in London in 1870 after having apprenticed with his father Edwin Henry Ward, also a taxidermist. This company became the industry leader in quality for over a century. The heart of its business was always hunting trophies, but the contributions that were made to museums and exhibitions worldwide from as early as 1871 are extensive. Ward developed his business to involve much more than taxidermy. He was someone from whom one would solicit advice and purchase equipment when arranging large-scale hunting trips, as he was a trophy hunter himself (Jones 2015: 227). Theodore Roosevelt bought equipment from Ward in preparation for his Africa expedition and purchased his taxidermy services afterwards (Morris 2003: 27–37).

The company Rowland Ward—Rowland Ward Ltd. since 1890—is especially interesting in a trophy hunting context, as Ward originated the tradition of careful measurement of weights and dimensions of the specimens he mounted. Measurements are of course required for hunters who wish to compare trophies, so in 1892, taxidermy merged with trophy hunters' desire for record-keeping in the book *Horn Measurements and Weights of the Great Game of the World*. The book was published by Rowland Ward Ltd. and authored by Rowland Ward himself.

Ward's trophy taxidermy business reached its peak around the time of Rowland Ward's death. Customers included many state leaders and royals, and the prices were set accordingly. A pair of mounted tigers, for instance, would cost about the same as a two-bedroom house at the time. Taxidermy was extremely fashionable, and this in turn promoted hunting: "The popularity of taxidermy and the social esteem curried by elite practitioners brought kudos to hunting as an exercise in imperial adventuring, scientific collection, marksmanship, and honorable masculinity" (Jones 2015: 234). Like Theodore Roosevelt in Africa, many hunters collected not only for themselves but also for museums. Some royals shot so many animals that their donations or sales to museums numbered in the hundreds of big animals. The majority of them passed through Ward's on their way to the displays.

The First World War had little effect on Ward's business, but the Second took its toll, and business slowed down even further afterwards. This was due partly to competition from the American taxidermic industry, capitalizing on the legacy of notable taxidermists like Carl Akeley, and partly to increasingly fewer large game animals in the world and growing negative attitudes towards shooting them. Also, a lot of taxidermy business came from India, and this ended with India's independence from Britain in 1947. Still, Ward's remained the taxidermy world leader until the 1960s (Morris 2003: 42–48).

The taxidermist side of Rowland Ward Ltd. closed shop in the 1970s. Trophy hunting in Africa had recovered after the Second World War but declined again in the sixties and seventies. One of the death blows to the business was the ban in 1977 of trophy hunting in Kenya. Kenya was responsible for three quarters of Ward's African business, and the company had opened a branch there, which now had to close. Other factors were previous customers switching to wildlife photography due to antihunting sentiments, museums no longer needing (or being able to afford) taxidermy, and a general ire at taxidermy because of its association with trophy hunting (Morris 2003: 63–65).

The publishing side of Rowland Ward Ltd. has been American-owned, South-African-owned, and American-owned again since the taxidermy business ended. They now publish with their sister-company, Safari Press, "Publisher and Seller of Fine Quality Hunting and Firearms Books." The flagship publication is still *Rowland Ward's Records of Big Game*.

In the late nineteenth century, North America took over Britain's role as the leading taxidermic nation and has held it since. American taxidermists organized themselves as *The Society of American Taxidermists* and got in touch with a wide customer base via mail order catalogues, which contributed to its increasing popularity. It is also important to note that trophy taxidermy was by then much more coveted in North America than in Britain, due to domestic American deer hunting. In Britain, a larger part of the customer base simply sought curiosities and household decorations on a much smaller scale (Morris 2010: 298–310).

Another factor contributing to taxidermy's spread and quality in North America was *Ward's Natural Science* (WNS). WNS was founded in 1862 by Henry Augustus Ward (unrelated to the Wards of London) and adopted best practices from Europe and gathered taxidermists from all over North America to service museums and educational institutions with specimens and knowledge. This provided the profession in North America with a continuity of taxidermic knowledge and experience. It ensured that when taxidermists needed to mount exotic animals, they would be able to seek help from someone with experience and/or gain the experience that others would later need. Many of the great taxidermists of North America in the nineteenth century learned their craft at WNS and would later put their skills to use at famous museums (Morris 2010: 311–313).

Two of the most notable taxidermists trained at WNS were William Temple Hornaday and Carl Ethan Akeley.[5] Hornaday was responsible for two of the most famous dioramas in museum history: the "Battle in the Treetops" diorama from 1883 with two orangutans and the "American Buffalo Group" from 1888. Both were displayed at the Smithsonian where he was the chief taxidermist (Asma 2001: 42).

Akeley was initially fired from WNS because he spent too much time trying to improve and reinvent taxidermy and too little time on his tasks. He was later reemployed at a much higher salary when his methods proved superior. He was also incidentally the inventor of the concrete spray (shotcrete) and had many more patents both related and unrelated to taxidermy. Akeley was a frequent visitor in Africa and met with Theodore Roosevelt on Roosevelt's expedition in 1909. He dedicated much of his time to gorilla (*Gorilla gorilla*) conservation in his later years.

As scientific, popular, and tasteful as taxidermy had become in the 19th until the mid-twentieth century, by the late twentieth century, it had become useless, unfashionable, and reviled. Television allowed everyone to see what living wildlife looked like, so there was less reason to display dead wildlife in museums. Museums

---

[5] Both men have names that are associated with ammunition but in both cases it is because of someone else who had a similar name: Parker Ackley, rather than Akeley, (who invented a lot of different cartridges) and Joyce Hornady (whose company Hornady ammunition co-invented the 6.5 mm Creedmoor cartridge).

at this time often chose to discard even well-preserved taxidermy rather than storing it as historical relics. Much other taxidermy has been dismantled and sold off in pieces, as some parts of taxidermy—like rhino horns for TCM or feathers of rare birds for fly fishers—were suddenly worth a lot of money. Bird mounts were often turned into birdskins because mounts take up a lot more space in museum storages (Morris 2010: 352–353). Morris remarks that

> by the late 20th century, general disapproval had set in. Displays of dead things began to be considered vaguely unhygienic and disturbing. Massed ranks of animal trophies were seen as vulgar and associated with reprehensible activities such as hunting and shooting, an affront to the prevailing conservation ethos at a time of diminishing wildlife. This approach may even extend to modern taxidermists, one of whom told me that he refuses to mount hunting trophies as he disapproved of the motives for preservation. *"I love animals",* he said, explaining that he wanted to preserve them *"as an act of admiration, not as a form of triumphalism".* (Morris 2010: 353, emphasis in original)

As Morris indicated, taxidermy had fallen out of favour not only because of technological advances but also because of changing attitudes to what is proper and respectful to do to a dead animal. In the 19th and early twentieth century, it was considered important to utilize every part of the animal. Neither Roosevelt nor his contemporary naturalists considered the dignity of an animal when making use of its carcass. A dead elephant was a dead elephant, and to discard an elephant's foot when it could be displayed as an umbrella stand was simply wasteful.

Rowland Ward was for many years the world leader in turning animal parts into more or less useful items:

> Ungulate feet could be presented as candlesticks, table lamps or doorstops for example. […] Hippopotamus feet were also on offer as biscuit barrels, with zebra and buffalo feet made into bookends. Other trophies could be turned into cigarette boxes, penholders, ashtrays, paper weights and picture frames, usually by the addition of appropriate metal fittings. (Morris 2003: 130–131)

This, however, is not just a fad of the past. Matthew Scully gives a more recent example:

> At the Skin and Bones boutique, a Houston firm specializing in curious and trinkets, a giraffe-skin day planner goes for $189, a zebra-skin purse for $149, impala-skin gun cases for $249, giraffe rugs for a few hundred, golf bags crafted of silver and finest Cape buffalo hide for a grand and up. (Scully 2002: 49)

Currently, taxidermy is controversial at best, and taboo at worst. Melissa Milgrom explains that when taxidermy goes out of fashion, it does not simply become uninteresting; it becomes positively reviled, which means that much of it is destroyed (Milgrom 2010). Morris, however, argues that museums should be ensuring that a record of past generations is preserved instead of engaging in "retroactive censorship" (Morris 2010: 360). If we are to not display anything offensive from the past, why are historical museums displaying gruesome instruments of torture, cannibalism, and war crimes? Why are art museums displaying paintings, sculptures, and installations depicting death, dismemberment, or worse? Taxidermy is somehow in a no-man's land of not quite art and not a respectable craft, and the mounts have often been treated accordingly.

Taxidermy is in less of a crisis now than it was at the end of the twentieth century. More collectors are showing interest in old taxidermy of high quality, and, thanks to technological advances, new taxidermy is more realistic than ever before (Morris 2010: 374–378). According to DataUSA and Manta there are now about 9000 taxidermy businesses, employing half a million people in the US (DataUSA 2019; Manta 2019).

A few representatives of post-War American taxidermy are worth singling out: the Jonas brothers and the Klineburger brothers. The latter are also important to the history of safari clubs and trophy hunting tourism.

The five Jonas brothers, Coloman, John, Guy, Leslie, and Louis Paul migrated to USA from Hungary in the early 1900s with some taxidermy experience from Europe. They started three taxidermy companies over the coming years. Coloman Jonas was the oldest brother, the most experienced taxidermist, and the first to leave Hungary. He set up the Jonas Brothers of Denver, Colorado in 1908 with help from his brother John. The business was a success from the start, and he soon brought over his remaining brothers to work for him. The Jonas Brothers of Denver became the premiere taxidermy studio in USA in the 1920s and 30s. Eventually, either because of the company's success or brotherly disagreements, all four of Coloman's brothers left the Denver business to pursue their own goals. Guy started the Jonas Brothers of Seattle in 1937, John started the Jonas Brothers Studio of New York, and Louis Paul, also in New York, focused on museum taxidermy work, leaving hunting trophies to John. Leslie had an engineering background, and it is unclear if he left the taxidermy industry or merely played a smaller role in the New York business. The three taxidermy studios bore the same name and did some promotions and brochures together but were essentially competitors. John Jonas died in 1950, shortly after selling the Jonas Brothers Studio of New York to his employee Steven Horn. Horn retired in 1985 and sold the studio to Carl and Mary Pepi, who now run the company as "Jonas Brothers – The Pepi Family Studios." The Seattle Studio was sold to the Klineburger brothers in 1954. Coloman Jonas died in 1969 but the taxidermy studio in Denver is still in operation. Part of the manikin-making business (Jonas Supply Inc.) was sold to the McKenzie Taxidermy Supply company in 2013 (Asma 2001: 11–13; Klineburger 2010: 36, 45–52; Edwards 2013; Jonastaxidermy 2019).

The Klineburgers, Gene, Bert, and Chris, grew up in the Mule Mountains of south-eastern Arizona. Bert Klineburger started working for Guy Jonas in Seattle in 1948 and Chris came on as a summer worker in 1949. Both had a lifelong interest in taxidermy and improved quickly. Chris started Klineburger Taxidermy by himself in Tucson soon after, which his partner took over in 1951. Chris and Bert then both worked for Guy Jonas again until they bought the Seattle studio in 1954. They had their brother Gene take up the part of tanning skins for the customers' hunting trophies, allowing for faster service than other taxidermists. The Klineburgers expanded the taxidermy business to Anchorage, Alaska in 1956, and successfully branched into the fur business there. In 1962 the Klineburgers expanded to Uganda with the establishment of the Jonas Brothers of Africa, a taxidermy studio in Kampala

(Klineburger 2010: 5–67; 227–232). This foothold allowed them to start arranging African hunting trips that same year, and later all over the world, as Phil Dougherty explains in one essay:

> From Uganda the Klineburgers branched out and began handling bookings for hunts worldwide -- initially in Africa, India, South America, and the South Pacific, which were the few places outside North America where hunting was then open to sports enthusiasts. But later in the 1960s other countries, such as Iran and Mongolia, began to open their doors to hunters. This trend continued into the 1970s and 1980s, with even more countries worldwide allowing hunting. The Klineburgers spent considerable time in Asia and Africa from the mid-1960s through the early 1990s developing various programs, referring clients, and participating in numerous hunts themselves. Chris remarked that he tended to spend more time in Asia, while Bert favored handling the company's operations in Africa. The Klineburgers sold Klineburger Brothers Worldwide Travel in 1991. (Dougherty 2007)

The Jonas Brothers taxidermy studio in Seattle that launched the Klineburger rise to fame in 1954—later renamed Klineburger Taxidermy—closed in 1996. The Klineburger brothers themselves were among the world's most experienced trophy hunters, amassing some of the largest and most complete trophy collections of especially wild sheep from all over the world. John Mitchell (1981) notes that Bert Klineburger seemed to be in fierce competition with Herb Klein (1901–1974), an oil tycoon from Dallas and renowned hunter, about who would be the greatest hunter in Boone and Crockett's *Records of North American Big Game*.

A few of taxidermy's fringe cases are worth mentioning here at the end. The first is human taxidermy, which is fringe because it is exceedingly rare and somewhat taboo, and the others are the methods of plastination and freeze-drying, which are fringe because they are not technically taxidermy at all but merely newer, alternative, and still pretty uncommon methods of preservation.

Human taxidermy has an expectedly sparse history once embalming is disregarded. Wax figures have instead been the custom for modelling human likeness. Wax figures, aside from being considerably less morbid, do not require any earthly remains and can portray humans who are still alive. Morris notes just a few historical examples of human taxidermy, and even those were apparently hard to root out. Museums are loath to admit that human taxidermy has been—or much less still is—in their possession.

An alternative to taxidermy that is sometimes used for the immortalization of pets is freeze-drying, described here by Rachel Poliquin:

> The process involves freezing the animal in a vacuum chamber. Frozen moisture is slowly extracted from the animal in a gaseous state, leaving the tissue, bones, and all internal organs intact and unaltered. Once all moisture (the source of all organic decay) is removed, the animal is returned to room temperature, perfectly and eternally preserved. The process is not quick. Small animals require anywhere from six to ten weeks, while large dogs may take as long as six months. But advocates claim that the procedure produces results vastly superior to the traditional method of rebuilding the creature almost from scratch. In addition, freeze-drying allows the owner to choose the exact pose in which the pet will be immortalized, and—perhaps best of all—many owners are comforted by the thought that their pets have not been skinned. With the artistic interpretation and craftmanship of taxidermy eliminated, freeze-drying gives the impression that here is the whole pet as it once was, complete with

bones and guts. As with traditional taxidermy, eyeballs are replaced with glass replicas. (Poliquin 2012: 209)

While it is possible to freeze-dry a hunting trophy, and some companies do offer this service, it is very rarely done. Pets, as Poliquin notes above, may be freeze-dried because the owner wishes to capture the pet's expression and demeanour as it was in life. Hunted game are not pets with whom hunters have had a personal and intimate relationship, so there is typically no such need. Taxidermy of commonly modelled animals has also become so lifelike that there would be little point in doing something much more expensive.

Another way of preserving human or animal remains is known as plastination. Plastination could almost be described as the opposite of taxidermy. While taxidermy is the skin, reshaped around an artificial interior, plastination is the original interior with the skin completely removed. The technique was developed in 1977 by anatomist Gunther von Hagens. He has since 1995 displayed the technique in the Body Worlds exhibition, touring the world with tremendous success and equally tremendous—and predictable—controversy. The main exhibition features plastinated humans who donated their bodies to the project. The name of the whole venture, *Body Worlds*, is also the name of the human exhibition. A second educational exhibition since 2010 is *Animal Inside Out*, which features more than 100 specimens of plastinated animals, including a giraffe, a gorilla, and even an elephant.

3.2 Fish and Birds

While we usually do not confuse fishing with hunting in everyday speech, the precise distinction is less straightforward than we might assume. One might think that fishing is just used to describe the "hunting" of aquatic animals.[6] Not all hunting of aquatic animals is called fishing, however. The hunting of seals and whales is more commonly called hunting than fishing, yet the terminology of "seal fishery" and "whale fishery" has traditionally been used about the sealing and whaling industries in Canada. Although most would probably say that they "hunt" seals and whales (or go sealing or whaling), Canada's Marine Mammal Regulations state that every person "who fishes for seals for personal or commercial use shall land the pelt or the carcass of the seal" (SOR/93-56 2022). Conversely, in EU's *Statistical classification of economic activities in the European Community*, the catching of whales is categorized as a fishing activity while the catching of seals and walruses is a hunting activity (Eurostat

---

[6] This does not necessarily involve the killing of said animals though, as exemplified by the increasingly common practice of catch and release fishing. Fish caught on hooks or in nets can often be released with little to no harm done, depending on the species and the details of the method. Hunting, due to the lethal weaponry normally used, usually involves killing. An exception to this is the fringe-practice of recreational hunting with a tranquilizer gun, commonly known as "darting safaris" in the academic literature. That fishing generally involves non-lethal methods of "taking possession" of the animal is presumably the reason why we usually say that we "catch" a fish but "kill" a deer if we are successful, even if the fish is also killed after being caught.

2008). Could fishing and hunting then more easily be distinguished by defining fishing as only those activities that involve fish? Not really, it is still confusing; if you catch fish by swimming underwater and shooting them with a harpoon gun or speargun, it is commonly referred to as both underwater hunting and as spearfishing. If you shoot fish with a bow and arrow from land, it is called bowfishing. All this said, the most common and familiar types of fishing to most of us, i.e., with hook and line or with nets and trawls, are generally only referred to as fishing.

Not only is it difficult to cleanly separate fishing from hunting, the practices of collecting trophies from these two activities can be the same, and a great variety of trophies are collected from fish and other aquatic animals. Fish trophies are sometimes mounted as taxidermy, and the method here is not that different from taxidermy methods otherwise, as they were described in Chapter 3.1. While you remove the skin from a deer in order to mount it, however, the process for a fish can more suitably be described as removing the insides from the skin, and it must also be kept moist during the whole process. A cut in a straight line from head to tail on one side of the fish (the worst-looking side, facing the wall) allows for cutting up the spine and removing and scraping out meat and entrails. An outline of the fish can be drawn on paper or card before this step to estimate the size of the piece of foam or other material to insert into the fish. This manikin is of course of a much simpler shape than is used for most non-fish taxidermy and may not need to be fashioned by a professional taxidermist. Clay can be used on top of this simple manikin and inside the head of the fish to fill out cavities (Browne 1893: 173–187). Full mount taxidermy fish like these have sometimes been displayed in glass cases, made to simulate their natural environment, like the dioramas discussed in Chapter 3.1. Another way to do the fish taxidermy, often used for larger fish like tuna, sharks, and marlin is to just mount the head, in the same way that more conventional mammal taxidermy can be either a full mount or a shoulder mount.

Many kinds of trophies from fish and aquatic animals exist in addition to taxidermy. Full skulls are common, and from fish with particularly impressive jaws and teeth like sharks or pike, the jaws along with teeth are often kept alone. For whales, scrimshawed bones and baleens have been common; for lobsters and crabs, some will keep the largest and most impressive claws; for turtles, the shell; for tuna, the tails and tail fins dried and mounted.

For fishermen who want to eat their catch but also keep a full-size trophy, at least one carpenter and fisherman turned artist, Adam Ashdown from Dunsborough in Western Australia, offers to paint your fish with black ink (water-based, so you can wash it off and consume the fish afterwards) and then make an imprint of it on paper, following a traditional Japanese artform called Gyotaku (Honeybone 2020).

All these physical mementos aside, and mainly due to the ever more popular tradition of catch and release angling, a major difference between the trophy practices for fish and trophy practices otherwise is that photos, be they purely on a screen or also printed, of the hunters/fishermen holding their fish are commonly the only trophies that are kept. For catch and release anglers who want more than just a trophy photo, 3D-models based on the photos and made with techniques such as 3D-printing are options offered by more and more taxidermists.

3.2 Fish and Birds

Moving on to birds, as discussed in Chapter 3.1, the first taxidermic manuals from the mid-16th century were manuals for how to mount birds. There have been two main ways to preserve a whole bird with taxidermy; one is the full mount, and the other is the birdskin. A birdskin is the skin, head, feet, and feathers of a bird, removed from the body, stuffed, and sewn back together. It is not mounted in any kind of pose, as it is filled with soft stuffing only. The mount also has wiring inserted inside for manipulation into a lifelike pose but can be reverted to a birdskin for storage, as has been practiced in museums.

The light weight and smaller size of birds compared to fish, and especially compared to large mammals, makes the full taxidermy mount a more common and practical option for birds than for other animals. Mounts very often feature the birds in flight or in other lifelike positions, but a trend for bird trophy taxidermy that we do not seem to see in other trophy taxidermy is the mount that is supposed to look like a dead bird. An example of this could be two or three ducks, mounted together to look like they are hanging by their feet from a string, as one might carry them on the way home from a successful hunt. The birdskin seems to be a less preferred method for displaying a hunting trophy than the mount, perhaps because it does not in fact display the bird very well. Skulls or beaks alone appear similarly rare. Heads with the skin and feathers still attached, and tailfeathers are sometimes mounted either in a combination or separately as trophies. The tail feather display mounted on a plaque for this purpose seems particularly common for wild turkey (*Meleagris gallopavo*) and wood grouse (*Tetrao urogallus*) and sometimes it will be a mounted display of fanned-out tailfeathers as well as fanned-out wings. Hunters may also save the feet of turkeys as trophies if the spurs are particularly impressive. A practice that is common with the hunting of Eurasian woodcock (*Scolopax rusticola*) is to keep only a pin feather as a trophy, traditionally displayed in a hat. These feathers are also sought after by artists for use as very delicate paint brushes. Woodcock with rare colorings also make for popular trophies.

Generally speaking, many species of hunted birds have individual feathers that are particularly beautiful and saved by hunters. An attractive feather is after all a simple trophy that takes up almost no space and requires no effort or money.[7] Still, trophy keeping habits vary for bird species that are commonly hunted, with waterfowl hunters often seeming to take a particular interest in collecting a diverse set of hunting trophies from different waterfowl species (see Chapter 4.5.6 for more about trophy hunting as collecting). A practice that is common for waterfowl hunting in the US is to keep the bands from birds that have been banded by scientists for research purposes. This band is normally too small to see when you shoot the bird but is a nice surprise for the hunter who retrieves the bird. In exchange for reporting the identity of the bird (identified by the code on the band) to a federal agency, the hunter will receive information by e-mail about the age of the bird, when and where it was

---

[7] A couple of common idioms stem from the keeping of bird feathers as hunting trophies. There is an expression in at least Danish and German but perhaps other languages too, referencing not to "decorate oneself with borrowed feathers," which means that one should not take credit for the accomplishments of others. Another idiom refers to "a feather in your cap" as an accomplishment or something to be proud of, as one would be of a hunting trophy feather.

banded, etc. This makes banded birds extra interesting, and the bands are considered valuable trophies by bird hunters.

3.3 Skulls, Antlers, Skins, and Other Trophies

The most common of all hunting trophies from big game is the half-skull mount. This style of hunting trophy is also known as a European mount, or Euro mount. The taxidermic shoulder mount has traditionally been the favoured hunting trophy method in North America, while half-skull mounts are better known from medieval castles in Europe, hence the moniker. The half-skull mount has also become very popular in USA now though, and it is today for most hunters the preferred method of preparing hunting trophies from most deer. While taxidermy generally requires the services of a professional taxidermist, the half-skull mount is something hunters can do themselves. Many hunters still send the task to a taxidermist though, who will charge about a fifth to a third of the price of shoulder mount taxidermy. The more exceptional the trophy, the more likely a hunter will be to pay for a shoulder- or even full taxidermy mount, so trophies from Africa or Asia or other places where the hunting is more expensive are more often mounted through taxidermy, though trophies from lions and other species may also be made as skull mounts. Additionally, if one can afford the exorbitant prices of trophy hunting abroad in the first place, the price difference between taxidermy and a half-skull mount is probably less likely to matter.

A half-skull mount is the upper part of the skull, mounted on a wooden board. Previous common practice was to screw the skull onto the board, but more commonly today, the board comes with pegs or a clip to hang and fasten the skull on, so marring the skull with screws is avoided. The trophy consists of the skull and antlers or horns along with the upper jaw with teeth. It may also, more common for roe deer (*Capreolus capreolus*) for example, be sawed over somewhere above the upper jaw, or sometimes through or above the eye sockets, so only a smaller part of the skull remains. One may even find examples where only enough sliver of skull to connect the antlers remains. This is sometimes the preferred method for record-sized antlers, as antlers can thus be weighed for trophy scoring purposes without also weighing the skull. The most common practice is to include the whole upper jaw with teeth. The boards used for the mounting are of the same kind as for most taxidermy head mounts but vary in sizes and designs, depending on the species and size of the specimen. Some taxidermists have unique board designs, so connoisseurs can sometimes identify the taxidermist based on the backboard.

To make the half-skull mount, one might boil the head for a couple of hours at a low heat to soften up all the skin and meat on it. This can be gradually scraped off the skull while it is cooking and afterwards. Once all the skin and meat has been removed from the skull surface, a water hose or an air compressor can be used to flush or blow the remaining brain matter, eyes, cartilage, and other soft tissue out of the skull and nasal cavities. It is also possible to clean a skull with maggots. After washing and soaking the cleaned skull in soap water, a bleaching of it by extended drying in the sun or with for example peroxide can turn it from a yellowish colour into the whiter shade that is often preferred.

What shoulder mounts and half-skull trophies of deer and other ungulates have in common is that their objective quality as trophies all derive from the size and weight of the animal's antlers or horns. John Wootters wrote in *Hunting Trophy Deer*, that

> [a]lthough hunters are always interested in the body size of deer and often brag about taking an exceptionally heavy buck, the *trophy* part of a deer is what grows upon his brow—his antlers. (Wootters 1977: 23, emphasis in original)

Considering the importance of antlers and horns to trophy hunters, a few clarifying words about them are perhaps timely as well. Antlers are the pronged projections on the heads of males (and, in reindeer, females) of the Cervidae family, better known as deer. Horns are found usually on males in the Bovidae family, which, among other species, comprises sheep, goats, bison, buffaloes, antelopes, and gazelles. The main differences are that antlers are made entirely of bone[8] and are shed and regrown annually, while horns have a bony base that is sheathed in keratin. Horns are never shed but grow a little bigger each year. The North American pronghorn antelope (*Antilocapra americana*) is the exception to this, shedding and regrowing its horn sheaths annually like deer. A broken set of horns are broken for life, while broken antlers are only broken for a season. Horns are usually a single pair of spikes, but between species (and even between individuals within a species) there may be considerable variation in overall shape (curved or straight), relative and absolute length, and whether horns are smooth, or lightly or heavily ridged. For species with ridged horns, older individuals, and males, will likely have more prominent ridges and this can be a factor in determining the value of the individual as a trophy, with heavy ridging often more desirable. Antlers can also have a wide variety of shapes and number of points between and within species, with the massive, wide, palmate antlers of a moose contrasting with the simpler, smaller three-pronged antlers of a roe deer for example. Antlers grow slightly larger each year than they did the previous year. While they get heavier and longer, and develop more points over the years, the pattern is predetermined and builds on the same shape as in previous years. If one is familiar with the deer in an area, one can therefore recognize individuals based on the shape and size of their antlers from season to season, or even tell which animal a shed pair of antlers came from. Antlers can to some extent be used to determine a deer's age but not precisely, and less so as the animal gets older. The number of points on an animal's antlers are however the common way of indicating the size and quality of a trophy animal. A 10-point whitetail is a whitetail deer with 5 points on each side, for example.[9] Other quality factors such as symmetry are often considered secondary to the number of points, but for scoring purposes, some record books will have a ranking for both typical and non-typical antlers.

---

[8] They are also covered in a layer of what is called antler velvet during growth, which sheds off upon maturity.

[9] There are a few confusing exceptions. A 6-point elk for example does not mean 3 points on each side but 6, so 12 in total.

Lacking branching, horns are more straightforward in terms of overall form. Length and circumference are broadly related to age. In South Africa, game breeders will sometimes breed for desirable qualities in some species. Typically, these qualities are length and symmetry, notable especially in the kudu, a spiral-horned antelope much desired as a trophy species. Other factors that are sought after might include a greater width between horn tips, thicker horns and, in wildebeest, larger "bosses," the base of the horns that grow and start to meet across the top of the skull. Game breeders also select for horn length, symmetry, width and boss thickness in the cape buffalo. Buffalo are a very dangerous and often difficult species to hunt on foot, sometimes living up to its various nicknames that include "black death" and "widow-maker." The breeding of buffalo with different horn shapes (including how "hooked" they are and how far down they curve), sizes (overall length, width between tips, and girth) and thicker, more pronounced bosses is a large part of the more intensive game breeding operations that operate in some parts of South Africa.

For a precise estimation of age, an examination of the teeth is necessary, but that is of course a post-mortem exercise. A rough estimate can be made based on the wear of the molars, but some diets wear the teeth down faster than others, so this method can be misleading. A more accurate result is obtained by the cementum annuli method, where one of the front incisors is sliced so year rings can be counted, similar to how one might determine the age of a tree.

Whitetails and red deer have maximum lifespans in the wild of roughly ten and twelve years, respectively. Their bodies and antlers tend to be fully matured when the animal is five to seven years old for whitetails, and around eight or nine for red deer. If an animal lives longer than that, which is rare even in unhunted populations, the antlers tend to decrease in size and may start to become deformed. Most wild deer die from hunting, disease, injury (including from other deer) or vehicle collisions before full maturity of their antlers.

Tusks of elephants, walruses, pigs, and hippos are perhaps the most noteworthy of hunting trophies that do not fall under the categories of antlers and horns described above. Some of these animals can also be prepared as taxidermy, but it is less common. Elephants, rhinos, and hippos are of course particularly expensive to have mounted and require a lot of room for display. Skins or pelts may be hunting trophies of animals with exceptionally beautiful, impressive, or soft skin or fur. Tiger skins from India were hugely popular in Europe in the first half of the twentieth century, and zebra and giraffe skins are also iconic trophies because of their unique patterns.[10] Pelts, sometimes with the head attached, have been a common way to turn various bear species into trophies, but lions and other big cats are also sometimes seen prepared like this.

An honourable mention here at last is the trend known as "fauxidermy," which is usually a fake animal's head, made to look like traditional shoulder mount taxidermy

---

[10] Zebra skins are available for anyone to buy in many outlets in South Africa. In Johannesburg's O.R, Tambo airport for example it is possible to purchase zebra skins, as well as impala and springbok skins, and other artefacts that would be considered a "trophy" in other circumstances. Helsinki airport has several outlets selling reindeer skins.

or a euro mount. It often comes in the form of stuffed plush toy animal heads of elephants or moose or deer. When they are not plush toys, they may be made of wood, metal, plastic, porcelain, or any other material, fashioned to look like a hunting trophy head. They are often very artistic or fun in their impression and embellishments and never meant to look lifelike. They are not hunting trophies of course, as they were never a real animal, but they are an interesting way of giving an impression of the hunting trophy in a humorous and perhaps ironic manner.

3.4 Trophies and Record Books

Hunter and author James Swan wrote the following about his first hunting kill:

> The first wild game I ever shot was a sora rail, which is a relatively small marsh bird about the size of a quail with long spindly legs. I remember well that we had a celebration that night, baking the rail with a strip of bacon around it. The bird was smaller than a game hen, but to me it was a trophy that I will never forget. It was a milestone in life, a milestone in growing up. (Swan 1995: 192–193)

The first kill is probably memorable to most hunters. Swan demonstrates here that anything can be a trophy in the eyes of a hunter, and any trophy can be a great trophy, subjectively speaking. Yet unless Swan also kept something from his sora rail, it does not fit the definition of trophy that we are using here, which requires keeping something as a souvenir. John Wootters thought along the same lines as Swan:

> I've used the terms "trophy buck" or "trophy deer" several times, without a definition of the word "trophy." The reader may be surprised to find that I'm perfectly satisfied with the dictionary definition: a memento of a personal achievement. That dictionary says nothing about number of points, inside spread, or dressed weight, and I say that there is no single set of such specifications which can cover all regions, all hunters, and all situations. Under some circumstances, even a spike buck[11] might qualify as a legitimate trophy, as in the case of a youngster's first kill. (Wootters 1977: 11)

While great trophies can certainly be defined subjectively, there are also objective definitions of trophies. These definitions are for the purposes of trophy books and trophy record keeping and can serve as a way of quantifying trophies for recognition and, for some hunters, competition. A deer's hoof, the tail of a squirrel, the first sora rail killed, or even a spike buck's antlers are not trophies for the trophy books.

Trophy record books are colloquially referred to as "the books," or perhaps the singular version, which often refers to either Boone & Crockett's record book or Safari Club International's record book, depending on the context. In Africa the reference will also still often be to Rowland Ward. Hunters may for example ask their professional hunter (PH) or their guide whether a particular animal will get them "in the book." The record books differ in accepted species, geographical coverage, levels of prestige, accepted trophy collection methods, fair chase standards, methods

---

[11] "A male deer typically in its second year with unbranched antlers on both sides" (Merriam-Webster).

of measurements, level of information required for submission, etc. Many national and lesser trophy hunting organisations and regional scoring systems exist, so we will cover only what we consider the most important trophy hunting organisations and their record keeping systems.

Chronologically and in order of their appearance in this chapter, they are: Rowland Ward Ltd. and the *Rowland Ward's Records of Big Game* (*RW Records*), Boone & Crockett (B&C) and the *Boone and Crockett Records of North American Big Game* (*B&C Records*), Pope & Young (P&Y) with their *Bowhunting Big Game Records of North America* (*P&Y Records*), the International Council for Game and Wildlife Conservation (CIC) and its Trophy Evaluation System, and Safari Club International (SCI) and the *SCI Record Book of Big Game Animals* (*SCI Records*). We finish with a brief discussion of the age-related measuring system developed by Erongo Verzeichnis and implemented in Namibia.

The scoring systems are similar in many ways, and we will not discuss in details the technical differences in scoring methods. Our intention is not to replicate a guide for hunters on how to measure trophies or get their trophies accepted for record books, but rather to provide an overview of the most famous systems and a general understanding of their history and how they work. We will cover how *RW Records* works in some detail first, and the others mostly where they differ in significant ways. Wherever comparisons are made between the systems' handling of individual species, we will single out the whitetail deer as an example.

Aside from covering hunting clubs with trophy scoring systems, this chapter also serves as an account of species commonly hunted for trophies.

3.4.1 Rowland Ward

In the latter half of the nineteenth century, as previously covered, Europeans increasingly hunted the world and especially Africa, bringing their trophies back to Europe and often writing books about their adventures. Personal journals, personal record keeping, and even personal paintings and sketches from hunting trips became hugely fashionable. Every proper hunter kept a hunting diary; the bird hunter kept a record of his numbers of kills, and the hunter of big game kept a record of the anatomical dimensions of his kills (MacKenzie 1988: 33–35).

This inevitably developed into the first official scorekeeping system, collating such records from many hunters into one published work. The first big game trophy record book was published by Rowland Ward in 1892 as *Horn Measurements and Weights of the Great Game of the World: Being a Record for the use of Sportsmen and Naturalists*. The second edition of this book was published in 1896, now titled *Rowland Ward's Records of Big Game* (*RW Records*). It has kept this title ever since in a series that is now on its 30th edition from 2019.

Horn Measurements and the first editions of *RW Records* contained very few North American trophies, as it was too expensive to ship trophies across the Atlantic for official measurement (Nesbitt and Reneau 2016: 3). The 9th edition of *RW Records* from 1928 was the last edition for many years to include North American and European game. The 10th edition from 1935 recorded only Asian and African species, and the 11th edition from 1962 was solely dedicated to Africa. This was the case for all the following editions until the 22nd in 1989, which once again included species from all over the world. With the 29th edition from 2017, *RW Records* once again only contains African species and this is also the case for the 30th edition from 2019. The 31st edition, coming in 2023, will once again cover the whole world.

About 190 different species of animals from many parts of the world were entered in *Horn Measurements*, and for each of them were listed the trophies with their dimensions, the specimen's habitat (if such information was available), and the owner of the trophy. Many species also had short biographies, illustrations, or other notes. Common trophy species filled several pages of entries in the book, while rarely hunted species had just one or two entries. The measurements in *Horn Measurements* were almost all taken by Rowland Ward himself, as the specimens passed through his taxidermy shop. While measurements in *Horn Measurements* were already more-or-less standardised for deer and antelope species (as Ward worked on many of these), the entries for lions, elephants, rhinos, bears and others of the more uncommonly hunted species were at the time still scarce, and Ward included whatever information he had.

Over the first few editions of *RW Records*, the book became a famous and important reference book for naturalists, hobbyists, and hunters alike. Prestige in having one's name recorded in the famous book soon followed, and it turned into the trophy hunting record book that it is known as today.

RW Records now has minimum scores for entry into their record keeping database for about 190 species from Africa and about 75 from North- and South America.[12] Strict minimum requirements have been established and grown increasingly harder to meet. The methodology for measuring the animals has also been refined over the years. Great care is put into ensuring that trophies are scored correctly, using the method forms available on the Rowland Ward Website.[13]

For antlered species, the required measurements typically include circumference at the base of one of the main beams, the spread of the antlers from tip to tip of the

---

[12] Most of the African species are antelopes, which also includes wildebeest. Shorter categories are buffaloes, cats (wild cat, leopard, lion, caracal, cheetah, serval), civets and genets, crocs, deer, elephants, goats and sheep, hippos, and hyenas. Most of the American species are deer, separated under the headings of elk (4 species), moose (3), caribou (5), white-tailed deer (4), mule- and black-tailed deer (3), South American deer (4), and Brocket deer (4). Additional categories include but are not limited to cats (puma, southern puma, jaguar, ocelot, Canadian lynx, bobcat), bears (polar, grizzly, Alaskan brown, American black, and Andean), canids (wolverine and wolf), American plains bison, sheep and goats (Dall's sheep, Stone's sheep, Rocky Mountain bighorn sheep, desert bighorn sheep, and Rocky Mountain goat), muskox, walrus, and crocodile. Additionally, there are categories of "introduced" game, which include fallow deer for Africa and red deer for America.

[13] Most antelopes are scored and reported with measurement method and form 7-a, which covers the greatest amount of species of any of the methods. Antlers on the other hand, being much more diverse in their shapes than horns, have evolved in their measuring methods for each of elk, moose, caribou, white-tailed deer, mule deer, South American deer, and brocket deer, so depending on these categories, one has to submit a form 1-a, 4-a, 2, 1-d, 1e, 1-h, or 1-j, respectively.

two beams (usually the most forward or inward facing tips), outside spread (how wide the whole rack is, usually measured at the outside edges of the two main beams, as the tips tend to curve inwards), inside spread (distance from main beam to main beam, measured at the widest point), and the number of tips. Antlers are for some deer species[14] further divided into separate scores for "typical" and "non-typical." Typical refers to the conventional shape and size of antlers, while non-typical scores any uncommon shapes and growths.

Horned species, including rhinos, require measurements of length (in straight line and along the curves for species with curved horns like the kudu), circumference at the base, and tip-to-tip width where relevant. Scoring for all categories of species has today reached the same standardisation as antelopes and deer had from the outset. Pigs, hippos, walruses, and elephants are scored by the length and circumference of their tusks (methods 5 and 6). Bears, lions, leopards, cheetahs, hyenas, and a handful others are now recorded with the length and width of the skull only (method 18). Crocodiles and alligators are scored by their total length (method 19). Measurements are today only accepted if they are done by an officially appointed Rowland Ward Measurer.

Aside from the species-specific measurements, the method forms also require the hunter to supply personal information, the measurer's name and signature, the animal's country and area, how it was obtained (rifle, muzzleloader, handgun, bow/crossbow, or picked up), and whether it was obtained from an enclosed area. If it was from an enclosed area, the size of the area needs to be supplied. Certain species, and all entries vying for a top-10 placement in *RW Records*, require photos.

By signing the method form, one acknowledges to have "hunted or acquired this trophy in an ethical, lawful, fair-chase manner, and according to Rowland Ward Ltd.'s Guiding Principles" (Rowland Ward Ltd. 2018a). The Rowland Ward Guiding Principles (2018b) is a two-page document that promises Rowland Ward Ltd.'s commitment to habitat protection, promoting understanding between indigenous people and the fauna that surrounds them, sound conservation practices, and fair-chase hunting methods. Most of the document concerns the organisation's commitment to fair chase principles, and how animals killed otherwise will not be accepted in the record book. It expresses an understanding that fencing is often necessary in modern wildlife management to prevent human-wildlife conflicts, and that trophies from animals behind fences are therefore accepted, provided that the enclosures

> promote self-sustaining, breeding populations that can feed themselves from naturally occurring vegetation and prey without continual supplemental feeding by humans. They shall provide enough acreage and vegetation that animals can easily hide from humans and predators alike, and they must offer a hunting scenario whereby the outcome of obtaining a certain animal is by no means guaranteed. Animals that are released solely for the purpose of hunting them shortly thereafter will not be accepted for entry into the record book. Any animal shot in an enclosure that lacks adequate food and acreage is not eligible for entry into the record book. (Rowland Ward 2018b)

---

[14] North American white-tailed, Coues' white-tailed, Central American white-tailed, South American white-tailed, mule deer, sitka black-tailed, and Columbia black-tailed.

Introduced species, also known as "exotic," are accepted for entry in the book, only if they are from unfenced areas and if a hunting license is required for them. Animals shot from vehicles are not accepted, and neither are animals shot with aid from helicopter, airplane, or drone, regardless of local game laws.

The method forms, as mentioned, have an option to tick "picked up" when answering how the trophy was obtained. The guiding principles correspondingly specify that "Rowland Ward's record book honors the animal, and as such no distinction is made based on how the animal was obtained" (Rowland Ward 2018b).

3.4.2 Boone and Crockett

While the database of *RW Records* does cover North America, Europe, and Asia, its focus is on Africa, and Africa is the only continent listed in recent editions of *RW Records*.

Boone and Crockett Records of North American Big Game (B&C Records) only covers North America. It is widely considered the most prestigious of the trophy books and record keeping systems for North American hunting, and it is certainly the most restrictive. Chapter 2.8 already remarked on the very important role that Theodore Roosevelt and George Bird Grinnell personally played for North American conservation at the end of the nineteenth century, and as well on the role that B&C played with the Yellowstone Park Protection Act, the Lacey Act, and the establishment of national parks and forest reserves.

B&C accomplishments are inextricably connected to the individual accomplishments of its members in other roles, which have already been covered. This includes Roosevelt's as US President, Pinchot's as chief forester, Grinnell's as *Forest and Stream* editor, and even those of later B&C members like Hornaday and Leopold. We shall therefore expand on B&C's history and understanding of fair chase only briefly here, before discussing the record book.

Article I.

This Club shall be known as the Boone and Crockett Club.

Article II.

The objects of the club shall be—

1. To promote manly sport with the rifle.
2. To promote travel and exploration in the wild and unknown, or but partially known, portions of the country.
3. To work for the preservation of the large game of this country, and so far as possible to further legislation for that purpose, and to assist in enforcing the existing laws.
4. To promote inquiry into and to record observations on the habits and natural history of the various wild animals.
5. To bring about among the members interchange of opinion and ideas on hunting, travel and exploration; on the various kinds of hunting rifles; on the haunts of game animals, etc.

Article III.

> No one shall be eligible for membership who shall not have killed with the rifle in fair chase, by still-hunting or otherwise, at least one individual of one of the various kinds of American large game.[15] (Roosevelt and Grinnell 1895: 439–440, emphasis in original)

These were the first three articles of the constitution of B&C in 1895. The club was gradually conceived by Roosevelt and Grinnell from the year 1885 after Grinnell's review of Roosevelt's *Hunting Trips of a Ranchman* led to their close friendship. Knowing that they shared concerns over the state of North American big game and forest conservation, Roosevelt proposed forming the club to Grinnell and a group of his sportsmen friends in 1887. The club was officially founded in 1888 (Grinnell 1913a; Reiger 2001: 146–151). Membership was limited to 100 and, as per the constitution, members had to be hunters. Roosevelt believed that the practical mindset of the hunter and conservationist was necessary for saving wildlife. He did not believe in preservation; both forests and wildlife needed a value and a use. The conservation idea was initially, as Grinnell confirmed in 1923, not entirely altruistic: "We wanted the game preserved, but chiefly with the idea that it should be protected in order that there might be good hunting which should last for generations" (Grinnell in Cutright 1985). This is referred to as the selfish stage of American conservation by Krausman and Mahoney:

> The idea of game protection occurred in three stages. The first stage was selfish whereby sportsmen wanted to reduce wanton killing to increase hunting opportunities for themselves. This stage was followed by a sentimental or aesthetic stage. People saw wildlife disappear before their eyes and felt compelled to do something about it, leading to the third or economic stage. At that stage, society began to feel that by spending money on wildlife, we could preserve it for tomorrow's generation. (Krausman and Mahoney 2015: 747)

While the idea of respectful hunting requiring physical effort dates to Ancient Greece at least, and the idea of sportsmanship for conservation purposes in North America dates to the late eighteenth century's "code of the sportsman," one of B&C's most important contributions has been to revitalize and promote the essence of these ideas broadly as "fair chase." B&C defined fair chase as "the ethical, sportsmanlike, and lawful pursuit and taking of any free-ranging wild, native North American big game animal in a manner that does not give the hunter an improper advantage over such animals" (B&C 2016: 1; see also Chapter 6). While this may at first have been motivated by an anthropocentric desire to safeguard hunting opportunities, it instilled in many American hunters an idea of restraint that helped vulnerable North American big game recover from almost certain extinction. One central event relating to fair chase, which B&C refers to as the "The Birth of Fair Chase and the Teddy Bear," occurred in 1902, when Roosevelt went black bear hunting in Mississippi. Roosevelt had no luck finding a bear, but his guide did, and he succeeded in having it cornered and tied to a tree. He then summoned Roosevelt, so that he may shoot it. Roosevelt

---

[15] In the 1913 version of the constitution, this requirement had changed to "at least one adult male individual of each of three of the various species of American large game", underscoring that B&C was indeed not a club for preservationists or even non-hunting conservationist (Grinnell 1913b: 501).

of course found it all very unsportsmanlike and refused to shoot the bear. The news about this hunting trip and Roosevelt's bear spread to many who had not heard about such a code of ethics in hunting before (B&C 2019a).

For B&C contributions to conservation in North America since the end of the nineteenth century, we have this summary by Krausman and Mahoney:

> Over more than a century since the B&C was founded in 1887, its members have been instrumental in the creation and expansion of Yellowstone National Park, Glacier National Park, Denali National Park, and Sheldon Antelope Range; prominent organizations including the Bronx Zoo (renamed the Wildlife Conservation Society), American Ornithological Union, Camp Fire Club of America, the National Association of Audubon Societies (renamed the National Audubon Society), American Bison Society, American Game Protective and Propagation Association (renamed the Wildlife Management Institute), Save the Redwoods League, National Wildlife Federation, American Wild Fowlers (renamed Ducks Unlimited), Natural Resources Council of America, National Bison Range, National Key Deer Refuge, National Collection of Heads and Horns, Wetland's Reserve Program, and Roosevelt's Elkhorn Ranch; key legislation including the Lacey Act, Reclamation Act, National Wildlife Refuge System Act, Migratory Bird Act, Migratory Bird Hunting Stamp Act, and Pittman-Robertson Act Legislation. The B&C has also been responsible for the development of a scoring system for big game, numerous key conferences, and the B&C Professor's Program, as well as publications and numerous other policies and actions that have shaped conservation and management in North America. (Krausman and Mahoney 2015: 748)

B&C's trophy scoring system mentioned above took its first steps in 1895, three years after Rowland Ward published *Horn Measurements*. At a sportsman's exhibition in Madison Square Garden, Theodore Roosevelt and other B&C members began measuring trophies and assigning rudimentary scores. This developed into the first simple scoring procedure for big game in 1902, when B&C also organized its first official "Committee for Measurement of Trophies" with members Theodore Roosevelt, Caspar Whitney, and Archibald Rogers. The purpose of this committee was to establish a uniform method for scoring North American big game. In 1906, B&C and this committee—now called the "Committee on Big Game Measurements"—published *Big Game Measurements: Game Book of the Boone and Crockett Club*. This was not a record book but rather a description of B&C's initial methods for measuring different big game species. It was also a limited edition and now only two original copies are known to exist. It was reprinted in 2011 with a run of 150 copies (B&C 2011; Buckner 2011; Nesbitt and Reneau 2016: 4).

B&C began systematically collecting data on maximum sizes attained by various North American species of big game in the early 1930s. This led to the first edition of *B&C Records* in 1932 with the second following in 1939. Many of the trophies measured in these early editions came from the National Collection of Heads and Horns that Hornaday had been putting together between 1906 and 1922.[16] B&C obtained the rest of the measurements from museums and trophy hunting collections all over the world (Archives West 2011; B&C 2013a; Nesbitt and Reneau 2016: 3–4).

---

[16] In 1922 this collection was put on display at the New York City Bronx Zoo, later to be displayed at the Buffalo Bill Historical Center in Cody, Wyoming. Today it is at the Wonders of Wildlife National Museum and Aquarium in Springfield, Missouri (Nesbitt and Reneau 2016: 3).

The first methods described in the 1906 *Big Game Measurements* and used in the 1932 and 1939 editions of *B&C Records* were very simple. Following the 1939 edition there were two competing scoring systems, advocated by long time members Grancel Fitz and James Clark, but all development of these systems stopped with the beginning of WWII and would not resume until after the war (Nesbitt and Reneau 2016: 4). In 1947, B&C began hosting an event called the North American Big Game Competition, overseen by the B&C Records of North American Big Game Committee, established in 1935. During this event, noteworthy trophies would receive awards based on the simple measurement techniques that had been employed so far and on the verdicts of a panel of judges. The subjectivity involved in scoring, however, convinced B&C of the need for a more objective system, which would allow hunters to estimate trophy scores themselves. A committee charged with developing this new system was formed in 1949, and B&C adopted the revised scoring system in 1950. This system—in many ways a compromise between the competing systems developed by Fitz and Clark—is still being used today. The new scoring was used for the first time in the third edition of *B&C Records* in 1952, for which many of the entries were obtained at the Big Game Competitions (Archives West 2011; B&C 2013a).

The North American Big Game Competition was held annually from 1947 to 1952, every other year until 1968, and every third year ever since. The name changed to North American Big Game Awards in 1971 to emphasize that the purpose was to celebrate exceptional animals, and not to stage a competition between trophy hunters. The 31st Big Game Awards were held in 2022 in Springfield, Missouri.

In 1973, B&C began sharing the responsibilities for record keeping and the Big Game Awards events with the National Rifle Association (NRA). This lasted until 1980, when B&C once again assumed full responsibility for both the triannual events and the record keeping. In 1985, B&C began publishing the *How to Score North American Big Game* official measurer's manual with scoring instructions. This manual is now on its 5th edition from 2021. From every Big Game Awards event follows a publication of a *Big Game Awards* book. This publication contains the trophies accepted for the Big Game Awards event, and therefore covers individual three-year periods. This is a publication separate from and in addition to the *B&C Records*, which contains the all-time records and therefore has higher minimum requirements (Archives West 2011; Nesbitt and Reneau 2016: 6–7).

The *B&C Records*, with a new edition every six years, is as of June 2022 on its 14th edition, published in 2017. The entire trophy database can also be searched and browsed online, following an access purchase. The scoring system remains essentially unchanged since its adoption in 1950 and has also been fully adopted by Pope and Young and partly (using it for North American game) by the International Council for Game and Wildlife Conservation, to be presented shortly. Covering only North America, the number of species accepted into *B&C Records* is lower than in *RW Records*, numbering at only 38 different big game categories. But entries for each species can number in the thousands. In 2016, there were over 46,500 accepted trophies in the B&C Awards Program (Nesbitt and Reneau 2016: 4–5; Spring 2016).

There are more than 3000 entries for the typical whitetail deer, and around 500 of them fall in the narrow range between 171 inches and the 170 inches minimum score.

The following categories of big game are listed in *B&C Records*, and separate scoring sheets are available on the B&C website for each: Bears (black, grizzly, Alaska brown, and polar), cougar and jaguar, walrus (Atlantic and Pacific), typical and non-typical American elk, Roosevelt's elk and tule elk, typical and non-typical mule deer and blacktail (mule deer, Columbia blacktail, and sitka blacktail), sheep (bighorn, desert, Dall's, and Stone's), typical and non-typical whitetail deer and Coues' deer, moose (Canada, Alaska-Yukon, and Shiras), caribou (mountain, woodland, barren ground, Central Canada barren ground, and Quebec-Labrador), pronghorn, bison, Rocky Mountain goat, and musk ox.

Entry in *B&C Records* requires the submission of a basic information- and trophy-scoring sheet, pertaining to one of the categories mentioned above. These sheets closely resemble the scoring sheets for *RW Records*. The two-page species-specific sheet lists the pertinent measuring instructions and minimum scores for *Big Game Awards* and *B&C Records*. The required information is (for deer in this example) the number of typical and non-typical (abnormal) points, their individual and collective lengths, tip to tip spread, greatest spread, lengths of beams, and beam-circumferences at various points. All these factors are then boiled down to a single trophy score, and it must all be measured—"with a 1/4-inch wide flexible steel tape to the nearest one-eighth of an inch" (B&C 2019b: 2)—and calculated by one of the roughly 1400 official B&C measurers (Spring 2016), whose details also need to be provided. An additional requirement for *B&C Records* is the submission of a "Hunter, Guide, and Hunt Information Sheet" and an "Entry Affidavit for all Hunter-Taken Trophies."

The Hunter, Guide, and Hunt Information Sheet requires information about the guide and the taxidermist, the exact location of the hunt and whether the property is private or public, duration of entire hunting trip and when the animal was shot,[17] mode of transportation in and out, whether motor-powered vehicles were used, what kind, and how, shot distance, detailed information on firearm/bow or bullet/arrow and scope used, and weather conditions.

About "picked up" trophies, the general policies of the Records Committee of B&C specify as such:

> Trophies in the records book with hunter listed as "Picked Up," include specimens that were found dead by hunters or hikers, and animals that have died from natural or unnatural causes, such as old age, severe winters, falls, car accidents, lightning, fence entanglements, drowning accidents, etc. "Picked Up" trophies also include animals that were illegally taken but are now held in trust for the public by game and fish departments. In the latter case, the poacher's name will not be listed in the book to avoid giving him credit for the kill. "Picked Up" trophies are an integral part of the Club's records-keeping activities. The fact that they were not taken by a hunter is irrelevant because the Club recognizes the trophy, and because they give a more complete picture of the successes of conservation efforts. (B&C 2017)

On page two of the Information Sheet, B&C asks for age data:

---

[17] If the trophy was picked up or acquired in any other way that does not involve killing the animal, a separate narrative document is required.

In an effort to increase the relevancy and value of the Club's records program, Boone and Crockett Club wishes to obtain age data for your trophy, if possible, even though age data is not required. Records keeping is more relevant to big game managers if B&C can supply them with age data. Age data also helps substantiate the contention that most trophy-sized animals are well past prime breeding age. (B&C 2015a)

An offer is also made that B&C will determine the age of the trophy, if a relevant premolar (bears/cougars) or incisor tooth (all other categories except walrus) is attached to the submission.

The third document required for submission—the Entry Affidavit—asks whether (and if so, how) dogs were used in the pursuit or harvest. The hunter's signature also confirms adherence to B&C Fair Chase principles, including that none of the following methods or conditions were involved:

I. Spotting or herding game from the air, followed by landing in its vicinity for the purpose of pursuit and shooting;
II. Herding or chasing with the aid of any motorized equipment;
III. Use of electronic communication devices (2-way radios, cell phones, etc.) to guide hunters to game, artificial lighting, electronic light intensifying devices (night vision optics), sights with built-in electronic range-finding capabilities (including smart scopes), drones/unmanned aerial vehicles (UAVs), thermal imaging equipment, electronic game calls or cameras/timers/motion tracking devices that transmit images and other information to the hunter;
IV. Confined by artificial barriers, including escape proof fenced enclosures;
V. Transplanted for the purpose of commercial shooting;
VI. By the use of traps or pharmaceuticals;
VII. While swimming, helpless in deep snow, or helpless in any other natural or artificial medium;
VIII. On another hunter's license;
IX. Not in full compliance with the game laws or regulations of the federal government or of any state, province, territory, or tribal council on reservations or tribal lands (B&C 2019c).

In addition to all aforementioned documents, high quality photos and copies of hunting licence/big game tag/salvage permit are also required for all trophies.

B&C Records, to sum up, is characterized by elaborate fair chase standards and attention to detailed information about the trophies within. The B&C trophy scoring system is frequently used and referred to in wildlife management science. Green and Stowe (2000), for example, explain how trophy deer management generally means managing to maximize B&C scores; Bartoskewitz et al. (2007) used B&C scores to determine effects of dietary copper and zinc concentrations on whitetail antler growth; and Strickland et al. (2013) show that since antler quality is used to assess characteristics of whitetail herds and B&C scores are the best indicator of antler quality, wildlife managers may benefit from simpler models for predicting B&C scores. The MSUDeerLab unit of Mississippi State University likewise developed and publish on their website equations for quickly and informally estimating B&C scores for whitetail deer, arguing that

3.4 Trophies and Record Books

[t]he Boone & Crockett Scoring system [...] has become the "gold standard" for measuring white-tailed deer antlers. Although many people have tried, no scoring system has surpassed the Boone & Crockett (B&C) system for recognition and popularity. (MSUDeerLab 2013)

Last—but to many hunters certainly not least—*B&C Records* is characterized by exclusivity and prestige. Wootters' opinion here from 1977 illustrates this well, and minimum scores have only increased since then, even though the accepted entries for trophy book species number in the thousands of specimens:

> The ultimate achievement in trophy deer hunting is, of course, a head which is listed in the Boone and Crockett Club Records of North American Big Game, but it must be understood that this is *not* the definition of a trophy head which will apply through this book. To begin with, the Boone and Crockett minimum score for entry is, in my opinion, ridiculously high. The best statistics available for the last few seasons, as I write this, suggest that not more than one whitetail rack out of each *one million* bucks killed in the U.S. qualifies as a B&C record. I'm not certain what percentage of the antlers taken of any species of deer should be recognized in a records-keeping system, but I'm pretty sure that it should be more than the top .000001 percent!. (Wootters 1977: 12, emphasis in original)

Further emphasizing the prestige aspect of having a trophy recorded in *B&C Records* is the memorabilia that one may purchase to show off the accomplishment. The B&C *Trophy Memories Catalog* sports, for example, B&C score chart plaques, engraved bronze coins, and customized belt buckles, complete with trophy category depiction, location, harvest date, and trophy score. B&C trophy rings are also available for purchase, featuring engraved species relief, B&C score, awards period, and a wide selection of metals, stones, and elk ivory (B&C 2015b).Such merchandise would seem to directly and robustly challenge B&C's claim that trophy scoring is about the animal and not the hunter.

B&C has two awards in addition to *B&C Records* called the Sagamore Hill Award, and the Special Sagamore Hill Award, named after Roosevelt's New York residence from 1885 until his death. Both come with the Sagamore Hill Medal. They were created in 1948 and are awarded by the Roosevelt family. This medal is the highest award given by B&C. The Sagamore Hill Award has been presented 17 times for "outstanding trophies worthy of distinction" and the Special Sagamore Hill Award 8 times for "distinguished devotion to the objectives of the club" (B&C 2019d).

3.4.3 *International Council for Game and Wildlife Conservation*

CIC[18] was conceived in 1910 at the first International Hunting Exhibition in Vienna, during which work also began—inspired by the 1896 second edition of *RW Records*—on a more elaborate scoring system than RW's. Hungarian hunting journalist Herbert

---

[18] Fritz Heje Hansen is part of Denmark's national delegation in CIC, where he is also a Senior Trophy Judge. He has worked with trophy measurement and in hunting organizations for over 30 years. We rely heavily on personal communication between him and NB as a source for all of Chapter 3.4.3.

Nadler described a scoring formula for red deer for the first time in 1925 (to become known as the Nadler formula). In 1928, Louis Károlyi, a Hungarian count, Maxime Ducrocq, a French lawyer and president of the French St. Hubert Club for deer hunting, and others of their friends organized an international conference in the Slovak town of Nové Zamky with the intention of forming an international hunting council. The conference resulted in the "Declaration of Nové Zamky," stipulating the specifics of an international hunting council, which was founded in 1930 in Paris at the first general assembly (CIC 1977, 2019a, 2019b, 2019c).

The Nadler formula was revised in 1934, 1937, 1952 (the Madrid formula), 1954, and 1955. Some of the revisions included adopting formulae from the third edition of *B&C Records* from 1952. From 1955 a "Trophy Valuation in the CIC" working group, headed by Kenneth Whitehead, worked on new system. This was finally presented at the fifth International Hunting Exhibition in 1971 in Budapest. Whitehead was the author of all formulae for Asian and African game. A revised, unified scoring system was completed with the publishing of the first version of *The Game Trophies of the World, International Formula for the Measurement and Evaluation of Trophies* in 1977, which is also referred to as *The Handbook* or The *CIC Red Book* (CIC 1977: 12).

The Handbook, however, lacked clarity of rules, and this led to different interpretations of the measurement methods in different countries over time. The executive committee of CIC therefore decided in 2009 that it was time to revise the rules. The result, however, was finally considered too fundamentally different from the Handbook. One important issue that had not been taken into consideration was that while different countries had interpreted the Handbook in different ways, these interpretations were now pretty well established over many years, and trophy measurements based on them were stored in national databases and often used in national research. CIC did not need something fundamentally different but rather more stringent versions of the old rules, leaving no—or at least less—room for misinterpretations. With this now in mind, the revision work continued in the hands of the main office in Hungary and some European specialists, and a new Handbook version was presented in 2014 at the 61st General Assembly in Milan. It was henceforth made available to CIC members, Senior Trophy Judges, and certified CIC measurers. The new Handbook has been gradually revised since 2014. It no longer exists in a complete book format, but rather—as is the case with the scoring sheets used by the other organisations discussed in this chapter—as an online collection of separate sheets for separate species. The advantage is that these can be individually altered and replaced without the need to publish an entirely new book.

An important difference between CIC and the other trophy scoring organisations in this regard is that the new CIC Handbook is under strict copyright protection and unavailable to the public. It is part of the CIC Trophy Evaluation System and stored in the Trophy Evaluation Database, which is only accessible by official CIC Trophy Measurers and Senior Trophy Judges.[19] What also differentiates CIC from

---

[19] We are grateful to Fritz Heje Hansen for allowing NB to read his copy of the Handbook for our research purposes.

the other organisations is that there is no record book. What merits the organisation's inclusion in this chapter nonetheless is that trophies of a high enough quality may instead receive either an international CIC gold medal, awarded by CIC, or—in the case of Denmark, Germany, Norway, UK, and Finland—a CIC officially recognized national medal, awarded by these countries' CIC-delegations or national hunting organisations. An international "Grand Prix" award also exists for truly exceptional trophies, scored by at least three Senior Trophy Judges from different countries.

CIC's trophy measurement system incorporates the methodologies of both B&C and Rowland Ward. The evaluation parameters and regulations of B&C are implemented for all North American game. For all African game, CIC has adopted the RW method of measuring horn lengths and base circumferences, which also means that trophies scored by CIC methods are eligible for *RW Records* (CIC 2018).

CIC's measurement system stands out from the rest by incorporating weight and beauty points in its trophy scores for game outside North America and Africa. Beauty is an especially interesting factor to discuss here, as it is undeniably a more subjective factor than size and weight. It could also be questioned what the relationship is between the beauty of a trophy and the genetic desirability of a specimen as seen from a wildlife management perspective. Beauty points are for the relevant species added to a trophy score after scoring for size and weight. Two of the most commonly hunted European species are red deer and roe deer, and these are examples of categories where beauty points are added. This addition necessitates much more thorough measurement instructions, so while most scoring sheets are 2–4 pages, the scoring sheets for red deer and roe deer are ten pages each. Beauty is measured differently for each species; red deer beauty is measured on five parameters (colour, pearling,[20] condition of tine tips, formation of bay tine,[21] and formation of crowns), and roe deer is measured on six (colour, pearling, coronets, tine tips, "regularity, symmetry and form," and length of tines). Using red deer as an example, for the colour, light grey yellowish or artificially coloured gives 0 points, grey to medium brown 1 point, and dark brown to black 2 points. For the pearling, smooth or light pearling gives 0 points, normal and medium pearling gives 1, and good pearling on the whole antler gives 2 points. The tips of the tines may add 0 points if they are dull or rotten, 1 if they are pointed but dark in colour, or 2 if they are rounded to a point, polished ivory-coloured, and contrasting. The formation of bay tines and formation of crowns refers to how the antlers split into tines on each side and how long the tines are (CIC 2018).

Just like B&C, CIC places great emphasis on symmetry in its measurement system, penalizing what is commonly referred to as "abnormal" trophies. The rationale is that highly asymmetrical antlers, large and heavy as they may be, represent genetic traits that are undesirable from the wildlife management perspective of maintaining a healthy population. The goal is that bucks with abnormal antlers are killed off by hunters as soon as the abnormal antlers are noticed and before they have had much

---

[20] Pearling is the formation of tiny knobbly lumps on the antler shaft, most prominently just above the brow.

[21] The first antler split is the brow tine and the second is called the bay tine.

time in which to spread their genes. Bucks with good genes (very symmetrical antlers) should be allowed to live and breed and spread their good genes until they are old and have developed very large antlers. The CIC-medals for large and symmetrical trophies would thereby be rewarding the patience of hunters and sound scientific management of wildlife populations.

Another part of CIC's system that is meant to contribute to scientific management and scientific research is the inclusion of many parameters on the scoring sheets that are not part of the actual scoring of the trophy. Taking red deer as an example again, the "measurement parameters" are the antler dimensions and weight and the "beauty" additions, and these are mandatory for having the trophy measured. Under "supplementary information," mandatory for gold medal level trophies and optional but strongly encouraged otherwise, one may supply such additional scientific information as skull dimensions and the animal's age based on teeth abrasion (CIC 2018).

The scoring sheets of the *Handbook* also include an affidavit box that the hunter must sign, and with this signature acknowledge that CIC's guidelines on fair chase have been followed. These Fair chase principles of CIC are on the surface similar to those of B&C:

6.4. **A trophy will be ineligible** for measuring with the **TES**, if:

6.4.1. the animal has been hunted by means and methods which are not in full compliance with the wildlife laws or regulations of the respective national hunting authority, government, or communal authority;

6.4.2. the animal was herded or chased with the aid of any motorized equipment;

6.4.3. the hunter, whilst hunting, used artificial lighting, thermal imaging equipment and electronic light intensifying devices except where legally permitted;

6.4.4. any pharmaceuticals were used on the animal before or during the hunt;

6.4.5. the hunter used artificial feeding stations and baiting except where legally permitted;

6.4.6. the hunted animal had been translocated from breeding to shooting areas exclusively for the purpose of "hunting" the animal. (CIC 2016: 5, emphasis in original)

It is worth noting, however, that CIC stipulates the ineligibility of trophies acquired with the use of the methods described in 6.4.3 and 6.4.5 only "except where legally permitted," whereas B&C bans these methods regardless of their legality. It does appears, in fact, that 6.4.3 and 6.4.5, ending in "except where legally permitted" are superfluous, as 6.4.1 already states that wildlife laws and regulations must be followed. This makes the CIC fair chase principles somewhat less ambitious than the B&C equivalent.

The signing of the affidavit further acknowledges that the trophy in question does not "come from an animal that was raised in a breeding centre, breeding farm or comparable establishment or was transported from one of these into a hunting area" (CIC 2018). See Chapter 6.4 for further explanation and discussion of canned hunting.

Some of the scoring sheets cover one species only while others cover a whole range of species, sharing methodology for measurement. While red deer and roe

3.4 Trophies and Record Books

deer have their own sheets, for example, the sheet for "carnivores" covers all hunted species of wolves, dogs, bears, and cats, as these are all simply measured by the length and width of the skull. Antlers and horns come in many more variations than skulls and will often need their own method of measurement. Information about categories of species and what constitutes a trophy in these categories may be found in article 1 of CIC's rules and regulations:

> 1.2. Trophies are defined as the antlers, horns, tusks, skulls, skins, furs, etc. of game species which occur in a wild state as part of self-sustaining populations. Trophies in the context of the CIC Trophy Evaluation System (TES) are defined as:
>
> 1.2.1. The horns and antlers of Cervidae, Bovidae, Antilocapridae, Rhinoceridae;
>
> 1.2.2. The tusks of Suidae, Tayasuidae, Elephantidae, Hippopotamidae, Moschidae, Tragulidae, Odobenidae;
>
> 1.2.3. The skulls and skins[22] of certain Felidea, Ursoidea, Canidae, Castoridae. (CIC 2016: 1, emphasis in original)

CIC is not a membership club in the same way as B&C or SCI. Anyone can become a member of SCI or an associate member of B&C. These memberships require only an annual subscription. CIC membership is not meant or suited for individual hunters. Members and partners of CIC are mostly national states, institutions of the United Nations, and NGOs. FAO, UNEP, UNESCO, and IUCN are examples. Individual members are generally either experts or representatives from the national delegations. There are about 1600 members of CIC, which includes both individuals, institutions, and states. Membership confers votes in the general assembly. It is through cooperation with its members that CIC supports or participates in science and does conservation work. CIC does not conduct research independently. In addition, the Trophy Evaluation Database is still relatively new and based on voluntary membership contributions of data, so its usefulness for research purposes has so far been limited. CIC's contribution to science is the CIC Technical Series, which has been published in cooperation with FAO since 2008. The technical series covers research topics such as *Best Practices in Sustainable Hunting*, *Developing Sustainable Wildlife Management Laws in Western and Central Asia*, and *Contribution of Wildlife to National Economies*.

3.4.4 Pope and Young

Like *B&C Records*, the P&Y *Bowhunting Big Game Records of North America* (*P&Y Records*) covers only North America, but it is further limited to animals taken with bow and arrow. As one needs to get considerably closer to shoot an animal with bow and arrow than with a rifle, and because there are far fewer bowhunters

---

[22] Skins, according to Fritz Heje Hansen, are no longer part of the CIC Trophy Evaluation Database, as measuring them accurately has long been considered too difficult. Species where the skin would previously be considered are now measured by the skull instead.

than rifle hunters, minimum requirements are lower for *P&Y Records* than for *B&C Records*. A typical whitetail deer, for example, must (as of 2022, but these numbers rarely change) meet a minimum score of 170 inches for entry in *B&C Records*. The minimum score for *P&Y Records* is 125 inches. Comparing these scores is easy, as the P&Y scoring method and measuring instructions are word-for-word copies of B&C's.

P&Y was founded in 1961 and named after Saxton Pope and Arthur Young. In the years 1912–1916, Pope and Young learned archery hunting and bow- and arrow-making from Ishi (meaning "man" in the language of his tribe, the Yahi). Ishi is considered the last Native American to grow up and live an adult life without any contact with European settlers. They encountered Ishi in 1911 but, lacking the natural immunities to the settlers' common diseases, he died of tuberculosis in 1916. Pope and Young continued improving their archery methods and equipment after his death. They published numerous books and even made a film about archery hunting. They killed (with special permissions) grizzly bears in Yellowstone with bow and arrow, something unheard of at the time, and even killed lion, gazelle, wildebeest, and eland in Africa and bears, giant moose, and Dall's sheep in Alaska. Both men are—along with Fred Bear who was inspired by Young and also made archery his business from the thirties onwards—considered fathers of modern bowhunting (P&Y 2010).

Bowhunting gained in popularity in the twenties and thirties, initially supported by the fair chase ethics pushed by B&C, by enthusiastic bowmaker Aldo Leopold, and by the fame of actor Errol Flynn, starring as Robin Hood in the 1938 movie. The bow was recognized as a legal hunting weapon for the first time in 1930 in Wisconsin, and more states would follow in the forties. Bowhunters and bowhunting organizations realized, however, that they would not gain real popularity or acceptance without the establishment of a club and a records program, like the ones established by Theodore Roosevelt. Glenn St. Charles, vice president of the National Field Archery Association and head of the Association's Hunting Activities Committee, was from 1956 to 1958 instrumental in the first systematic collection of archery hunting trophy information. Heavily inspired by B&C and with permission to use their scoring system, he defined the rules for trophy scoring and record book acceptance for bowhunters. The records program was introduced in 1958 and quickly became a success. An organization separate from the National Field Archery Association was needed to handle the records keeping program, however. Thus, the Pope and Young Club, "dedicated to scientific records keeping, sound conservation practices, quality hunting and fair chase" was established in 1961 (P&Y 2010).

The requirements for getting a hunting trophy in the *P&Y Records* are that the animal must have been killed:

a. entirely by the use of the bow and arrow (as defined by the Pope and Young Club),
b. in complete compliance with the controlling state/provincial game regulations,
c. in complete compliance with the Club's Rules of Fair Chase,
d. and must meet or exceed the corresponding minimum score requirement for that category (P&Y 2011).

3.4 Trophies and Record Books

The materials required for submission are a scoring sheet with measures taken by an official P&Y or B&C measurer (as the systems are the same), a fair chase affidavit, and photos.

The fair chase affidavit states—similarly to the one used for *B&C Records*—that fair chase does not include the taking of animals under any of the following conditions:

1. Helpless in a trap, deep snow or water, or on ice.
2. From any power vehicle or power boat.
3. While inside escape-proof fenced enclosures.
4. By "jacklighting" or shining at night.
5. By the use of any tranquilizers or poisons.
6. By the use of any power vehicle or power boats for herding or driving animals, including use of aircraft to land alongside or to communicate with or direct a hunter on the ground.
7. By the use of electronic devices for attracting, locating or pursuing game or guiding the hunter to such game, or by the use of a bow or arrow to which any electronic device is attached with the exception of lighted nocks and recording devices that cast no light towards the target and do not aid in rangefinding, sighting or shooting the bow.
8. Any other condition considered by the Board of Directors as unacceptable (P&Y 2015).

It also queries whether the animal was recovered on the same day as it was hit and requires a detailed account of its recovery if it was not.

The second page of the fair chase affidavit requests information about the hunt. This includes weather conditions, shot distance (for each arrow if more than one was required), shot angle (broadside, rear quartering, front quartering, rear, front, above), where the arrow struck the animal (chest, paunch, rump, leg, head, other), how far the animal ran before being recovered, and hunting style (bait, stalk, still, tree stand, ground blind, calling, dogs). Equipment information required includes type of bow (longbow, recurve, compound), manufacturer, model, draw weight, type of quiver (hip, back, bow), type of arrow (wood, carbon, aluminium, hybrid), arrow's model, manufacturer, length, and weight, type of broadhead (generally being fixed blades, removable blades or expanding blades), type of sight and optics, method of release (use of release aid or not), rangefinder, clothing, boots, and more.

P&Y Records—unlike the other record books, and naturally since it is specifically for bowhunting—does not accept entries that were picked up or obtained in other ways. Species covered are mostly the same as in *B&C Records*.

The prestige of having a trophy entered in *P&Y Records* is not as centred around sheer size and antler symmetry as it is for the other record books. It is instead a combination of the trophy qualities and of how the trophy is evidence of the additional woodcraft and hunting skill required to kill with bow and arrow. Bow and arrow, however, is a broad definition. *P&Y Records* accepts kills made with longbows, recurve bows, and compound bows but not with crossbows.

Longbows and recurves are simple bows. The longbow is the most basic and long (up to as long as the person using it) of the "Robin Hood-style" bows. It is shaped like a half-moon and usually made of laminated wood. The recurve bow is shorter and has tips that extend forward (re-curve) and are pulled back under tension when the bow is strung, adding extra force and speed to the arrows. Recurve bows, being shorter, tend to be much lighter and more field-portable than longbows. Both kinds of simple bows require a very strong arm to pull back (to "draw") and lots of practice before they can be used for hunting big game as well as lots of practice to maintain the skills. The compound bow is another beast entirely.

Compound bows have a very advanced system of cam-shaped pulleys that provide a mechanical advantage to the archer, making the bow easier to draw than a traditional bow of comparable power. The design also allows for what is called "let-off." This feature means than once the bow is drawn back past a certain point, the cams rotate and allow the archer to hold the string back using only around 15–30% of the effort that a traditional bow would need. This "let-off," combined with the additional power of the set-up, means that an archer has a better chance of making a good shot (since the bow is easier to hold at full draw, and to aim steadily despite being usually heavier than a traditional bow) and of that shot penetrating the target with sufficient power. Some compound bows may to the untrained eye not even be recognized as bows; John Mitchell wrote that "most of America's two million bow hunters come equipped with a device that looks like something William Tell might have invented, in consultation with the National Aeronautics and Space Administration" (Mitchell 1981: 43). Compound bows are far more advanced today than they were when Mitchell was writing, and they allow for a wide variety of high-tech gadgets and modifications (including arrow quivers, stabilizers, and sighting systems).

The compound bow is looked down upon among some bowhunters. They argue that one of the common reasons for bowhunting in the first place is to return to the traditional roots of hunting. This purpose appears defeated if the bow is even more high-tech and expensive than a rifle would have been. Mitchell continues:

> To anyone who has never pulled a bowstring, compounded or otherwise, bow hunting is a puzzlement. Here are men and women who would seem to be reaching directly for Paleolithic vibrations, for hunting in the purest, most aesthetic form. Yet here are all these pulleys and cables, these alloys and epoxies, forever getting in the way of the pristine experience. (Mitchell 1981: 43–44)

If the bow—additionally—becomes too effective, the purpose of bowhunting because it is more challenging is defeated as well. Hunting author David Petersen disdains the "high-tech, mechanically assisted missile launchers," by which he refers to crossbows and compound bows. He argues that

> while killing with a rifle or high-tech compound bow demands only a well-tuned instrument, a modicum of practice, and steady hands, making meat with unaccoutered sticks and strings requires a whole panoply of exacting skills, including pinpoint arrow accuracy (necessitating regular practice), wide-ranging woodcraft, physical and mental fitness, and a passion for your

prey extending far beyond just hunting it; all the best bowhunters I know are avid amateur naturalists as well: nature hunters.[23] It comes with the territory. (Petersen 2010: 180–181)

Bowhunting is a particularly interesting case study in a fair chase discussion. It is widely considered a method of hunting with greater emphasis on fair chase than rifle hunting because the prey has a greater chance to escape. The issue is, however, that common interpretations of fair chase also emphasize the importance of "one shot, one kill" and quick and painless deaths. Yet sometimes the limitations that hunters impose on themselves to increase the challenge and give the animal a greater chance of escaping with its life—such as eschewing bullets in favour of arrows—also risk increasing wounding rates. We will return to this dilemma and some other features of fair chase hunting in Chapter 6.

3.4.5 Safari Club International

Several big game hunting organizations came and went in North America between the forties and the seventies. Notable hunting clubs included the Shikar Safari Club in 1952 with a membership criterion of having been to at least two hunting trips in India or Africa, the Mzuri Safari Club in 1958 (Mzuri Wildlife Foundation since 1969), and Game Conservation International in 1967, responsible for some of the first international hunting conventions. Game Conservation International held these large conventions every other year, with the Mzuri Wildlife Foundation hosting it in the alternate years. These conventions provided the model for how SCI would be arranging conventions in the future. In 1971 the Safari Club of Los Angeles, under the leadership of famous trophy hunter C.J. McElroy, spun off from The Southern California Safari Club, also founded by McElroy in 1966. The ambitions did not fit either of the names, however, so the Safari Club of Los Angeles changed its name to Safari Club International in 1972. The focus at the time was on conventions, and SCI took over convention hosting from Game Conservation International and the Mzuri Wildlife Foundation in 1973. SCI has since been hosting conventions every year with recent statistics showing up to 20,000 visitors, attending the different talks and events, and browsing the 2000 or so booths that sell guns and equipment, auction hunts, display taxidermy, and other hunting-related activities and paraphernalia. The 2022 convention was held in Las Vegas, which has become the regular venue in recent times.

The early development and efforts of SCI were greatly boosted by the Klineburger brothers (see Chapter 3.1) who, thanks to their thriving businesses that began with taxidermy in 1954, had become quite famous and well-connected in the world of trophy hunting. The Klineburgers came up with the idea of opening local SCI chapters, and SCI was soon represented in Los Angeles and Chicago. Chris Klineburger started the Northwest chapter in 1974 (covering Oregon, Washington, Idaho, Canada,

---

[23] "Nature hunters" is a reference to a categorization of hunters based on their motivations by Stephen R. Kellert (1978). We will return to Kellert and these categories in Chapter 4.3.

and Alaska), which spawned ten additional chapters (Klineburger 2010: 102–112). Both Bert and Chris Klineburger are in the SCI Hall of Fame. SCI today has over 180 chapters worldwide and over 50,000 members. It is by far the largest, richest, and most influential hunting organisation in the world. It is also, since 1985 and after a long legal battle and many appeals, recognized as a 501(c)(3) tax-exempt charity organisation (Scully 2002: 78).

SCI does its conservation work mainly through the Safari Club International Foundation (SCIF), which, according to their figures, has provided over $75 million to conservation work since the year 2000. SCIF works globally and their work includes supporting and conducting science, educational and humanitarian service programs, and donations to other organisations and individuals (SCIF 2019). Aside from the direct contributions to conservation initiatives by SCIF, SCI argues that all hunters contribute locally to the conservation of the hunted species and support local communities through trophy hunting fees. The extent to which this is true is discussed in Chapter 5.2.

The slogan of SCI is "First for Hunters," and the slogan of SCIF is "First for Wildlife." Everything under the SCI/SCIF umbrella that has to do with records keeping, trophy hunting awards, and trophy scoring belongs in SCI. The scoring system of SCI came about in 1977, under the leadership of McElroy, who felt that *B&C Records* was too restrictive and placed undue emphasis on the spread and symmetry of antlers. SCI's scoring system was designed to allow more hunters to get their names in a record book. For comparison, Ted Kerasote gives the example of a Dall Sheep shot by famous trophy hunter Robert (Bob) Kubick.[24] It ranked 126th in *B&C Records* but 45th in *SCI Records* (Kerasote 1994: 92).

SCI Records comes in a published and an online version, and the first noteworthy difference between it and the other record books is that you need to be an SCI member to submit a trophy for the book. Anyone can submit trophies and have them accepted in the other systems discussed in this chapter, regardless of membership. Members of SCI have access to the SCI Online Record Book, a very extensive database with a user-friendly interface and many options for searching and browsing. However, full access to all features, which includes access to individual hunting reports and browsing/searching hunters' names, requires having an entry accepted in the book, being an official SCI measurer, or being a hunting guide or outfitter. With just basic membership access, the record book is anonymous, meaning that for each species, one can view a ranked list of all the entries with their trophy scores, method of kill, hunting company/guide, location, and date taken, but without the names of hunters.

As of May 2019, there were 886 different "species names" in the book. Most of these are separate species. Some are the same species but separated by continent or by categories that include introduced, feral, estate/free-range, typical/non-typical, darted (for some pumas, jaguars, and rhinos), or even airplane/dogsled hunted (for polar

---

[24] Robert Kubick rose to a net worth of over US$60 million from real estate development during the Alaska oil boom in the 1970s. He spent most of his time trophy hunting and was the winner of the 32nd Weatherby Award with about 270 species in *SCI Records* and a place in the SCI hall of fame (Kerasote 1994: 92–95). He went bankrupt in the nineties and was sentenced to five years in prison for tax fraud. All of his trophies have since been auctioned off (Rose 2011).

3.4 Trophies and Record Books

bears). The European fallow deer (*Dama dama*), as an example of a species represented by separate categories, is divided into South Pacific, South Pacific free range, North America Introduced, North America Introduced free range, Africa Introduced, South America, and South America free range. White-tailed deer are separated into more than 50 such categories. Two hundred and thirty-six of the total 886 categories are deer.

Ninety-nine categories have no record book entries and about 550 have fewer than 100 entries. The top 10 of most trophy entries have from 2526 (common nyala) to 4546 (southern impala). The highest trophy score for southern impala is 70 3/8 while the lowest is 46 5/8, but there are so many trophies with identical dimensions that these 4546 trophies only have 165 different scores, with low-score entries having most of these duplicate scores, as was also the case for B&C. 3727 of the southern impala entries were shot with a rifle, 667 were shot with a bow, 78 with a handgun, 40 with a crossbow, 34 with a muzzleloader, and only 5 are registered as "picked up." This is a typical distribution of methods, but of course some species are hard or impossible to obtain with certain methods. Only a few have killed elephants with bow and arrow, for example, though the fact that anyone has is a testament to the power of a modern compound bow coupled with a large broadhead arrow. Interestingly, elephants were often killed with simple longbows used by the Waliangulu of Tsavo East. The penetration afforded by their large, powerful bows was augmented by arrows tipped with the poisonous extract of the Acokanthera bush (Anderson 2022).

Numbers of entries for some of the more iconic species are African lion (1534), African leopard (1979), African elephant (820), rhinos (separate categories added up: 958), European red deer (367), and North American grizzly bear (576). Leopard and lion are ranked 17 and 22 respectively, sorting by most entries. Looking at the distribution across continents, 202 of the 886 categories are from Africa, 23 from Africa (introduced), 220 from Asia, 99 from Europe, 109 from North America, 34 from North America (estate), 98 from North America (introduced), 50 from South America, and 51 from South Pacific. Estate refers to hunting animals behind a fence, also commonly, but not always unambiguously, referred to as canned hunting (see Chapter 6.4).

Clicking a category yields—in addition to the anonymized book records and required minimum scores for different levels of awards—data on the species and its distribution, taxonomy, endangered status, and occasional remarks about hunting advice, what to look for in a good trophy, or other miscellaneous information. It also reveals a "Where do I rank?" -feature, informing about placement on the record list if one were to submit a trophy with a given score. There is also an "Eligible Rewards" feature, revealing which World Hunting Awards a species is eligible for. An African lion, for instance, was eligible in May 2019 for the following World Hunting Awards: *African 29, African Big Five, Animals of Africa, Cats of the World, Dangerous Game of Africa, Global Hunting Award, Hunting Achievement Award, Predators of the World, Top Ten Award, Africa Big Five (Bow), African 29 (Bow), Animals of Africa (Bow), Cats of the World (Bow), Dangerous Game of Africa (Bow), Global Hunting Award (Bow), Hunting Achievement Award (Bow), Predators of the*

World (Bow), and the *Top Ten Award (Bow)*. More about the World Hunting Award system shortly.

To have a trophy accepted in *SCI Records* it must be scored by a certified SCI measurer and obtain the minimum score for the relevant category. The scoring sheets are of a single page and have most in common with the sheets used by *RW Records*, as the B&C sheets require much more information and a fair chase affidavit. For white-tailed deer, for example, the *RW Records* method 1-d corresponds to the *SCI Records* methods 17-T and 17-NT (typical and non-typical), but with two noteworthy differences. The first is that the sheet for *RW Records* enquires if the animal comes from an enclosed area, and if so, how many acres/hectares, whereas the *SCI Records* sheet asks only whether the animal was "Free-ranging," without further definitions. The second is that one may, on the SCI scoring sheet, tick an anonymous submission box, which means that a Record Book ID number will replace the hunter's name in the *SCI Records* (SCI 2018a). With the signing of the scoring sheet, one acknowledges that

> to the best of my knowledge, I took this animal without violating the wildlife laws or ethical hunting practices of the country or province in which I hunted. I also certify that, to the best of my knowledge, the laws of my country have not been violated by my taking or importing this animal. (SCI 2018a)

This brings us to the question of SCI's official positions on hunting ethics and fair chase. Information about this is given in the official 2017 *Policies and Procedures Manual*. Article 22.1.1 of this document concerns the "Hunter's Code of Ethics," and states the following:

> Recognizing my responsibilities to wildlife, habitat and future generations, I pledge:
>
> To conduct myself in the field so as to make a positive contribution to wildlife and ecosystem;
>
> To improve my skills as a woodsman and marksman to ensure humane harvesting of wildlife;
>
> To comply with all game laws, or the spirit of fair chase, and to influence my companions accordingly;
>
> To accept my responsibility to provide all possible assistance to game law enforcement officers;
>
> To waste no opportunity to teach young people the full meaning of this code of ethics;
>
> To reflect in word and behavior only credit upon the fraternity of sportsmen, and to demonstrate abiding respect for game, habitat and property where I am privileged to hunt. (SCI 2017: 29)

Article 22.1.3 then asserts about the fair chase standard for captive animals, that

> SCI opposes the hunting of all animals bred in captivity unless the hunting operator, or an association of hunting operators, can demonstrate that the hunting meets SCI's fair chase standards (see 27.5.2 (iv) and (v)). (SCI 2017: 29)

Article 27.5.2 specifies that trophies are only accepted for submission if the animal was not shot or hazed from an aircraft. About animals shot on hunting estates, articles

3.4 Trophies and Record Books

27.5.2 (iv) and (v), referenced above, specify further that entries are only accepted, if

(iv) in the case of an animal taken on a hunting estate, the hunter was in compliance with the following policy:

A. The animals hunted must have freely resided on the property on which they are being hunted for at least six months, or longer.
B. The hunting property shall provide escape cover that allows the animals to elude hunters for extended periods of time and multiple occurrences. Escape cover, in the form of rugged terrain or topography, and/or dense thickets or stands of woods, shall collectively comprise at least 50% of the property.
C. The animals hunted must be part of a breeding herd that is a resident on the hunted property.
D. The operators of the preserve must provide freely available and ample amounts of cover, food and water at all times.
E. Animals that are to be hunted must exhibit their natural flight/survival instincts.
F. No zoo animals, exhibited animals or tame animals are to be hunted.
G. Hunting methods employed cannot include driving, herding or chasing animals to awaiting hunters.

(v) in the case of an animal taken on a hunting estate in North America, the conditions in (iv), above, apply and in addition:

A. Every effort must be made to utilize all meat commonly consumed from a taken animal;
B. There may not be any hunting or selling of hunting rights to a specified animal. (SCI 2017: 37–38)

It is relevant to notice about articles 22.1.1 and 27.5.2 that the passages about complying with game laws and not shooting from an aircraft or using an aircraft for hazing are the only parts (excluding iv and v, dealing specifically with hunting estates) that are concrete and verifiable. B&C's fair chase affidavit, in comparison, has a long list of specific conditions under which animals killed will not be accepted for trophy records, including not accepting anything that was confined by artificial barriers at all. B&C refers to this as canned shoots (B&C 2013b). Some SCI records come from what they call estate hunting, and some award categories in their World Hunting Awards Program (*Introduced Animals of North America* and *Introduced Animals of Africa*) are estate hunting, as the species are unrepresented in the wild. SCI announced in February 2018, as an exception to their support for estate hunting, that lions bred and hunted in captivity (a greatly criticized industry in South Africa) will no longer be accepted in *SCI Records*, and that SCI will not support or promote the captive breeding of lions (SCI 2018b). As uncovered by the Humane Society, however, there was still widespread promotion and selling of captive lion hunts during both the 2019 and 2020 SCI conventions (Associated Press 2019; Coleman 2020) (lion breeding and canned lion hunting will be covered further in Chapter 6.4).

SCI is of course interesting because of its size and influence, but what more than anything else sets SCI apart from the other organisations discussed in this chapter is the elaborate and extensive SCI *World Hunting Awards Program*. This cornucopia of awards honours the hunters who can document the harvesting of species in many categories from all over the world, and who can demonstrate various different ways of harvesting them.

The *World Hunting Award Field Journal* summarizes the awards categories as follows:

SCI Inner Circles: This award program offers a variety of leveled (sic) awards for any type of big game hunter. Inner Circle awards contain 5 different levels per award: Copper, Bronze, Silver, Gold, and Diamond levels.

SCI Milestone Awards®: Milestone Awards do not have levels. There is only one way to achieve a Milestone Award and that is to obtain all the required animals for that award.

SCI Continental Awards: This award program was designed to challenge those hunters who want to travel to all the different continents and hunt.

SCI Pinnacle of Achievement Awards, Zenith Award and Crowning Achievement Award: This award program takes into account how many Inner Circles and Milestone Awards you have achieved. The more Inner Circles and Milestone Awards you achieve the higher you progress in these awards.

SCI World Hunting Award Ring: The World Hunting Award Ring has been achieved by an exclusive group of SCI members who have excelled in the World Hunting Awards program. If a member reaches the World Hunting Award ring they are presented a complimentary custom-crafted ring made of 14-karat white or yellow gold, onyx and 21 point diamonds with six stones.

SCI World Conservation and Hunting Award: This Award is the highest achievement an SCI member can receive in the World Hunting Awards Program. Less than 60 SCI members to date have received this prestigious award (SCI 2018c: 2).

The continental awards are the following: *African 15* (38), *European 12* (55), *South America 8* (25), *South Pacific 8* (19), *North American 12* (44), and *Asia 8* (114). The first numbers indicate the minimum number of species that a hunter needs to have harvested from a given area to receive the award and the numbers in parentheses indicate the total numbers of species. To receive the *African 15* continental award, for example, a hunter needs to check the boxes and provide trophy scores for at least 15 of the 38 African species of game. Some of the continental awards have lower requirements for bowhunters. The *European 12* award, for example, requires 12 of the 55 species if taken with a rifle but only 6 of the 55 if taken with a bow (SCI 2018c: 3–6).

The SCI Milestone awards are more diverse in their requirements, some of them also with lower requirements for bowhunters. These awards are *Alternative Methods 24*,[25] which requires the harvesting of 8 of any species with muzzleloader,[26] 8 with

---

[25] This award does not count towards most of the cumulative awards, discussed shortly.

[26] Muzzle-loading firearms require the user to load the propellant powder and the bullet down the bore of the barrel (i.e. from the muzzle, or open end of the barrel) rather than using a brass cartridge, which packages propellant and bullet together. To ignite the propellant requires a primer that ignites

3.4 Trophies and Record Books

handgun, and 8 with bow/crossbow; *African 29* (38); *Dangerous Game of Africa*, which requires the harvesting of 5 of these 7: African lion, African leopard, African elephant, African rhinoceros, African buffalo, Hippopotamus, and Nile Crocodile; *North American 29* (40); *White-tailed Deer of the World*, requiring 5 of 19 kinds of white-tailed deer; *African Big Five*, requiring all 5 of African lion, African leopard, African elephant, African rhinoceros, and African buffalo; *North American Caribou*, 5 of 6; *Elk of North America*, 3 of 3; *North American Deer*, 4 of 8; *North American Wild Sheep*, 4 of 5; *Wild Turkey Milestone*, 5 of 8; *European Deer*, 9 of 21; *Moose of the World*, 4 of 9; *Indigenous Animals of South America*, 4 of 11; *Bears of the World*, 5 of 13; *Cats of the World*, 4 of 9; and *European 25 Milestone* (29) (SCI 2018c: 7–11). Some of the SCI Milestone awards were previously called "Grand Slams" and are among many trophy hunters still referred to either as such or as "Super Slams" (SCI 2015). Among these, the *North American 29*, known as the Super Slam of North American Big Game, and the *North American Wild Sheep*, known as the Grand Slam of North American Wild Sheep, are widely considered some of the most prestigious milestone awards.

The Inner Circle Awards are like the milestone awards, except for their tiers (copper, bronze, silver, gold, and diamond) and generally longer lists of accepted species. These awards are *Antlered Game of the Americas, Antlered Game of the World, Animals of Africa, Animals of Africa (Bow), Top Ten Award, Hunting Achievement Award, Global Hunting Award, Ringed-Horn Antelopes of Africa, Ringed-Horn Antelopes of Africa (Bow), Animals of North America, Animals of North America (bow), Mountain Game of the World, Multiple Methods, Animals of Europe, Introduced Animals of North America, Wild Oxen of the World, Desert Game of the World, Wild Sheep of the World, Pygmy Antelopes of Africa, Spiral-horned Antelopes of Africa, Animals of the South Pacific, Red Deer/Wapiti of the World, Chamois of the World, Wild Pigs and Peccaries of the World, Animals of Asia, Gazelles of the World, Ibex of the World, Predators of the World, Introduced Animals of Africa, Animals of South America*, and *Wild Goats of the World* (SCI 2018c: 12–33).

Here we start to see some of the awards for having accumulated other awards or record book entries. A diamond *Hunting Achievement Award*, for example, requires having the trophies of 125 different species accepted in the record book, and a diamond *Top Ten Award* requires 15 or more top-ten record book trophies.

The *SCI Pinnacle of Achievement Award, SCI Zenith Award*, and *SCI Crowning Achievement Award* are where the accumulation of awards really escalates. The *SCI Pinnacle of Achievement* award has four pinnacles, the 4th of which requires five gold and one diamond Inner Circle as well as three Milestone awards; the *SCI Zenith Award* requires the 4th pinnacle of the *SCI Pinnacle of Achievement*, plus six milestone awards, and ten inner circle awards (two silver, six gold, two diamond);

---

when the firing pin hits it. This then causes the propellent to ignite, essentially exploding in a controlled manner and propelling the bullet down the barrel. The primer is built into the base of a cartridge, but it must be loaded into the breach of a muzzleloader. A modern muzzleloader is very different from an antique musket and is more than capable of the power and accuracy required to kill deer cleanly. The downside against a rifle using cartridges is that a second, follow-up, shot is greatly delayed because of the loading process.

and the *SCI Crowning Achievement* requires the 4th pinnacle of the *SCI Pinnacle of Achievement*, plus nine milestone awards, the *Hunting Achievement Award* on Silver or higher or the *Animals of North America* on gold or higher, the *Animals of Africa* on silver or higher or *Animals of North America* on gold or higher, and 13 additional inner circle awards on copper or higher (SCI 2018c: 39).

The last two noteworthy world hunting awards are the highly cumulative *World Hunting Award Ring* and the *World Conservation and Hunting Award*. The *World Hunting Award Ring*—a "custom-crafted ring made of 14-karat white or yellow gold, onyx and 21-point diamonds with six stones" (SCI 2019)—was introduced in 1995 and was, for about ten years, the ultimate world hunting award, received by only 95 people. It requires that SCI members "have achieved and purchased 11 SCI Milestones, including the North American 29 and the Africa 29, 17 SCI Inner Circles at the Diamond level and the Fourth Pinnacle of Achievement and the Crowning Achievement" (SCI 2019).

In 2005, however, an even greater award was introduced:

> To achieve the World Conservation and Hunting Award, members must achieve and purchase all 15 Milestones, the diamond level of 25 of the 27 Inner Circles,[27] the fourth Pinnacle of Achievement, Zenith and the Crowning Achievement. (SCI 2019)

Aside from the world hunting awards described over the last few pages, SCI also has an assortment of lifetime achievement awards. These are the *C.J. McElroy Award*, the *Diana Award*, the *Hall of Fame Award*, the *International Hunting Award*, the *Pathfinder Award*, the *Professional Hunter Award*, and the *SCI & Cabela's Young Hunter Award*. These are awarded for various levels of hunting excellence combined with contributions to conservation, hunters' interests globally, and the SCI organization, or to exceptional female hunters (*Diana*), professional hunters, 15–21 year old hunters (*Young Hunter*), or to hunters who have had to overcome physical challenges or disabilities (*Pathfinder*) (SCI 2019).

These award systems have always put SCI under attack, not only from antihunters but also from hunters with a different hunting attitude. David Petersen, for example, wrote in his book *Heartsblood* that

> [f]or those eager trophy "collectors" unwilling to abide by even the foundation-level ethics of B&C and P&Y, there's always Safari Club International (SCI). Despite its much-touted "global wildlife conservation agenda," I see a whole lot bad in SCI and damned little good. In fact, Safari Club International and its members have the onerous distinction of embodying most of the attitudes that antihunters love to hate. When I hear someone attempt to defend SCI, I know I'm hearing a fool. (Petersen 2010: 48–49)

At the very least, Petersen is right about SCI representing everything that antihunters love to hate. The system of cumulative awards and a history of high-ranking members of SCI breaking laws and behaving unethically make the antihunting project easy.

---

[27] There are in fact 17 Milestone awards, but the *Alternative Methods 24* does not count towards this award, and the *European 25* Milestone is newly added. There are also 31 Inner Circle awards and not 27, so these descriptions of award requirements are not fully updated.

3.4 Trophies and Record Books

SCI's founder and first president, C.J. McElroy, who has hundreds of trophies in *SCI Records*, was among the first of such cases. He was forced to resign in 1988, according to prominent member and later President William (Bill) Quimby, because of persistent rumours that McElroy ignored hunting laws and behaved highly unethically. Peter Laufer tells the story of Sonny Milstead, former president of SCI's Arkansas/Louisiana/Texas chapter, who was convicted for illegally shooting a Bengal tiger, resting under a tree on a hunting ranch in Texas (after having just legally shot a resting lion) (Laufer 2012: 49–51).

Famous speechwriter Matthew Scully reports on a story of then-president of SCI Skip Donau and two SCI colleagues in Mozambique in 1998. They shot five elephants there at a time when elephant hunting and the export of ivory had been illegal in Mozambique since 1990. A further three lion, five buffalo, one eland, three kudu and six impala were shot, and several animals had been found dying of gunshot wounds in the area following the hunt (Scully 2002: 69–71).

When Ted Kerasote did his research for *Bloodties*, he travelled to Siberia in 1989 on a two-week hunting trip with prominent SCI hunters Robert Kubick, Ali Üstay, Vance Corrigan, Paul Asper, and Donald Cox. His book provides some of the best anecdotal evidence of the trophy fervour of high-ranking SCI members. Robert Kubick, according to hunting guides, was an overly competitive killer who got his trophies at all costs, except when he had to leave behind a Jentink's duiker in Africa, because it is endangered and illegal to import into the US (Kerasote 1994: 101, 171); Donald Cox shot his trophy sheep in Siberia after being dropped off near the sheep and having the helicopter haze it back towards him, while a guide got out of the helicopter to shoot at the sheep with a submachine gun (Kerasote 1994: 130). Paul Asper was sentenced to two and a half years in jail and almost US$200,000 in fines for—among other felonies—violating the Endangered Species Act by killing and illegally smuggling eight endangered species of animals into the US. He had also attempted to smuggle the skin of a Nile crocodile, a leopard inside the skin of an elephant, and a bald eagle inside a mountain lion hide into the US. He told his taxidermist to list the eagle as a goose (Kerasote 1994: 169–170).

One month after shooting his Siberian sheep with the aid of a helicopter and secondary machine gunner, Donald Cox was honoured with the Weatherby Award. The Weatherby Award is one of the most prestigious awards in hunting, given to those who—apart from having obtained trophies from almost everything there is to hunt in the world—have shown "dedication to the principles of fair chase and ethical hunting" (Weatherby Foundation 2019).

David Petersen references investigative journalist, hunter, and writer for *Audubon* Ted Williams about SCI's attitude towards the ESA:

> Safari Club International … is a tireless opponent of the Endangered Species Act [ESA]… At the 1996 Outdoor Writers [Association of America] Conference in Duluth, I sat with several hundred of my fellow scribes while then-president of SCI, John Jackson, droned on about how awful was the ESA and how it inconvenienced him and his fellow members in their quest to acquire the biggest horns. It was excruciatingly embarrassing for every genuine conservationist in the room. (Ted Williams, quoted in Petersen 2010: 49, brackets in original)

WildCRU's Andrew Loveridge and Zeke Davidson went to the SCI convention in Reno, Nevada in 2005. They were there to present their findings that excessive and poorly managed trophy hunting near Hwange was causing a worrying decline in the lion population, and that temporarily halting hunting efforts could allow the population to recover. Loveridge comments on the reaction to their research here:

> At the end of the meeting, one of the hunters summed up the mood with a comment I will always remember. He said, "Well, maybe hunting does have an effect on lion populations, but we can't let the anti-hunters see any of these results." We were left with the sense that they thought we were pretty cheeky coming to the SCI Convention, the bastion of all things hunting, to criticize how hunting was managed. Our impression was that inconvenient scientific data was not going to get in the way of business. And business was what they were there for. (Loveridge 2018: 96–97)

He continues later:

> There were other conservationists at the convention, mostly biologists funded by SCI, and several were attending a meeting on lion conservation organized by SCI. We were conspicuously not invited to attend. [...] Our perspective on the management of lion conservation aimed to be entirely objective; we were collecting data to support wildlife-management decision-making. But the fact that our research had precipitated a hunting ban, even a temporary one, put us very much in the Anti camp. (Loveridge 2018: 99)

One statement from SCI explains the "first for hunter" experiences of Loveridge and Davidson at the SCI convention quite well; on February 4, 2020, SCI posted on their official YouTube channel a commercial called "Triggered," featuring video clips of angry antihunting demonstrators. The video description reads as follows:

> Nobody's more pro-hunting than SCI. No wonder the antis hate us so much. From opening more public lands to hunters and protecting hunters' rights to fighting anti-hunting legislation, nobody does more for hunters. Period. We're the radical pro-hunting solution to the radical anti-hunting problem. No matter what, where or how you hunt, we defend and promote your right to do it. Triggering anti-hunters is just icing on the cake. Join the fight to save hunting. Join SCI. (SCI 2020)

The implications of the sentence "No matter what, where or how you hunt, we defend and promote your right to do it" are clearly cause to question the conservation motivation as well as the ethical and legal position of SCI with respect to hunting. Would they defend your right to kill a gorilla with a harpoon gun in a zoo? "No matter," it seems they would. Of course, we aren't suggesting that SCI actually mean what they have said, but the blanket wording is interesting. We would suggest that such a "no matter what" statement is reflective of a highly entrenched and defensive position—or, as the description says, a "radical" solution.

3.4.6 Erongo Verzeichnis for African Game Animals

The "Erongo Verzeichnis for African Game Animals" (EVAGA) and its Age-Related Measuring System (ART) are a limited record book and trophy scoring system for

African game animals that have been under development since 2010 by Erongo Verzeichnis (EV), headed by Namibian professional hunter Kai-Uwe Denker.[28] In 2019, the Namibia Professional Hunting Association and EV agreed that the ART should be implemented in Namibia. This means that the top ten of trophies obtained by the Hunting Association's members, which has hitherto been scored with the SCI method, will now be scored with the ART method, developed by the EV working group.

Erongo Verzeichnis was also the name of a magazine that EV published between 2010 and 2018 to gather materials for EVAGA and ART. The EVAGA record book was published in German in January of 2020 and an English version is expected.

The physical book itself is not a record book in the traditional sense. There will be some recordings of trophies, but the intention was to make a record book that is more about wildlife, ecology, habitat information, and articles than just trophy records.

The ART—inspired by the acknowledgement that harvesting males that are in their prime or immature might remove strong genes from populations too early—aims to divide animals into age classes of "immature," "prime," and "past prime." The size of the trophy would still be an important factor, and trophies are scored in a way that is similar to that of the other scoring systems. However, the measuring points of "past prime" trophy scores are multiplied by 1.12, so older animals with similar quality of genetics as animals in their prime would rank higher in the records. "Immature" trophies are disqualified from entry altogether. The system intends to exclude penalties for flaws like broken or highly worn tips of horns and tusks, as such imperfections mostly penalize past-prime animals.

Since 2015, Erongo has also been offering certifications to wilderness hunting areas if they live up to the Erongo Recordbook principles, which are an integral part of the EVAGA. This includes (1) that game animals live within the natural historical habitat of their species; (2) that they are able to evade capture (only unfenced areas and fenced areas larger than 1000 km^2 are eligible); (3) that at least one of the species of lion, leopard, cheetah, African wild dog, and spotted hyena lives in the area and regulates game populations; (4) that only fully mature animals are harvested, and that animals past their prime are valued higher. Twenty hunting areas in Namibia, one in Tanzania, and one in Mozambique have at the time of this writing received certifications for this system.

3.5 A Critical Look at Trophy Records

There is genuine concern among hunters that emphasis upon trophy heads is seriously misguided. Clubs such as Safari Club International and record-keeping organizations such as Boone and Crockett, and Pope and Young, though initially conceived to enhance appreciation for the game animal itself, now tend to emphasize hunter success. The hunter's name

---

[28] Chapter 3.4.6 is largely based on personal communication with Kai-Uwe Denker.

goes into the record book and the success of his hunt, based upon the size of the game taken, is celebrated. (Vitali 1990: 70)

Having covered the major trophy scoring organisations and record keeping systems, it is time to take a critical look at their purposes and effects.

Are the purposes of record keeping systems to incentivise the selective and sustainable harvest of old males that have long since passed on their good genes, to honour exemplary specimens of these big game species, and to inform science and wildlife management efforts by facilitating information about genetics, habitats, ethology, and ecology? Or is their real, but unstated, purpose to provide venues for competition and bragging rights, in which hunters can show off their personal exploits and be rewarded and admired for the quality and quantity of their kills? And, regardless of purpose, what kinds of behaviour do trophy records encourage in hunters?

Jim Posewitz formulated the ethical problem with trophy record keeping very accurately here:

> Hunting is not a contest between humans. Trophy scoring and big game contests come perilously close to, and sometimes cross, the line of proper ethical practice. In other words, trying to take a trophy to get your name in a record book is taking a fine animal for the wrong reason. Contests between hunters that require killing animals should be prohibited. Trying to kill the "big buck" to win a contest or a monetary prize also represents pursuing and killing wildlife for the wrong reasons.
>
> The idea of hunting trophy animals and the preservation of wildlife have a deep and common root in our history. One of our nation's oldest and proudest conservation organizations has done both well for over a century. There is value in keeping trophy records, and they should include information about the animal, the land that produced it, and the wild nature of the habitat that sustained it. Over time, valuable information about a species and the land is collected and preserved. Displaying the name of the hunter, however, may no longer be necessary. (Posewitz 1994: 97–99)

The last paragraph—being mostly a reference to B&C—gives us a place to begin this discussion.

3.5.1 Do Record Books Honour Animals or Hunters?

In the foreword to the 4th edition of B&C's *How to Score North American Big Game*, Frederick King writes that.

> [t]he Club will keep the records of native North American Big Game for the purpose of providing a vital conservation record of excellence in specimen quality and linking it with both excellence in habitat management and wildlife management practices that produce such specimens and excellence in hunting skills demonstrated by the taking of such specimens in a manner that fully complies with the Club's Fair Chase Code. (King 2016: viii)

The "excellence in specimen quality" part is indeed true. As Wootters said in 1977, only one out of about a million whitetail deer racks would score high enough to be a B&C trophy. The idea is that only a truly exceptional animal gets to appear in

3.5 A Critical Look at Trophy Records

the book, and only a truly exceptional hunter gets to put his name next to it. Prentiss Gray, author of the 1932 first edition of *B&C Records*, stated herein that

> developing a measuring system to pay tribute to outstanding big game trophies and natural history was a way to preserve our natural heritage before the hunter or trophy owner passed away and their relatives discarded or gave away the prized trophies. (Gray in Nesbitt and Reneau 2016: 4)

The stated purpose of *RW Records*, similarly, is to honour the animal. This is the reason for accepting picked up trophies in the records. However, we see at least three viable arguments why record books serve primarily to honour the hunters and not the animals they hunt.

The first argument has to do with anonymity; a hunter gets to put their *name* next to an animal that is represented only by a number. We do not often honour our fellow human beings anonymously, or at least not individuals. Humans have names and are recognized as individuals, and when we honour people, we do so by name. We may honour groups of people anonymously, but this is usually out of convenience. We may for instance honour our fire fighters, military, nurses, or teachers as groups, but if naming them individually was feasible, we would do that instead. We rarely honour individuals anonymously. We mostly do so if we do not know the identity of the individual or if the individual has requested to remain anonymous. Anonymous donors may be thanked and honoured for their donation to something, or we may honour someone who has performed a heroic act and then disappeared. Yet we would first try to uncover the unknown individual's identity, so we could honour them by name. This all points to the same conclusion, which is that we honour individuals by name whenever possible. By being named next to a record, hunters are clearly being honoured, animals—represented only by a number—are not.

Yet what are the alternatives to current practices in trophy record keeping? Wild animals do not have names, so to honour hunter and prey equally our only options would be to honour hunters anonymously, or with an identifier corresponding to the identifier of the animal. Since we honour trophy animals by the size of their skulls, antlers, or horns, should we perhaps honour hunters by their height? Their IQ? Their annual income? Hunters would be unlikely to cherish the idea of their identity being reduced to a single and arbitrary score. People want to be honoured, with a unique identifier; they want to be named and to be "in the book."

The alternative is the anonymous record book, but that would just be a collection of long lists of dimensions of trophies, a location where each trophy was obtained and perhaps some other information such as hunting method and date. Granted, hunters could still use this as a guide to see where exceptional trophy animals are harvested, but would many hunters buy such a record book or care about having their trophy accepted in it? There seems to be little in it for the hunter if the trophy book is anonymous.

The second argument that record books appear to not honour animals to the same extent as hunters is that record books only honour dead animals. When a hunter's trophy is accepted into a record book, the hunter is generally still alive. For old entries in a record book, the hunter may be alive or not, but the animal, necessarily, is not.[29] When we honour a human, it is often to show gratitude to that person, and a person being honoured is likely to be happy and proud of the recognition. If we honour someone posthumously, as we often do with famous people shortly after they pass away, then the family and friends of that person are likely to feel happy and proud. The rest of us tend to enjoy the trip down memory lane that usually follows. We can use such honouring therapeutically, if we feel sad about the person's passing. When a famous singer or movie star dies, we often hear their songs or watch their films relentlessly for some time afterwards. Someone usually benefits, in other words, when we honour people posthumously. When we honour a trophy animal in a record book, on the other hand, who stands to gain from this honour other than the hunter who killed it? The animal itself is not around to feel proud of its accomplishments, and its relatives in the wild are not likely to be informed of the score that the trophy obtained.

This of course brings us neatly to the third argument, which is that the whole idea of being honoured is entirely nonsensical to wildlife to begin with. This alone may make the first two arguments redundant. Hunters in a record book know they are being honoured and why; the animals, even if they were not dead, have no idea. Nor do animals have any concept of the human value of honour. Humans alone have decided what is worth honouring about animals. The honouring of dead trophy animals would be arrogantly anthropomorphic, even if it were true that record books are about honouring animals and not hunters. Hunting trophies are kept as souvenirs and certificates of accomplishments for the benefit of hunters. If the Predator film series were authentic, a series where aliens hunt humans and other species for trophies, would the humans killed, skinned, and prepared for skull mounts feel honoured by this process? The idea seems foolish.

With SCI's *World Hunting Awards*, we rid ourselves of all pretence of honouring animals. Each trophy animal is now nothing more than a step on the way to an award, which is a step on the way to another award, all in honour of the hunter. Each animal need not even be exceptional or even necessarily a wild animal. It just needs to be of a specific species, and maybe die in a specific place or to a specific weapon, allowing the hunter to check a box on a piece of paper.

For SCI, we also find strong indications in *SCI Records* that they are purposely catering to the ego of hunters. In the description of the European red deer in SCI's online record system, for example, they write:

> For the most memorable experience, the sportsman should probably make his hunt in the mountains of eastern Europe during the roar. If all goes well, he could find himself surrounded by roaring stags that are some of the finest in the world, and see antlers that are unaffordable for most sportsmen.

The phrasing "antlers that are unaffordable for most sportsmen" plays into the common criticism of trophy hunting; that it is just a spending competition. SCI does not appeal here to the hunters who want to spend days testing their skills and patience

---

[29] Rare cases with for example darted rhinos excluded.

against the finely-honed instincts of old and exemplary specimens of animals. They appeal to shooters who want to be surrounded by trophies that their peers could not afford. Ted Kerasote, on that note, quotes trophy hunter Vance Corrigan in *Bloodties*: "Today getting the biggest Grand Slam is nothing more than an award for who's spent the most money" (Kerasote 1994: 135).

3.5.2 The Consequences of Record Books

To identify further examples of the detrimental effects of record books, we need only browse B&C's General Policies, as listed on the B&C website. Many of the policies are direct responses to trends in hunters' behaviour (which could fairly be called cheating) and thus indicative of the unintended consequences of *B&C Records*. The "altered trophies" policy quoted here, for example, is in response to attempts to modify hunting trophies:

> Trophies that have been tampered with to gain an advantage obviously are not eligible for entry into the records books. Examples of trophy tampering include the deliberate removal of abnormal points from typical racks to increase a trophy's score and/or the addition of antler or horn material. If any points are deliberately removed from antlers of any trophy by breaking or sawing them off, that trophy is not eligible for entry in the Awards Programs and/or records books. Deliberate modification of horns by adding to horn length or artificially increasing circumference measurements is also grounds for disqualification of a trophy. (B&C 2017)

A related policy addresses antlers and horns that are altogether fake. If there is any doubt, then a small sample of the horn is collected to verify that it is biological material.

A notable case that illustrates these problems was reported in 2014 in the magazine *North American Whitetail* and other hunting periodicals. The story was about a renowned taxidermist and trophy hunter who was very successful in collecting P&Y and B&C trophy deer (seemingly as many as 24 in *P&Y Records*) with a bow. He used the fame that comes with such success to market his blog and merchandise and gain sponsorship and commissions as a writer. One day, someone pointed out (with photographic evidence) that the antlers on one of his trophy deer (as seen in a trophy photo, where he stands above a dead deer) were identical to another set of antlers, which was traded in 2008. The common consensus in the community, which subsequently cast doubt on all his trophies, was that he (being a skilled taxidermist) had removed the skull cap and antlers from the deer that he had just killed and replaced them with the trophy sized antlers in question. These antlers had never been scored, so all he needed to do was pretend they came from this newly killed buck and have them measured. He left the hunting community and abandoned his social media presence following the controversy (Hart 2014).

Two other B&C policies concern the buying and selling of trophies. One pertains to trophy dealers, who may want trophies measured to increase the price that they could sell them for, and the other to internet sales:

The information used to advertise big game trophies for sale on internet sites is frequently misleading and fraudulent. Unofficial B&C scores, claims of "new B&C World's Records," and unauthorized reproductions of the Club's copyrighted score charts are commonly used. Such blatant statements as, "You can have your own B&C trophy in the records book if you buy this one because it hasn't been accepted by B&C" have been seen. (B&C 2017)

Other policies are that B&C Measurers should work purely pro bono, as otherwise: "Measurers could feel obligated to "find" an inch or so for trophies scoring near the minimums if paid by trophy owners for their services" (B&C 2017) and that official B&C Measurers should ask hunters if their trophy has already been scored, to prevent "score shopping," i.e. getting the same trophy measured by multiple measurers in an attempt to get a higher score (B&C 2017).

While the target audience for admonitions such as these are clearly the bad apples among hunters and not representative of all hunters who seek entry in *B&C Records*, it is still fair to interpret these policies as a recognition of what some hunters will do to get their name in B&C Records. After all, there must have been enough experiences with bad apples to warrant putting the policies in writing.

Gerhard Damm's recognition of the CIC scoring system's negative consequences is even more direct:

> In Europe, the CIC formula system is occasionally being misused for individual glorification and an unhealthy competition between European hunting countries to "produce" the largest antlers.[30] Moreover, the CIC scoring system has distinct flaws, like beauty and penalty points, which may be construed as being given or subtracted on an understandably subjective basis.
>
> A consequence of this development was that the tape measure and "shopping lists" with animals and specific trophy sizes unfortunately became part of many hunts. A good number of hunting managers, professional hunters and landowners succumbed to the "record temptation." Ultimately, the trend encouraged genetic manipulation of game animals, canned shooting, high-fenced killing grounds supplied by "breeder facilities" and abetted the killing of immature animals and of those which are essential for breeding. (Damm 2008: 7)

Damm further criticizes the argument that record books contribute valuable biological data to science. He points out that the formulas give weight to anthropomorphic factors like beauty and colour (in CIC's case), but that species-specific biological information is generally missing, that geographical data is not specific enough, and that important biological data about the animal's body and its behaviour/activity as it was shot, is missing (Damm 2008: 8).

Some examples of questionable and illegal behaviour of hunters in the upper echelons of SCI have already been provided. Kerasote's *Bloodties* offers further evidence of what hunters will do to get their name in *SCI Records* or—especially— complete some of the *World Hunting Awards*. Trophy hunter Robert Kubick said the following in conversation with Kerasote:

> One of the bad things about trophy hunting and record books is a person comes all this way and they don't get something big enough to qualify and then they're unhappy. [...] It's

---

[30] "Producing" trophy animals and "genetic manipulation" refers to the practice of artificially breeding animals specifically to produce the largest possible antlers and then provide food that further contributes to excessive antler growth. This practice is a major ethical issue and largely a consequences of trophy record books. We will return to this discussion in Chapters 5.4 and 6.4.

3.5 A Critical Look at Trophy Records

perfect until he takes out the tape and measures the horns. It doesn't score high enough, it's three inches too small, and he doesn't want it. (Kerasote 1994: 126)

Fritz Heje Hansen adds that in his many years of experience measuring trophies for Danish hunters for CIC, it is often bragging rights that drive the desire to have trophies measured. Trophy scores that fall just short of the medal range are usually accepted with expressions of disappointment.

Trophy animals being too small can lead to unethical behaviour by hunting guides as well as hunters. Notably so if guides expect hunters to have the record book mindset. Here Kerasote visits the home of trophy hunter Paul Asper:

As we walk into the main hall I pause again in front of the photo of him with the Dall sheep. From behind me he says, "When I walked up to that ram my guide told me, 'We'll bury it.' He thought I was hunting for the biggest one on the mountain and he was disappointed at its size. But this was a wonderful sheep! Thirty-eight inches. 'Hell no,' I told him. 'I'm keeping it.'" He wags his head. "That's what the book does to some people". (Kerasote 1994: 153)

Kerasote also quotes hunting legend and Weatherby award winner Jack O'Connor as saying that the creation of the Grand Slams of hunting was a mistake:

"All this whoopla is a very sad thing," he said. "I wish Grancel Fitz[31] hadn't started it all. The old-timers hunted sheep because they loved sheep, because they loved to be up on those high windswept ridges where they shared the sheep pastures with the sheep, the grizzly, the hoary marmot, the soaring eagle. When they brought back a ram trophy, they were not seeking honor and prestige—they were bringing back memories of icy winds fragrant with fir and balsam, of the smell of sheep beds and arctic willow, of tiny, perfect alpine flowers, gray slide rock, velvet sheep pastures. The old-timers had sheep and sheep country in their blood."

Not so with today's hunters, O'Connor went on to write. "They're after glory ... and the sooner they can get the tiresome business over with and slap those ram heads on a wall, the better they like it." One man collected his Grand Slam in thirty days; one had collected ten Grand Slams. Some men were so desperate to accomplish the feat that their collection had "not a single head that was a trophy—all little rams from five to seven years old that when they were knocked off had useful lives before them." Worse. O'Connor alleged that the desire to complete a Grand Slam had created a "tremendous amount of lying, poaching, and cheating". (Kerasote 1994: 96)

Additional indications of what hunters will do for record book rankings, or just more impressive trophies in general, come from taxidermists. The North American National Taxidermists Association has a code of ethics, which in the version on their newest website (www.nationaltaxidermists.com) is rather vague and inspirational, but their previous 11-point code of ethics can still be found on their old website (www.taxidermy.net). Points 6, 7, and 8 about hunting trophies give a bit of insight into requests that taxidermists must be getting from unethical trophy hunters:

---

[31] Grancel Fitz was one of the most important figures in the development of B&C's scoring system. He coined the "Grand Slam" name in a 1948 article about killing all four varieties of North American sheep. The original and biggest Grand Slam is the Grand Slam of North American Sheep from 1977. Grand Slams are pillars of SCI's World Hunting Awards program.

6. I will make every effort to fully acquaint my clients with the true potential of their trophies to avoid possible disappointment with the finished product.

7. I will refuse to alter or falsify trophy characteristics. I will replace broken portions of a trophy only at the express direction of the client.

8. I will not accept as work any trophy or material that appears to be in violation of legal statutes or commonly defined rules of sportsmanship. (Taxidermy.net 2000)

Two other serious problems with record books is that they can lead to trophy poaching and to the artificial breeding of freakshow trophy bucks. See Chapter 5.7 for a discussion of trophy poaching and Chapters 5.4.1 and 6.4 for discussions of breeding for trophies.

Is it then, as Posewitz suggests, time to squash the greatest of evidence that record books honour hunters more than their prey and perhaps put an end to the awful behaviour of some record book trophy hunters by striking hunters' names from the books? With anonymous record books there can be no doubt that the purpose of the record list is to honour the exceptional animals and keep a useful scientific database. However, as David Petersen argues, if the established record books scrubbed hunter names from their books, then new ones with even lower ethical standards would probably appear instead. The unethical hunters—who are the problem—would just use the new record books instead and continue behaving unethically (Petersen 2010: 52).

In lieu of this optimal solution, Erongo Verzeichnis' ART could be a small step in the right direction for trophy scoring. If ART becomes a success, then perhaps, in time, the established record books could adopt similar approaches and nudge trophy hunters towards an interest in trophy age over trophy size. While this would not solve all problems with unethical behaviour encouraged by record books, it might address concerns about trophy hunting's potential for negative selection for genetics, as discussed in Chapter 5.4.

3.6 The Meanings of Hunting Trophies

How hunting trophies are perceived and interpreted, and why they themselves as objects are controversial, are important issues to consider in terms of how trophy hunting is framed and discussed. An important source for the next section is Rachel Poliquin's excellent chapter about hunting trophies in her book, *The Breathless Zoo—Taxidermy and the Cultures of Longing*.

3.6.1 Dead Nature

There was a British comedy show about ten years ago called *That Mitchell and Webb Look*, starring comedians David Mitchell and Robert Webb. In one of their sketches they are dressed up as two Nazi soldiers on a battlefield; Webb's character is talking about how the Communists will soon be beaten, but Mitchell's character is distractedly thinking about something else. He finally asks Webb if he has ever noticed the badges on their caps: "They've got skulls on 'em…" he says, and proceeds to ask: "Hans… Are we the baddies?" They sit down and talk about the significance of the skulls on their caps and how a skull—apart from maybe a rat's anus—is probably the worst symbol they could have chosen. Skulls only make you think of things like death, cannibals, beheading, and pirates. The sketch does a good job of illuminating the negative connotations of skulls, and why someone represented by skulls must be the bad guys.

Trophy hunters are in a similar situation. The trophy hunter is inextricably connected to hunting, the killing of an animal, and to the hunting trophy. Without it they would not be a trophy hunter. If we had to come up with a symbol to represent the trophy hunter, it would have to be the hunting trophy. How can those whose symbol is a skull, who produce and keep skulls as souvenirs, and who hang skulls on their walls as decorations possibly not be "the baddies"?

Granted, hunting trophies are not always skulls. Only the half-skull or full-skull mount are technically and visibly skulls. Taxidermy used to be a skull with skin on it but is today usually the animal's skin shaped around an artificial mannikin. Still, a decapitated head, the classic "head on a wall" that adorns many a Scottish highland hotel lobby and African safari lodge bar, is symbolically not much better than a skull.

The question that we need to ask is the one that Poliquin asks here:

> Abhorrence is understandable in vegetarians, animal rights activists, and any community that believes that there is no excuse for killing an animal. But for meat eaters and anyone who owns leather shoes, bags, or furniture, the sharp negative reaction to taxidermy is less straightforward. What is the difference between a steak, a belt, and a trophy? In all cases an animal has been killed, dissected, and reshaped, and yet a distinction exists in contemporary opinion. What makes hunting trophies so particularly objectionable? (Poliquin 2012: 151)

We suggest a fourfold answer to Poliquin's question.

First, hunting trophies invoke our fear of death.

Thoreau wrote the following in an undated entry in one of his early journals:

> I hate museums; there is nothing so weighs upon my spirits. They are the catacombs of nature. One green bud of spring, one willow catkins, one faint trill from a migrating sparrow would set the world on its legs again. The life that is in a single green weed is of more worth than all this death. They are dead nature collected by dead men. I know not whether I muse most at the bodies stuffed with cotton and sawdust or those stuffed with bowels and fleshy fibre outside the cases. (Thoreau 1837–1846: 464)

He refers to museum taxidermy here, but of course hunting trophies are the same "dead nature," collected by eventually dead men. Like Thoreau, many people are uncomfortable around hunting trophies because they are a kind of *memento mori*. Memento mori in Latin means "remember (that you have) to die," and a memento mori, according to the Oxford dictionary, is a "warning or reminder of the inevitability of death, esp. a skull or other symbolic object."

The fear of death is pervasive in society. Jan E. Dizard provides telling examples here, and relate them to the reputation of hunters:

> Think of the peculiar status those who deal with death occupy in society. The coroner and medical examiner do not enjoy the highest status within the medical profession. Morticians occupy a curious standing in the community that is not much relieved by being renamed "funeral director." We have moved our slaughterhouses farther away from our centers of activity and further from our consciousness, and even within the "meat processing" industry, clear lines are drawn between those who do the actual killing and those who ready various animal parts for our consumption. Hunters enter a similarly charged symbolic arena in which their hands become bloodstained. No wonder hunters evoke decidedly mixed responses. (Dizard 2003: 34–35)

Even graveyards, with no dead people in sight, are places that many consider unnerving. The fear of graveyards is connected to irrational fears of dead people coming back to life, which has always fuelled horror stories. Even the words death and dead are unpleasant to some people, who might opt to use euphemisms like "passed away" or "passed on" about death or someone who died. With these words we signify, whether to comfort ourselves or because we really believe it, that the dead merely pass to some other mode of existence rather than no existence at all.

Despite these qualms, we enjoy fictional death. We like murder mystery books and horror movies. Shows like CSI, which focus on solving murder cases, are extremely popular, and action movies and videogames contain so much death that we are liable to become desensitized to each individual occurrence. The screen provides us with a comfortable distance from the death and dying depicted, and relegates it to the realm of fantasy. Being around actual death, as we are when we look upon hunting trophies, is different. A hunting trophy is evidence that real death has occurred, and to kill a large animal yourself—as a hunter has done to produce a trophy—is unthinkable to many in the West, who have nothing to do with the killing of the animals most of us consume. The hunting trophy is a very visceral and concrete reminder of death that we cannot just turn off with a remote control and dismiss as fiction.

Second, and closely related to hunting trophies invoking our fear of death, the hunting trophy goes against western cultural and religious traditions surrounding death.

The hunting trophy represents a way of displaying death that is not customarily used for humans in western cultures. If we were to save something of our dead grandmother to remember her by, we would save favourite possessions of hers, photos of her, or perhaps her ashes in an urn. We would never save her stuffed head or her skull and display it on our wall. Ashes are about the furthest we go in terms of saving physical parts of a person, and some people may even find human ashes disturbing. Religious and cultural traditions dictate that there are certain ways in which we can treat a dead body and certain ways we cannot. Decapitated heads fall squarely in the "cannot" category. Poliquin here relates our discomfort with trophies to our discomfort with heads:

> When most people think of taxidermy, they think of hunting trophies. In a sense, trophies epitomize everything that is disliked about taxidermy in general. All taxidermy makes death overt, but just heads are decidedly deader. The eerie sensation of immortally motionless liveliness becomes even more eerie with trophies. Any fantasy of reanimation can only be a nightmare of a disembodied head. (Poliquin 2012: 151)

3.6 The Meanings of Hunting Trophies

Lewis Aiken—in his famous book, *Dying, Death, and Bereavement* (2001)—connects our fear of death (thanatophobia) with the emergence of funerary rites and customs. Death, he writes, has always been an emotional experience that reminded everyone of their own mortality—something they could not even comprehend. Thanatophobia contributed to the emergence of funerary rites, which provided people with the illusion that death was to some extend under their control (Aiken 2001: 126). Phillip Walker notes that there are signs of early funeral rites already some 600,000 years ago, and that 50–100,000 years ago, these practices "had evolved into elaborate rituals that involved painting bodies with red ochre and including food or animal remains with the body as offerings" (Walker 2000: 4).

For more recent documentation we may look to ancient Egypt, where embalming was practiced in the belief that the soul would eventually return to the same body, or to ancient Greece where—at first—nothing was allowed to mar the dead body and—later—the body had to be burned to set the soul free (Pine 1981). Most civilizations have had customs surrounding the treatment of the dead, and they have often involved religious or spiritual ideas of an afterlife; either the body needed to be burned to free the soul to go elsewhere, or the body had to buried intact, so it could be used again. A mutilated body would also be a mutilated body in the afterlife, or it might be prevented from entering the afterlife altogether. Walker argues that the integrity of the corpse and its importance for the afterlife dominated medieval discussions of the body; an intact corpse went to heaven and a damaged corpse went to hell. This made anatomical studies of the dead troublesome, as resurrection was considered impossible if the integrity of the corpse had been compromised (Walker 2000: 4–5).

The strength of these religious views waned from the Renaissance onwards but are still important to many people, and we retain laws against the desecration of dead bodies. Poliquin relates this to our discomfort with hunting trophies:

> A degree of respect is always due the dead. Parading or desecrating dead bodies is as morally reprehensible as violating the living. That is, just because we may eat an animal, this does not mean that we want to see its death memorialized. (Poliquin 2012: 152)

Since hunting trophies are usually skulls or taxidermy shoulder mounts, they go against these ingrained traditions of either disintegrating the body entirely or preserving it intact. To create either of these trophies, the animal must be decapitated, and decapitation is a practice that has not traditionally played a role in western funerary customs. On the contrary, decapitation has often been reserved for defeated enemies to humiliate them or display dominance. More on that subject shortly.

Hunting trophies are also never "laid to rest," as we describe it when human remains are properly disposed of. A common belief in many cultures and religions is that it is sacrilegious to exhume the dead or in other ways disturb burial sites, because they are "resting places" for the dead. In archaeology, we see an example of this in what is called "the reburial issue." When archaeologists have excavated the bones of peoples such as the Indigenous peoples of America and Australia these have often been sent to museums around the world. Increasingly, in the last fifty or so years, the descendants of these peoples have demanded the return of their ancestors' remains in order to put them properly to rest (Goldstein and Kintigh 1990). Animals that become hunting trophies are never properly laid to rest. This may serve as another contributing factor to the sense of unease around them.

Third, hunting trophies symbolize dominance and conquest.

The reasons that we would not commemorate grandmother by mounting her head on the wall are not only that it would be a much too unnerving reminder of death and mortality (hers and our own) or that western common practice is to not preserve human beings in any form more physical than ashes. It is also that a decapitated head commonly symbolizes dominance and victory over the deceased.

A closer comparison than grandmother to the hunting trophy would perhaps be the deceased pet. We love our pets and sometimes even consider them family members, so we apply—albeit to a lesser extent—human funerary rites and customs to them. Our feelings about pet taxidermy resemble what our feelings would be about human taxidermy. Pets are sufficiently anthropomorphized and familiar to be treated almost as humans in death, or as Christina Colvin writes about the stuffed pet category of taxidermy:

> Because of our familiarity with these humanized animals, we understand the invasive processes of taxidermy as violations to pets' bodily integrity. Even more potentially disturbing about pet taxidermy is our familiarity with pets' personalities and, by extension, the seeming absence of said personality in the still, taxidermic body. (Colvin 2016: 68)

It was quite commonplace to have shoulder mount taxidermy made of deceased pets in the nineteenth century, but the trend did not last, because taxidermy just does not capture the spirit of animals that we knew personally. The facial expressions of taxidermy are not the ones that we know from when our pet was still alive (Poliquin 2012: 209).

This argument goes for people as well, of course. A taxidermist could capture grandmother's uniquely lovable character to an even lesser extent than they could capture Fido's. Her expression would almost assuredly be grotesque and frightening to look upon. If pet taxidermy is unsettling and unseemly, human taxidermy would be much more so.

A common plot in legends and fairy tales from all over the world is that a hero must slay a monster or enemy of some kind, and the hero almost invariably must collect the head as proof of its demise. The severed head thus becomes the symbol of the victory. Yet the finest examples of decapitated human heads symbolizing dominance and conquest are probably found in historical accounts of human head-hunting. This has occurred in many cultures all over the world. Walker explains that collecting skeletal remains, scalps, heads, and other body parts as trophies, to obtain their power, and for religious purposes can be considered nearly a cultural universal (Walker 2000: 9).

Chacon and Dye write in *The Taking and Displaying of Human Body Parts as Trophies by Amerindians*, that "[t]he removal of heads, scalps, eyes, ears, teeth, cheekbones, mandibles, arms, hands, fingers, legs, feet, and sometimes genitalia for use as trophies by Amerindians was an ancient and widespread practice in the New World" (Chacon and Dye 2007a: 7). Most human body parts have been potential trophies

3.6 The Meanings of Hunting Trophies

somewhere in the world at one time or another. Chacon and Dye provide a very thorough historical overview (2007a: 8–21) of all manner of taking and displaying human trophies. We list below some of their examples, specifically of human heads as trophies.

The first skulls with signs of scalping appear from 600,000 years ago in Ethiopia and skull caps turned into drinking cups from 25,000 years ago have been found in France and Spain. Scalped skulls from Denmark and China from 4500–4000 BC and many more up through the Bronze and Iron Age have also been uncovered. Scythian warriors from 500 BC brought the heads of their killed enemies to their king while they kept the skin of the heads as trophies. The Celts in 800 BC to 100 AD attached the heads of their enemies to the necks of their horses and would embalm the heads of prominent enemies for celebratory displays. Scalped skulls and coins depicting severed heads have been found from Ancient Rome, where human skulls would also be used for drinking. Genghis Khan's grandson Hulagu killed a large part of the inhabitants of Baghdad in 1258 and constructed a pyramid of their skulls. Similarly, a Muslim ruler in India around the year 1500 constructed a tower of the skulls of his enemies. William Wallace's head and severed limbs were prominently displayed by the English in 1305. Many other human heads were hung out for display on London Bridge for several hundred years, and heads of the Irish would adorn military camps under Elizabeth I. Padre Miguel Hidalgo called for Mexican independence in 1810 and was decapitated by the Spanish, who displayed his head in a cage for a decade. English officers were photographed with the decapitated heads of Chinese pirates in Hong Kong ca. 1900. And Japanese warriors who did not follow the Bushido code commonly collected heads as proof of kills (Chacon and Dye 2007a: 8–15).

It is worth emphasising that the motivations for taking and displaying human trophies are usually very different from hunters' motivations for collecting trophies. Most of the examples above involve the taking of human trophies for the purposes of intimidation or deterrence, which is clearly never a motivation for keeping a hunting trophy. The comparisons drawn here are not intended to imply that hunting trophies and human trophies are the same or serve the same purposes—although they can be. The point is that a history of human heads being used to display dominance and conquest may negatively influence our perception of hunting trophies as well, because we come to view them in the same light.

Van der Kroef provides an interesting example of how head-hunting among many New Guinean tribes in the early twentieth century was driven and necessitated by the need to collect the victims' names:

> A man's real name, however, prized beyond all others, is the head-name (*koei-igis*) the name of the victim whose head has been acquired. The head-name is a badge of honor, with whom the bearer is completely identified. To exchange one's head-name with a friend is therefore the highest honor which one can bestow. For every head-name given to a child means that its father, uncle or other immediate relative has travelled into the interior, and has undergone the dangers that accompany slaying the victim whose head-name is bestowed on the child. The social pressure behind the acquisition of such a name is so great that a child is derided and, indeed, feels inferior unless he possesses his own head-name. Among the Boetinese the child will frequently carry the lower jaw or chin of the victim whose head-name he bears sewed in a little cotton bag around his neck, and mothers keep their children under control

by threatening to take this prized possession away from them unless they behave. (Van der Kroef 1952: 222–223)

An interesting difference between these human trophies and most hunting trophies is that the trophy is not the decapitated head of the enemy per se, but the name that comes with it,[32] which may only be acquired via the physical head.

Staying on the subject of names but getting back to the hunting of animals, killing to earn a name has been a feature of lion hunting for the Maasai:

> In Maasai culture the first warrior to spear a lion in a successful hunt is given a name that represents the characteristics of both the warrior and the lion he speared. A lion name conveys upon the young warrior recognition and prestige amongst his community and peers. A warrior with a lion name feels that something great has happened to him. When the successful warrior brings the lion's mane and tail back to his manyatta (his home in the community) to be put on display, he is treated and celebrated as a hero. Other warriors who don't yet have their lion names yearn to have this same recognition and so dream about the day that it will be their turn to bring home the lion trophy. (Lion Guardians 2012)

Here one might make a comparison to record-book trophy hunting, discussed further in Chapter 4.5.7. Some hunters may desire trophies not for the trophies themselves but because a trophy is needed for a name in a record book.

Downs describes here some reasons for head-hunting practices in Indonesia:

> Heads were always taken in combat when possible, but raids were often made for the express purpose of obtaining them. A head was necessary for ending the mourning for an important person, and for the consecration of a temple. Young men took heads to prove their bravery to girls and it was sometimes done to prove one's innocence with regard to some accusation or other. Apart from such specific grounds, however, there was the general one that the well-being of the village as a whole and the success of its crops depended on the taking of heads. (Downs 1955: 41)

These heads seemingly served more diverse and general purposes than heads did in the New Guinean culture described by Kroef. Other Melanesian and Australasian examples are the Iban warriors of Borneo, who were not considered ready for marriage before they had collected a human head, and the Maori warriors who would save the heads of detested chiefs for insults. They would also use the heads of their enemies in a bowling game, as explained by one Maori chief:

> The heads of slain chiefs were cut off and piled in a heap, and the head of the principal chief was placed on the top of that heap. Then we took other heads and threw them at the pile of heads. The one head at the top of the pile made a fine target to throw at. This is an ancient game; it was practiced by our ancestors although stones were often used instead of human heads to throw at the pile of heads. We continue our game until all the heads were quite crushed. (Chacon and Dye 2007a: 17)

---

[32] One might here question how the headhunter knows the name of the head's previous owner. One way is to spy on the enemy long enough for someone to mention his name, another method, according to Kroef, is even more deceptive: "The hunter will induce his victim to eat with him, or chew some betel with him or give him other evidences of friendship. In the ensuing conversation, names and places of origin are usually mentioned. The hunter will then suddenly attack his victim, and take his head" (Kroef 1952: 225). One could imagine a measure of hesitation here when your new friend from another tribe asks for your name.

3.6 The Meanings of Hunting Trophies

Taking human heads as trophies has been at least as prevalent in North- and Latin American history as in the rest of the world, as demonstrated by the examples throughout Chacon and Dye. The reasons for taking trophies have been many, including spiritual, cultural, strategical, and practical, but many of the examples above show the direct relation of trophy heads to displays of power and dominance:

> Individuals in tribal and chiefdom societies sought recognition as great warriors based on their personal achievements through the possession of human trophies. The acquisition, public display, and hoarding of trophies were each vital for the creation, maintenance, and retention of both tribal and chiefly positions of leadership. (Chacon and Dye 2007b: 635)

To prestige:

> Obtaining an enemy trophy resulted in the enhancement of the trophy taker's prestige, and in some cases the prestige of the victim would be conferred on the trophy taker. (Chacon and Dye 2007b: 633)

And to proof of fighting capabilities:

> Among the chiefdom societies of Colombia, "They stored their war trophies in special structures or sanctuaries. They also displayed the skulls of their enemies on stakes they erected in front of their residences and hung other human trophies inside their homes along with other war spoils as tokens of their bravery and victory in warfare. They bedecked themselves with necklaces of human teeth" (Redmond 1994:36). (Chacon and Dye 2007b: 635)

The serial killer who keeps some form of memento from his victims is a classic trope in crime literature, cinema, and television. A popular TV-show that ran from 2006 to 2013 (and was revived for a new season in 2021) was called *Dexter*. The ambiguous protagonist of the show was Dexter Morgan, played by actor Michael C. Hall. He worked as a blood spatter analyst for the police by day and killed criminals by night to satisfy his urge to kill—his "dark passenger." His trophies were a blood sample from each victim.

The trophy kept by the serial killer is a deeply personal (and sometimes sexualized) item that must be kept secret, lest the killer get caught. One famous example of the serial killer collector is Ted Bundy, who displayed the heads of some of his victims in his apartment, but there are many other such examples (Hickey 2015: 198–200; 206–207). Serial killer trophies seem to resemble hunting trophies more than other human trophies do. Whereas many human trophies serve practical purposes of ritual or intimidation, serial killer trophies and hunting trophies mainly have personal significance.

Finally, the **fourth** factor that might contribute to negative perceptions of hunting trophies is just the obvious fact that the hunting trophy is inextricably associated with hunters and the activity of hunting. Poliquin writes that trophies provoke negative emotions not just because they are evidence of humans killing animals, but because they are evidence of the *desire* of humans to kill animals (Poliquin 2012: 152). If someone is already opposed to trophy hunting and trophy hunters, and unable to comprehend how someone could desire to kill, then this doubtlessly influences their attitude towards hunting trophies. One can hardly, after all, approve of hunting trophies without at least appearing to approve of hunting.

3.6.2 Achievement, Tribute, and Gratitude

The previous section featured rather grim interpretations of the symbolism and connotations of hunting trophies and reasons for disliking them. Yet we must assume that the hunters who own them think otherwise of them.

Uncovering what hunting trophies mean to hunters is probably a more challenging task than uncovering what they mean to nonhunters or antihunters, since the relationship should be more complex. While nonhunters or antihunters may feel strongly about hunting trophies, their emotions are likely to be simpler because they are unlikely to be shaped by the same kind of personal narratives. The head of a deer on a wall in a hunter's home may greatly disgust the antihunter, but it is also in all likelihood just a random dead deer. To the hunter who killed that deer, it is one that the hunter—even if only for a minute—had a very personal and emotional relationship with. The hunter was an active participant in the story of its death, and that makes it special to the hunter. Garry Marvin writes that

> [h]unting trophies as material objects are primarily markers of what developed between the hunter and the hunted, the experiences of the hunter prior to the death of the animal. Their significance does not principally reside in what they are in the present (although they do have such significance in terms of their taxidermied qualities) but rather in their power to evoke the past, a past that can be relived by the hunter and by only that hunter. Such a collection is more than an assembly of individual trophies; it is also a complex site of recollection at which animals that had no connection with each other when alive are brought together because of the manner of their deaths. (Marvin 2011: 203–204)

This is not to say that all hunting trophies are powerful and deeply significant memento of the hunt. As Garry Marvin points out, not all trophies have the same meaning, nor are they equally memorable:

> Not all [hunting trophies] are valued equally or in the same manner. Some may be trophies in the original, military sense of spoils of battle and symbols of both victory and defeat on the hunting field. Some may be valued in and of themselves as the biggest, rarest, or the most difficult to hunt. Others [...] are sites of memory that both invoke and evoke the hunter's journey out to the animal and the return with it. Not all hunters make the journey in the same manner, for the same purpose, or for the same experiences. (Marvin 2011: 205)

For a hunter whose primary motivation is meat for the freezer, the killing of average animals may become routine. Hunting trophies from most hunts may be made and kept mostly out of habit and tradition. NB talked to one Danish hunter and wildlife manager who has killed more than 200 roe deer. He related that he once promised himself that the day he could not remember the circumstances of all his kills by looking at his trophies, he would stop hunting. He then admitted that he had not kept that promise; he still hunts and keeps trophies, even though he has forgotten how many of the hunts went down. AH has spoken with a deer manager who keeps the cartridge cases of the ammunition used for every deer he has shot as a kind of trophy, with some details written onto it, like species, date, location, sex, age and so on. This is a more complex trophy relationship, since the trophy itself is simply a piece of metal that acts as a surface on which to record key details (and aides memoire) of hunts. Nonetheless, since the case itself was involved, intimately, in the death of the animal, it still carries some personal significance, albeit perhaps less than a head or set of antlers from the animal.

But hunting trophies are often very powerful mementos. If a hunter after many years finally succeeds in outsmarting one particularly impressive and cunning animal, then the story of that trophy might fill a book, and the feelings involved could be very complicated. There could be—and all at the same time—feelings of happiness, sadness, grief, regret, pride, pity, determination, resignation, love, and more. It is also possible that the hunter who finally had a chance to kill this animal would be unable to do so, as it would be almost like killing an old friend. The trophy from such an animal would be a very meaningful and, to the hunter, priceless possession. Dizard writes that

> [k]illing an animal is an emotionally complicated and conflicted act. Everyone who has even casually read the literature of hunting has encountered descriptions of how the successful hunter, at the moment of the kill, is bathed in a flood of conflicting emotions—triumphal elation alternates with feelings of remorse, and both are mixed with awe and humility. (Dizard 2003: 125)

Previous executive director of Wildlife Forever, Barry Vorse, wrote this in a letter to Forrest Wood, Jr.:

> Do I feel remorse when I kill an animal? Yes, I do. I even talk to the animals that I kill sometimes. I discussed this very topic with two very deep-thinking, thoughtful hunters who are acquaintances. Both of them admitted to doing the same thing. (Wood 1997: 20)

James Swan argues that the conflicting and complicated emotions of hunters lead to that paradox of hunters loving the animals that they also kill. He writes that if hunters let themselves feel the full range of complicated emotions associated with killing, they will come to feel a deep reverence for the animals they hunt (Swan 1995: 239).

But the complicated and ambivalent feelings hunters may have about their quarry arise even before the animal is killed. The Spanish philosopher José Ortega y Gasset suggested in his *Meditations on Hunting*, which is perhaps the foremost of all works on hunting philosophy, that the good hunter is never entirely sure about the morality of his intended action:

> More than once, the sportsman within shooting range of a splendid animal, hesitates in pulling the trigger. The idea that such a slender life is going to be annulled surprises him for an instant. *Every good hunter is uneasy in the depths of his conscience when faced with the death he is about to inflict on the enchanting animal.* He does not have the final and firm conviction that his conduct is correct. But neither, it should be understood, is he certain of the opposite. (Gasset 1942: 88, emphasis in original)

The trophies on the walls of hunters' homes generally reflect these complicated feelings.

Regardless of the complicated feelings hunters may have about their hunting trophies, they are surely less influenced by the previously discussed negative associations that trouble many nonhunters. Hunters are unavoidably more accustomed to killing and death than the average nonhunter, and so less troubled by any association with mortality. Most urban nonhunters who are comfortable eating meat would be uncomfortable killing the animal that provided the meat. Many are uncomfortable even if the meat they pick up in the supermarket looks too much like an animal. A chicken with its head still attached is, to many people, a too stark reminder that you cannot consume meat without sanctioning killing and consuming dead animals.[33] Hunters also associate the hunting trophy with death, but the hunter *knows* that there is no eating meat without killing. The nonhunter also knows this, but the hunter understands it in a more personal and visceral way. The hunter does not compartmentalize it in some far-off corner of the brain where knowing something is dissimilar from being meaningfully aware of it, accepting it, and even embracing it. Hunters cause death and accept death as a part of life. The trophy being a memento mori is not a burden on the hunter; hunters already live with death and surround themselves with it.

What seems to separate hunters from nonhunters in their interpretation of hunting trophies is that many nonhunters—and most likely all antihunters—interpret hunting trophies as triumphant celebrations of death, while most hunters see their trophies as celebrations of life. Poliquin explains the former interpretation here:

> Obviously, hunting trophies stand apart from most other sorts of mementos, with the exception, perhaps, of tombstones: the material existence of both signifies loss of life. However, even here a sharp distinction exists. If tombstones are commissioned from grief, trophies are commissioned from triumph. Animals have been vanquished and broken, and their fragmented forms accentuate the raw fact that trophies are souvenirs of termination. And it is this celebratory display of deliberate death that inclines most nonhunters to see trophies as nature destroyed, not nature displayed. (Poliquin 2012: 150–151)

This interpretation does not fit many hunters. The notion that the hunting trophy symbolises a "vanquished and broken" animal, celebrating "deliberate death" is simplistically judgemental. As an illustration of hunters' interpretation of their trophies, we may instead compare the hunting trophy to funeral parties. In some cultures, funerals are joyous parties with happiness and dancing and singing. Obviously, the purpose of these funerals is to celebrate the life of the deceased and not to celebrate their death. To many hunters, similarly, hunting trophies are celebrations of the animals' lives and not—as the antihunters see them—celebrations of killing

---

[33] This is partly cultural; in Asia it is more common to prepare fish and poultry whole, for example, and to eat the brains and feet of poultry and the meat on the head of a fish (the tender jaw muscle is often the prize part). In western countries it is common to only eat the meaty parts, like breasts and filets, disassembled. The species of animals that we eat vary from culture to culture as well of course. Few in the western world would be comfortable eating dogs, because dogs in western culture are pets and not food, yet there is no reason other than custom and culture why eating a dog—rearing and method of killing assumed to be the same—should be morally different from eating a pig, which is generally considered the more intelligent of the two. See for example Melanie Joy's *Why We Love Dogs, Eat Pigs, and Wear Cows* (2010) or Hal Herzog's *Some We Love, Some We Hate, Some We Eat* (2010) for more on this topic. Also, it is technically not true today that you cannot eat meat without sanctioning the killing of animals. Meat can be grown from cells in a laboratory. We are however still at least some years away from this being an affordable consumer option.

and death. As Morris Jones says about a sheep trophy on his wall: "The only tribute I can pay it now is to watch it for the rest of my life: remembering" (Jones 2012: 44).

Lastly, trophies might be a kind of penitence to some hunters. Hornaday provided this as a reason to hunters for keeping trophies: "If you must go and kill things, save their heads and mount them as atonement for your deeds of blood. They will give pleasure to you and your friends long after you have hung up your rifle forever" (Hornaday in Jones 2015: 264). Dan Crockett, similarly, here describes how an owl's feather became what we might call his penitence trophy:

> I was hunting alone through a mix of gooseberry tangles, sycamores, and pine oaks on a friend's farm in southeast Kansas. Having lucked into the scant, ephemeral woodcock migration, I was lathered by the rare opportunity. Twice that morning, woodcock whirred up almost at my feet. Both times, brush veiled their flight and I didn't fire. Then a small bird lifted and went lilting toward a screen of cedars.
>
> Doubt descended immediately after the shot, and I walked over to find the delicate body of an eastern screech owl. Numbed, I laid down the gun and knelt beside the bird. After constructing a burial platform of sticks and bark in the low branches of a redbud, I arranged the owl in a bleak semblance of flight. Walking out of the woods, I turned back and withdrew one of the owl's variegated flight feathers. I keep it to remind me that done in haste, or without respect, hunting can deliver the most profound regret conceivable. (Crockett 1996: 240–241)

From this melancholy interpretation of trophies, we move to something quite the opposite, which is the sometimes humorous or kitschy decorative role of some hunting trophies, when they cease to be personal mementos.

3.6.3 Trophies as Interior Decoration

When a hunter dies, trophies in the hunter's possession generally lose much, and potentially all, of their original significance. Even if the narrative of how each animal died was shared by the hunter, the emotions tied up in the hunting experience is lost, because only the hunter felt them. The trophies do not necessarily lose all value though when the hunter dies. From being representations of a personal hunting narrative, they can instead—if inherited by family members for example—become memories of the hunter who died. They can also become objects that say something about their new owner's sense of style or humour, or they can lend a certain feeling to a room. They may go—in other words—from being personal mementos with mainly sentimental value to items of interior decoration with mainly monetary value, not very different in purpose from a designer chair or a piece of art.

There is an increasingly lucrative market for second-hand trophies. When a hunter dies, it is often the case that no one wishes to inherit the trophies. Either the relatives are not hunters and do not want them, or they *are* hunters and consider hunting trophies something deeply personal, so they may not want them. In either case, trophies are often auctioned off or sold wholesale to a retailer. Eventually they may end up in homes or establishments of someone who wanted them purely for their aesthetic value. This atmosphere, this certain style that someone wants to communicate, can according to Poliquin be "either a flaunting of all things crass and backwater or a Victoriana aesthetic" (Poliquin 2012: 164).

Although most exceptional trophies are sold through specialized channels, a search for hunting trophies on eBay reveals a few things about this interior decoration stage in the lifetime of a hunting trophy.

First, it reveals how much people are willing to pay for hunting trophies as interior decoration. The price ranges, while widespread, seem to depend mostly on antler/horn size (most listings are deer and most of the rest are antelopes, bulls, or sheep) and overall condition. Roe deer half-skull mounts are about US$50–150, steer skulls with horns about 130–300, half-skull red deer or whitetail start at 150 but can reach 1000 or more. Shoulder mount taxidermy may be as low as 100 for poor taxidermy with small antlers and reach over 5000 for a record book deer, a bison, or similar large species/specimen of high quality. The online market for shed antlers is seemingly much larger than the market for hunting trophies, presumably because these are easier to come by and can also appeal to those opposed to hunting.

Second, it reveals something about why people want such trophies. Relevant search words are often included in a listing's title on eBay, so the title may indicate what interested parties search for, other than "hunting trophy." Two examples of eBay hunting trophy listing titles in June 2019 are "Whitetail 4 PT [4-point antlers] Buck Deer Antler Hunting Mount Trophy Cabin Decor Man Cave" and "Red Deer Antler, Taxidermy, Hunting, Log Home, Rustic, Knife, Trophy, Handmade." The atmosphere that someone looking for antlers might want to create should be clear from these keywords.

As hunting trophies become decorative items, they cease being mementos of great personal significance. Their physical attributes, atmosphere, and aesthetics become more important than what they once meant to a hunter. What is interesting is the likely response of other people to them. A deer head in your house will almost certainly elicit a very different response from visitors if you have shot it yourself (a trophy) than if you bought it on eBay (a decorative item), but the object is the same, and the deer is still dead.

3.7 The Trophy Photo

Finally, we need to discuss the trophy photo. The trophy photo is a photo of the trophy animal, and usually of the hunter as well, more often than not holding the weapon used to kill the animal. It is commonly taken on-site, shortly after the kill.[34] The purposes of trophy and trophy photo are similar because both are meant to

---

[34] In some rare cases, trophy hunting may not involve killing. Rhinoceros hunting, for example, sometimes involves shooting the rhino with a tranquilizer dart and then posing with it for a photo. In these cases, unless a mould has been made of the rhino while it was unconscious, with which an artificial trophy is later created, the trophy photo *is* the hunting trophy. Trophy photos of all the killed birds are also the only trophies for many kinds of bird hunting.

3.7 The Trophy Photo

commemorate the hunting experience and the animal. Trophy photos that feature the hunter, however, are also about commemorating the hunter.

Kalof and Fitzgerald did a study in 2003 of 792 hunting trophy photos from 14 different hunting magazines. They came to some conclusions about gender and ethnicity in trophy hunting, which we will return to in Chapter 4, but here are some of their other general findings:

Ninety percent of the photos surveyed were taken in the animal's natural environment where it was just killed, and just over half of the photos were of deer.[35] The 792 photos altogether featured 803 humans, 95% of which were male. Almost all held a hunting weapon, or had it propped up against the animal. The animals were often manipulated in such a way as to give an impression of life or natural behaviour. This is common practice, and hunting guides will usually be proficient in positioning animals to produce the classic "trophy" photo. The wounds were carefully hidden, and care was often taken to properly show the full size of the animal. In practice, blood trails down the side of an animal will often be washed away or disguised with dust. A trophy photograph is usually a post-mortem manipulation that, intentionally, disguises much of the reality of the kill. Exit wounds that pass through the gut cavity, for example, may cause parts of the gut to fall out. In such a case, the trophy photo would be taken from the entry-wound side. These manipulations were generally less important for antlered or horned species, as focus was on the antlers and horns. Predators like leopards, cougars, and bobcats were often held up vertically by the hunters to better show their relative size. The hunter very often had a hand placed on the animal. Kalof and Fitzgerald interpret this as a symbol of dominance over and possession of the animal (Kalof and Fitzgerald 2003). Most hunters would probably interpret this more positively as affection, appreciation, and pride.

There is a very credible argument to be made that the trophy photo is more important to antagonistic trophy hunting discourse than the hunting trophy. Without the trophy photo, the only visual evidence of trophy hunting would be the hunting trophy. However, as hunting trophies reside in the homes of hunters, they are generally only seen by antihunters in occasional trophy room photos, which are much rarer than the trophy photos taken right after a hunt. Without the trophy photo, antihunters (or those inclined to become so) would have heard about the activity but possibly never seen evidence of it. Hatred of trophy hunting is arguably ignited and fuelled by the trophy photos, and even more so when this imagery features dead specimens of highly anthropomorphized species (discussed in Chapter 7). Without photos of dead charismatic animals alongside smiling hunters, it would be much harder for news of animals killed by trophy hunters to attract enough attention to go viral. Far fewer people, consequently, would be aware of trophy hunting and the controversy surrounding it. Without photographic evidence, one might simply not know that there was anything there to be upset about. Trophy photos fuel the anger towards

---

[35] This is going to depend a lot on the magazines that one chooses to survey. Magazines may be dedicated to deer hunting, hunting in Africa, bowhunting, etc. There would not be many antlered species in a magazine about African hunting.

trophy hunters, which seems to fuel the opposition to trophy hunting, as we discuss in Chapter 8.

Berger and Milkman (2012) concluded in a study of articles in *New York Times* that while articles evoking positive emotions go viral more often than those evoking negative emotions, the articles that overall generate *stronger* emotions (specifically high-arousal emotions such as anger and not low-arousal emotions such as sadness) have a greater viral potential. Trophy photos undeniably generate very strong emotions, and while they do evoke sadness for the animals, they seem to evoke a far greater portion of high-arousal anger towards hunters. Other research (Fein and Spencer 1997; Peters and Kashima 2007; Guadagno et al. 2013) indicates that viral potential is increased if the emotions generated by forwarding something are likely to be shared. A story about trophy hunting with trophy photos, then, is likely to go viral if it is likely to generate the same strong anger in the person receiving it as in the person sharing it. The shared emotional experience increases closeness in an in-group. The presence of an out-group (hunters), for whom hatred is shared, further bolsters positive perceptions of the in-group and boosts self-esteem.

Overall, trophy photos are almost always the emphasis in social media posts about trophy hunters. Any accompanying description seems irrelevant to most social media commenters. Descriptive texts are often just some negatively worded variation of the sentence "look at what this hunter did." This is despite the photo's context (what was shot? where was it shot? how fast did it die? what is its conservation status on a species/population basis? what happened to the meat? how much was paid? where did the money go?) being very important to whether the hunt was ethical or not. Social media comments—when they are not just general expressions of rage such as "I hope he gets eaten by a lion" or just strings of angry emoticons—generally reference the photo. Comments are frequently either about the cuteness, beauty, or majesty of the animal or insults related to the physical appearance of the hunter. The "fat white slob" stereotype is one of the most common. Overweight hunters in trophy photos are sure to get lots of these comments. See Chapter 8.1.1 for an analysis of trophy hunting antipathy on Twitter, based on hunting photos.

That the trophy photo is an enabler of resentment is also indicated by the negative comments that certain behaviours or attitudes of hunters in trophy photos elicit. If the hunter is smiling, for example, this tends to receive many negative comments about how someone can be happy killing such beautiful animals. Comparisons to serial killers are commonplace, as are comparisons to paedophiles.

Even worse than smiling in a trophy photo is when the hunter is depicted doing something or posing in a way that is perceived as disrespectful to the dead animal. A notable example featured a baboon that a hunter had killed and then dressed up in a hat and sunglasses and positioned in the driver's seat of a car, so it looked like it was driving with one arm out the window. Another famous example, which will be familiar to anyone who frequents places where trophy photos are being shared for the purpose of shaming, features six white men, standing behind a dead rhino with an American flag draped over it. This one is used as a stock photo for many antihunting articles. American hunter Rebecca Francis posted a picture of herself on social media, lying down, looking up, and smiling beside the neck of a giraffe that she had just killed. This prompted Ricky Gervais—whose critical tweets about trophy hunting are often referenced by the media—to tweet "What must've happened to you in your life to make you want to kill a beautiful animal & then lie next to it smiling?" (BBC 2015).

Child and Darimont (2015) argued that smiles are closely related to hunting success. Based on an analysis of 2791 trophy photos, featuring smiling hunters, they found out that there were more genuine smiles[36] in photos where a hunter was displaying a large and/or carnivorous prey than in photos where a hunter displayed a smaller and/or herbivorous prey. It makes sense, in this light, that trophy photos featuring large and charismatic species face a stronger backlash online, as not only do these species elicit more empathy than smaller and less charismatic ones, but they also contain more genuinely smiling hunters, which further infuriates hunting critics and animal lovers. Of course, these animals are also better known; everyone can identify and probably knows something about a lion, but an impala or blesbok are far less recognisable (see Chapter 7).

Sometimes in photos, in obvious attempts to display dominance and conquest as opposed to wittiness, the hunter will stand with a foot placed on the back of the dead animal or (for carnivores) kneel and pry the animal's mouth open to show its teeth. These poses equally seem to provoke more anger than when hunters in trophy photos adopt a more sombre attitude. In these cases, it is interesting to note that the trophy photo may have a different significance, one more akin to dominance and conquest, than the trophy, which may be more of a memento, imbued with emotional significance related to the challenges of the hunt and the experience of the hunter. This further emphasises the point that "trophies," in all their forms, are complex to interpret and understand.

References

Aiken, Lewis R. 2001 (First 1985). *Death, Dying, and Bereavement*, 4th ed. Mahwah: Lawrence Erlbaum Associates, Inc.
Anderson, Martin. 2022. *Galana: Elephant, Game Domestication, and Cattle on a Kenya Ranch*. Redwood City: Stanford University Press.
Archives West. 2011. Boone and Crockett Club Records, 1888–2009—Historical Note. Archives West. http://archiveswest.orbiscascade.org/ark:/80444/xv46765 (accessed 29 January 2019).
Asma, Stephen T. 2001. *Stuffed Animals and Pickled Heads*. New York: Oxford University Press.
Associated Press. 2019. Safari Club Event Vendors Sold Products Made from Threatened Wildlife. Voice of America News. https://www.voanews.com/a/safari-club-event-vendors-sold-products-made-from-threatened-wildlife/4750014.html (accessed 18 April 2019).
B&C. 2011. 1906 Measurer's Manual Now Available!. Boone and Crockett Club, March 29. https://www.boone-crockett.org/news/featured_story.asp?area=news&ID=101 (accessed 30 January 2019).
B&C. 2013a. *The Boone and Crockett Club: A History (1887–2013)*. Missoula: Boone and Crockett Club.

---

[36] Genuine smiles activate the muscles around the eyes, whereas fake smiles activate only the muscles around the mouth.

B&C. 2013b. Boone and Crockett Club Position Statement—Canned Shoots. Boone and Crockett Club, June 11. https://www.boone-crockett.org/about/positions_CannedShoots.asp?area=about&ID=6B455080&se=1&te=1 (accessed 18 April 2019).

B&C. 2015a. Hunter, Guide, and Hunt Information Sheet. Boone and Crockett Club. https://www.boone-crockett.org/pdfs/HGH_2015.pdf (accessed 12 February 2019).

B&C. 2015b. Boone and Crockett Club Trophy Memories Catalog. Boone and Crockett Club. https://www.boone-crockett.org/pdfs/2015-16%20Trophy%20RecognitionCatalog.pdf (accessed 13 February 2019).

B&C. 2016. The Boone and Crockett Club on Fair Chase. Boone and Crockett Club. https://www.boone-crockett.org/pdf/On_Fair_Chase.pdf (accessed 23 January 2019).

B&C. 2017. General Policies of the Boone and Crockett Club's Records Committee. Boone and Crockett Club. https://www.boone-crockett.org/bgRecords/records_policies.asp?area=bgRecords (accessed 21 February 2019).

B&C. 2019a. The Birth of the Fair Chase and the Teddy Bear. Boone and Crockett Club. https://www.boone-crockett.org/huntingEthics/Birth_of_fair_chase.asp?area=huntingEthics (accessed 29 January 2019).

B&C. 2019b. Boone and Crockett Club—Official Scoring System for North American Big Game Trophies—Typical Whitetail and Coues' Deer. Boone and Crockett Club. https://www.boone-crockett.org/pdf/SC_whitetail_typical.pdf (accessed 27 February 2019).

B&C. 2019c. Entry Affidavit for All Hunter-Taken Trophies. Boone and Crockett Club. https://www.boone-crockett.org/pdfs/EntryAffidavit_2019.pdf (accessed 12 February 2019).

B&C. 2019d. Sagamore Hill Award. Boone and Crockett Club. https://www.boone-crockett.org/bgRecords/records_sagamore.asp?area=bgRecords (accessed 7 May 2019).

Bartoskewitz, Marc L., David G. Hewitt, Jamie C. Laurenz, John S. Pitts, and Fred C. Bryant. 2007. Effect of Dietary Copper and Zinc Concentrations on White-Tailed Deer Antler Growth, Body Size, and Immune System Function. *Small Ruminant Research* 73: 87–94.

BBC. 2015. Extreme Hunter Has 'No Regrets' After Killing a Giraffe. BBC Newsbeat, April 15. http://www.bbc.co.uk/newsbeat/article/32319210/extreme-hunter-has-no-regrets-after-killing-a-giraffe (accessed 19 June 2019).

Berger, Jonah, and Katherine L. Milkman. 2012. What Makes Online Content Viral? *Journal of Marketing Research* XLIX: 192–205.

Bostock, Stephen St. C. 1993. *Zoos and Animal Rights—The Ethics of Keeping Animals*. London and New York: Routledge.

Browne, Montagu. 1893. *Practical Taxidermy: A Manual of Instruction to the Amateur in Collecting, Preserving, and Setting Up Natural History Specimens of All Kinds. To Which Is Added a Chapter Upon the Pictoral Arrangement of Museums*, 2nd ed., Alfred Bradley, 170. Strand, WC: London.

Buckner, Eldon L. "Buck." 2011. Introduction. In *Records of North American Big Game*, ed. Jack Reneau, Justin Spring, and Eldon L. "Buck" Buckner, 13th ed. Missoula: Boone and Crockett Club.

Chacon, Richard J., and David H. Dye. 2007a. Introduction to Human Trophy Taking: An Ancient and Widespread Practice. In *The Taking and Displaying of Human Body Parts as Trophies by Amerindians*, ed. Richard J. Chacon and David H. Dye, 5–31. Springer.

Chacon, Richard J., and David H. Dye. 2007b. Conclusions. In *The Taking and Displaying of Human Body Parts as Trophies by Amerindians*, ed. Richard J. Chacon and David H. Dye, 630–653. Springer.

Child, K.R., and C.T. Darimont. 2015. Hunting for Trophies: Online Hunting Photographs Reveal Achievement Satisfaction with Large and Dangerous Prey. *Human Dimensions of Wildlife* 20 (6): 531–541.

CIC. 1977. The Game Trophies of the World—International Formula for the Measurement and Evaluation of Trophies. International Council for Game and Wildlife Conservation, Version 1.0.

CIC. 2016. CIC Trophy Measuring Rules and Regulations. International Council for Game and Wildlife Conservation, September 20. http://www.cic-wildlife.org/wp-content/uploads/2017/02/Trophy_rules_Revision_September_20_2016.pdf (accessed 12 March 2019).

References

CIC. 2018. CIC Handbook for the Evaluation and Measurement of Hunting Trophies. International Council for Game and Wildlife Conservation, Edition 09-2018 (CIC—Division of Applied Sciences).

CIC. 2019a. The Handbook. International Council for Game and Wildlife Conservation. http://www.cic-wildlife.org/trophy-evaluation/the-handbook/ (accessed 13 March 2019).

CIC. 2019b. The Beginnings. International Council for Game and Wildlife Conservation. http://www.cic-wildlife.org/who-we-are/the-beginnings/ (accessed 13 March 2019).

CIC. 2019c. CIC Museum Palárikovo. International Council for Game and Wildlife Conservation. http://www.cic-wildlife.org/divisions/culture/core-working-fields/cic-museum-palarikovo/ (accessed 13 March 2019).

Coleman, Justine. 2020. Undercover video finds trophy-hunting convention vendors selling captive-bred lion hunts. *The Hill*, February 12. https://thehill.com/policy/energy-environment/482783-undercover-video-finds-trophy-hunting-convention-vendors-selling (accessed 5 January 2021).

Colvin, Christina. 2016. Freeze-Drying Fido: The Uncanny Aesthetics of Modern Taxidermy. In *Mourning Animals—Rituals and Practices Surrounding Animal Death*, ed. Margo DeMello, 65–71. Michigan State University Press.

Crockett, Dan. 1996. All Birds Flying. In *A Hunter's Heart—Honest Essays on Blood Sport*, ed. David Petersen 1997 (First 1996), 237–245. New York: Henry Holt and Company.

Cutright, Paul Russell. 1985. *Theodore Roosevelt: The Making of a Conservationist*. Urbana and Chicago: University of Illinois Press.

Damm, Gerhard R. 2008. Recreational Trophy Hunting: "What Do We Know and What Should We Do? In *Best Practices in Sustainable Hunting—A Guide to Best Practices From Around the World*, ed. Rolf D. Baldus, Gerhard R. Damm, and Kai-Uwe Wollscheid, 5–11. CIC Technical Series Publication No. 1. http://www.cic-wildlife.org/wp-content/uploads/2012/12/Technical_series_1.pdf (accessed 16 May 2019).

Datausa. 2019. Taxidermy. Datausa, March 28. https://datausa.io/profile/cip/010508/ (accessed 28 March 2019).

Dizard, Jan E. 2003. *Mortal Stakes—Hunting and Hunters in Contemporary America*. Amherst and Boston: University of Massachusetts Press.

Dougherty, Phil. 2007. The Klineburger Brothers and the High Lonesome Ranch (Sammamish). HistoryLink.org, November 16. https://www.historylink.org/File/8306 (accessed 27 March 2019).

Downs, R. 1955. Head-Hunting in Indonesia. *Bijdragen Tot De Taal-, Land- En Volkenkunde* 111 (1): 40–70.

Edwards, Ken. 2013. The Jonas Legacy Lives On. Ken's Corner, April 5. https://www.taxidermy.net/ken/?p=861 (accessed 28 March 2019).

Eurostat. 2008. NACE Rev. 2—Statistical Classification of Economic Activities in the European Community. Eurostat Methodologies and Working Papers. https://ec.europa.eu/eurostat/documents/3859598/5902521/KS-RA-07-015-EN.PDF (accessed 14 October 2022).

Fein, Steven, and Steven J. Spencer. 1997. Prejudice as Self-Image Maintenance: Affirming the Self Through Derogating Others. *Journal of Personality and Social Psychology* 73 (1): 31–44.

Gasset, José Ortega Y. 1942. *Meditations on Hunting*, trans. Howard B. Wescott, 1985. New York: Charles Scribner's Sons.

Goldstein, Lynne, and Keith Kintigh. 1990. Ethics and the Reburial Controversy. *American Antiquity* 55 (3): 585–591.

Green, Deborah, and Johnny P. Stowe, Jr. 2000. Quality Deer Management: Ethical and Social Issues. *Human Dimensions of Wildlife: An International Journal* 5 (4): 62-71.

Grinnell, George Bird. 1913a. Brief History of the Boone and Crockett Club. In *Hunting at High Altitudes—The Book of the Boone and Crockett Club*, ed. George Bird Grinnell, 433–491. New York: Harper and Brothers.

Grinnell, George Bird. 1913b. Constitution. In *Hunting at High Altitudes—The Book of the Boone and Crockett Club*, ed. George Bird Grinnell, 500–503. New York: Harper and Brothers.

Guadagno, Rosanna E., Daniel M. Rempala, Shannon Murphy, and Bradley M. Okdie. 2013. What Makes a Video Go Viral? An Analysis of Emotional Contagion and Internet Memes. *Computers in Human Behavior* 29: 2312–2319.

Hart, David, 2014. NAW TV: Marc Anthony Talks About Buck He Allegedly Faked. *North American Whitetail*, May 20. https://www.northamericanwhitetail.com/editorial/illinois-hunter-marc-anthony-scrutiny/263001 (accessed 16 May 2019).

Herzog, Hal. 2010. *Some We Love, Some We Hate, Some We Eat: Why It's So Hard to Think Straight About Animals*. HarperCollins (e-books; Reprint Edition).

Hickey, Eric W. 2015 (First 1991). *Serial Murderers and Their Victims*, 7th ed. Boston: Cengage Learning.

Hölscher, Tonio. 2006. The Transformation of Victory into Power: From Event to Structure. In *Representations of War in Ancient Rome*, ed. Sheila Dillon and Katherine E. Welch, 27–48. New York: Cambridge University Press.

Honeybone, Ellie. 2020. Meet the WA Fisherman Turning Dead Trophy Fish into Detailed Works of Art. ABC News, September 11. https://www.abc.net.au/news/2020-09-11/dunsborough-artist-creating-gyotaku-art-using-dead-fish/12649164 (accessed 27 October 2022).

Jonastaxidermy. 2019. The Jonas Brothers Story. Jonas Brothers Taxidermy, March 28. https://jonastaxidermy.co/jonas-brothers-story/ (accessed 28 March 2019).

Jones, Allen Morris. 2012 (First 1997). *A Quiet Place of Violence—Hunting and Ethics in the Missouri River Breaks*. Montana: Bangtail Press.

Jones, Karen R. 2015. *Epiphany in the Wilderness—Hunting, Nature and Performance in the Nineteenth-Century American West*. Boulder: University Press of Colorado.

Joy, Melanie. 2010. *Why We Love Dogs, Eat Pigs, and Wear Cows—An Introduction to Carnism*. San Francisco: Conari Press.

Kalof, Linda, and Amy Fitzgerald. 2003. Reading the Trophy: Exploring the Display of Dead Animals in Hunting Magazines. *Visual Studies* 18 (2): 112–122.

Kellert, Stephen Robert. 1978. Attitudes and Characteristics of Hunters and Anti-Hunters and Related Policy Suggestions. United States Department of the Interior Fish and Wildlife Service. https://catalog.data.gov/dataset/attitudes-and-characteristics-of-hunters-and-anti-hunters-and-related-policy-questions-197 (accessed 10 September 2018).

Kerasote, Ted. 1994. *Bloodties—Nature, Culture, and the Hunt*. New York: Kodansha International.

King, Frederick J. 2016. Foreword—More Than Just Methods To Measure Trophies. In *How to Score North American Big Game—Boone and Crockett Club's Official Measurer's Manual*, 4th ed., viii–ix. Missoula: Boone and Crockett Club.

Klineburger, Chris R. 2010. *Gamemasters of the World—A Chronicle of Sport Hunting and Conservation*. Skyhorse Publishing.

Krausman, Paul, and Shane P. Mahoney. 2015. How the Boone and Crockett Club (B&C) Shaped North American Conservation. *International Journal of Environmental Studies* 72 (5): 746–755.

Laufer, Peter. 2012. *No Animals Were Harmed—The Controversial Line Between Entertainment and Abuse*. Guilford: Lyons Press.

Leopold, Aldo. 1968 (First 1949): *A Sand County Almanac, And Sketches Here and There*. New York: Oxford University Press.

Lion Guardians. 2012. Miterienanka: It's Not Just Another Name. Lion Guardians, January 2. http://lionguardians.org/miterienanka-its-not-just-another-name/ (accessed 2 July 2021).

Loveridge, Andrew J. 2018. *Lion Hearted: The Life and Death of Cecil & the Future of Africa's Iconic Cats*. New York: Regan Arts.

MacKenzie, John M. 1988. *The Empire of Nature—Hunting, Conservation and British Imperialism*. Manchester and New York: Manchester University Press.

Manta. 2019. Taxidermists in the United States. Manta, March 28. https://www.manta.com/mb_35_B32BB232_000/taxidermists (accessed 28 March 2019).

Marvin, Garry. 2011. Enlivened Through Memory—Hunters and Hunting Trophies. In *The Afterlives of Animals—A Museum Menagerie*, ed. Samuel J. M. M. Alberti, 202–217. Charlottesville and London: University of Virginia Press.

References

Milgrom, Melissa. 2010. *Still Life—Adventures in Taxidermy*. Boston and New York: Houghton Mifflin Harcourt.

Mitchell, John G. 1981 (First 1980). *The Hunt—A Provocative Exploration of Hunting in North America*. Penguin Books.

Morris, P.A. 2003. *Rowland Ward—Taxidermist to the World*. Ascot: MPM Publishing.

Morris, P.A. 2004. *Edward Gerrard & Sons—A Taxidermy Memoir*. Ascot: MPM Publishing.

Morris, P.A. 2006. *Van Ingen & Van Ingen—Artists in Taxidermy*. Ascot: MPM Publishing.

Morris, P.A. 2010. *A History of Taxidermy: Art, Science and Bad Taste*. Ascot: MPM Publishing.

Moyer, John W. 1953. *Practical Taxidermy*. New York: The Ronald Press Company.

MSUDeerLab. 2013. Estimating Boone and Crockett Score. Mississippi State University. MSUDeerLab. http://msudeerlab.com/mobile/estimatebooncrockettscore.asp (accessed 13 February 2019).

Nesbitt, William H., and Jack Reneau. 2016. Records Keeping—A Brief History. In *How to Score North American Big Game—Boone and Crockett Club's Official Measurer's Manual*, 4th ed., 3–7. Missoula: Boone and Crockett Club.

Nielson, Kristina P. 1983. The "Tropaion" in the "Aenid." *Vergilius* 29: 27–33.

P&Y. 2010. Our History, Our Heritage, and Our Heroes. Pope and Young Club. https://pope-young.org/images/uploads/files/HistoryHeritageHeros.pdf (accessed 14 February 2019).

P&Y. 2011. Entry Procedure. Pope and Young Club. https://pope-young.org/images/uploads/files/entry_procedure_r1-2011.pdf (accessed 19 February 2019).

P&Y. 2015. Fair Chase Affidavit. Pope and Young Club. https://pope-young.org/images/uploads/files/fair_chase_affidavit.pdf (accessed 19 February 2019).

Peters, Kim, and Yoshihisa Kashima. 2007. From Social Talk to Social Action: Shaping the Social Triad With Emotion Sharing. *Journal of Personality and Social Psychology* 93 (5): 780–797.

Petersen, David. 2010 (First 2000). *Heartsblood: Hunting, Spirituality, and Wildness in America*. Washington and Covelo: Island Press/Shearwater Books.

Phillipps-Wolley. 1894. Chapter XV—Hints on Taxidermy, etc. In *The Badminton Library of Sport and Pastimes—Big Game Shooting*, ed. His Grace the Duke of Beaufort, K. G., Vol. II, 413–420. London: Longmans, Green, and Co.

Pine, Vanderlyn. 1981. The Care of the Dead: A Historical Portrait. In *Death and Dying—Challenge and Change*, ed. Robert Fulton, Eric Markusen, Greg Owen, and Jane L. Scheiber, 272–278. San Francisco: Boyd & Fraser.

Poliquin, Rachel. 2012. *The Breathless Zoo—Taxidermy and the Cultures of Longing*. Pennsylvania State University Press.

Posewitz, Jim. 1994. *Beyond Fair Chase: The Ethic and Tradition of Hunting*. Helena and Billings: Falcon Press.

Reiger, John F. 2001 (First 1975). *American Sportsmen and the Origins of Conservation*, 3rd ed. Corvallis: Oregon State University Press.

Roosevelt, Theodore. 1885. *Hunting Trips of a Ranchman*. New York: G.P. Putnam's Sons.

Roosevelt, Theodore, and George Bird Grinnell. 1895. Constitution of the Boone and Crockett Club. In Hunting in Many Lands—The Book of the Boone and Crockett Club, Theodore Roosevelt and George Bird Grinnell, 439-441. New York: Forest and Stream Publishing Company.

Rose, Eve. 2011. Kubick gets 5-Year Sentence in Fraud Case (Feb. 28, 1998). *Anchorage Daily*, April 18. https://www.adn.com/alaska-news/article/kubick-gets-5-year-sentence-fraud-case-feb-28-1998/2011/04/18/ (accessed 15 May 2019).

Rowland Ward Ltd. 2018a. Method Forms. Rowland Ward Ltd. https://rowlandward.org/measurements/ (accessed 2 January 2019).

Rowland Ward Ltd. 2018b. Conservation and Hunting—Guiding Principles of Rowland Ward Ltd. Rowland Ward Ltd. https://docs.wixstatic.com/ugd/180d91_091bbe1ac8d74bc89160bef2b4928061.pdf (accessed 17 September 2019).

SCI. 2015. World Conservation & Hunting Award. Safari Club International. http://member.scifirstforhunters.org/static/WHA/docs/15/World%20Conservation%20Hunting%20Award.pdf (accessed 3 April 2019).

SCI. 2017. Safari Club International Policies and Procedures. Safari Club International, April 19. https://www.safariclub.org/sites/default/files/2018-10/Policies_and_Procedures_Manual_as_of_April_19_2017_rev_0.pdf (accessed 4 April 2019).

SCI. 2018a. Safari Club International Method 17-T Entry Form. Safari Club International, May. https://www.safariclub.org/sites/default/files/2018-10/17%20-%20t.pdf (accessed 9 April 2019).

SCI. 2018b. SCI Adopts Policy On Captive Bred Lions. Safari Club International, February 2. https://www.safariclub.org/news/sci-adopts-policy-captive-bred-lions (accessed 18 April 2019).

SCI. 2018c. Safari Club International Record Book—World Hunting Award Field Journal. Safari Club International. https://www.safariclub.org/sites/default/files/2018-10/world%20hunting%20award.pdf (accessed 3 April 2019).

SCI. 2019. World Hunting Awards. Safari Club International. https://www.safariclub.org/world-hunting-awards (accessed 4 April 2019).

SCI. 2020. 30 sec Television Commercial: "TRIGGERED". Safari Club International, February 5. https://www.youtube.com/watch?v=KXSqyZhQ5Lg (accessed 23 February 2021).

SCIF. 2019. About. Safari Club International Foundation. http://safariclubfoundation.org/about/ (accessed 6 April 2019).

Scully, Matthew. 2002. *Dominion—The Power of Man, The Suffering of Animals, and the Call to Mercy*. New York: St. Martin's Press.

SOR/93-56. 2022. Marine Mammal Regulations (SOR/93-56). Government of Canada—Justice Laws Website. https://laws-lois.justice.gc.ca/eng/regulations/sor-93-56/page-3.html (accessed 14 October 2022).

Spring, Justin. 2016. Introduction. In *How to Score North American Big Game—Boone and Crockett Club's Official Measurer's Manual*, 4th ed., x–xi. Missoula: Boone and Crockett Club.

Strickland, Bronson K., Phillip D. Jones, Stephen Demarais, Chad M. Dacus, Jocephus R. Dillard, and Harry Jacobson. 2013. Estimating Boone and Crockett Scores for White-Tailed Deer from Simple Antler Measurements. *Wildlife Society Bulletin* 37 (2): 458–463.

Swan, James A. 1995. *In Defense of Hunting*. San Francisco: Harper.

Taxidermy.net. 2000. Taxidermists Code of Ethics as adopted by the National Taxidermists Association. National Taxidermists Association, August 18. https://www.taxidermy.net/nta/code.html (accessed 16 July 2019).

Thoreau, Henry David. 1837–1846 in Bradford Torrey, ed. 1906. *The Writings of Henry David Thoreau—Journal*. Boston and New York: Houghton Mifflin and Company.

der Kroef, Van, and M. Justus. 1952. Some Head-Hunting Traditions of Southern New Guinea. *American Anthropologist* 54 (2): 221–235.

Vitali, Theodore. 1990. Sport Hunting: Moral or Immoral? *Environmental Ethics* 12: 69–82.

Walker, Phillip L. 2000. Bioarchaeological Ethics: A Historical Perspective on the Value of Human Remains. In *Biological Anthropology of the Human Skeleton*, ed. M. Anne Katzenberg and Shelley R. Saunders, 3–39. Wiley-Liss, Inc.

Weatherby Foundation. 2019. What It Takes to Win the Weatherby Hunting and Conservation Award. The Weatherby Foundation, April 18. https://www.weatherbyfoundation.com/about-us/winning-the-weatherby-award/ (accessed 18 April 2019).

Wood, Forrest, Jr. 1997. *The Delights and Dilemmas of Hunting—The Hunting versus Anti-Hunting Debate*. Lanham, New York, and London: University Press of America, Inc.

Wootters, John. 1977. *Hunting Trophy Deer*. New York: Winchester Press.

Chapter 4
Hunters and Antihunters

If you examine social media posts where antihunters upload and comment on trophy photos, then you will get the following impression of hunters: they are unintelligent and uneducated. They are overweight, alcoholic rednecks. Hunters are bad at shooting and do not care when they wound animals. They are cruel; they enjoy killing animals and watching them suffer. Hunters are psychopaths or have other mental illnesses. Hunters have no empathy. Hunters hate nature and love destroying it. Hunters are impotent and have tiny penises that they compensate for with big guns. They all voted for Trump, and they are all racists.

Occasionally—maybe one in twenty or thirty comments—someone who hunts or wants to defend hunters will pop up. They will fire off one of the go-to arguments: that hunters are conservationists; or that anyone who eats meat but is against hunting is a hypocrite. But trophy photos are frequently reposted in antihunting or animal rights forums that are dedicated to the sharing of such material, such as closed Facebook groups, so in these cases, the comments defending hunters are exceedingly rare, as anyone who shows an inclination towards a pro-hunting or even neutral stance is banned from participating. Videos or news of hunters dying from animal attacks or in hunting accidents are met with cheers and celebrations in these groups.

Hunting forums can be more balanced, but not by much. There generally does not seem to be the same volume of hatred towards antihunters in hunting forums as there is towards hunters in antihunting forums. This makes sense of course; antihunting forums are dedicated specifically to hating hunting and hunters, while hunting forums are dedicated to the hunting activity and various aspects of it. Antihunters are just one such aspect. There is also a tendency towards more nuance and diversity on hunting forums since many different hunting issues are discussed. Discussions on some subjects can nonetheless get heated. Predators like wolves, bears, and mountain lions are a particularly divisive and sensitive subject for hunters. Other subjects that divide hunters include hunting over bait, trapping, and hunting contests, as well, of course, as trophy hunting.

Yet there is little room for disagreement when it comes to antihunters. Antihunters are usually demonised simply as "antis," as if their very nature is to just be *against*.

© The Author(s), under exclusive license to Springer Nature Singapore Pte Ltd. 2023
N. Bichel and A. Hart, *Trophy Hunting*,
https://doi.org/10.1007/978-981-19-9976-5_4

Antis, according to many hunters, know nothing about nature or wildlife and what they do know they learned from Bambi and the Lion King. They live in big cities with their cats and tiny pampered dogs, and the closest they get to experiencing nature is these pets. Antihunters think that meat grows on Styrofoam trays in the supermarket. They do not have a natural relationship with life and death. Antihunters are overly emotional and sentimental. They are liberal "snowflakes," or part of the "woke brigade."

While antihunting and animal rights forums do not differ much—the attitudes underpinning them are quite clearly defined and generally uncompromising—hunting forums can vary greatly. They may seek to attract different types of hunters or be dedicated to different aspects or kinds of hunting. The closer you get to hunting forums catering for what you might call the trophy hunter stereotype, the more unbendingly hostile towards antihunters they tend to get. In the types of hunting forums where for example the hunting of captive-bred lions in enclosures is accepted, anyone who questions the ethics of anything that any hunter does is branded an "anti." Hunters frequenting such forums are as uncompromising as the antihunters are in their respective forums. Loveridge makes mention of the types we refer to here:

> To many in the hunting industry, there are two kinds of people: those who hunt and those who do not. Anyone who expresses an opinion that does not support hunting, is termed "an Anti." This is short for anti-hunter but could equally mean antichrist for all the venom spewed at critics of hunting. (Loveridge 2018: 99)

The point we are making is that many hunters and antihunters live their online lives in echo chambers.[1] Very rarely are their opinions challenged and, when they are, the in-group[2] stands ready to pull back anyone who starts to doubt or question. Social media relationships are important to many people. Questioning the status quo of the in-group comes with the risk of losing these relationships. Life is easier and the right answers are simpler when the world is black and white, and there is a lack of grayscale in both hunting and antihunting forums.

4.1 Hunter Demographics

For hunting to occur there needs to be a hunter, and as we've seen already, much of the ire directed towards trophy hunting tends to be focussed less on the actions and more on the actors. With that in mind, it becomes essential to understand more about hunters if we are to understand more about trophy hunting and the various reactions to it. Fortunately, we can turn to some excellent resources to learn more about hunter demographics and motivations. It is important to mention, however, that

---

[1] Echo chambers are environments in which someone encounters only opinions that they already agree with. Their views in such environments are only reinforced and never questioned. Twitter is argued to have a particularly strong echo chamber effect. See e.g. Colleoni et al. (2014).

[2] In social psychology, a person's in-groups are groups that the person belongs to and identifies with, and out-groups are groups that the person does not identify with.

demographic-, statistical-, and research data about hunters is to our knowledge widely more available for USA than for anywhere else in the world. European countries, in contrast, collect hunter- and hunting data differently or (often the case in Eastern Europe) scarcely at all (Vicente et al. 2018). This Chapter 4.1 on hunter demographics and to some extent the following Chapter 4.2 are therefore heavily skewed towards US hunters.

One of the most recent and exhaustive sources of information on general demographics and statistics for hunting and hunters in the United States is the *National Survey of Fishing, Hunting, and Wildlife-Associated Recreation* (hereafter, the *National Survey*), conducted roughly every five years since 1955 by the U.S. Fish & Wildlife Service (USFWS). We will refer to the report from 2018 and its 2019 addendum, *Recruitment and Retention of Hunters and Anglers: 2000–2015*. Another very relevant source of data is the Responsive Management survey research firm in Virginia, USA. *The Sportsman's Voice—Hunting and Fishing in America* (2010) by Duda et al. is one of their flagship publications and an important source of information for this chapter.

The *National Survey* reports that a national average of 4% (11.5 million) of the US population of 16 years or older hunted in 2016. This was the lowest hunting participation in USA in at least 25 years, down from 6% (13.7 million) in 2011 and record highs in 1960 and 1975 of 9%.[3] As mentioned at the end of Chapter 2, however, there are indications that US hunting recruitment has increased—at least temporarily—in the face of the Covid-19 pandemic, with many states reporting a "modest to massive spike" in hunters in 2020 (Brown 2020).

Duda et al. (2010: 57–60), comparing different statistical sources, report that probably around 14 million, but perhaps as many as 27 million, Americans had hunted in any given year between 2000–2006. They also note that numbers might be higher if respondents in general were asked if they *consider* themselves hunters, rather than if they *had hunted* in the previous year. For comparison, FACE reports around 6.7 million hunters across 36 European countries in 2016, which is only 1–2% of the total population (FACE 2016: 6).

US hunting participation is lowest at 2–3% on the East and West coasts, roughly corresponding to the Pacific and Eastern time zones, and highest in the Mountain and Central time zones where participation rates in some states is as high as 8%. Participation rates increase with distance from big cities, but urban Americans still account for many more hunters in total than rural areas (8.9 million vs. 2.6 million). Ninety-four percent of all hunters had hunted big game (defined here as deer, elk, bear, and wild turkey) within their own home state in 2016 and 14% had hunted big game in another state. Thirteen percent hunted exclusively on public lands, 64% exclusively on private lands, and 21% on both private and public lands. Out of the total

---

[3] Note that the USFWS numbers exclude anyone below the age of 16 and that they only indicate how many went hunting in a given year. If someone who identifies as a hunter happened to not hunt in 2016, they will not be included in USFWS numbers for that year. These numbers are therefore most likely considerable underestimations of the total number of hunters in the US. Kellert (1978: 5) reported that 37% of the American population at the time of his study had hunted at some point in their life.

11.5 million who hunted, 9.2 million hunted big game for a total of 133 million days, and 8.1 million of them hunted deer for a total of 115 million days (14 days a year per hunter). This makes deer hunting by far the most common form of hunting in the United States. Of US deer species, the whitetail is by far the most hunted. According to 2007 figures, 78% hunted whitetail deer and 23% hunted wild turkey. Elk and mule deer (*Odocoileus hemionus*) were hunted by only 7% and 5% respectively (Responsive Management and National Shooting Sports Foundation 2008: 12).

Rifles are used by about two out of three hunters, shotguns are used by a little more than half, and archery equipment is used by about a quarter.[4] Total hunting expenditures were $26.2 billion in 2016, 49% of which was spent on equipment, 35% on hunting trips, and 16% on "other." Only about 3% was spent on hunting licenses, stamps, tags, and permits. 57% of the total amount was spent on big game hunting.[5] The hunting industry produces around $5 billion in annual federal taxes and—as Duda et al. note (2010: 164)—"sportsmen help to redistribute dollars from the wealthier urban and suburban areas to rural areas" (Duda et al. 2010: 64–65, 161–164; USFWS 2018: 24–26, 28–29, 34; USFWS 2019: 4).

Of the 11.5 million or so who hunted in 2016, 90% were male and 10% female. Recruitment of new male hunters was down 31% from 2000 to 2015 while recruitment of new female hunters was up by 32% in the same period. Women who hunt are generally less active (in days per year) than men and more likely to hunt with family or friends. The decrease in total number of hunters is mainly due to recruitment for children being down 51% from 2000 to 2015. Four percent of the population hunted in the age brackets between 18 and 44 and 6% between 45 and 64, dropping to 3% at 65 and older (Duda et al. 2010: 76–77; USFWS 2018: 30–31; USFWS 2019: 6, 9).

Participation rates in hunting increased with household income until $100,000 per year. In the income brackets ranging from less than $20,000 up to $35,000, participation was at 2%, which rose to 7% in the $40,000 to $100,000 brackets. Brackets above $100,000 were at 5%. For education levels, hunting participation for those with 11 years of education or less was at 3%, while brackets with more than 11 years of education were at 5%. Most remarkable of these demographics is perhaps the distribution by race; USFWS reports that as many as 97% of hunters were white in 2016, which also explains why changing US racial demographics contribute to the decline in hunting participation (USFWS 2018: 31–33).

According to Dizard (2003: 43–46), American hunters are generally more politically conservative than nonhunters. Almost half of male hunters identify as conservatives, which is true for about a third of male nonhunters,[6] but unlike the average

---

[4] Based on 2008 statistics (Duda et al. 2010: 65). Numbers add up to more than 100% because many hunters use more than one weapon. Archery has increased in popularity since 2008 at the expense of the other categories.

[5] The USFWS survey covers only domestic hunting, so spending on trips and equipment on hunting trips abroad is not included.

[6] Dizard compares males with males here because—as he rightly points out—when about 90% of hunters are male, a comparison between hunters and nonhunters irrespective of gender would be comparing mostly men to an even split between men and women. The results would then show differences between men and women muddled up with differences between hunters and nonhunters.

conservative, they are as supportive as nonhunters of the federal government limiting individual freedoms to protect the environment. Male hunters are (or were in 2003; these percentages have probably changed) more likely than male nonhunters to disapprove of homosexuality (72% versus 57%) and to be anti-abortion.

Responsive Management similarly reports that

> hunters more commonly consider themselves conservative on issues (44%) rather than moderate (29%) or liberal (9%) (another 18% don't know or declined to answer), and they more commonly are Republican (35%) than Democrat (21%) or without a major party affiliation (28%) (15% were undecided or declined to answer). (Duda et al. 2010: 77)

Backing up these observations, the *OpenSecrets.org Center for Responsive Politics* reports that SCI donated $24,000 to six Democratic congressional candidates and $431,000 to 123 Republican congressional candidates in 2018 (OpenSecrets 2019).

The demographics and statistics based on USFWS and Responsive Management surveys and summaries pertain to American hunters and hunting in general. It is difficult to single out trophy hunting from all of this, as that can cover almost no one or almost everyone, depending on the definition. However, sources that might give us an impression of hunters with special interests in trophies are the readership bases of B&C's *Fair Chase* magazine and SCI's *Safari Magazine*. Both organisations periodically release "Media Kits" to showcase for advertisers who they can reach by advertising in their respective publications. These kits are a useful source of demographic information about the target audience.

Fair Chase subscribers in 2018 were 97% male and 3% female. The most represented age brackets were 55–64 with 33% and 65–74 with 27%. Only 6% had a household income less than $40,000 a year, 14% had $40,000 to $60,000, 31% had $60,000 to $100,000, 25% had $100,000 to $150,000, and 28% had more than $150,000. Education was more diverse than income but leaning towards higher education levels, with 30% having a bachelor's degree and 26% a graduate degree. Subscribers' areas of residence generally corresponded with where the USFWS National Survey indicated the highest numbers of hunters (B&C 2018: 10).

A couple of differences between the USFWS national survey respondents and the B&C *Fair Chase* readership are clear; *Fair Chase* readers are older and wealthier. Yet the difference between the average American hunter and the hunter with special interests in trophy records and record books becomes even clearer when we look at the demographics for SCI's *Safari Magazine* readership. The *Safari Magazine* is distributed to all members of SCI. The readership demographics and details that SCI emphasizes to advertisers are so telling that we shall simply insert the "Our Readers" section of the 2019 SCI Media Kit in its entirety:

This still leaves the issue of US hunters being about 97% white, which is not the case for average nonhunters and which is not accounted for here.

Income • An overwhelming majority of our readers either own their own businesses, or are professionals or executives. • Average annual income is $200,000+ • 40%+ earn in excess of $250,000 • Approximately 10% earn $750,000+ • Nearly 7% earn more than $1 million • 85% hold a four-year college degree	**Activities** • Members spend on average of 37 days a year hunting - 21 of those are spent outside the U.S. • Members spend an average of nearly $44,000 a year on hunting • Members spend an average of nearly $10,000 a year on hunting/shooting-related equipment • Nearly 60% own ammunition reloading equipment • Members own an average of 36 firearms each
Lifestyle • SCI members own an average of 1.8 Homes, two cars, two trucks/SUV, two ATVs • 38% own one ranch or farm • 38% own tractors and nearly half of those own an average of two tractors • 11% own planes and hold pilots' licenses • Members spend an average of $7000 a year on vehicle accessories • 27% own a fishing boat • 5%+ own yachts • 22%+ smoke cigars • 88%+ consume alcohol	**Generous Giving** • Each member contributes on average $50,000 a year to various charitable causes • More than 90% contribute to political campaigns
Travel • Members travel on average more than 100 days a year and spend an average of $61,000 • Of those who travel above economy class when flying, nearly 33% travel first class, while 63% travel in business class	**Reading Practices** • An impressive 88% spend an average of three hours reading each issue of SAFARI Magazine • Nearly 82% share our publications with others for a potential readership in excess of 100,000
Buying & Investing Habits • Members spend on average $46,000+ a year on jewelry and furs • 75% hire the services of a professional broker or investment firm to manage their wealth • More than half of the members collect art, wine, stamps, or coins as investments • Nearly 14% collect antique firearms • Almost 20% collect antique duck decoys	

It is almost as if the *Safari Magazine* readership was designed to fit the Great White Hunter stereotype. This media kit is no longer available online and the later version (SCI 2020) is more subdued with less emphasis on private yachts, furs, cigars, and jewellery. Most noteworthy points from the 2019 Media Kit are probably the very high annual incomes, the frequency and duration of hunting trips abroad, and the staggering numbers of firearms owned. Matthew Scully reported similar statistics in 2002, though average incomes were significantly lower:

> The average Safari Club member owns eleven rifles, six shotguns, five handguns, and a bow. He spends $14,000 a year on hunting, compared to $1500 for the average American hunter, for an annual half a billion dollars spent by the entire membership. Two-thirds hunt more than twenty-six days a year, and a quarter of them more than fifty days, typically in fenced ranches and game parks. Half of them [...] have hunted in Africa at least once. Over 50 percent of members report an annual income exceeding $100,000, as against 6 percent of hunters nationwide. (Scully 2002: 53)

And here is one of Andrew Loveridge's impressions from the 2005 SCI convention in Reno:

> In amongst the taxidermied hunting trophies were the trophy wives, dripping in expensive jewelry, tottering after their aging Actaeons. The wealth on display was astonishing, especially for two lowly biologists straight out of Africa. This was a place where animals, at least the lives of animals, were bought and sold on a colossal scale. (Loveridge 2018: 98)

These impressions and the apparent demographics of the most trophy-interested segment of American hunters bring to light the theme of trophy hunting and privilege.

4.2 Hunting, Privilege, and Social Schisms

From the kings of Ancient Egypt and the citizen athletes of Ancient Greece to the European nobles in the Middle Ages and the politically influential gentlemen sportsmen of the late nineteenth century USA, hunting has historically been a privilege. Illegal killing of wild animals for meat, poaching, may always have been there for anyone to risk engaging in, but legal and "proper" hunting and "fair chase" has been defined and carried out by the rich and powerful. The hunt itself, the definition of poaching, and even the language and rituals surrounding the hunt (see Chapter 2.6 especially), have consistently been instruments and symbols of wealth and power. Ortega y Gasset wrote that

> [f]rom all the revolutionary periods in history, there leaps into view the lower classes' fierce hatred for the upper classes because the latter had limited hunting—an indication of the enormous appetite which the lower classes had for the occupation. One of the causes of French Revolution was the irritation the country people felt because they were not allowed to hunt, and consequently one of the first privileges which the nobles were obliged to abandon was this one. In all revolutions, the first thing that the "people" have done was to jump over the fences of the preserves or to tear them down, and in the name of social justice pursue the hare and the partridge. And this after the revolutionary newspapers, in their editorials, had for years and years been abusing the aristocrats for being so frivolous as to ... spend their time hunting [...]
>
> Argue, fight as much as you like, over who should be the privileged ones, but do not pretend that squares are round and that hunting is not a privilege. (Gasset 1942: 29–30)

American hunters are, as we have now uncovered, about 90% male, 97% white, about four times as likely to be conservative as liberal, likely to be against abortion and homosexuality, and mostly—for obvious reasons—gun owners and pro-guns. Given this mix of characteristics it seems reasonable to assume (although without directly applicable statistics) that hunters are also far less likely to not be (at least openly or knowingly) cis-gender and heterosexual than the average American. Carol J. Adams argues specifically that men's social hunting encourages homophobic behaviour:

> With hunting, bonds develop between men who hunt together, but they must have an outside target to justify their male bonding, otherwise their heterosexuality could be called into question. Besides conversing in a language that assumes women's object status, heterosexual hunters must devalue homosexual men to assure themselves that *they* aren't gay. (Adams 2003: 93–95, emphasis in original)

Lastly, while the average American hunter is not particularly rich, the average $200,000 + income of SCI members certainly does speak to the stereotypical trophy hunter's affluence. In most of Europe, where hunting is generally an expensive hobby, it is also associated with a certain level of wealth.

What the attributes of being white, male, heterosexual, and wealthy all have in common is that they are all associated with privilege. In this light, one cannot but wonder to what extent modern opposition to hunting is about more than just hunting itself. Robin Hood's poaching of the King's deer was as much, or more, about defying those in power as it was about venison, and so perhaps today's criticism of hunting is also as much about sticking it to the privileged as it is an objection to hunting itself? Indeed, as shown in Chapter 2.11, the general opposition to hunt*ers* seems to be much stronger than to hunt*ing*.

Antihunters, being often female, non-white, non-wealthy, and liberal students (see Chapter 4.4) tend to be the opposite of the mostly conservative hunters in many ways. They are often anti-guns, less religious or not religious at all, pro-choice, and either very accepting of the LGBTQ community or part of it themselves (Duda et al. 2010: 44–45).

Analysing the controversy over hunting as a representation of not just hunting and antihunting stances but other societal tensions as well, we find there are five areas worth paying special attention to. These are gender, race, wealth, guns, and God.

4.2.1 Gender

If the controversy over hunting were an expression of only one broader societal tension, then it would be between men and women. This is not only because hunting is traditionally more of a male activity, but also because the hunting activity is commonly compared to sex. Antihunting, as we have seen, is closely associated with feminism in academic literature, where men's objectification of, domination over, or even rape of women is likened to hunters' relationship with nature and wildlife. Adams (2003) and Luke (1998, 2007) are prominent advocates of such views in addition to the authors discussed in Chapter 2.10.

A lot of American hunting merchandise speaks in favour of this kind of interpretation and sexual objectification of women. Google searches yield plenty of hunting-related bumper stickers, decals, coffee mugs, T-shirts, and other merchandise available for online ordering. The following are examples currently available that fit the

4.2 Hunting, Privilege, and Social Schisms

hunting/sex theme. Most of these are bumper stickers or car window stickers with lewd imagery to accompany the text:

- I like my deer like I like my women—horny.
- Tail Chaser.
- Chase-n-tail.
- Show me your rack.
- I love a big rack.
- Big Rack Hunter.
- Size does matter.
- Givem' the shaft [for bowhunting].
- Master Baiter [featuring a barrel—a reference to hunting over bait and, of course, to masturbation].
- Deer hunting is like sex—Size, equipment & technique are all important.
- Poontang Hunting Club [with a naked woman—poontang is North American slang for female genitalia, or a woman regarded only for sexual gratification]
- Nothing butt hunting [featuring antlers and a female butt].
- Show me your honkers [a reference to ducks as well as breasts].
- Just blowin my wad [slang for ejaculation].
- Blow my load [ditto].
- Warning—Bowhunting makes me horny [including a naked woman].
- Hunting whitetail with huge racks [including a naked woman with antlers].
- Booty Hunter [featuring a naked woman with crosshairs on her butt].

Or perhaps the ultimate bumper sticker for hunters who think the innuendo is too subtle in the examples above:

- I LOVE GUNS TITTIES & HUNTING.

The sexual-, macho-, and misogynistic themes are expectedly met with counter-thematic antihunting merchandise, intended to emasculate:

- Hunting—How men with small penises feel macho.
- Hunts are for cunts.
- How ya like my meat now? [featuring a hunter being sodomized by a buck].

A subset category of merchandise is intended for female hunters. This category generally emphasizes either the femininity and sexuality of women in a hunting context or plays on the prejudice that women cannot or do not hunt or shoot:

- Girls make hunting look good.
- Shoot like a girl.
- Girls hunt too.
- Backwoods Princess.
- Pretty in Pink—Dangerous in Camo.
- Camo Barbie.
- Yes I'm a Daddy's Girl—But Daddy taught me how to shoot.

These examples, and most others, use the word "girl" instead of "woman," which further underscores the point that power disparity and the men = subjects and women = objects theme is central to the sex/hunting analogy. These slogans relate closely to a trophy hunting stereotype that seems to have appeared in recent years and features prominently on Google's image search for "trophy hunting." These female hunter stereotypes tend to be very young and attractive. They wear lots of make-up, very revealing outfits (sometimes just a bikini), and strike sexy poses with trophy animals (or sometimes just holding and drawing a bow). The tabloid *Daily Star* reported in 2016 on this fad with the article *Meet the girls with blood on their hands as #girlswhohunt craze hits new levels.* Prominent representatives of the trend are Tess Thompson Talley, Kendall Jones, Heather Strickland, Rebecca Francis, and Melissa Bachman. Interestingly, these women seem to infuriate antihunters even more than male trophy hunters. As argued by Gjerris et al. (2016: 60), this may be because antihunters (who are often women and feminists) think such idiocy as hunting may be expected of men, whereas women should know better. A female hunter is like a betrayal, and it becomes even worse if the female hunter is also willingly objectifying herself by appealing to the sexual fantasies of men with a gun fetish.

While these *#girlswhohunt* seem to strongly emphasize their femininity and sexuality on social media, they also seek to portray themselves as very competent and knowledgeable hunters, generally adopting the conservation rhetoric in defence of their activities. The profile of these women only partly matches with the general tendency of portraying women as helpless and bewildered that Kalof and Fitzgerald observed in their study of trophy hunting photos:

> Women comprised only about 5% of the adults exhibited alongside a trophy kill, and these images had some interesting features. First, in about 30% of these images men were also present, making it unclear exactly who was to get credit for the kill. Second, while almost all of the male hunters were photographed with their weapons, only 55.6% of the images with females included weapons. Third, women were often represented in gender stereotypical ways, such as striking poses of confusion or helplessness, which was illustrated by a picture of a woman with a weapon in her hand, looking down with bewilderment at a dead moose that she presumably felled. In addition, photographs of women often focused on appearance, comfort or clothes, such as the picture of a smiling woman kneeling next to a freshly killed antelope; the picture caption read: "Even on this 1987 antelope hunt, I'd figured out that jeans were better on bottom than the fleece pants that went with this jacket". (Kalof and Fitzgerald 2003: 116–117)

While not all the stereotypes of women are the same, it is important to note that they are all *some* kind of female stereotype. It may be women as sex objects, women as young, girly, fragile princesses, women as overly dominating wives (and men as their children), women as out of place in a hunting situation and therefore confused and clueless, or women as poor shots. The latter is strikingly akin to the "women are bad drivers" stereotype. While unfounded, this stereotype does perhaps foster a healthy humility among aspiring female hunters that is notably absent in aspiring male hunters, who are often dangerously overconfident:

> Firearms instructors and outdoor educators are virtually unanimous in noting how much less guarded and defensive women are than men about their naïveté when it comes to guns and

hunting. Somehow, men feel as if they should know this stuff intuitively. Women do not. (Dizard 2003: 165)

The over-the-top crassness of merchandise that uses the hunting/sex analogies suggests the possibility that much of it may be used ironically or to deliberately provoke antihunters and feminists. It is as if to say "you think hunters are primitive slobs? I'll show you primitive slob!" It may also be a way of doubling down in the face of criticism to show that some hunters take pride in backwoods culture. Supporting this claim is a range of hunting merchandise, again mostly car decals and bumper stickers, with slogans that underscore the stereotypical connection between hunting and drinking, which most ethical hunters otherwise try hard to disprove and dissociate from:

- Shootin Deers n Drinkin Beers That's How I Roll.
- All I care about is hunting—and like maybe 3 people—and beer.
- Beer Season [featuring a can of beer with antlers].
- A bad day of hunting turns into a good day of drinking.
- Elkaholic.
- Rackaholic.

Of course, bumper stickers and similar merchandise are often designed to be excessively in-your-face, and maybe only a certain type of people buy bumper stickers to begin with. What you get from analysing such slogans is certainly not a fair representation of the average hunter's mindset. Yet these slogans do indicate a pervasiveness of the hunting/sex theme among a certain subset of hunters as well as associations between hunting and antifeminism on one hand, and feminism and antihunting on the other.

A final example of these associations comes from an interview by Dizard with a hunter who claimed to not be a feminist. Dizard found out that he simply did not want to be associated with what is usually correlated with feminism:

> I couldn't resist prodding this "I'm not a feminist" statement a bit. Since his wife was far better educated than he was and he was actively encouraging his daughters to become outdoorswomen, his disclaimer sounded a bit hollow. Jim explained that what he meant was that he disagreed with what he saw as an antifamily stance of feminists. Even more troubling to him was his sense that feminists were allied with gun-control forces as well as antihunting and animal rights causes. (Dizard 2003: 60)

4.2.2 Race and Wealth

If "male" defines the hunter, then "white" defines him even more. Ninety-seven percent of US hunters being white makes hunting a relevant target for those who wish to challenge white privilege. Hunting tourism in Africa is very easily understood as the white man's exploitation of Africa's people and natural resources. A long history of colonial trophy hunting in Africa where white hunters each relied on hundreds of indigenous servants—and still singlehandedly killed more animals than all of them could eat—only furthers this narrative. So too does the blatantly racist autobiographical hunting literature by colonial hunters of the nineteenth and early twentieth century, Theodore Roosevelt included.

Kalof and Fitzgerald point out that trophy photography also furthers the racial narrative. In their analysis of trophy photos, they found that less than 2% of the 792 trophy photos analysed featured minority men and none of them featured minority women. Sixty percent of the minority men seemed to be guides or assistants to white hunters. None of them ever held a weapon. Only one non-white male out of all 792 photos was photographed alone with his kill.

A simple but important privilege to emphasize is wealth, but here we come to a case where the ascription of privilege depends heavily on our definition of trophy hunting. The Great White Hunter stereotype is certainly wealthy, yet this world of the SCI record book and award systems featuring species from across the globe is very far from the world of the rural American meat hunter who only hunts in his own state but still cares greatly about the antlers of the deer he kills. The hunter who travels to Africa to kill the Big Five is surely privileged. The rural meat hunter may be male, he may be a white conservative, he may be heterosexual and cisgender, and he may bring a trophy home, but on account of increasingly often also being uneducated, poor, and unemployed, he probably does not consider himself privileged, nor do we.

Finally, he is also often (since he hunts locally) the kind of hunter that nonhunters encounter, which does not always favour the reputation of hunting or hunters, as noted here by Dizard:

> [P]eople who hunt are not always the best of ambassadors. Even though most hunters are now city or suburban dwellers themselves, most, as we have seen, have rural roots and most are working or lower middle class. Only eleven of the thirty-seven hunters I interviewed had graduated from college. While this compares favorably to the percentage of college graduates in the population as a whole, it is well below the average educational level of the people who are moving into the countryside to find their own little patch of Eden. Although hunters and the urbanites who have moved to the country might have rubbed shoulders uneventfully in the hustle and bustle of the city, class differences show up almost immediately in exurbia and make some hunters' efforts at communication awkward and even counterproductive. (Dizard 2003: 167)

4.2.3 Guns and God

Lastly in this discussion of societal tensions between hunters and antihunters, we turn to two passages from two texts, which are very important to many American hunters.

One is the Second Amendment of the Constitution:

> A well-regulated Militia, being necessary to the security of a free State, the right of the people to keep and bear Arms, shall not be infringed.

And the other is Genesis 1:26 in the Bible:

4.2 Hunting, Privilege, and Social Schisms

> And God said, Let us make man in our image, after our likeness: and let them have dominion over the fish of the sea, and over the fowl of the air, and over the cattle, and over all the earth, and over every creeping thing that creepeth upon the earth.

Guns and God. The former enables hunters to kill animals and the latter gives them the right.

Organisations in the USA like The Christian Deer Hunters Association are testament to a link between protestant conservatism and hunting that goes back to colonial times. This is not necessarily to say that religion makes people hunt, but hunters are predominantly conservative with traditional conservative values and beliefs, one of which is Protestantism. Robert Young—seeking an explanation as to why protestants are far more likely to own guns than those of other faiths and the non-religious—showed in 1989 a strong correlation between hunting and Protestantism: 43% of protestants hunted compared to 23% of non-protestants. The correlation could be explained, however, by Protestants being considerably more likely to have grown up in the rural South, where guns and hunting are also an integral part of rural culture (Young 1989).

A few examples may serve to illustrate the connections between trophy hunting and the belief that God gave man dominion over animals. In the 2017 documentary movie *Trophy*, we follow the hunting exploits of American trophy hunter Philip Glass as he works on collecting his Big Five (elephant, buffalo, lion, rhino, leopard) and kills an elephant, a hippo (*Hippopotamus amphibius*) (for lion bait), and a lion. He explains here why it is right for him to do so:

> The Bible says he gave man dominion over all the animals. And that dominion comes with responsibility, but it also means it's the right to use. And so I think that is a big part of it, and it's a big part of appreciating God's creation. And some people think, "Well, how can you go out and shoot God's creation?" That's a totally false statement, false point of view. God said we have dominion over the animals. That means we can do what we choose with them. It's a very powerful statement that's in the Bible. [...]
>
> I think it does make it more special for me as a believer to go out there and pursue these animals and know that God placed them. When I put my hand on that lion, I can promise you… at that moment, as with all of my life, anybody that believes in evolution is a complete fool. I just don't understand how people can't understand that God raised that animal into existence. (Trophy 2017: 01:30:28)

SCI's Chris Klineburger dedicated his book *Gamemasters of the World* to God and demonstrates an attitude to evolution similar to that of Philip Glass above:

> There are many theories of evolution, but if one opens his eyes to the millions of living things, he would see they could not start from amoebas without a profound Divine guidance—a Creator. Man evolved late in the process as a predator of sorts. He developed skills and weapons for hunting and fishing and even learned how to domesticate living things. He became the most important player in the Balance of Nature. If removed, there is no "Balance." Sport hunting and fishing is a continuance of man's inbred heritage from his primitive ancestors and is an important modern role in the dominion over the earth that God gave us. (Klineburger 2010: Front matter)

And John Mitchell quotes former president of the Michigan Bear Hunter's Association Carl T. Johnson:

> Now, the Bible tells us that man should have dominion over the fowls of the air and the cattle that roam the hills and the game and the wild animals. And I believe that, because man is superior. He should manage the wild animals just as he does a herd of cattle. (Mitchell 1981: 59)

The dominion idea also comes into play regarding canned hunting, to be further discussed in Chapter 6.4. Ian Michler from the famous *Blood Lions* documentary about lion breeding and canned hunting states in Ashcroft (2020) that the Calvinist beliefs held by most lion farmers in South Africa are that man has dominion over all animals and is entitled to do with them as he wants.

The connection between hunting and a favourable view of guns is obvious, and nowhere is it more so than in the close relationship between SCI and the NRA. The NRA presents hunting as one of its main arguments and causes, and SCI is heavily involved in lobbying for gun rights. About half of NRA's members are hunters and 30% of American non-NRA gun owners are hunters (Parker 2017).

Critics of hunting are commonly critical of American gun laws as well and some of them see hunters simply as "gun nuts" who only hunt because they enjoy killing or shooting at moving targets. John Mitchell quotes wildlife biologist David Klein about the public perceptions of guns and its relation to how hunters are perceived:

> In the public eye, wrote Klein, "the gun has increasingly come to represent the tool of criminal violence and the weapon of seemingly pointless wars which degrade humanity; it is seen as a direct agent of death or destruction. When large elements of society are turning against expressions of overt aggression, the gun, whether in the hands of the criminal, the soldier, or the hunter, symbolizes aggression; therefore the hunter by association becomes a primary focus for anti-aggression appeals". (Mitchell 1981: 94)

The close relationship between hunting and the NRA unfortunately pushes some hunters into supporting political agendas with which they do not wholly agree but must support if they want to keep hunting. Here is an example from Dizard's interviews:

> [Peter] belongs to the NRA and is an alternate on his rod and gun club's pistol team. He's not wild about the NRA's politics or their growing intransigence to any and all efforts to control access to exceptionally lethal weapons, he has no problem with tightened registration and licensing regulations, but he noted that the NRA is the only organization protecting what he believes is his right as a lawful person [...] to own firearms. [...] Peter is somewhere on the moderate to liberal end of the political spectrum and has no sympathy for the right-wing agenda of many supporters of the NRA. In a sense, he feels trapped into supporting the NRA. But he is determined to see guns and hunting persist. (Dizard 2003: 58)

With guns and God, we get to the core of what defending hunting is about for American hunters, when that defence is about more than just defending hunting itself. Defending hunting means defending core *rights*: the God-given right to hold sway over animals, and the constitutionally ensured right to bear arms. When antihunters attack hunting, they may not realize that some hunters do not just see it as an attack on that activity; they see it as an attack on their most fundamental rights and beliefs. As Joe Bageant writes in *Deer Hunting with Jesus*, the rural, poor, uneducated American harbours "long-buried anger at those faraway people who seem to govern the world:

4.2 Hunting, Privilege, and Social Schisms

city people, educated city people who win and control while the rest of us work and lose" (Bageant 2007: 128). When the educated liberals attack hunting, the defence may come with a fervour and unwillingness to compromise that is hard for antihunters to truly comprehend, because they thought that all they attacked was someone's hobby. But hunting is what defines some Americans more so than their job or their fatherhood or anything else. Duda et al. provide an example of this as well:

> During a focus group on hunting participation, a woman said that her husband had suffered a severe stroke to the extent that he could not remember his children, He did, however, one day ask his wife to get a new hunting license for him. When she said that he could no longer go hunting, he replied, "if I can't go hunting, who am I?" (Duda et al. 2015: 68–69)

"I am a *hunter*, and I don't apologize to anybody," Matthew Scully writes in reference to this attitude (2002: 67). Variations of this sentence are commonly found on hunting forums. An attack on hunting is not just an attack on what these hunters *do*, it is an attack on everything they *are*. Scully later continues from the floor of an SCI-convention:

> The important thing is not to let a lot of outsiders start laying restrictions on things. Then you're fiddling with basic rights, above all the constitutional right to firearms, as the National Rifle Association is here to remind us at its booth. Nothing unites like a common enemy, and they've all got one here: Give an inch to the enviros and Bambi lovers, all those urban types who know little about firearms and even less about wildlife, and it won't end there. It won't end with elephants or giraffes or the Big Five or wolves in Arizona or bison near Yellowstone. Before long it'll be deer. If anyone in this convention hall feels uneasy about any display, they keep it to themselves. Here, the mildest qualm or fugitive doubt is heresy. Let them take away our helicopters and next it'll be our guns. (Scully 2002: 54)

The hunting debate in general but in USA especially has little room for compromise. Concede anything and you are on a slippery slope to conceding everything. We have seen this attitude ourselves repeatedly in our engagements with both hunters and antihunters. As Forrest Wood, Jr. also writes:

> In the heat of the battle in the hunting/anti-hunting controversy, hunters defend too much and anti-hunters attack too much. Hunters are afraid that if they give an inch, all will be lost. Their argument is the slippery slope argument: if you admit that any form of shooting is undesirable, or if you agree that hunting should not be done at this time and at this place, you will end up with no hunting at all. (Wood, Jr. 1997: 10)

All this said, USA is different from Europe on some key issues here. The gap between hunters and antihunters is wider than it is in Europe. European hunters are generally on the right or centre-right of the political spectrum, but they hold less strongly to the values associated with conservatism in USA. Far fewer hunters in the generally more secular Europe see hunting as a God-given right, and guns are mostly just tools for hunting. They are not the symbols of freedom that they are in USA. The defence of hunting is not a defence of core rights to the same extent as it is in USA. Besides, many conservatives, especially in Northern Europe, would be considered liberals in USA. There are simply more issues that hunters and antihunters can conflict over in USA, and the disparity of positions on these issues is more extreme. It is also much less common for European hunters to rely on hunting for meat in the way that many of the rural poor in USA do.

That said, what they call hunting and shooting in the UK for example—i.e. fox hunting from horseback and driven or rough shooting at birds—are still deeply rooted rural traditions and rituals as well as important sources of rural employment. Criticism of driven grouse shooting in particular is seen as an attack on the rural identity and way of life itself and on a set of deeply emotionally resonant practices that tie the fabric of the countryside together and situate people in their communities.

For some, of course, hunting is just an important hobby and a hobby of theirs is all that you would be taking away, but wherever there is hunting, there are also many for whom it is much more than just a hobby.

4.3 Hunting Motivations

Psychology professor Geoffrey Beattie writes in *Trophy Hunting—A Psychological Perspective* (2020: 79) that "direct research focussing on the personality characteristics of trophy hunters is still waiting to be carried out." Beattie's book does not provide that required research, and neither have any other studies to date. Lacking psychological data about trophy hunters, the best that we can do to get know hunters is probably to study their motivations.

Americans, as we have seen, generally think that hunting for food is acceptable while hunting for trophies is not. But what does it mean to hunt "for" something? If you ate what you killed, did you automatically hunt for food? What if you ate what you killed and kept a trophy? Did you hunt for food or for a trophy?

The word "for" must imply motivation, so if you hunt for food, you are motivated by the meat that the hunt will hopefully yield. Most cases are not so simple though. Most deer hunters who save trophies of their kill, and take trophy photos, will also eat the meat and enjoy the experience, the challenge, the natural environment, the connection with nature and participation in the food chain, and either the company of others or the solitude. When hunters have so many motivations, what can we really say that they hunt *for*?

A 2015 study among California hunters conducted by Responsive Management reported that survey respondents, when presented with five options for the primary motivation to hunt, chose, in order of popularity: sport or recreation (32%); meat (30%); the company of family and friends (21%); closeness to nature (15%); and a long way behind, in last place, the trophy (1%)[7] (Duda et al. 2015: 181).

Eliason (2012) summarized other studies that come to similar conclusions and specifically surveyed the importance of hunting trophies for resident and non-resident

---

[7] It should be noted that especially in the case of surveys conducted over the telephone, which was the case for part of this 2015 study, respondents might be less likely to admit it if the trophy is their primary motivation for hunting. The trophy is probably the primary motivation for very few, but we suspect 1% could be too low.

elk hunters in Montana. His results were that 55% of resident hunters said the trophy was of no importance to them, 34% indicated that it was moderately important, and 11% said that the trophy was extremely important. For non-resident hunters, these percentages, respectively, were 34%, 50%, and 16%, expectedly indicating a greater interest in trophies among hunters who travel out of state for hunting. Many of the hunters, resident and non-resident, who had marked trophies as "extremely important" commented in Eliason's survey that because they would only shoot an exceptional trophy animal, they were satisfied with coming home from almost every hunt without having fired a shot.

Dizard's interviews led him to similar overall conclusions about the importance of trophies:

> To be sure, any hunter would be thrilled to see, much less to bag, a deer with an impressive set of antlers, but no one I interviewed had an impressive rack (or the equivalent measure in other game species) as his or her preeminent goal. Indeed, when the subject of trophy hunting came up, most expressed distaste for it. (2003: 97–98)

If hunting "for" something is to make any sense, then it must refer to the primary motivation for hunting. Luckily, while there are often several motivations for hunting, it is often possible to identify one primary motivation for hunting, and that is very rarely the trophy.

Even now, the best research that has been done towards characterizing different kinds of hunters and hunting motivations is probably by Stephen R. Kellert. Kellert published extensively about American attitudes towards animals and wildlife in the seventies and eighties. Most relevant of these publications is the *Attitudes and Characteristics of Hunters and Anti-Hunters and Related Policy Suggestions* from 1978. This study was part of a series based on in-depth interviews and questionnaires conducted over a 3-year period. Kellert developed a typology of nine basic attitudes towards animals, refined and increased to ten in Kellert and Berry (1980). Table 4.1 below combines Tables 23 (attitude definitions) and 24 (estimated percentages of the American population, behavioural expressions, and values/benefits) from Kellert and Berry (1980: 42, 44).

Not all typologies are equally relevant or useful to consider, so we will focus on what Kellert described as the three primary hunting attitudes (1978: 8), which are still widely referenced and used today: the *Naturalistic*, the *Utilitarian*, and the *Dominionistic*, underlined in Table 4.1. These attitudes, in a hunting context, roughly correspond to the hunting motivations of nature, meat, and sport, respectively (Kellert 1978: 3–5, 8; Kellert and Berry 1980: 42–45).

Another famous and quite similar study about hunting motivations was done by Decker et al. (1984). The three categories of motivations that they found for hunters were *Appreciative*, *Affiliative*, and *Achievement*. Appreciative corresponds to the Naturalistic motivation from Kellert, to be discussed shortly, indicating appreciation of nature. Affiliative is a social motivation; an affiliative hunter hunts mainly because of the opportunity to share the experience with others. This fits many Dominionistic hunters in Kellert's terminology, as these are more interested in people and less interested in wildlife than Utilitarian and Naturalistic hunters. It also fits the social

Table 4.1 Kellert's typology definitions and typology of basic wildlife values

Value Type	Definition	Estimated % of American Population Strongly Oriented Toward the Attitude	Common Behavioral Expressions	Most Related Values/Benefits
Naturalistic	Primary interest and affection for wildlife and the outdoors	10	Outdoor wildlife-related recreation—backcountry use, nature birding, and nature hunting	Recreational
Ecologistic	Primary concern for the environment as a system, for interrelationships between wildlife species and natural habitats	7	Conservation support, activism and membership, ecological study	Ecological
Humanistic	Primary interest and strong affection for individual animals, principally pets	35	Pets, wildlife tourism, casual zoo visitation	Companionship, affective
Moralistic	Primary concern for the right and wrong treatment of animals, with strong opposition to exploitation or cruelty toward animals	20	Animal welfare support/membership, kindness to animals	Ethical, existence
Scientistic	Primary interest in the physical attributes and biological functioning of animals	1	Scientific study/hobbies, collecting	Scientific
Aesthetic	Primary interest in the artistic and symbolic characteristics of animals	15	Nature appreciation, art, wildlife tourism	Aesthetic

(continued)

4.3 Hunting Motivations

Table 4.1 (continued)

Value Type	Definition	Estimated % of American Population Strongly Oriented Toward the Attitude	Common Behavioral Expressions	Most Related Values/Benefits
Utilitarian	Primary concern for the practical and material value of animals or the animal's habitat	20	Consumption of furs, raising meat, bounties, meat hunting	Consumptive, utilitarian
Dominionistic	Primary interest in the mastery and control of animals, typically in sporting situations	3	Animal spectator sports, trophy hunting, animal training	Sporting
Negativistic	Primary orientation an active avoidance of animals due to dislike or fear	2	Cruelty, overt fear behavior	Little or negative
Neutralistic	Primary orientation an active avoidance of animals due to indifference	35	Avoidance of animals	Little or negative

Source Kellert and Berry 1980: 42, 44

motivation of rituals and social identity, discussed in Chapter 4.5.2. The Achievement motivated hunter could be considered mostly identical to the Dominionistic hunter, but with some elements of the Utilitarian hunter mixed in. Achievement can be about bagging the largest animal or making the most difficult shot, which are clearly Dominionistic trait, but an achievement can also be to claim a large deer for the freezer, which is characteristic of the Utilitarian hunter (Decker et al. 1984; Messmer and Enck 2012).

4.3.1 The Naturalistic Hunter—Nature

The Naturalistic Hunter—or Nature Hunter[8]—constitutes an estimated 10–20% of American hunters (Kellert 1996: 71). They are characterized most of all by a desire to engage with nature as an active participant; a predator in its natural element, trying to take down its prey. Marty Kheel's "holy hunter" archetype would fit in the category of the Naturalistic Hunter, as feelings of an almost spiritual connection with nature are often part of this hunting attitude.

The attitude embodies

> [an] interest in experiencing human nature at its most fundamental evolutionary level. It is the desire for reaffirmation of basic relationships between humans and the natural world which, of course, are separated today by incredible barriers of civilized, industrial culture. Hunting, from the naturalistic perspective, provides an awareness of man's hunting origins, as well as a reminder of the basic interflow of energy and materials involved in the interdependence of all living things. (Kellert 1978: 11)

And also:

> The naturalistic hunting attitude is focused about a strong interest and affection for wildlife and the opportunity the hunt provides for immediate, first-hand contact with wild animals in their natural settings. The prey animal is even perceived as an object of strong affection, respect and, at times, reverence. (Kellert 1978: 10)

Being a participant in nature rather than an observer of nature is central to understanding the Nature Hunter. Some hunter-authors use the metaphor of being an actor in a play versus watching a play. Ortega y Gasset captures well the essence of being a participant in nature through hunting:

> If we want to enjoy that intense and pure happiness which is a "return to Nature," we have to seek the company of the surly beast, descend to his level, feel emulation towards him, pursue him. This subtle rite is the hunt.

> When one is hunting, the air has another, more exquisite feel as it glides over the skin or enters the lungs, the rocks acquire a more expressive physiognomy, and the vegetation becomes loaded with meaning. But all this is due to the fact that the hunter, while he advances or waits crouching, feels tied through the earth to the animal he pursues, whether the animal is in view, hidden, or absent. (Gasset 1942: 123)

The Nature Hunting attitude has received much attention from hunter-authors and scholars seeking to defend hunting ethics with more than just the conservation argument. David Petersen, Valerius Geist, Ted Kerasote, Mary Zeiss Stange, Allen Morris Jones, Jim Posewitz, Tovar Cerulli, Aldo Leopold, James Swan, and many more would all probably describe Nature Hunting in much the same terms as simply "ethical hunting." All would probably fit the category of nature hunter themselves.[9]

---

[8] We use "naturalistic hunter" and "nature hunter" interchangeably. The original word used in Kellert (1978) was naturalistic, but he has later (1996) used nature hunter in the same way, and later writers like Petersen (2010) and Marvin (2010) have also used the term nature hunter to describe the same hunting attitude. Petersen (2010) also refers to this as the "spiritual" hunter, and Swan (1999) as well emphasizes the religious/spiritual connection with nature felt by many nature hunters.

[9] Nature hunter authors tend to also write very eloquently about nature hunting, so we try to keep paraphrasing to a minimum in this part of the chapter.

4.3 Hunting Motivations

Some authors emphasize the spiritual connection with nature and sense of deep purpose that Nature Hunters feel when they hunt (like Gasset 1942; Swan 1995; Petersen 2010). Some emphasize the instinctual desire to engage with prey as a predator, as we did when hunting was a matter of survival, and the way our perceptions change while hunting (like Thoreau 1882; Gasset 1942; Leopold 1968; Jones 2012). There is a duality in the Nature Hunting attitude, which is absent in the others: The Nature Hunter is both civilized and aware of an instinct that draws them towards a natural, pre-civilized way of life. Thoreau wrote:

> I found in myself, and still find, an instinct toward a higher, or, as it is named, spiritual life, as do most men, and another toward a primitive rank and savage one, and I reverence them both. (Thoreau 1882: 226)

And Ortega y Gasset refers to the same instincts here:

> Do not forget that man was once a beast. His carnivore's fangs and canine teeth are unimpeachable evidence of this. Of course he was also a vegetarian, like the Ovidae, as his molars attest. Man, in fact, combines the two extreme conditions of the mammal, and therefore he goes through life vacillating between being a sheep and being a tiger. (Gasset 1942: 103)

A Quiet Place of Violence—Hunting and Ethics in the Missouri River Breaks (1997) by Allen Morris Jones is one of the best works on hunting ethics since Gasset's *Meditations*. In it, he reframed some of Gasset's thoughts in a very comprehensible and relevant way.

Essentially, *A Quiet Place of Violence* may be boiled down to a defence of Nature Hunting. Jones accomplishes this by separating *Project* from *Process*. A project is a goal, and the project of any hunter is to kill an animal. The process is the way in which the project is carried out; in other words, how the goal is reached. Hunting, Jones argues, is ethical only if the purpose of the project is to align oneself with the process:

> If the purpose of hunting is only to kill an animal, then the process is moot; we contain the technological ability to kill all animals. The *project* may be to kill an animal, but the project is useful only to the extent that it allows us to orient ourselves to the process. (Jones 2012: 40, emphasis in original)

Jones uses an analogy of mountain climbing to explain: The project of a mountain climber is to reach the top of the mountain. This project may be completed if the mountain climber flies to the peak in a helicopter. The project has then become meaningless, however, because the truer purpose of the mountain climbing had to do with the process. It had to do with accomplishing the goal in a certain way and being in a certain kind of accord with the mountain. And so it is with hunting:

> If we're hunting only for the project, only to reach the peak, then it becomes possible to reduce the relationship back to one of subjects and objects. If the animal is *only* a trophy to be taken, then we've ignored the process by objectifying the animal.
>
> The animal must never be a means to an end. If you want to acquire a trophy so badly that you're willing to do anything to get it—if you let the project take precedence over the process—then the animal has been reduced to an object, a means to an end. [...]

> If you are exclusively a trophy hunter, your new project does not return you to the natural order, it places you further beyond it.
>
> The process must take precedence, if only to temper the project. (Jones 2012: 72–73, emphasis in original)

Aldo Leopold would probably call Nature Hunting a man-earth experience:

> [T]here is value in any experience that reminds us of our dependency on the soil-plant-animal-man food chain, and of the fundamental organization of the biota. Civilization has so cluttered this elemental man-earth relation with gadgets and middlemen that awareness of it is growing dim. We fancy that industry supports us, forgetting what supports industry. (Leopold 1968: 178)

But hunting must strike a balance. If the project requires no stealth, endurance, speed, hard work, patience, or even luck, or in other words nothing that resembles the natural behaviour and skillset of the animal predator, then immersion in the process has failed. The project is no longer hunting but is just killing. This is a criticism often levelled against activities that come under the banner of canned hunting, and can also be applied to hunting over bait, where the hunter simply has to sit still in a blind and kill the animal, often at close range, as it feeds. Yet the project cannot be too hard either. It must be a *valid* project.

Setting out to kill a deer with bare hands may not be hunting either, because this project—persistence hunting aside—is not valid. It is too hard and not realistic. Setting out to kill a lion with nothing but a knife is not hunting either, it is combat or, arguably, suicide. This balance is discussed further in Chapter 6 when considering notions of "fair chase."

Nature Hunting must be fair chase hunting because the relationship between prey and predator must be something that immerses the hunter in the process. The predator must have a realistic but not guaranteed chance of killing its prey. We achieve this by limiting weaponry (bow instead of rifle for example) and limiting valid prey (a trophy male instead of a doe or spike buck). The hunting ethic that Jones develops is much like the sport hunting ethic that Gasset argued for:

> To the sportsman the death of the game is not what interests him; that is not his purpose. What interests him is everything that he had to do to achieve that death—that is, the hunt. Therefore what was before only a means to an end is now an end in itself. Death is essential because without it there is no authentic hunting: the killing of the animal is the natural end of the hunt and that goal of hunting itself, *not* of the hunter. The hunter seeks this death because it is no less than the sign of reality for the whole hunting process. To sum up, one does not hunt in order to kill; on the contrary, one kills in order to have hunted. (Gasset 1942: 96–97)

The last sentence here is probably the most quoted in all hunting philosophy literature. The hunt is what is important; as long as the project is valid (an intention to kill, and a realistic chance of doing so) and secondary to the process, then the hunt was a success, regardless of it ending with a kill or not. The intention to kill—the project—must be there, or it could never be real participation in nature. This is the meaning of "one kills in order to have hunted." Here, then, is also the reason Nature Hunters are almost universally unwilling to substitute the gun for the camera. If you carry a gun and plan to use it, you participate. If you carry a camera, you merely observe.

It is confusing that Gasset did not mean the same when he spoke of "sport" as Kellert and Jones do. The meaning of the word has evolved since Gasset's time. Kellert refers to "sport" as the objective of the Dominionistic Hunter rather than the nature hunter. Gasset, on the other hand, used "sportsman" in the early twentieth century sense of the word, when it referred more to "sporting," i.e. fair chase, and less to sport as competition. Gasset's sportsman compares to the Nature Hunter of Kellert, and the hunter who values process over project of Jones. Whatever we may call this hunter, Kellert, Gasset, and Jones all refer to the hunter who tempers his project to the process to participate in nature in a way that returns him or her to the natural order.

Given the characteristics outlined above it is unsurprising that Nature Hunters are often bowhunters. They are also more likely than other hunters to use a recurve or longbow than a compound bow. The added challenge of a bow—necessitating stealth, physical exertion, woodcraft, and other behaviour natural to animal predators—brings the Nature Hunter closer to the original role of predator. It immerses the hunter deeper in the process. However, this *must* be the motivation for bowhunting for the nature hunter. Walter Palmer used a bow when he killed Cecil. Presumably, according to Loveridge (2018: 214), it took him 11 h to kill Cecil after first wounding him, because if he had finished Cecil off with a rifle, it would not have counted as a bow-kill for record book purposes. It is obvious that despite using a bow, Walter Palmer is no Nature Hunter. The hunter who uses a bow to get his name and a bow-hunted trophy in a record book has let his project take precedence and reduced the animal to a means to an end. So too has the hunter who bow hunts in order to brag about his fieldcraft or marksmanship with a bow.

4.3.2 The Utilitarian Hunter—Meat

As we have seen, many hunters in rural USA hunt to make ends meet. These are the unprivileged hunters who have been looked down upon by those who have defined "proper hunting" since Xenophon or earlier. The Utilitarian Hunter—the meat hunter—accounted, according to Kellert, for four out of ten hunters in 1996 (1996: 72). The naturalistic and ecologic values of the Nature Hunter are, in the Utilitarian Hunter, overshadowed by the material, nutritional importance of wild game. There is not necessarily an absence of complicated emotions in the meat hunter, or any dislike of animals, but there is often a more indifferent attitude towards animals than we would see in the nature hunter. Any feelings or moral considerations about the hunt or the animals come second to simple utilitarian considerations. Wildlife is first and foremost a material benefit.

It is, as such, a less complicated hunting attitude than that of the nature hunter, which Kellert describes as philosophical and complex. It is sometimes perhaps more comparable to raising farm animals for meat than to Nature Hunting. For the Utilitarian Hunter, the project is more important than the process, but the project may be less to kill an animal and more to put meat on the table. The cheap and available process just so happens to be hunting. If there was an easier and even cheaper process that lead to the completion of this project, then some Utilitarian Hunters might not even hunt. While hunting is an end in itself to the Nature Hunter, the hunt is a means to an end to the Utilitarian Hunter.

Utilitarian Hunters are likely to embrace the wildlife management rhetoric, which compares animals to crops. The surplus of crops may be harvested for human benefit, and so may the surplus of deer. This rhetoric is often disdained by Nature Hunters and actively despised by antihunters. David Petersen, for example, writes that "[these] euphemisms have since infiltrated both outdoor journalism and wildlife management, insidiously reducing such lovely and lively creatures as deer, elk, and grouse to the level of turnips" (Petersen 2010: 43). Still, the lines also blur between these hunting categories. Edward Abbey was a Utilitarian Hunter like his father, who gave up hunting when his family no longer relied on the meat. He finishes the essay *Blood Sport* with these utilitarian thoughts:

> Now I have railed against the sportsman hunter long enough. I wished only to explain why first my father and then I have given up hunting, for the time being. When times get hard again, as they surely will, when my family and kin need meat on the table, I shall not hesitate to take that old carbine down from the wall and ramrod the spider out of the barrel and wander back once more into the hills. (Abbey 1996: 16)

Yet his opinion of the wildlife management rhetoric reflects the dispositions of David Petersen and other Nature Hunters:

> Where did the ugly term "harvesting" come from? To speak of "harvesting" other living creatures, whether deer or elk or birds or cottontail rabbits, as if they were no more than a crop, exposes the meanest cruellest, most narrow and homocentric of possible human attitudes toward the life that surrounds us (Abbey 1996: 16).

Conversely, Aldo Leopold is best described as a Nature Hunter, but he was also a game management professional who referred to deer as harvestable crops. In truth, to answer Abbey's rhetorical question, Aldo Leopold may largely be the one to blame for this language.

A 2020 New York Times article, *Deer Overpopulation Meets Its Match: Women Who Hunt*, provides an interesting example of the utilitarian hunting motivation, which perhaps inflates the recorded number of female hunters a little:

> For some time, the only Long Island deer story involving a woman was the tale of Maria Caccese. Thirty-five years ago, she was honored at the annual hunter's banquet on Shelter Island for bagging the biggest buck of the year. Her speech was short: "Thank you very much. I shot it from my kitchen window in my pajamas."
>
> Everyone knew that Ms. Caccese's husband shot that deer, and that she only agreed to hunt with him so he could use the extra tag to fill the family's freezer with venison. (Robey 2020)

The utilitarian attitude is closely associated with the American rural frontier tradition and relative rural poverty; hunted meat is important for economic and nutritional reasons, as with Edward Abbey and his father. European settlers in America were for several centuries waging a religious and survivalist war against both wilderness and wildlife. An aesthetic and moral appreciation for wildlife developed later in the cities, where wilderness ceased being a threat to survival. So, as Kellert states, "the pioneering perception of wilderness still underlies the utilitarian hunting tradition today" (Kellert 1978: 15).

For poor and rural Utilitarian Hunters whose job security is not always guaranteed, hunting may also be important psychologically. In Dizard's interviews, he encountered several hunters who relied on hunting for survival during periods of unemployment. They commonly felt proud of, and took solace in, their ability to survive and provide for their families the old-fashioned way. Equally, they felt disdain for those urbanites who would be unable to do so (Dizard 2003: 90–95).

4.3.3 The Dominionistic Hunter—Sport

The Dominionistic attitude, according to Kellert and Berry (1980), accounts only for an estimated 3% of the population. The Dominionistic Hunter, however—the sport hunter in the "hunting for sport" sense and not the traditional "hunting in a sporting way" sense—accounts for perhaps one hunter out of three, according to Kellert (1996: 73). Jan Dizard also came across sport hunters in his interviews and wrote as follows about them:

> Among the avid are a few who take their hunting to another level of preoccupation. These are the men and women for whom the challenge inherent in hunting takes on a competitive edge. Unlike George Dunnerston, who didn't keep score, these hunters do keep score, literally and figuratively. They strive for perfection. They have fun, and time in the woods or at the shooting range is recreation as it is for other hunters, but Charlie Rock, Jon Harcourt, and Cal Jones hunt with an uncommon intensity and focus. They are competitors who prepare for the hunt in much the way an athlete would train for a game. Jon and Cal compete with themselves—they want to improve over their past performance. There's always the prospect of a bigger deer, and there is always the hope of making a more precise shot. (Dizard 2003: 113–114)

While far from all Dominionistic Hunters are in it just for the trophies, it is, not surprisingly, among the dominionistic subset of hunters that we find the ones who are. Kellert writes:

> Although the specific pursuit of a trophy is far from necessary, and in fact relatively infrequent, trophy hunting usefully illustrates aspects of the dominionistic hunting attitude. The hunted animal represents a kind of trophy even in the absence of obtaining a large rack of some big game animal. Its conquest and defeat implicitly represent mastery and dominance. The actual big game trophy hunter simply takes the sport, the competition, the display of skill, and the emphasis on achievement and success to the logical dominionistic hunting extreme. (Kellert 1978: 16)

Other characteristics of the Dominionistic Hunting attitude are displays of masculinity and very low levels of compassion and curiosity about animals. Hunting is important for the Dominionistic Hunter as a way of signalling "aggressiveness, strength, courage, boldness, physical endurance and hardiness" and for satisfying the desire to exert "physical domination over an opponent" (Kellert 1978: 16). The gun "represents more than just a means to an end; in many cases, it is the end in itself" (Kellert 1978: 17).

Dominionistic Hunters are likely to be inflexible about their trophy demands, as the biggest trophies supposedly represent the greatest challenge. They may also be likely to resist shooting does, even when wildlife management requires it. There is little to no masculinity to be displayed in the killing of a female to the Dominionistic Hunter.

When antihunters claim that hunters in general just hunt because they enjoy killing, it seems to be mostly untrue. Nonetheless, there are hunters (and more than the hunting community will admit) for which it *is* true. Hunting for the enjoyment of killing is a motivation that most hunters would probably wish did not exist in any hunter, because its mere existence in some hunters taints the reputation of all hunters, perhaps more so than any other motivation. Yet its existence is very difficult to deny. There are too many examples of animal killing thinly disguised as hunting—some canned hunting for a start—which could be motivated by little else if not the pleasure derived from killing animals or using animals for target practice. However, even this motivation may not be simple. AH has spoken with a hunter, for example, who hunted vervet monkeys for the utilitarian purpose of vermin control (as is the case in part of South Africa), but they also happened to really enjoy killing monkeys. They would make the kills more challenging by shooting from greater range, but they would still only shoot monkeys if there was a genuine need for vermin control. Thus, although they were driven by a motivation to kill, they would not wish to shoot a monkey in a cage, or just any monkey. The target had to be legitimised (as vermin) and the killing made challenging before it became acceptable to act on a desire to kill it.

4.3.4 Motivation Demographics and Prevalences

According to Kellert (1978: 20), 47.5% of hunter-respondents fit the profile of Utilitarian Hunters, 36.8% dominionistic hunters, and 15.7% nature hunters. These attitudes are associated with clear demographic differences.

Utilitarian Hunters, who were disproportionately 65 + years of age and in the lowest income bracket, were more likely than other hunter types to reside in rural areas, to have had farming-related occupations, and to have grown up on a farm. They were, like Dominionistic Hunters, generally against protecting predator animals that prey on farm animals. Contrary to Dominionistic Hunters, the Utilitarian Hunters generally disagreed with the idea of hunting for sport. Utilitarian hunters "reported far more frequent formal religious participation" (Kellert 1978: 23), underscoring the connection between meat hunting and religious conservatism.

The attitudes of Nature Hunters were reflected in more time spent outdoors and strong interests in engaging with wildlife of all kinds, including insects. Nature

Hunters have the highest nature knowledge scores while Dominionistic Hunters have the lowest. Nature Hunters were also more likely to be female (though females were still a minority) and far better educated than Dominionistic and Utilitarian Hunters (Kellert 1978: 20–21).

Nature Hunters revealed stronger empathy for animals and interest in animal welfare than most others in the study. This extended also to pets, for which Utilitarian and Dominionistic Hunters had much less affection. Nature Hunters were significantly younger than Utilitarian Hunters and particularly highly represented in the younger than 30-bracket. They also tended to live in towns of 10–50,000 population rather than very urban or very rural environments.

Kellert found in 1996 that "[o]lder people generally express significantly less interest and affection—and more dominionistic and utilitarian values—toward animals and nature than do young adults" (Kellert 1996: 50). Gender is also a good indication of attitudes with men being more utilitarian than women (moderately high vs. moderately low), a little more naturalistic (mean vs. moderately low), and much more dominionistic (moderately high vs. very low). The much lower dominionistic scores for women might explain why they constitute about 10% of hunters but only 3% of the *Fair Chase* readership. Men express greater interest in wildlife and the outdoors, more support for hunting and fishing, and possess more knowledge and less fear of wildlife than women. In contrast, women are more emotionally attached to animals (domesticated ones especially) and express greater humanistic and moralistic sentiments toward nature than men (Kellert 1996: 50–53).

Education levels reflect attitudes as well. The lowest educated are almost the opposite on every measure of the highly educated. Respondents with high school education or less scored low or very low on the naturalistic scale while college graduates and those with graduate education scored moderately high (with no significant differences between different college majors). The trend was the same for knowledge scores and for ecologistic, humanistic, moralistic, and scientific values: the higher the education level, the greater the likelihood of being a Nature Hunter. Conversely, the lowest educated scored highly on utilitarian, dominionistic, and negativistic values, where the college educated or above scored very low. Income proved almost irrelevant (Kellert 1996: 54–56). Unsurprisingly, the variation between urban and rural Americans is large. Rural residents showed more utilitarian, pragmatic, and less protectionist attitudes. Urban residents tend to Muir-esquely romanticize nature and

> view as irrelevant and sometimes contemptible the practical dependencies on mastering wild living resources. In their eyes, nature may be viewed simplistically as large, attractive, and charismatic species or perhaps as pristine wilderness whose innocence and grandeur are debased by the hands of humans. For many urban wildlife advocates, practical considerations of animal harvest, control, and utilization are often viewed as unappealing and even repugnant. […] Ironically, many city residents barely recognize nature as an integral part of their urban lives—either from the perspective of practical benefits derived from environmental exploitation or in the natural complexity encountered even in the largest cities. (Kellert 1996: 59)

Dominionistic Hunters were the least interested in wildlife and the least nature oriented. Possibly related to their interest in guns and shooting, sport hunters were the most likely to have been in the armed forces. Very high negativistic scale scores indicated low interest and affection for animals among sport hunters. As Kellert notes, "one could speculate that sport hunters were more people- than animal-oriented, engaging in the activity largely for the social satisfactions derived from competition, companionship, shooting and displays of masculinity" (Kellert 1978: 22).

Concerning the differences between hunters and nonhunters, Kellert reported that

> [o]ne of the most striking findings was the much greater naturalistic attitude of hunters compared to nonhunters. Hunters generally revealed far more interest in the outdoors, and more affection and desire for contact with wildlife. Hunters reported significantly greater desire for exposure to wilderness areas and living in proximity to wildlife […]
>
> Additionally, hunters reported significantly greater affection for pets and generally more recreational involvement with animals. Thus, although hunters killed wild animals for food, recreation or other purposes, they appeared to be characterized by substantially greater affection, interest and lack of fear of animals than those who had never hunted.
>
> Hunters did reveal significantly greater scientistic perceptions of animals and, relatedly, had far higher knowledge-of-animals scores. (Kellert 1978: 6–7)

Pro-hunting nonhunters represented 21% of the sample. This group was characterized by low naturalistic and humanistic scale scores and considerably less interest in and affection for animals than both hunters and antihunters. Negativistic scale scores were higher for this group than for any other group examined, and many questions revealed a high degree of alienation and detachment from animals. Kellert thus argued that the pro-hunting stance of some nonhunters may indicate an indifference about the treatment of animals more than a positive attitude to hunting per se (Kellert 1978: 23–24).

A much newer study by Ebeling-Schuld et al. (2017) used the terminology of *Appreciative*, *Affiliative*, and *Achievement*, adopted from Decker et al. (1984) to assess the prevalence of motivations (or in this study, "satisfactions") among hunters. The authors collected data from three online hunting forums, totalling around 380,000 members. Four-hundred and fifty-five hunting stories from the forums were analysed for occurrence indicators of the three identified satisfactions. Three-hundred and seventy-nine of the stories involved ungulate hunting and 76 involved carnivore hunting. The authors found that achievement was by far the dominant satisfaction expressed in the hunting stories. In stories where multiple satisfactions were identified, achievement was dominant in 81% of the ungulate stories with affiliation at 11% and appreciation at 8%. Achievement was dominant in 86% of the carnivore stories with affiliation at 12% and appreciation at only 3%. In stories where only one satisfaction was identified (40% of the stories), this satisfaction was achievement in 97% of the ungulate stories and 100% of the carnivore stories (Ebeling-Schuld et al. 2017).

What this research could indicate is that there are far more dominionistic/achievement hunters than affiliative or appreciative. However, it could also just indicate that dominionistic hunters are by far most likely to write hunting stories on hunting forums and that nature/appreciative hunters are very unlikely to do so. As noted by the authors, online forum users might be neither broadly representative of the population of interest nor behave in ways that characterize typical behavior. In our case, hunting forums might elicit posts by hunters who are particularly achievement oriented and emphasize their achievements more in relation to other aspects of a hunt. (Ebeling-Schuld et al. 2017: 527)

Considering that Dominionistic Hunters are motivated by the competitive aspect of hunting—the sport—it seems probable that they would be more likely to want to show off their accomplishments to other hunters by writing stories about their successes. A hunting story, in this sense, might just be compared to the hunting trophy or the hunting photo as a display of hunting achievement.

Ebeling-Schuld et al.'s research might also indicate that the achievement satisfaction is much easier to write about, or that achievements are what hunters think other hunters want to read about. It is presumably easier for the average hunter to describe the process of hunting and killing an animal than it is to describe the feelings that being in nature evoked. This naturally leads to an emphasis on achievement satisfactions and a de-emphasis on appreciative satisfactions.

4.4 Antihunters

As might be expected, some characteristics of antihunters are simply the opposite of hunters. The antihunter, for example, is more often urban than rural. Duda et al. (2010: 44) state that in 2006, nationwide, 70% of urban residents, 72% of suburban residents, 80% of residents in small cities or towns, and 89% of rural residents approved of hunting in general. This, of course, makes sense as rural residents are much likelier to hunt or to know a hunter, and knowing a hunter is the single most important indicator of hunting approval for nonhunters. The antihunter is also more likely to be female. Twenty percent of females disapprove of hunting compared to only 13% of males. Females are also more likely to prohibit hunting on their land and to express a desire to join antihunting organisations. Differences are even more marked when motivational specifications like "hunting for recreation" are included, with more than twice as many women as men disapproving (Duda et al. 2010: 45). Antihunters are more likely to be well-educated than hunters. Fifty-one percent of those with no college education strongly approve of hunting, declining to 43% among holders of a bachelor's degree and 40% of those with a postgraduate degree. Young persons are more likely to be antihunters than older persons, and the same goes for persons of colour. One of the lowest approval ratings of hunting (25%) is found among full-time students, who are often also young, urban, and more likely to be female. Antihunters are also much more likely to be liberal than conservative. This is reflected on social media in the way that hardcore pro-hunting advocates often use derogatory names for liberals about anyone who displays antihunting sentiments.

The conclusion of Duda et al. is that being 18–24 years old, female, Hispanic or Latino, and a full-time student are all characteristics that are strongly correlated with disapproval of hunting. Living in a large city or urban area is of a medium correlation strength.

Antihunters as a combined group represented 29% of Kellert's studied sample. They were—unsurprisingly—mostly characterized by very high moralistic scores. Their moralistic concern extended far beyond hunting to many kinds of human exploitation of animals. Antihunters tend to see hunting as just one out of many examples of animal mistreatment. This may tell us something about what is at stake for the hunter compared to the antihunter. For the hunter—who presumably does not also fight bulls or conduct medical experiments on animals—hunting is the only activity that matters. If antihunters should succeed in taking away hunting, many hunters will lose their identity. Some utilitarian hunters may even lose their livelihood. If antihunters fail, then they may accept the loss but move onto an equivalent (for them) win against factory farmers the next day. Hunting, one might say, is a battle for antihunters but the whole war for hunters.

Humanistic scores were also high for antihunters but did not approach their moralistic scores. Dominionistic and utilitarian scores were expectedly very low, while ecologistic scores were relatively high. No significant trends were found in their naturalistic or knowledge of animals scores (Kellert 1978: 24–27).

Twenty-four percent of hunters fell into the "high knowledge" of wildlife category while 17% of antihunters did. Kellert found that while high knowledge hunters and high knowledge antihunters of course thought very differently on some key moral questions, they were surprisingly similar demographically and regarding their positive attitudes towards animals. Notable demographic differences that remained were that the high knowledge antihunters were significantly more often female and unmarried (Kellert 1978: 38–40).

Distinguishing between types of antihunters, Kellert found that three groups represented the antihunting stance: a humanistic, a moralistic, and a much smaller ecologistic group. The humanistic antihunter is characterized by a strong emotional identification with animals. Humanistic antihunters were particularly closely attached to pets. Their concern for the emotions and feelings of pets extended to a general concern for the pain and suffering experienced by all animal individuals. The emphasis on the welfare of individuals—particularly large, charismatic mammals—was very strong with humanistic antihunters. They were correspondingly characterized by an absence of concern for species or habitats. Humanistic antihunters consistently viewed hunting as nothing but killing for the fun of killing or a desire to inflict suffering. Humanistic antihunters demonstrated profound incomprehension as to why someone would kill "innocent," "beautiful," and "kind" animals (Kellert 1978: 28–29). This marks them as the most relevant antihunters to keep in mind for Chapter 7 where we discuss the role of anthropomorphism for antihunting sentiments.

The moralistic antihunting stance is more intellectually sophisticated. While the humanistic antihunting stance is based on emotions and gut feelings, the moralistic stance relies on moral reasoning. Aside from Peter Singer and Tom Regan, moralistic antihunters might also base their arguments on Joseph Wood Krutch, Cleveland Amory, or Albert Schweitzer. Moralistic antihunters tend to view wildlife management as the invention of sport hunters. They argue that the purpose of game management is to artificially create an overpopulation of game to kill rather than to keep naturally growing populations in check. Moralistic antihunters are also concerned about hunters, perhaps using the Kantian argument that hunting (which they consider cruelty to animals) may lead to cruelty towards humans. Some respondents even expressed that their antihunting stance was more about hunting's effect on humans than the consequences for animals. For many moralistic antihunters, hunting is associated with—and not much different from—other forms of violence in society like war or violent crime. Some even think that the desire to hunt is a psychopathic tendency (Kellert 1978: 29–34). It is extremely common to see antihunters causally label all hunters as "psychopaths" or "psychos," although the label is much more commonly attached to hunters they regard as trophy hunters, an identification often assisted by the depiction of those hunters in a trophy photo (see Chapters 3.7 and 8.1).

The ecologistic antihunting stance assumes that hunting leads to the endangerment and extinction of the hunted species, which in the modern world is of course rarely a valid claim. John Mitchell writes the following about the ecologistic antihunting stance, which he associates with the "hyperbolic" antihunter:

> As for the third type of anti-hunter, I use the modifier "hyperbolic" advisedly, for his argument is richly steeped in extravagant overstatement. In claiming that hunters upset the natural balance by causing species extinctions, the hyperbolist plays somewhat the same game as the moralist: He resorts to the lineup. He places the sport hunter next to the market hunter of yore and hopes that the audience won't notice the difference. (Mitchell 1981: 80–81)

The ecologistic attitude also leans towards a criticism of conservation hunting on the grounds of anthropocentrism.

The final type of antihunter, merely of our own observation, can be termed the activist antihunter. One might compare the activist antihunter to the football hooligan. The football hooligan is not always very interested in football. Some football hooligans are just interested in fighting, shouting, and being a part of a boisterous group of likeminded individuals who seek confrontations with other groups of hooligans from opposing teams. Supporting the team is an excuse for these activities, and so is the sport. Whatever team this hooligan supports might be replaced with any other team at random, and it would make little difference. The hooligan's sport could have been tennis if tennis had a hooligan culture. Whether the hooligan's team wins, loses, or ties is not necessarily important; any result can be made to work as an excuse for violent confrontation. This hooligan lives for the collective identity of being a hooligan, and hooliganism is the point.

The activist antihunter is similar. To the activist antihunter, activism is a lifestyle, and hunting is simply an outlet. The hunting activity may have been chosen by happenstance. The activist antihunter could just as easily be (and in practice may well also be) anti-fascist, anti-nuclear, anti-guns, anti-fur, or anti- something else (or pro-something for that matter), as long as it involves the collective identity of a group that engages in activism and confrontation.[10] The activist antihunter is characterized by not having a great amount of knowledge of hunting or wildlife. Some

---

[10] We do not claim that needing to belong to a group is a characteristic exclusively of those who become antihunters. As we will discuss in Chapter 4.5.2, belonging to a group of hunters and finding security in established hunting traditions is a strong motivator for some hunters as well.

vital basic propaganda is picked up and some slogans are memorized, but since the activist antihunter is not particularly interested in the subject of hunting per se but merely interested in the lifestyle and image that follows from being an activist and part of a group, the knowledge of animals or hunting does not go very deep. The activist antihunter, like the hooligan, is drawn to the social community that exists among activists and the excitement and adrenaline rush of confrontational or even violent activism. Activities that appeal to the activist antihunter involve being part of a group of activists and direct confrontations with hunters, such as hunting sabotage, vandalism, or protests. To a lesser extent, the activities of some people online, especially on social media, can also be considered activist antihunting. Both authors have engaged on Twitter, for example, with prominent animal activists who are as likely to post antihunting statements (often wildly untrue and comfortably fitting Mitchell's hyperbolic antihunter) as they are to protest fur farming, whale killing in the Faroe Islands, or pet abuse. Such people are activists for what they perceive to be animal rights, and hunting falls squarely into this jurisdiction, but the subject is treated no less superficially than the many other animal rights issues that they embrace.

4.5 More on Hunting Motivations

Kellert presented most of his research on hunting motivations in sources from 1978 and 1996, and, while we have supplemented with more updated sources, the archetypes that Kellert developed are still widely recognized as holding true today. Of course, other authors indicate signs of other kinds of motivations that were perhaps less relevant in 1978 or 1996 than they are today. Some motivations are more recent variations of Kellert's attitudes, which deserve further attention. Some are altogether different or relate specifically to trophy hunting and so the rest of this chapter focusses on motivational discussions not covered by Kellert's work.

4.5.1 New Nature Hunting

There is a trend, or some connected trends, in many Western countries but perhaps especially in Northern Europe, of simplifying certain aspects of modern life.[11] *Local* is one of the keywords if you are part of these trends. The trendiness of localness is largely a reaction to global climate change and other environmental issues. Where before it was trendy to fly around the world and stay in expensive hotels, flaunting

---

Some people are hunters because of the social aspects of hunting in the same way that some people are antihunters because of the social aspects of antihunting activism.

[11] While predominantly a Western world trend, it is gaining followers elsewhere too. For example, one of China's most recently famous internet stars, (helped along by the Chinese leadership's attempt to promote traditional Chinese values) with over 70 million followers is Li Ziqi. She documents on social media her subsistence life of making food, clothing, furniture, etc. in a rural Chinese village.

material wealth, people who fly are now being shamed for their contribution to climate change. The same goes for people who drive big cars or perhaps even drive at all. The environmentally responsible holiday is to explore parts of your own country and to use environmentally sensitive transport. We visit family, we sleep in tents, and we ride bikes. If we go abroad, we try to go by train. Greta Thunberg has been a driving force behind this trend in the last few years, and since 2020—because of Covid-19 restrictions on international travel—it has become even more popular.

Taking an interest in local culture and history is popular too. Nordic mythology and paganism have seen a resurgence of interest in Scandinavia, and runic "Viking" tattoos have become more common. Restaurants adapt to the trend by experimenting with local ingredients instead of being responsible for additional carbon emissions by importing ingredients from abroad. New Nordic food has been very trendy for restaurants and home cooks in Scandinavia for the last ten to fifteen years. The focus of Nordic food is to use traditional, fresh, local, seasonal, and organic ingredients. The Danish restaurant Noma is a Nordic food restaurant, famous for using unusual local ingredients, combined and prepared in innovative ways. Knowing the source of one's food and how it was treated is also part of the local food trend. "Fra jord til bord" is the Danish expression, translating to "from soil to table." Similar trends are apparent throughout Europe and beyond, reflecting simultaneously the importance of knowing the provenance of the food we eat and topophilia (literally, the love of place), a description of the strong sense of place and identity that has come to the fore recently.

Time is another keyword. A desire to slow things down is a reaction to the fast pace of modern life. Everyone is always working and always stressed. Women especially seem to struggle, as they try to combine the perfect careers with being perfect wives and perfect mothers and keeping in perfect shape. When we get off work, we still need to be online and available. We must not miss anything. Working less, not answering work calls after working hours, and perhaps retiring early yields time for family and interests, for rest and meditation. Showing that one has time to spare is a reaction to trying to seem important by showing that one has no time to spare. Having time is a new way of showing off a life in balance.

Norway has become the capital of "slow TV." Slow TV is unedited and sometimes live television that shows long, uninterrupted, and uneventful programs such as (real examples) a train driving through the Norwegian landscape, a boat trip, or perhaps just someone knitting a hat. Denmark has some slightly less slow TV, but still following the trend of calm and simple, almost meditative television. Two of the most popular Danish television shows in recent years are *Bonderøven* (a partly affectionate and partly derogatory term for a farmer; it translates directly to peasant ass) and *Nak og Æd* (Kill and Eat). Bonderøven is about a Danish small-scale semi-subsistence farmer and everything that he does in the course of daily work on his very traditional and unmechanized farm. Nak og Æd shows a hunter and a chef (Jørgen Skouboe and Nikolaj Kirk), as they travel to somewhere in the world with a mission to hunt a specific animal and then cook and eat it. Characteristically "gentle" UK-based TV shows like the Great British Bake-off, and The Repair Shop are similar reflections of this change of pace.

Being *offline* is closely related to having time. Offline culture is a reaction to online culture. Online culture is being present on all the social media and using it to present a facade of a perfect life. Online culture is to always—phone in hand—be available and respond instantly to calls and messages. Online culture is checking the phone constantly out of fear of missing out (FOMO), or to check how many likes your photo yielded. Offline culture is to quit social media, and to not have to answer the phone if it rings or respond to messages instantly. By being offline, you gain time.

Nature is the final keyword. Closely related to the trendiness of everything local is the trendiness of the primitive and the natural. CrossFit gyms have replaced the advanced exercise machines of now "traditional" gyms with tractor tires, wheelbarrows, and ropes. Paleo-dieting, where one seeks to emulate eating habits of the Palaeolithic, has been trendy for many years, and so has raw food dieting. Foraging for mushrooms and berries in nature has gained enormously in popularity, following from both the nature- and local food trends.

Hunting checks all these boxes. It is a primitive activity that takes place locally and in nature.[12] It is offline, and it is a slow and meditative endeavour. In no small part thanks to the popularity of Nak og Æd, hunting has been on the rise for the last fifteen or so years in Denmark. Young people, students, and especially teenage girls are much more common among new hunters than they have been before. Some Danish efterskoler[13] have chosen to focus on nature- and hunting education and offer opportunities to obtain a hunting license. This introduces some children to hunting that do not come from hunting families and were unlikely to be introduced to it otherwise.

Hunting meets our desire to know how our food was produced. Just as we want to know exactly what transpired on a carrot's journey from the soil to our plate, we want to know how our meat ended there. When we buy meat in the supermarket, we know very little. We have some vague awareness that organic or free-range should equate to better animal welfare than conventional meat, but the rearing and treatment of livestock is generally a closed world to most of us. If we want exact traceability and to verify for ourselves that the animals we eat had a good life and a humane death, we can either take a very close interest in specific farms and their husbandry and slaughtering, or we can go hunting. Hunting fits very well into the lives of many of the urban and well-educated women who often have these concerns. Animal welfare thus becomes an important hunting motivation for many new female hunters in Denmark who do not think that eating meat is morally wrong (or at least not wrong enough that they are willing to give it up). 6.5% of new Danish hunters (11,587 of 177,103) in 2019 were female. While not as high a percentage as in USA, it is the highest percentage and total number that has been registered in Denmark. The recruitment of males is slightly down in recent years. Sweden and Norway began concerted efforts to recruit female hunters already in the nineties, long before Denmark (Olesen and Baagøe, 2009: 13). The result is that 14.5% of registered Norwegian hunters were

---

[12] We write locally because going on hunting trips abroad is obviously not part of this hunting trend.

[13] Directly translates to afterschools. An efterskole is a kind of voluntary, independent boarding school in Denmark, where kids can finish the last 2–3 years of their primary education.

female in 2018, and around 9% of Swedish hunters in 2017. And these percentages are only increasing.

As suggested by Gjerris et al. (2016), the growing number of Danish women who take up hunting may also be a feminist statement. Instead of criticising the male-dominated activity of hunting, some women choose to show that they can participate as equals. This is increasingly the case for social activities and professions that used to be dominated by men. These women are mentioned by Olesen and Baagøe (2009: 21, NB translation) as "women for whom the hunting license was a box that had to be ticked in a world where all options are available for women." Olesen and Baagøe (2009: 17) noted that as many as half of the new Danish female hunters in 2009 had a graduate or postgraduate education, while the national average was about one in four. New male hunters were typically well-educated as well. This is a stark contrast to the composition of hunters in 2000, where 40% had only primary schooling, i.e., until around 15 years of age.

The new Danish female hunters are in many ways the same as the nature hunters described by Kellert. They tend to have been children in a rural or suburban environment and to have family members who hunted or still hunt. Two thirds are introduced to hunting by either a husband or father and two thirds have a partner who also hunts (Olesen and Baagøe 2009). Yet their hunting is also part of societal trends that were not around when Kellert wrote about hunters in 1978. No nature hunters that Kellert described in 1978 were ever trendy in urban environments because they hunted. In these environments, being a hunter sends important signals, the most important of which is probably *time*.

4.5.2 Rituals and (Male) Identity

Kellert suggested that to some dominionistic hunters especially—characterized by their greater interest in humans than animals—the social interactions that are often a part of hunting may be more important than interacting with wildlife. For some hunters, the rituals, traditions, and social bonding are the primary motivations for going hunting altogether. This seems to increasingly be the case. Duda et al. (2010: 68–71) note that 9% of American respondents hunted primarily to be with family and friends in 1980, while 27% did so in 2006.

This motivation might be particularly important in a trophy hunting context, as there seems to be something inherently more social about big game hunting in the US than small game hunting or bird hunting, as noted by Decker et al.:

> Big game (i.e., deer) hunting is special in a social-psychological context and is a socio-cultural event for many hunters and their families/friends. (1984: III)

Nowhere is this special social status of deer hunting more visible than at deer camp. Deer camps or hunting camps are a North American phenomenon most of all, and "camp" is where hunting traditions are particularly important. Simon Bronner, in his research for *Killing Tradition*, asked a group of hunters in a Pennsylvania deer camp why they hunt. Their answer is a good portrayal of socially motivated hunting and the importance of tradition and rituals:

> A cascade of stock answers flowed: connecting with nature and the wild, the thrill of the hunt, getting away from everyday routine (and their wives). They knew it was not about providing food anymore, and one of the more vocal members got to the "heart of the matter": "When we were younger, we looked forward to the kill, the big shot, but I tell you now it's about the rituals and the traditions of camp; it's about being with these guys. And lately, we've been more into our rituals and traditions because they've been more important to us." Pointing to a board behind him displaying cut shirttails from the 1950s to the present, framed by a buck's head on top and photographs of the group gathered at deer camp, he said, "This is why we hunt," and the other men nodded their assent. (Bronner 2008: 34–35)

Swan, similarly, writes that

> [h]unting camps are today's versions of the secret societies of tribal hunting cultures, the social groups that men have formed for thousands of years to support the ethics of the hunt and adulthood as well as to have a good time. At their worst, they are drinking societies where little hunting is done and most of the time is spent drinking and playing cards. At their best, hunting clubs are a touchstone of meaning, magic, mystery, and service. (Swan 1995: 138)

While each group of hunters has their own traditions, Bronner notes that some are notably common among American hunters at hunting camps. Two of these are shirttail cutting, which is when the shirttails are cut off the shirt of a hunter who missed a shot at a buck[14] and "blooding" where someone who has just killed their first deer has its blood ceremonially smeared on their face. Others take the blooding ritual further and drink the deer's blood or eat part of its raw heart, liver, or testicles (Bronner 2004: 16).

Many hunting rituals may easily be related to the masculinity of the hunting activity. The blooding ritual is clearly a "becoming a man" ritual, while the shirttail cutting symbolizes a castration of the impotent hunter:

> When the shirttail is cut by a gang of men, the buttocks are being exposed, suggesting in the company of men that he is prone to homoerotic attack [...] Added to the symbolism is the act of cutting which leaves the shirt ineffective, and the man therefore symbolically castrated. Thus the offender is humiliated because he is being emasculated or feminized [...] By symbolically castrating him, his campmates are reminding him of the consequences of his impotence in the face of the buck. In the drama of the hunt, configured as a male combat ritual between two physically imposing, and therefore sexually, potent forces, the man needs to muster his phallic gun as the sign of prowess against the horned buck. Missing the shot undermines the phallic power of the gun for his fellows, since it is the main extension of the man given potent symbolism when loaded. In the ritual is a reaffirmation of the campmates' potency and manliness. (Bronner 2004: 23–24)

Rituals such as blooding and shirttail cutting are important rituals because of their powerful symbolism. They are also important because they set deer camp very distinctly apart from the hunters' normal lives. Rituals at deer camp are a way of emphasizing that where we are now is somewhere completely different from where

---

[14] Hunters typically wear flannel shirts with shirttails.

we are in our normal everyday lives, and that different rules apply. Setting deer camp apart from modern life via rituals in this way is important when the motivation for hunting is to "get away," which will be discussed shortly.

Outside of male hunters' ritualistic traditions and the interpretation of hunting itself as a symbol of male potency and sex, hunting has historically served as a tradition that creates and maintains masculine identities or as rites of initiation into manhood, as covered throughout Chapter 2. This has been evidenced at least since Ancient Greece where young men of standing were expected to hunt as training for war. African explorer-hunters like Cumming, Harris, and Selous and the members of The Shikar Club were masculine role models of their time (Mckenzie 2000), and Roosevelt believed that hunting was an essential source of manliness:

> The chase is among the best of all national pastimes; it cultivates that vigorous manliness for the lack of which in a nation, as in an individual, the possession of no other qualities can possibly atone. (Roosevelt 1893: 7–8)

Similarly, B&C sought "to promote manly sport with the rifle," as stated at the very beginning of its constitution. Even Thoreau—notoriously ambivalent about hunting and often very critical—said that a boy's education would be neglected with the absence of hunting (see Chapter 2.9).

Hunting serves the purpose of reinforcing masculine identity particularly for men in vulnerable positions who may struggle in the areas of economy, employment, and education. We touched on this subject in Chapter 4.3.2 about utilitarian hunters. To the American rural poor, the ability to provide for one's family through hunting in times of unemployment or other economic hardship is a source of pride. It provides at the very least the identity of "hunter" to fall back on, should other personal narratives fail. A hunter is a survivor. Lisa Fine—studying the historical importance of hunting for the identity and well-being of Michigan autoworkers—noted that

> [t]here is no question that many men who took (and take) to hunting did so because it was considered one of the few remaining authentically masculine and white, male-only activities left in an increasingly feminized and integrated world. (Fine 2000: 816)

This is not exclusively an American phenomenon. Linda Marie Bye, for example, discusses the importance of hunting for rural working-class men in Norway:

> In many rural Norwegian wooded regions, elk hunting plays a central role in the shaping of the masculine rural identity. [...]
>
> [T]raditional masculine activities like hunting, outdoor life, tinkering with engines, and 'hard drinking' have become increasingly important in the construction of male identity in rural areas when local boys cannot find work in their home town. In the arenas of the job market and public life where the local boys lose out to the local girls, traditional masculine leisure activities are the primary source for gender identity. (Bye 2003: 145–146)

Hunting's purposes of creating and maintaining masculine identities and providing venues for male bonding and ritualistic traditions rely on a distinct separation of hunting trip from everyday life and on the different rules that apply. This distinction allows the hunting trip to be a way to "get away" from everyday life, discussed next.

4.5.3 Getting Away

Hunting to get away is very reminiscent of the trends of simple living and reconnecting with nature, discussed above. One may hunt to get away from the stress of modern life. Gasset called this an escape from the human condition, and there is easily more human condition to escape from today than there was in Gasset's time:

> [M]an, projected by his inevitable progress away from his ancestral proximity to animals, vegetables, and minerals—in sum, to Nature—takes pleasure in the artificial return to it, the only occupation that permits him something like a vacation from his human condition. (Gasset 1942: 111)

He continues later:

> Man is a fugitive from Nature. He escaped from it and began to make history, which is trying to realize the imaginary, the improbable, perhaps the impossible. History is always made against the grain of Nature. The human being tries to rest from the enormous discomfort and all-embracing disquiet of history by "returning" transitorily, artificially, to Nature in the sport of hunting. (Gasset 1942: 121)

When we stopped hunting (for our survival) and started farming, we escaped from nature and found security in culture. Now that our survival is no longer at stake, we escape from culture to find meaning in nature by hunting again. We thus find something akin to Kellert's nature hunting motivation here with the notion of hunting to get away. Yet there are variations of getting away. One may hunt because of a desire to participate in nature, which is the motivation of the nature hunter. One may also hunt because of a desire to escape from culture, which is the motivation of getting away:

> Getting away was a recurrent theme in my interviews. For some, like Russ, hunting got him away from work. Tommy also valued getting away from the hassle of managing a business—employees who didn't show up for work, customers who never seemed satisfied, haggles with insurance companies, and mounting regulations of the hazardous materials he used in the garage gave him good reason to get away now and then. But it was also clear that, as for several others with whom I spoke, getting away also meant getting away from the responsibilities of being a father and husband. (Dizard 2003: 110–111)

Getting away from work seems common for men and women; getting away from a partner or children seems to be a characteristically male motivation. Women who hunt usually have husbands who hunt. Sharing hunting experiences tends to be important for both in these cases. Men who hunt much more rarely have a wife who also hunts and getting away from said nonhunting wife is not an uncommon motivation for men. Once again, we may look towards bumper stickers for thematic indicators. While the hunting as sex theme is the more common inspiration for such merchandise, another genre plays instead on the hunt being men's opportunity to escape from their oppressive wives:

- Every day I live with fear. But occasionally I leave her and go hunting.
- I love it when my wife lets me go hunting.
- ATTENTION—I love my wife—when she lets me go hunting.

- We interrupt this marriage to bring you the—HUNTING SEASON!

These all underscore the theme that the man is dominated by his wife but may, if he is so permitted, find sanctuary in his hunting. This idea of hunting being a man's sanctuary is reflected also in the hunting paraphernalia discussed in Chapter 3.6.3, where "man cave" was used as keywords to describe hunting trophies for sale on eBay. The narrative—be it true or not—behind the man cave is that the house is the woman's domain, and the man is being allowed to live in it by her good graces. Husbands being forced to sleep on the couch is a good example of this narrative.[15] The man cave gives a husband one place in the house where he can make all his bad decisions and implement his awful ideas of interior decoration without being overruled. His wife gets to relieve herself of all responsibility of that room's appearance and activities. She gets to be the generous wife for allowing him his little space (often a basement, a garage, or an attic). In his man cave, he can have big, ugly loudspeakers on the floor, a huge television with a gaming console, a home bar, and hunting trophies on the walls.

Hunting being a male activity is what has traditionally allowed hunting to be a man's sanctuary from his wife or family. As hunting trophies turn a basement into a "man cave" and therefore sanctuary, so does the gun and the word "hunting" turn an outdoors experience into a hunting trip, and therefore sanctuary. If it were a walk or a hike, then the wife could come along, but because it is hunting—and most women are not interested in hunting—a man can (try to) claim it as his sanctuary. Just so, we get hunters who bring along a gun for their hunting trip but no ammunition. An intention to kill animals is not necessary to create the sanctuary; only a façade that is believable enough to convince the wife is necessary. This may have an air of theatre to it, however, because the wife is probably perfectly aware of what is going on. She must notice that he never shoots anything. She allows her husband to maintain the illusion of hunting, just as she might allow a man cave in the house.

Hunting is, in effect, the "outdoor man cave." It is therefore to be expected that some men, those who need hunting as their man cave, would oppose the rise of women hunting. If a group of male hunters get invaded by just one wife, then the hunt is no longer a man cave. The humour, subjects of conversation, and rituals would not be the same if someone's wife was present. Letting one woman inside sets a precedent. A sanctuary is only a sanctuary if everyone agrees what it is a sanctuary *from*.

Of course, getting away is not only a social thing. Some hunt to get away from everything, including other people. Some hunt to seek out the solitude that is hard to come by in modern life. Hiking or nature watching can be excellent ways of seeking solitude, but they serve a different meditative purpose than hunting. If there is no other purpose to being in nature than simply being in nature then the mind is still occupied with everything that occupies it in day-to-day nonhunting life. Hiking is not the "vacation from his human condition" that Gasset referred to. If the meditative purpose is to rid oneself of all concerns and thoughts that relate to modern life, then

---

[15] A woman may of course be a victim in a household in much more serious ways but having to sleep on the couch seems to be a punishment reserved for men.

hunting serves the purpose better. When hunting, one is not just in nature for the sake of being in nature; one is in nature—to use the terminology of Morris Jones again—with a project. A long time may be spent there, just sitting still and still also thinking, but because one is also *hunting*, there is an awareness of and connection with nature and prey that is only there because of this project. This alertness—this *participation*—removes the hunter from modern life in a way that just observing nature would not. That is "getting away."

4.5.4 *The Conservation Justification*

Marty Kheel—recalling Chapter 2.10—called the hunter whose motivation for hunting was conservation the "holist hunter" and proceeded to criticize the motivation. We suggest replacing Kheel's criticism of the holist hunter with the argument that no such motivation even exists, which is why we call it a "justification" in the headline above. Hunters surely *defend* hunting with conservation rhetoric, yet no hunters can seriously claim that conservation is their primary motivation for hunting. It is certainly hard to imagine the hunter who, on a cold and rainy day before the break of dawn, and while really wanting nothing more than to stay in bed, goes out hunting anyway because of a sense of arduous duty to regulate game populations. That hunting may serve wildlife management purposes and fund conservation is a bonus and a justification. John Mitchell came to a similar conclusion about hunters whose motivational rhetoric relies on hunting as a public service to wildlife:

> Hunting man must arise in the cold before dawn, dress and arm himself expensively, travel long miles into the country, trudge steeply up hill and down vale, shoot unerringly, immerse hands and forearms in the steaming guts of the fallen beast, then wrestle the meat all the way home to the freezer. And all of this, for what? For the sake of the animal? For the advancement of wildlife euthanasia? It hardly seems possible. And it is not. (Mitchell 1981: 21)

The subject is also broached by Forrest Wood, Jr., who said that the question of why one hunts is often cause for misunderstanding between hunters and antihunters, because it may be interpreted in two ways: One interpretation refers to the motivation for hunting, and the other refers to the justification for hunting (Wood, Jr. 1997: 16–17). Conservation is a justification and not a motivation.

Hunters are often conservationists, and they may even work with conservation professionally, but when they go hunting, they are motivated by something else. For the trophy hunters who pay a fortune to hunt exotic game abroad, the conservation-as-motivation appears even more dubious. If someone claims to pay tens of thousands of dollars to go to Africa to shoot a lion mainly because of a desire to conserve lions, then why not just dispense with the hassle of travelling and hunting altogether and donate the money directly to lion conservation? It is, after all, usually not the shooting of a lion that contributes to conservation but the money that is paid for doing so. Paying $50,000 without shooting a lion usually does more for lion conservation than paying $50,000 for shooting one, and with a much-reduced carbon footprint.

4.5 More on Hunting Motivations

Kellert described hunting motivated by conservation as the "Ecologistic Hunting Attitude" and had the same conclusion:

> This hunting attitude, like the aesthetic, may be regarded as of relatively minor importance. While an ecologistic rationalization is often invoked to justify the need for hunting, it rarely operates as a personal motivational source for hunting. That is, while some may argue that hunting is a necessary form of game management in the absence of natural predators and in situations of serious overpopulation, few persons are motivated to hunt because they believe they are practicing essential wildlife management. (Kellert 1978: 18)

Yet conservation is very frequently mentioned by hunters if one asks them why they hunt. It is possible that hunters are so used to defending hunting that if anyone asks them why they do it, they immediately become defensive and want to justify hunting. It is also possible that they are embarrassed by their motivations—maybe they think their motivations sound selfish or, to borrow from antihunting rhetoric, psychotic—so they employ the justification in place of their motivation. Or maybe their motivation is too hard to explain or has not been fully thought through. It is probable that many hunters do not quite know or cannot quite formulate why they hunt. But they have all learned to say that hunting is conservation. In fact, the phrase "hunting is conservation," often said as "hunting *is* conservation," is so widely used in hunting contexts, forums, publications, and events that it would be almost impossible for a hunter to avoid it. "Hunting funds conservation" would be a far better framing in many contexts and would have the other advantage of sheltering hunters from the obvious retort that "hunting is not conservation," a statement that is trivial to prove correct in the vast majority of cases. In some cases, it is possible for the hunting itself, as opposed to the revenue it generates, to be conservation. For example, population regulation that prevents overgrazing, or the removal of low-fertility old males that prevent other males from breeding. However, such cases are exceptions and perhaps more accurately fall under the umbrella of culling. A hunter seeking out an old male antelope as a trophy is neither culling nor conserving; they are simply reducing the population by one. It is the fees they pay to hunt that have the potential to conserve, not their action of hunting.

Ann S. Causey indicates that both the hunter's argument for hunting (that it contributes to conservation) and the antihunter's argument against hunting (that killing is immoral) are in a sense irrelevant. Both sides are just trying to justify a position that is held regardless of the justification's validity (Causey 1989). In many if not most cases, the hunter who becomes convinced that hunting does not contribute to conservation would still hunt, and the antihunter who becomes convinced that their moral objection to hunting is illogical would still be against it.

The real motivations—the real reasons that some people are hunters and some are antihunters—are not so much reasoned as they are *felt*. No amount of rational argument is going to change the mind of the hunter or antihunter unless they also come to *feel* that their position is wrong. To borrow a famous quote, usually attributed to Irish poet and satirist Jonathan Swift (1667–1745): "Reasoning will never make a Man correct an ill Opinion, which by Reasoning he never acquired."

4.5.5 Costly Signalling

Signalling theory comes from evolutionary biology, where the traits of animals serve to send certain messages to others. Costly signalling is when an animal demonstrates an ability to bear a handicap. This "handicap principle" is commonly illustrated with the example of the peacock's tail. The feathers of a peacock are a practical disadvantage; they require a lot of nutrition to grow, and they are cumbersome if the peacock is attacked by a predator and needs to escape. What the peacock signals to the peahen with these feathers is that he has been able to shoulder the handicap of his large plumage and still survive. A weaker male would have succumbed to starvation or predators if it had grown the same plumage as a strong male. The peahen, seeking the strongest mate, therefore tends to seek out a male with an impressive plumage.

The costly signalling theory has commonly been applied to the study of hunter gatherer cultures' hunting of large prey species, which seem like they are not otherwise worth the effort or danger to hunt. It is also easily applied to the purchasing of luxury items in modern society. Exclusive, high quality and expensive items that are status symbols are known as Veblen goods (after economist Thorstein Veblen, who introduced the term "conspicuous consumption"). Paying large amounts of money for handbags, impractical shoes, and jewellery shows off wealth that is considerable enough to not only ensure survival but to allow for useless but conspicuous spending in addition. Recent attempts have been made to use the costly signalling theory to explain trophy hunting as well. In the ambitiously titled article *Why Men Trophy Hunt*, Darimont et al. (2017) argue that the costs associated with trophy hunting can be the very motivation for hunting. The authors argue that the hunting of large, dangerous, and costly (rarer and therefore more expensive) animals can be explained by the additional status associated with such animals in trophy award systems and among trophy hunter peers. By showing off the kill of a lion, one shows off the ability to absorb the costs of killing a lion. Historically, and in some subsistence cultures, part of this cost would be the danger involved, as Darimont et al. describe here:

> The Maasai peoples of eastern Africa themselves describe lion killing as a manhood ritual that awards prestige to the hunter who first spears the animal. Why is status awarded? Simply put, killing large, dangerous, and/or rare prey is difficult with high failure risks that impose costs on the hunter. Accordingly, successful hunts signal underlying qualities to rivals and potential allies. (Darimont et al. 2017: 2)

Rituals of manhood linked to the killing of lions is seen in other groups within lion ranges as well. For example, the Nandi, in the highlands of the Rift Valley, and the Ishenyi and Ikoma communities on what is now the western border of the Serengeti National Park had rituals of manhood that involved killing lions. A ritual called aghaso involved young men hunting lions or leopards, with the man killing the lion taking trophies and eating a portion of the animal's heart (Somerville 2019b).

For foreigners hunting in Africa, the cost is mainly the price paid for the hunt, which makes trophy hunting a simple demonstration (or signal) of affluence, just like purchasing jewellery or a flashy car.

4.5 More on Hunting Motivations

Finally, Darimont et al. argue that the show-off behaviour of hunters on social media and the large audience that they can reach make signalling theory even more relevant:

> Worldwide social media creates for trophy hunters a vast audience to which to boast. Signalling the costs of hunting are no longer restricted to carcass displays in small social groups. Men can now communicate an ability to absorb trophy hunting costs not only to their immediate social group but also—with the help of the Internet—to a global audience. (Darimont et al. 2017: 3)

An earlier proponent of a natural selection hypothesis very similar to costly signalling was Randall Eaton. His claim was that

> trophyism [the hunting for and displaying of trophies] has for thousands of years affected the reproductive success of males, and thus has been strongly influenced by natural selection. Males would benefit in competition for females if they could demonstrate and advertise their hunting prowess. Display of hunting prowess to other males could also affect the probability of a male being accepted as a hunting partner, thus increasing his fitness. A trophy usually communicates bravery or skill, both desirable traits for a hunting or warring associate. (Eaton 1978: 110)

This hypothesis involved the three predictions that (1) "Males Benefit in Competition for Females If They Advertise Hunting Prowess," (2) "Status Among Males Should Correlate with Success in Trophy Hunting," and (3) "Trophy Values Should Evolve According to Rank of Hunters Among Predators" (Eaton 1978: 110–112). Eaton finds evidence for his first two predictions in a series of examples from indigenous communities from across the world of how hunting prowess and trophy display lead to greater reproductive success and more partners or wives, as well as greater status among other males. For the third prediction, Eaton finds evidence, among other places, in the wearing of leopard skins and lion manes in African cultures, tiger skins in Eurasian cultures, and in what he calls *pseudo-trophyism*, which is the acquisition of trophies of dangerous predators through nonhunting means, so as to make oneself seem more intimidating and avoid being attacked by warring groups. Such "cheating" has in turn provoked strict procedures to catch the people who try to cheat and ensure that trophies are proper proofs of hunting success:

> If selection has so strongly favored trophyism as a social evaluator, cheating and attempts at detection by males are to be expected in any society in which trophies are displayed. Among [indigenous] hunting societies, testimony is insufficient—the prey must be located or presented (e.g., in the !Kung, Thomas, 1958). Among warring societies, claims of kills not observed or substantiated by a trophy, such as a scalp, are not recognized (e.g., in the Kurela, Matthiessen, 1962). Hunter organizations such as the Boone and Crockett Club act as detectors, demanding clear proof of the kill. I have personally observed extreme ostracism extended to hunters who buy their trophies. (Eaton 1978: 114)

For some hunters, their trophies cease to just be individual proofs of and mementoes from separate successful hunts; they become collector's items, discussed next.

4.5.6 *Collecting and Collections*

Despite indications that hunters motivated primarily by trophies are very few, it is necessary to discuss this motivation separately for two reasons. First, as mentioned, surveys about this could be misleading as hunters—even anonymously—may be unwilling to admit it if they are motivated by trophies. Hunters generally do not wish to tarnish the reputation of hunting, so they have an incentive to respond to surveys in ways that talk down the importance of trophies. Second, while hunters who seem motivated mainly by trophies may be very few in numbers, they appear disproportionately often in popular- and social media because they generate outrage. Outrage online generates site visits, and site visits generate money. Also, antihunters naturally seek to present trophy-motivated hunters as representatives of all hunters, because it is one of the least sympathetic motivations.

There are two kinds of trophy-motivated hunters that warrant special attention and, though they usually overlap, the motivation is slightly different. The first kind collects trophies not just for the value and meaning of the individual animals but for how the trophies constitute missing pieces, contributing to a trophy *collection*—the hunter-turned-collector, in other words, discussed shortly. The second kind, to be covered in Chapter 4.5.7, is one for whom trophies are not themselves the motivation but rather the only means (barring cheating) of obtaining that which really motivates, which is a name in trophy record books.

When the late nineteenth century and early twentieth century Europeans hunted in Africa and sent wagonloads of hunting trophies back to their respective home countries—commonly for existing museums or to start their own museums upon returning—they became some of the early collectors of hunting trophies. John MacKenzie said about the late nineteenth century trend of collecting exotic (mostly African) trophies, that

> [l]ike all collecting interests, it stimulated a desire to possess representative collections from specific areas and a desire for rarities to fill gaps. (MacKenzie 1988: 29)

"Filling gaps," is key to the analysis of hunting trophies as collections and trophy hunters as collectors.

Here is a comment by Robert Kubick, as he showed Ted Kerasote around in his trophy rooms:

> When you have taken as many big game animals as they have, as I have, some of these trips are not *hunts*, they're *collecting*. For instance, if you're going to hunt a bontebok, you'll find them in only about three places in South Africa, on ranches. And I've taken one—walking—but they're about as wild as these tame reindeer. It was brutal. But I wanted a complete collection, not from an ego standpoint, but to have most of the animals of the world. So I shot it. (Kerasote 1994: 129, emphasis in original)

The quote shows that filling gaps can be a motivation of the hunter-turned-collector. Kubick admits freely that his killing of a tame bontebok on a hunting ranch was collecting—not hunting.

But what are the factors that make someone a collector or a noncollector? What constitutes a collection? Most big-game hunters collect the trophies of their kills, but this does not necessarily make them *collectors* nor their trophies *collections*. Catherine Carey may assist in clarifying this statement:

> A noncollector would not value an additional unit to a collection any differently than the value of the marginal unit of the good in ordinary use. A collector would place value on both the marginal unit for its ordinary use and on its contribution to the collection. (Carey 2008: 338)

With this distinction, Kubick is clearly a collector, because his bontebok had value as a contribution to the completion of his collection and little to no value as a marginal unit. The bontebok had value because it filled a gap.

Susan Pearce defines a collection as

> a group of objects, brought together with intention and sharing a common identity of some kind, which is regarded by its owner as, in some sense, special or set apart. (Pearce 1995: 159)

And McIntosh and Schmeichel define a collector as

> a person who is motivated to accumulate a series of similar objects where the instrumental function of the objects is of secondary (or no) concern and the person does not plan to immediately dispose of the objects. (McIntosh and Schmeichel 2004: 86)

These definitions also clearly recognize Kubick as a collector and his "group of objects" as a collection, because he was intentionally bringing together a series of similar objects (trophies), sharing a common identity ("most of the [hunted] animals of the world").

Another important aspect of the collecting question concerns utility. McIntosh and Schmeichel note that the "function" of the objects is of secondary importance, while Carey speaks of "ordinary use." We find similar thoughts about utility in Baudrillard (1994: 8), who notes about collector's items, that

> any given object can have two functions: it can be utilized, or it can be possessed. The first function has to do with the subject's project of asserting practical control within the real world, the second with an enterprise of abstract mastery whereby the subject seeks to assert himself as an autonomous totality outside the world. The two functions are mutually exclusive. Ultimately, the strictly utilitarian object has a social status: think of a machine, for example. Conversely, the object pure and simple, divested of its function, abstracted from any practical context, takes on a strictly subjective status. Now its destiny is to be collected. Whereupon it ceases to be a carpet, a table, a compass, or a knick-knack, and instead turns into an 'object' or a 'piece'.[16]

---

[16] Though Baudrillard writes that the two categories of functions—utilized and possessed—are mutually exclusive, there are collector's items, for which we think the distinction could be at least blurry. There are people who collect sneakers for instance (called "sneakerheads"), who will pay tremendous amounts of money for rare and old or limited-edition sneakers. Yet they commonly also wear the sneakers (though maybe not the most expensive ones, and they probably do not go jumping in puddles with them). It seems hard to claim that a shoe, worn and used, does not have utility. It seems equally hard to suggest that they are not collector's items. One could perhaps interpret sneaker collecting in the same way as we interpreted the elephant's foot turned into an umbrella stand. The elephant's foot umbrella stand is a trophy and not a utility item because its *primary* function is that of trophy. As we stated, any old bucket can be an umbrella stand, so turning an

It is tricky to speak of the "function" or "utility" of a hunting trophy, as it seems to have none. Upon closer consideration though, there are a few possible functions of hunting trophies. These were mostly covered in Chapter 3.6. Most hunting trophies are collected and displayed because they are reminders of hunting experiences and of the animals that were killed; to evoke memories is their function. The function of a trophy may also be to honour or pay tribute to the animal, but the function of trophies can also, and perhaps more often than many hunters will admit, be to show off the hunting prowess or wealth of the hunter.

Regardless of the function of trophies on their own, these functions become secondary once trophies become missing pieces, needed to fill gaps in collections. This is when hunters, as in the example with Kubick, are collectors, and their trophies collections.

So, when is a trophy hunter a noncollector? Hunters are noncollectors if they do not consider that a new trophy has somehow furthered their "collection," which they also do not consider a collection in the sense that Kubick considered his trophies a collection. We can delineate the noncollection trophies owned by noncollector hunters with the word "assortment" to prevent confusion. Kubick's trophies constituted a collection because the trophies—like the bontebok—had value as contributions to the collection, and not just value as marginal units. Many hunters' assortments are exactly the sums of their parts because new trophies do not fill gaps. Each individual trophy is valued on its own and not as a contribution to the assortment, which is therefore not a collection.

Most hunters' assortments of trophies are not planned or collected with a goal in mind. As such, there are no gaps to fill. The average hunter does not hunt many different species of animals. Many hunt only one. Some are devoted to turkey hunting or duck hunting for example. Many are exclusively deer hunters, though they may hunt a few different species of deer. A typical Danish deer hunter hunts mostly roe deer and/or red deer, but fallow and sika (*Cervus nippon*) are also options. UK deer hunters also hunt these species as well as Reeves muntjac deer (*Muntiacus reevesi*) and Chinese water deer (*Hydropotes inermis*), though most focus on those species present in the areas they hunt. Many US hunters solely hunt whitetail deer. These hunters are usually noncollectors who own assortments of trophies. For example, eight roe deer trophies and three red deer trophies would rarely constitute a collection to the hunter, because there is no theme to guide the future of the collection. One more of either trophy would most likely do nothing towards "completing" the assortment and turning it into a collection.

This is not to say that eight roe deer and three red deer trophies *cannot* be a collection, or the hunter not a collector. If the hunter is trying to, say, collect trophies from different geographical regions, then the trophies become meaningfully different, providing gaps to be filled, and a way of furthering or completing the collection. He

---

elephant's foot into one is just an alternative way of displaying a hunting trophy. Similarly, one does not need a pair of sneakers that cost more than a car to protect one's feet. Their primary function is that of collector's item; the option of wearing them is just a convenient way of showing them off.

or she has then become a collector. However, most such hunters will not have the collecting mentality or think of their trophy assortments as collections.

Nor is it to say that an assortment must *necessarily* have an end-goal—a way of completion—to be a collection. Some scholars define a collection very simply as whatever the collector thinks it is (Pearce 1995: 22). Yet few who would think of themselves as collectors and of their assortment as a collection would be interested in just collecting many duplicates of a few different items.

The objects in a collection, according to Pearce (1995) and McIntosh and Schmeichel (2004), should be "similar" or share a "common identity," but they are generally not just duplicates. One roe deer trophy will normally not be considered different from another roe deer trophy in the sense that they contribute differently to the assortment. There would be little reason to collect another roe deer trophy if the hunter was a collector, which indicates that most hunters are not. Most Danish hunters for example are probably to be considered noncollectors, because they are opportunistic. They may desire the trophies, and they will often consider bigger trophies better than small, but they hunt for whatever big trophy they have good opportunity to hunt and not with serious consideration of how a trophy might fit in with whatever trophies they already possess. Collecting is generally not an opportunistic endeavour.

So, when does this typical Danish hunter cross over from being a hunter with an assortment of trophies to being a hunter-collector with a collection of trophies? Let us consider another hypothetical example. Some hunters may hunt all of roe deer, red deer, fallow deer, and (less commonly) sika deer, which are the four species present in Denmark. One could imagine a Danish hunter, having shot and collected trophies from roe deer, red deer, and fallow deer, wanting to also—*for this reason*—shoot a sika deer and collect its trophy, so as to have killed and collected trophies from all the Danish deer. Now there is a theme, which is all Danish deer species. There is a goal and a guideline for how to proceed to complete the collection, which is to collect a sika. The sika trophy, in Carey's words, will have value not just as a marginal unit, but as a contribution to the collection as well. This makes the hunter a collector and the assortment a collection.

Yet a collection with four different pieces is not much of a collection. This trophy hunter turned trophy collector is unlikely to be very passionate about collecting a sika as the fourth piece to complete the collection. Collectors who are very passionate (to the point where their passion sometimes seems like madness to the uninitiated) are usually collectors who have worked on a collection with many often rare and/or valuable pieces for many years. A hunter who seeks to win one of SCI's most prestigious World Hunting Awards, requiring the collecting of hundreds of different trophies, is likely very passionate about completing his collection because of the time, money, and effort invested in getting thus far.

Although most trophy hunters are not to be considered collectors, it seems reasonable to assert that most trophy hunters flirt with the collector's mentality from time to time. After all, this only takes a hunter considering that it might be fun to hunt something for a trophy that they do not already possess.

4.5.7 Record-Book Hunting

Where hunting truly meets collecting is in the world of record books. We have examined the dire consequences of record books in Chapters 3.4 and 3.5 already, but a few more notes on record books as a motivational factor are necessary.

Here, we adopt the terms record-book hunter and record-book hunting from John Wootters. Wootters—contrary to most trophy hunters today, preferring "big game hunter" or just "hunter"—proudly called himself a trophy hunter. He did so because he made a distinction between trophy hunters and record-book hunters:

> The record book, whoever keeps it, will always be controversial, and so will the idea of hunting expressly "for the book." Obviously, I heartily approve of trophy hunting, or I would not be writing this volume, but there's a distinct difference between trophy hunting and record-book hunting. The latter can—and often does—result in certain abuses, both legal and otherwise, of superb game animals. I personally know of trophies poached or shot illegally (at night, for example), purchased, and even faked by taxidermists which today rank high in the august listings of North American big game records.
>
> To me, there is a subtle but essential difference between pursuing a great buck for the joy of matching one's wits and woodsmanship against his and for the deep satisfaction of winning the match and possessing him, and doing it solely for the public acclaim among hunters. The former is trophy hunting, while the latter is record-book hunting. (Wootters 1977: 18)

More so than anywhere else, SCI's World Hunting Awards program is where collecting and record-book hunting come together. The World Hunting Awards encourage and require the completion of collections of kills. Here, recalling the definitions of a collector vs. a noncollector, each trophy has value almost solely as proof of a missing piece in a collection—a box to be ticked. The marginal value is low to non-existent, and the whole awards system is nothing if not for the public acclaim among hunters.

Record-book hunting and especially the World Hunting Awards are interesting and—while overlapping—subtly different from trophy collecting in a discussion of motivations, because getting one's name in a record book does not literally require the retainment of the physical trophies in a hunting trophy collection. It requires only the registration of the killing of each individual animal. The trophies invariably end up in a physical collection regardless, but they *need* not. Record-book hunting as a motivation may be interpreted as the collecting of proofs of achievement in the form of hunting awards more so than the collecting of physical hunting trophies.

References

Abbey, Edward. 1996. Blood Sport. In *A Hunter's Heart – Honest Essays on Blood Sport*, ed. David Petersen, 1997 (First 1996), 11–16. New York: Henry Holt and Company.

Adams, Carol J. 2003. *The Pornography of Meat*. New York and London: Continuum.

Ashcroft, Lord Michael. 2020. *Unfair Game: An exposé of South Africa's captive-bred lion industry*. Britback Publishing.

References

B&C. 2018. Fair Chase Media Kit 2018. *Boone and Crockett Club*. https://www.boone-crockett.org/pdfs/MK_2018_MediaKit.pdf (accessed 18 July 2019).

Bageant, Joe. 2007. *Deer Hunting with Jesus—Dispatches from America's Class War*. New York: Crown Publishers.

Baudrillard, Jean. 1994. The System of Collecting. In *The Cultures of Collecting*, ed. John Elsner and Roger Cardinal and trans. Roger Cardinal), 7–24. London: Reaktion Books.

Beattie, Geoffrey. 2020. *Trophy Hunting: A Psychological Perspective*. London and New York: Routledge.

Bronner, Simon. 2004. "This Is Why We Hunt": Social-Psychological Meanings of the Traditions and Rituals of Deer Camp. *Western Folklore* 63 (1–2): 11–50. Hunting and Fishing.

Bronner, Simon. 2008. *Killing Tradition—Inside Hunting and Animal Rights Controversies*. Lexington: The University Press of Kentucky.

Brown, Alex. 2020, December 14. The Pandemic Created New Hunters. States Need to Keep Them. *PEW Stateline*. https://www.pewtrusts.org/en/research-and-analysis/blogs/stateline/2020/12/14/the-pandemic-created-new-hunters-states-need-to-keep-them (accessed 29 June 2022).

Bye, Linda Marie. 2003. Masculinity and Rurality at Play in Stories About Hunting. *Norsk Geografisk Tidsskrift—Norwegian Journal of Geography* 57 (3): 145–153.

Carey, Catherine. 2008. Modeling Collecting Behavior: The Role of Set Completion. *Journal of Economic Psychology* 29: 336–347.

Causey, Ann S. 1989. On the Morality of Hunting. *Environmental Ethics* 11 (4): 327–343.

Colleoni, Elanor, Alessandro Rozza, and Adam Arvidson. 2014. Echo Chamber or Public Sphere? Predicting Political Orientation and Measuring Political Homophily in Twitter Using Big Data. *Journal of Communication* 64 (2): 317–332.

Darimont, Chris T., Brian F. Codding, and Kristen Hawkes. 2017. Why Men Trophy Hunt. *Biology Letters* 16: 1–3.

Decker, Daniel J., R. William Provencher, and Tommy L. Brown, Outdoor Recreational Research Unit. 1984. Antecedents to Hunting Participation: An Exploratory Study of the Social-Psychological Determinants of Initiation, Continuation, and Desertion in Hunting. *Public Attitudes Toward Wildlife and its Accessibility (W-146-R-8)*. Final Report.

Dizard, Jan E. 2003. *Mortal Stakes—Hunting and Hunters in Contemporary America*. Amherst and Boston: University of Massachusetts Press.

Duda, Mark Damian, Martin Jones, and Andrea Criscione. 2010. *The Sportsman's Voice—Hunting and Fishing in America*. Responsive Management/Venture Publishing.

Duda, Mark Damian, Martin Jones, Tom Beppler, Steven J. Bissell, Andrea Criscione, Patrick Doherty, Gregory L. Hughes, Claudia Reilley, and Alison Lanier. 2015. *California Hunters' Opinions on Deer Hunting, Season Structure, Hunting Regulations, and Deer Management*. Responsive Management.

Eaton, Randall. 1978. The Evolution of Trophy Hunting. *Carnivores* 1/1: 110–121.

Ebeling-Schuld, Alena M., and Chris T. Darimont. 2017. Online Hunting Forums Identify Achievement as Prominent Among Multiple Satisfactions. *Wildlife Society Bulletin* 41 (3): 523–529.

Eliason, Stephen L. 2012. A Statewide Examination of Hunting and Trophy Nonhuman Animals: Perspectives of Montana Hunters. *Society and Animals* 16: 256–278.

FACE. 2016, September 27. The Economic Value of Hunting in The EU. *Federation of Associations for Hunting and Conservation of the EU*. http://www.face.eu/sites/default/files/documents/english/economia_della_caccia_27_9_2016_en.pdf (accessed 23 July 2019).

Fine, Lisa M. 2000. Rights of Men, Rites of Passage: Hunting and Masculinity at Reo Motors of Lansing, Michigan, 1945–1975. *Journal of Social History* 33 (4): 805–823.

Gasset, José Ortega y. 1942. *Meditations on Hunting*. Translated by Howard B. Wescott, 1985. New York: Charles Scribner's Sons.

Gjerris, Mickey, Sune Borkfelt, Christian Gamborg, Jes Harfeld, and Sara Kondrup. 2016. *Jagt—Natur, mennesker, dyr og drab*. Aarhus: Klim.

Jones, Allen Morris. 2012 (First 1997). *A Quiet Place of Violence—Hunting and Ethics in the Missouri River Breaks*. Montana: Bangtail Press.

Kalof, Linda, and Amy Fitzgerald. 2003. Reading the Trophy: Exploring the Display of Dead Animals in Hunting Magazines. *Visual Studies* 18 (2): 112–122.

Kellert, Stephen Robert and Joyce K. Berry. 1980. Knowledge, Affection and Basic Attitudes toward Animals in American Society—Phase 3. *United States Department of the Interior Fish and Wildlife Service*. Washington: U.S. Government Printing Office 20402.

Kellert, Stephen Robert. 1978. Attitudes and Characteristics of Hunters and Anti-Hunters and Related Policy Suggestions. *United States Department of the Interior Fish and Wildlife Service*. https://catalog.data.gov/dataset/attitudes-and-characteristics-of-hunters-and-anti-hunters-and-related-policy-questions-197 (accessed 10 September 2018).

Kellert, Stephen Robert. 1996. *The Value of Life—Biological Diversity and Human Society*. Washington: Island Press/Shearwater Books.

Kerasote, Ted. 1994. *Bloodties—Nature, Culture, and the Hunt*. New York: Kodansha International.

Klineburger, Chris R. 2010. *Gamemasters of the World—A Chronicle of Sport Hunting and Conservation*. Skyhorse Publishing.

Leopold, Aldo. 1968 (First 1949). *A Sand County Almanac, And Sketches Here and There*. New York: Oxford University Press.

Loveridge, Andrew J. 2018. *Lion Hearted: The Life and Death of Cecil & the Future of Africa's Iconic Cats*. New York: Regan Arts.

Luke, Brian. 1998. Violent Love: Hunting, Heterosexuality, and the Erotics of Men's Predation. *Feminist Studies* 24 (3): 627–655.

Luke, Brian. 2007. *Brutal—Manhood and the Exploitation of Animals*. Urbana and Chicago: University of Illinois Press.

MacKenzie, John M. 1988. *The Empire of Nature—Hunting, Conservation and British Imperialism*. Manchester and New York: Manchester University Press.

Marvin, Garry. 2010. Living with Dead Animals? – Trophies as Souvenirs of the Hunt. In *Hunting Philosophy for Everyone: In Search of the Wild Life*, ed. Nathan Kowalsky, 107–117. Wiley-Blackwell.

McIntosh, William D., and Brandon Schmeichel. 2004. Collectors and Collecting: A Social Psychological Perspective. *Leisure Sciences: An Interdisciplinary Journal* 26 (1): 85–97.

Mckenzie, Callum. 2000. The British Big-Game Hunting Tradition, Masculinity and Fraternalism with Particular Reference to the 'The Shikar Club.' *The Sports Historian* 20 (1): 70–96.

Messmer, Terry A. and Jody W. Enck. 2012. Human Dimensions of Wildlife Use Management. In *Human Dimensions of Wildlife Management*, ed. Daniel J. Decker, Shawn J. Riley, and William F. Siemer, 2nd edition, 203–220. Baltimore: Johns Hopkins University Press.

Mitchell, John G. 1981 (First 1980). *The Hunt—A Provocative Exploration of Hunting in North America*. Penguin Books.

Olesen, Bodil og Jette Baagøe. 2009. Diana og Artemis's Døtre – Kvindelige jægere i Danmark. In *Årbog for Kvindemuseet i Danmark 2009*, 7–33. Kvindemuseet i Danmark.

Opensecrets. 2019, April 14. Safari Club International. *Opensecrets.org—The Center for Responsive Politics*. https://www.opensecrets.org/lobby/clientsum.php?id=D000000757&year=2017 (accessed 14 April 2019).

Parker, Ian. 2017, January 16. Killing Animals at the Zoo. *The New Yorker*. https://www.newyorker.com/magazine/2017/01/16/killing-animals-at-the-zoo (accessed 28 November 2018).

Pearce, Susan M. 1995. *On Collecting: An investigation into collecting in the European Tradition*. London and New York: Routledge.

Petersen, David. 2010 (First 2000). *Heartsblood: Hunting, Spirituality, and Wildness in America*. Washington and Covelo: Island Press/Shearwater Books.

Responsive Management and National Shooting Sports Foundation. 2008. The Future of Hunting and the Shooting Sports—Research-based Recruitment and Retention Strategies. Produced for the U.S. Fish and Wildlife Service under Grant Agreement CT-M-6-0. Harrisonburg, VA.

Robey, Charity. 2020, February 6. Deer Overpopulation Meets Its Match: Women Who Hunt. *The New York Times*. https://www.nytimes.com/2020/02/06/nyregion/deer-hunting.html?smid=nytcore-ios-share&utm_campaign=revuenewsletter&utm_medium=email&utm_source=Revue%20newsletter (accessed 18 March 2020).

Roosevelt, Theodore. 1893. *The Wilderness Hunter*. New York: G.P. Putnam's Sons.

SCI. 2020. Media Kit. *Safari Club International*. https://safariclub.org/wp-content/uploads/2020/10/2021_SCI_Media_Kit.pdf (accessed 14 February 2021).

Scully, Matthew. 2002. *Dominion—The Power of Man, the Suffering of Animals, and the Call to Mercy*. New York: St. Martin's Press.

Somerville, Keith. 2019. Lions and Humans: Conflict, Conservation and Coexistence. London: Routledge.

Swan, James A. 1995. *In Defense of Hunting*. HarperSanFrancisco.

Swan, James A. 1999. *The Sacred Art of Hunting*. Minocqua: Willow Creek Press.

Thoreau, Henry David. 1882 (First 1854). *Walden*, 19th edition. Boston: Houghton, Mifflin and Company.

Trophy. 2017. *Trophy* (film by Shaul Schwartz and Christina Clusiau). http://trophy.film/about.php.

USFWS. 2018. 2016 National Survey of Fishing, Hunting, and Wildlife-Associated Recreation. *U.S. Department of the Interior, U.S. Fish and Wildlife Service, and U.S. Department of Commerce, U.S. Census Bureau*.

USFWS. 2019. Recruitment and Retention of Hunters and Anglers: 2000–2015–2016 National Survey Addendum. *U.S. Department of the Interior, U.S. Fish and Wildlife Service, and U.S. Department of Commerce, U.S. Census Bureau*.

Vicente, Joaquín, Radim Plhal, Jose A. Blanco-Aguiar, Marie Sange, Tomasz Podgórski, Karolina Petrovic, Massimo Scandura, Anna Cohen Nabeiro, Guillaume Body, Oliver Keuling, Marco Apollonio, Ezio Ferroglio, Stefania Zanet, Francesca Brivio, Graham C. Smith, Simon Croft, Pelayo Acevedo, Ramon Soriguer. 2018. Analysis of Hunting Statistics Collection Frameworks for Wild Boar Across Europe and Proposals for Improving the Harmonisation of Data Collection, 1–33. *ENETwild Consortium/European Food Safety Authority (EFSA) Supporting publication*.

Wood, Jr., Forrest. 1997. *The Delights and Dilemmas of Hunting—The Hunting versus Anti-Hunting Debate*. Lanham, New York, and London: University Press of America, Inc.

Wootters, John. 1977. *Hunting Trophy Deer*. New York: Winchester Press.

Young, Robert L. 1989. The Protestant Heritage and the Spirit of Gun Ownership. *Journal for the Scientific Study of Religion* 28 (3): 300–309.

Chapter 5
Trophy Hunting and Conservation

To some extent we have covered conservation already. Chapters 2.7 and 2.8 summarized the history of game laws and national parks in USA and Africa and conservation as we understand it today. This began with American sportsmen in the late eighteenth century and was later picked up again and popularized by figures such as Pinchot, Grinnell, and Roosevelt in the late nineteenth century. Chapter 2.9 discussed transcendentalist perspectives on conservation and preservation, and introduced Aldo Leopold, the father of what we today call wildlife management science. Chapter 2.10 folded in animal rights and ecofeminist objectives to conservation arguments, while Chapter 3.4 noted some of the conservation work of famous hunting organisations. In Chapter 4.5.4 the notion of conservation as a motivation (or, more accurately, a justification) for hunting was addressed. But there is one question related to conservation, arguably more important than all others, which has not been covered yet. That question is: Does trophy hunting support wildlife conservation today?

5.1 Hunting- and Trophy Trade Regulations

It is important to make a clear distinction between the uncontrolled and unregulated hunting of the past, and modern-day trophy hunting. Historical, legal, market hunting for meat and skins, and legal, intensive elephant hunting for ivory for example are very different topics with only tangential links to modern-day trophy hunting. Likewise, the killing of lions and rhino in Africa to facilitate land clearance and agricultural expansion, with those species being treated essentially as vermin, is not modern-day trophy hunting. Neither is the unregulated, unlawful killing of animals that comes under the problematic umbrella of poaching, a primary conservation threat for a great many species. A major difference between trophy hunting and present-day poaching,

---

The original version of the chapter was published with incorrect information. A correction to this chapter can be found at https://doi.org/10.1007/978-981-19-9976-5_10

is that trophy hunting is legal. Illegal trophy hunting is, de facto, poaching and if the motivation to poach is the trophy then it becomes trophy poaching, but it is not trophy hunting. More complex to categorise is poaching that is motivated by the same trophy that would motivate a trophy hunter, but for different underlying reasons. Long, heavy elephant tusks, and large, heavy rhino horns are desired by both trophy hunters and poachers, but the poacher is concerned solely with the illegal market value of the product, largely dictated by weight. Another major difference between trophy hunting and poaching (and much of historical market hunting) is that trophy hunting is governed by one or more sets of laws and regulations. Domestic trophy hunting is subject to national laws, and trophy hunting tourism abroad is subject to international laws and their incorporation in national legislations of both producer countries (range countries) and consumer countries. While it is outside the scope of this book to explain all laws and agencies related to hunting activities, it will facilitate the understanding of how instruments of conservation work if one understands at least the most important of the legislative frameworks that incorporate them.

5.1.1 CITES

The Convention on International Trade in Endangered Species of Wild Fauna and Flora (CITES) was conceived in 1960 at the seventh general assembly of the International Union for Conservation of Nature (IUCN). Recognizing that species of conservation concern were often in high demand in countries other than their country of origin, the purpose of CITES was to restrict imports of certain species of plants and animals in accordance with the export restrictions of their countries of origin. The first draft of the convention text was finished in 1964, an initial list of species was ready in 1969, the final convention text was presented in 1973, and CITES was ratified and entered into force—initially for just a few parties—in 1975 (Wijnstekers 2011: 31–32).

As of June 2022, there are 184 contracting parties to CITES, which the 1973 initial text specified could only be states. This was changed in 2013 with the Gaborone amendment, and the European Union joined as the only non-country party in 2015. Most EU members were already contracting parties and CITES regulations have been integrated in, or surpassed by EU law since 1984.

The decision-making body of CITES is the contracting parties, together constituting the Conference of the Parties (CoP). The CoP meets approximately every three years. The implementation of CITES is overseen by a Standing Committee, and scientific advice is provided by an Animals Committee and a Plants Committee. Day-to-day operations are handled by the Secretariat in Switzerland, which is managed by the United Nations Environment Programme (UNEP).

With CITES, the world established that protecting the world's flora and fauna from harmful trade is not just the responsibility of countries in which species live, but a shared responsibility between those "producer" exporting countries and importing consumer countries. CITES is a recognition that preventing poaching, smuggling and unsustainable trading of species is a costly and difficult task, and that prevention would be unsuccessful unless consumer countries complemented producer countries' efforts to monitor and restrict international trade (Wijnstekers 2011: 31–32). This is important in a trophy hunting context, as the consumer countries (often USA or EU-members) tend to be better capable of shouldering the associated costs than producer countries, which are often (but not always) relatively poorer countries especially those in Southern and Eastern Africa.

CITES works by regulating international trade in certain species of wild flora and fauna, alive and dead through permits and certificates that must be presented before specimens are allowed to leave or enter countries. Each party to CITES designates a management authority and a scientific authority responsible for this (Wijnstekers 2011: 47). Despite "endangered" being in the CITES name, species covered by CITES are not necessarily endangered in terms of their IUCN Red List status.[1] The roughly 5,800 populations, species, and subspecies of animals and 30,000 of plants covered by CITES are separated into three appendices:

Appendix I contains

> all species threatened with extinction, which are or may be affected by trade. Trade in specimens of these species must be subject to particularly strict regulation in order not to endanger further their survival and must only be authorized in exceptional circumstances. (CITES 1983: Art. II)

These strict regulations include—most importantly—requirements that "[t]he export of any specimen of a species included in Appendix I shall require the prior grant and presentation of an export permit" (Art. III, Par. 2) and that "[t]he import of any specimen of a species included in Appendix I shall require the prior grant and presentation of an import permit and either an export permit or a re-export certificate" (Art. III, Par. 3). Export permits will only be granted if "a Scientific Authority of the State of export has advised that such export will not be detrimental to the survival of that species"[2] (Art. III, Par. 2a) and that "a Management Authority of the State of export is satisfied that an import permit has been granted for the specimen" (Art. III, Par. 2d). Import permits will only be granted if "a Scientific Authority of the State of import has advised that the import will be for purposes which are not detrimental to the survival of the species involved" (Art. III, Par. 3a) and if "a Management Authority of the State of import is satisfied that the specimen is not to be used for primarily commercial purposes" (Art. III, Par. 3c). "Primarily commercial purposes" is interpreted in CITES to not include hunting trophies.

---

[1] The IUCN Red List of Threatened Species, near universally called "the Red List," is an inventory that details the conservation status of species. It is a very common error to assume that a species' presence on the list indicates that it is in danger of imminent extinction, but even overabundant species are on the Red List; it just means that the conservation status of a species has been evaluated. The status of each species is assessed by an expert group and species are assigned to one of nine categories. These categories are: Not Evaluated; Data Deficient; Least Threatened; Near Threatened; Vulnerable; Endangered; Critically Endangered; Extinct in the Wild; and Extinct. To be considered "threatened" a species should have a status of Critically Endangered, Endangered or Vulnerable. For more information on the IUCN Red List, visit www.iucnredlist.org, where individual species' listings can also be accessed, providing invaluable data and information on conservation status and actions.

[2] This finding is called a Non-Detriment Finding and its details were specified and adopted in resolution 8.6.

Appendix II species are not necessarily endangered, but they are liable to become so, and their trade needs to be closely monitored. International trade in Appendix II species is subject to export permits, as described in Article IV, Par. 2. Requirements for Appendix II export permits are the same as for Appendix I export permits, except for the Article III, Par. 2d concerning verification that an import permit has also been granted. CITES does not require import permits for Appendix II species. Appendix II contains several family listings, in which all species in that family are Appendix II listed: Ursidae (all bears) and Felidae (all cats) are two such examples.

Appendix III species are added by request of parties that already regulate their trade for whatever reason and need cooperation from other CITES parties. Their trade is regulated by Article V, which is comparable to Articles III and IV.

The CoP alone may add and remove species from Appendices I and II or move species between them. All species may be subject to CoP-decided annotations and exemptions. This can mean, for example, that some populations of a species are Appendix I while other populations of the same species are Appendix II (a so-called "split-listing"), or that an Appendix I species may be traded in certain amounts and under certain conditions. Parties to CITES may also, within limited timeframes, mark species as reservations, which means that they are not bound by CITES provisions for those species. For example, Botswana, Eswatini, D.R. Congo, Namibia, South Africa, Tanzania, Zambia, and Zimbabwe have had the giraffe (*Giraffa camelopardalis*) reserved since CoP18 in 2019 when it was added to Appendix II.

CITES Resolution 17.9 recognizes

> that well-managed and sustainable trophy hunting is consistent with and contributes to species conservation, as it provides both livelihood opportunities for rural communities and incentives for habitat conservation, and generates benefits which can be invested for conservation purposes

and acknowledges

> that where economic value can be attached to wildlife and a controlled management system is implemented, favourable conditions can be created for investment in the conservation and the sustainable use of the resource, thus reducing the risks to wildlife from alternative forms of land use. (Resolution 17.9)

CITES therefore also recognizes that restricting international trade in trophy hunting species by including them in Appendix II or upgrading them to Appendix I may or may not benefit conservation. This is why CITES allows for trophy hunting quotas and exemptions made for trophy hunting. In cooperation with other organisations, CITES has published several reports about trophy hunting conservation successes, such as the hunting of bighorn sheep in Mexico and of ibex and markhor in Tajikistan. We will return to these examples later.

How trophy hunting fits into CITES legislation is not an uncomplicated matter. CITES Resolution 12.3 from 2002 defines hunting trophies as such:

> The term 'hunting trophy', as used in this Resolution, means a whole animal, or a readily recognizable part or derivative of an animal, specified on any accompanying CITES permit or certificate, that:

5.1 Hunting- and Trophy Trade Regulations

i) is raw, processed or manufactured;

ii) was legally obtained by the hunter through hunting for the hunter's personal use; and

iii) is being imported, exported or re-exported by or on behalf of the hunter, as part of the transfer from its country of origin, ultimately to the hunter's State of usual residence. (CITES 2002: 5)

Second relevant text is Article VII, Par. 3 of the convention, which states that "[t]he provisions of Articles III, IV and V shall not apply to specimens that are personal or household effects" (Art. VII, Par. 3). Hunting trophies, as outlined in Resolution 13.7, Annex 1, *Guidelines for interpretation of personal and household effects*, are considered personal/household effects and should therefore be exempt from Articles III through V. However, Article VII goes on to stipulate in Par. 3a and 3b that that the exemption shall not apply, if

(a) in the case of specimens of a species included in Appendix I, they were acquired by the owner outside his State of usual residence, and are being imported into that State; or

(b) in the case of specimens of species included in Appendix II:
  (i) they were acquired by the owner outside his State of usual residence and in a State where removal from the wild occurred;
  (ii) they are being imported into the owner's State of usual residence; and
  (iii) the State where removal from the wild occurred requires the prior grant of export permits before any export of such specimens. (CITES 1983: Art. VII)

As noted by Wijnstekers (2011: 217–221), Art. VII, Par. 3b is notoriously complicated to interpret and enforce, but a simplified explanation is that the article text aims to allow the export of specimens if they are part of a household move, but not if they are tourist souvenirs. As (a) and (b) usually apply to hunting trophies, hunting trophies are not generally exempt from Articles III through V and do need export permits (Appendix II) or import and export permits (Appendix I).[3]

However, as noted earlier, any appendix species may be subject to exemptions and annotations that increase or decrease restrictions. This applies to several trophy hunted Appendix I species that have export quotas, an instrument established with resolution 14.7. It is important to realise that these quotas are the maximum allowable hunted individuals, and in practice many of these quotas are not reached, or even approached. This is a common error (or deliberate attempt to misinform) made by individuals and organisations that oppose trophy hunting. For example, Botswana has a quota of 150 cheetahs (see below) but exported none in the period 2012–2022. South Africa has a quota of nine black rhino in 2020 but exported no more than five. To provide an idea of quota ranges, relevant exporting countries, and critical species, these were the Appendix I species that had trophy hunting quotas in 2020[4]:

Leopard (*Panthera pardus*), Appendix I: Botswana: 130, Central African Republic: 40, Ethiopia: 500, Kenya: 80, Malawi: 50, Mozambique: 120, Namibia:

---

[3] Resolution 13.7, par. 3 further stipulates some exemptions that may be relevant to hunting trophies in very rare cases of export to a neighbouring country, but we will not get into that.

[4] This data is extracted from Species+ (https://www.speciesplus.net/species), developed by UNEP and CITES.

250, South Africa: 150, Uganda: 28, Tanzania: 500, Zambia: 300, Zimbabwe: 500, D.R. Congo: 5.
Cheetah (*Acinonyx jubatus*), Appendix I: Botswana: 150, Namibia: 150.
African elephant (*Loxodonta africana*), Appendix I/II[5]: Botswana: 800, Namibia: 180, South Africa: 300, Tanzania: 100, Zimbabwe: 1000.[6]
Nile crocodile (*Crocodylus niloticus*), Appendix I/II[7]: Namibia: 25, Tanzania: 1600, Zimbabwe: 200.[8]
Black rhinoceros (*Diceros bicornis*), Appendix I: Namibia: 5, South Africa: 9.
Markhor (*Capra falconeri*), Appendix I: Pakistan: 12.
Hippopotamus (*Hippopotamus amphibius*), Appendix II: Tanzania: 1200.

A handful more Appendix I species and many Appendix II species are trophy hunted. These are not subject to quotas but instead to the import and/or export permits granted based on Non-Detriment Findings, as per the convention articles.

One of these is the African lion (*Panthera leo*), which is Appendix II except for the population in India on Appendix I.[9] The lion stands out, however, as it has an export quota in South Africa for lion products (bones, skulls, claws, teeth) from farmed (captive bred) lions, traded for commercial purposes.[10] Trophy hunted lions are in CITES subject to standard Appendix II treatment without quotas, but EU has restricted lion imports on its own volition, and many other countries have done so as well, including USA, as we shall see shortly.

5.1.2 USA

Most hunting inside the USA is managed on a state-by-state basis but under the provision of several federal acts, including the Lacey Act of 1900, the Migratory Birds Act of 1918, and the Endangered Species Act (ESA) of 1973. Likewise, some lands are under federal protection by the National Park Service as national parks or national monuments. Federal coordination is handled by the USFWS. State agencies generally require hunters to acquire hunting licenses, which are only valid for hunting in the state that they have been issued in. These licenses usually require that the hunter

---

[5] Elephants are Appendix I, except for populations in Botswana, Namibia, South Africa, and Zimbabwe, which are Appendix II for a limited number of purposes, one of which is trophy hunting (described in annotation 2 of the CITES appendices).

[6] These numbers all refer to tusks, so the 800 tusks for Botswana equate to 400 elephants, etc.

[7] Appendix II in Botswana, Egypt, Ethiopia, Kenya, Madagascar, Malawi, Mozambique, Namibia, South Africa, Uganda, the United Republic of Tanzania, Zambia, and Zimbabwe. Otherwise Appendix I.

[8] Numbers for Nile crocodiles vary a lot because they are a mix of wild and ranched specimens.

[9] Lions in India consist of a single Asiatic lion population of around 500 specimens in the Gir National Park.

[10] TCM traditionally used tiger bones, but since the use of tiger bones for TCM has been banned in China, lion bones have been sought out as an alternative in recent years.

5.1 Hunting- and Trophy Trade Regulations

completes the local state's version of a hunter education and safety course. Multiple hunting licenses for different states may be held by the same person, but they are generally more expensive to obtain for out-of-state residents.

What one may hunt after getting a hunting license also depends on the state. Most states require hunting "tags" in addition to licenses for big game hunting, which are permissions to kill one specimen of a specific species, sex, and age, with a specific weapon, and in a specific area. These tags are used by the local agencies to manage game populations and are one of the most basic ways of doing so. More tags for whitetail deer does (and in general) may be issued, for example, if the local wildlife management agency records that whitetail populations have grown unsustainably large. If the situation is the opposite, where hunting pressure needs to be decreased and there are more willing hunters than available tags, then tags may be issued in a lottery. For popular hunts with only a few available tags, winning this lottery may be like... winning the lottery. Some tags are once-in-a-lifetime opportunities and the hunter that wins one cannot re-enter the lottery in future. Other tag lotteries have time-out limitations whereby hunters cannot enter again for a specified number of years. Tags can vary in price from a few dollars to thousands. Licenses and tags are of course subject to hunting seasons, which again may vary from state to state, from species to species, and from method to method. Archery season is generally longer than rifle season, and in some cases a muzzleloader season extends the hunting opportunities further. Most big game hunting and hunting in general takes place in the fall and over winter. Hunting- and fishing licences, tags, stamps, and taxes on hunting and fishing equipment are the main sources of funding for state agencies, wildlife management, and conservation in USA. Sportsman-generated funds on average comprise more than 75% of state wildlife conservation budgets.

USA is the biggest importer of hunting trophies in the world. The difference in scale between the United States and the rest of the world is shown in Fig. 5.1 which shows the CITES top ten of countries with the most hunting trophy imports for the years 2013–2022.

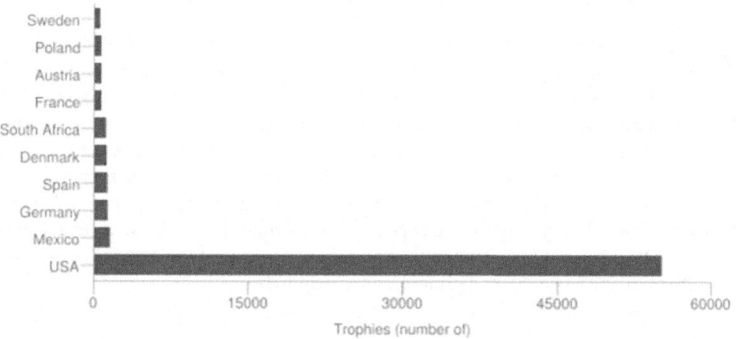

Fig. 5.1 2013–2022 CITES trophy imports by country as reported by exporting countries (https://tradeview.cites.org/en/overview)

Most trophies imported into the United States come from Canada, which accounts for the most commonly imported hunting trophies of snow goose (*Anser caerulescens*), mallard (*Anas platyrhynchos*), and black bear (*Ursus americanus*). Snow goose and mallard are not CITES-listed and therefore not represented in Fig. 5.1. 205,742 trophies were imported from Canada between 2009 and 2014 while 196,941 were imported from South Africa. Then follows Namibia with 41,345, Zimbabwe 22,581, New Zealand 20,782, Tanzania 14,651, Mexico 7282, Argentina 7206, Zambia 5197, Botswana 4198, and Mozambique 4126. Of CITES-listed species imported, the black bear is by far the most numerous on the top-ten with 32,100 trophies imported between 2013 and 2017. The sandhill crane (*Antigone canadensis*, also from Canada) is second at 10,112, and only then do we find mostly African species, starting with lion at 2645 trophies, chacma baboon (*Papio ursinus*) (2552), mountain zebra (*Equus zebra*) (2148), leopard (*Panthera pardus*) (1525), red lechwe (*Kobus leche*) (1506), blackbuck (*Antilope cervicapra*) (1096), Nile crocodile (*Crocodylus niloticus*) (1076), caracal (*Caracal caracal*) (1016), African elephant (*Loxondonta africana*) (984), vervet monkey (*Chlorocebus pygerythrus*) (926), grey wolf (*Canis lupis*) (873), and hippopotamus (*Hippopotamus amphibius*) (839) (Sheikh 2019: 7–8).

CITES is implemented in US legislation mainly in sections 8 and 9 of the ESA, and the ESA is implemented in US legislation in Title 50, Chapter I, Subchapter B, part 17 of the Code of Federal Regulations, which is easily browsed online. CITES is detailed in Code of Federal Regulations, part 23, and §23.74 describes what hunting trophies are and which rules apply to them. The USFWS is designated as both the management- and scientific authority (ESA sections 8A(b) and (c)), overseeing the US implementation of CITES.

The ESA aims to protect species by granting them a status of either "Endangered" or "Threatened" in accordance with ESA section 4(a). These categories are similar in function to Appendices I and II of CITES. However, as Sheikh and Corn (2016) lay out, ESA's Endangered listing generally offers better protection than the CITES Appendix I listing, as exemption from import restrictions requires proof of benefits to species survival rather than just proof of absence of harm:

> The protection of species listed as endangered under the ESA and listed in Appendix I of CITES is not equivalent. CITES allows for the trade in endangered species, *if* trade is not detrimental to the survival of the species. The ESA requires that importing endangered species have a net result of enhancing the survival of the species. For example, cheetahs are Appendix I species under CITES and listed as endangered under the ESA. Under CITES, some countries have been allowed to participate in the limited trade of sport-hunted cheetahs, because it was shown that this trade would not be a detriment to the population. Under the ESA, however, the FWS has not issued permits for importing sport-hunted cheetah trophies, because their take has not been shown to enhance the survival of the species in their range country. (Sheikh and Corn 2016: 10, emphasis in original)

5.1 Hunting- and Trophy Trade Regulations

The ESA also offers more comprehensive consideration of all threats than CITES, which is focused on threats that relate to trade (Sheikh and Corn 2016: 9).

Species categorized as Endangered, according to §17.21(b), may not be imported or exported, but this is subject to §17.22 exceptions. One of these possible exceptions is if USFWS concludes that the taking and importing of a species is estimated to contribute to its conservation. In this case, under the authority of section 10 of the ESA, an "enhancement of survival" permission to import a hunting trophy of an endangered species may be granted. As Sheikh and Corn note (2016: 10), however, requests for hunting trophy imports of endangered species are routinely denied by the USFWS. From 2013 to 2017, only 2.8% of CITES mammal trophies imported to the US were Appendix I. 4.8% were Appendix III and 92.4% were Appendix II (CITES trade data). It is important to remember that most hunting trophies are of species that are not listed on CITES Appendices, which can make them difficult, or in some jurisdictions, impossible to account for. For example, a hunter from the UK who kills an impala, kudu, wildebeest, hartebeest, gemsbok (*Oryx gazella*) or many other species in South Africa and brings back the horns or skin as a trophy, all perfectly legally, need not (and in practice usually will not) be recorded on any database, anywhere.

Species categorized as Threatened are by default offered the same levels of protection as endangered species, but the room for exemptions and special rules is wider. This is allowed via ESA section 4(d), of which one part states that

> [w]henever any species is listed as a threatened species pursuant to subsection (c) of this section, the Secretary shall issue such regulations as he deems necessary and advisable to provide for the conservation of such species.

"4d rules," as they are commonly referred to, are mainly implemented through Code of Federal Regulations §§17.40–17.48. For trophy hunting, the most relevant section is §17.40, which covers mammals.

CITES Appendix II species are according to §17.8 exempt from import permits unless otherwise stated in §17.40. One of the Appendix II species that the 4(d) rule offers more protection in the ESA than it has in CITES is the lion. In legislation that had been underway since at least 2011, on December 23, 2015 the USFWS included the two subspecies of the African lion in the ESA, as reported in the federal register 80/246:

> We, the U.S. Fish and Wildlife Service (Service), determine endangered status for the lion subspecies *Panthera leo leo* and threatened status for *P. l. melanochaita* under the Endangered Species Act of 1973.[11]

---

[11] If this seems confusing when the ESA's treatment of lions is compared to the CITES treatment of lions, it is because the ESA differentiates between threatened and endangered lions based on the ranges and population sizes of the two subspecies of lions (*Panthera leo melanochaita* and *Panthera leo leo*), whereas CITES differentiates between Appendix II and I lions based on their geographic distribution (Africa and India). These distinctions are not functionally the same, as *leo leo* populations exist in both India and West-/Central Africa. The Indian population of *leo leo* is also sometimes—e.g. by IUCN—taxonomically referred to as the Asiatic lion (*Panthera leo persica*). IUCN estimates that there are around 23,000–39,000 *leo melanochaita* worldwide, but

This listed *Panthera leo leo* (sometimes called the northern lion) and *Panthera leo melanochaita* (the southern or black-maned lion) in §17.11, which contains the list of endangered and threatened wildlife, similar to the CITES Appendices. The special regulation for *P. leo melanochaita* in §17.40(r), effectively affords all the protections of endangered species to all lions:

> Lion (*Panthera leo melanochaita*). (1) General requirements. All prohibitions and provisions of §§ 17.31 and 17.32 apply to this subspecies. (2) The import exemption found in § 17.8 for threatened wildlife listed in Appendix II of the Convention on International Trade in Endangered Species of Wild Fauna and Flora (CITES) does not apply to this subspecies. A threatened species import permit under § 17.32 is required for the importation of all specimens of *Panthera leo melanochaita*.

Consequently, all imports of lion trophies have been subject to USFWS import permits since January 2016. These, as described, are subject to very strict requirements and routinely denied so US lion trophy imports have declined sharply since.

Other trophy hunted species that receive greater protection under the ESA than under CITES include African savannah elephant, white and black rhinos, argali sheep, polar bears, cheetahs, and several species of deer, gazelles, and antelopes.

5.1.3 Europe

We will not go into legislative details for individual European countries as the implementation of CITES in EU regulations generally covers the significant international trade in trophy hunted species for Europe. Since EU is a single market without systematic border controls, CITES requires that its standards as a minimum be implemented uniformly by the EU member states.

The basic EU wildlife regulation that implements CITES for EU nations is *Council Regulation (EC) No 338/97 of 9 December 1996 on the protection of species of wild fauna and flora by regulating trade therein* (EC338/97). This regulation

> lays down the provisions for import, export and re-export as well as internal EU trade in specimens of species listed in its four Annexes. It provides for procedures and documents required for such trade (import and export permits, re-export certificates, import notifications and internal trade certificates) and it regulates the movement of live specimens. It also sets out specific requirements for Member States to ensure compliance with the Regulation and to impose adequate sanctions for infringements.
>
> The Regulation also establishes a number of bodies at EU level, i.e. the Committee on Trade in Wild Fauna and Flora, the Scientific Review Group and the Enforcement Group, all of which consist of representatives of the Member States and are convened and chaired by the European Commission. (EC 2020)

---

less than 25,000 seems more realistic (see Chapter 5.2.1). The number of *leo leo* seems to be around 500–2,000.

5.1 Hunting- and Trophy Trade Regulations

The four annexes are EU's version of the CITES appendices and they can be found in EC338/97 or in *Commission Regulation (EU) 2019/2117*, which amends and updates the annexes of EC338/97. They are called annexes so as to avoid confusion between them and the CITES appendices.

The specifics of how CITES is implemented are written in *Commission Regulation (EC) No 865/2006* (EC865/2006). Labels, permits, and other implementation items are located in *Commission Implementing Regulation (EU) No 792/2012 of 23 August 2012*.

In short, as described in Article 3 of EC338/97, Annex A includes Appendix I species, certain Appendix II and III species where EU has adopted stricter measures than CITES, and some non-CITES species. Annex B includes the rest of Appendix II and some non-CITES species. Annexes C and D include Appendix III and other species for the purposes of aligning CITES implementation with other EU legislation such as the Habitats Directive and the Birds Directive.

Article 4.1 of EC338/97 specifies that import permits are required for Annex A species, and Article 4.2 specifies that import permits are required for Annex B species. Export permits are always required. This makes the default EU implementation stricter than CITES minimums, which only requires export permits for Appendix II species. Import permissions for Annex B species are easier to obtain, however, and a positive evaluation regarding the non-detrimental effect of importing an Annex B species "shall be valid for subsequent imports as long as the abovementioned aspects have not changed significantly" (Art. 4.2(a)). Hunting trophies from Annex B species should as such require import permits, but this is changed by EC338/97, Article 7.3, which exempts personal effects and household items from import permit requirements. EC865/2006, Article 57.1 further explains that the derogation in EC338/97, Article 7.3

> shall only apply to specimens, including hunting trophies, if they meet one of the following conditions:
>
> (a) they are contained in the personal luggage of travellers coming from a third country;
> (b) they are contained in the personal property of a natural person transferring his normal place of residence from a third country to the Community;
> (c) they are hunting trophies taken by a traveller and imported at a later date.

Hunting trophies usually meet condition (c) above and are therefore generally exempt from import permit requirements. This applies only to hunting trophies from species in Annex B, however, as stipulated in Article 57.2. In addition, the following six Annex B species are subject to stricter protection than they receive in CITES and do require import permits (following Article 57.3a and annex XIII in EC865/2006): Southern white rhinoceros, hippopotamus, African elephant, argali sheep, lion (*P. leo melanochaita*),[12] and polar bear.

---

[12] The Asiatic lion (*P. leo persica*/*P. leo leo*) is Annex A and therefore, because of Annex XIII to EC865/2006, treated the same as *P. leo melanochaita*.

While EU's trophy import regulations are already stricter than CITES requirements, some EU member states have unilaterally implemented stricter domestic measures than required by both CITES and the EU. France has banned the import of lion trophies and the Netherlands have banned imports of over 200 species, including white rhino, hippo, cheetah, polar bear, and lion. No longer a part of the EU, the UK is considering a blanket ban on all trophy imports, although not currently on exports. Australia has also banned African lion trophy imports (Ares 2019: 8).

5.2 Hunting and Conservation in Africa

In terms of media, government and public attention, trophy hunting is most closely linked to African species hunted in southern and eastern African countries, including South Africa (the primary hunting destination in southern Africa and where captive bred, "canned," lion hunting attracts attention), Zimbabwe (home of "Cecil"), Zambia, Namibia, Botswana, Mozambique and Tanzania. Trophy hunting takes place in around or just over half of Africa's 54 countries but most of it on a very small scale. It is currently banned in Angola, Congo, Gabon, Ghana, Kenya, Malawi, Mauritania, Niger, Nigeria, Rwanda, and South Sudan. Traditional hunting and illegal hunting presumably take place everywhere. Botswana banned trophy hunting in 2014 but lifted the ban in 2019, including the hunting of elephants, a decision that has attracted considerable negative attention from some quarters but was taken in response to increasing elephant populations and human-elephant conflicts. It is likely to also resume trophy hunting of lions in the near future (Cooney et al. 2017; Sheikh 2019).

An estimated fifty million international tourists visit Africa each year, of which only 20,000 are hunters. Despite the low percentage of tourists being hunters, approximately 1.5 million square kilometres of Africa are protected and maintained for hunting purposes. This exceeds the area covered by national parks. Tourist hunters killed around 105,000 animals per year in 2009, which included 640 elephants, 3800 buffalos, 600 lions, and 800 leopards (IUCN 2009; Booth and Chardonnet 2015: 10–11). Wildlife tourism across Africa—pre-Covid-19 statistics—contributes $US29 billion annually to African economies and employs 3.6 million people. Out of this, trophy hunting is estimated to contribute $US217 million (Lindsey et al. 2020). The Timbavati Private Nature Reserve in South Africa, which hosts both photographic and hunting clients, calculated that 24,000 photographic tourists brought in just under a third of the income that just 46 hunters provided (Timbavati 2022). The carbon footprint of photographic tourism there, and the disturbance to wildlife, is several orders of magnitude greater than hunting, and yet pulls in far less revenue.

Most species of mammals that can be legally hunted are part of SCI's World Hunting Awards program, so the *World Hunting Awards Field Journal* is once again useful for exploring which species are of interest to trophy hunters. In this case "Animals of Africa" is the relevant award to look up. Almost 200 different species—most of them antelopes—are hunted for trophies in Africa. The Humane Society—obtaining data about trophy imports from USFWS's Law Enforcement

5.2 Hunting and Conservation in Africa

Management Information System (LEMIS) database for a report in 2016—showed that impala, wildebeest, greater kudu, gemsbok, springbok, bontebok, plains zebra, desert warthog, hartebeest, common warthog, common eland, and nyala were the African species with most imports into the United States from 2005 to 2014. Only after these do we find the first of the Big Five, which is the African buffalo. For comparison, according to this report, 58,423 impala trophies and 17,214 buffalo trophies were imported to the United States in 2005–2014. Lions, leopards, and elephants had between 4000 and 6000 registered US trophy imports in the same period (Humane Society 2016). Sheikh (2019) corroborates these statistics, but it should be noted that as with CITES, interpretations of LEMIS data are often flawed and uncertain. In the years 2009–2014, South Africa exported 196,941 trophies to the US, Namibia exported 41,345, Zimbabwe 22,581, and Tanzania 14,651 (Sheikh 2019). While European hunters tend to focus a little less on southern African countries and a little more on East- and West Africa than American hunters, this is a pretty fair representation of African trophy exports for the primary range countries in the period.

Despite its prevalence, we do not hear much about the trophy hunting of antelopes in the news or on social media. One reason is that most of these hunted species are very common and therefore not listed in the CITES Appendices. This makes trade volumes difficult to track, as CITES (or any other body, usually) does not monitor their imports or exports. Researchers and activists often look to the CITES trade database for documentation of the international trade in trophies. Since trade in common species is unmonitored, CITES species tend to be overemphasized in the trophy hunting discourse.[13] Antihunting campaigns often give the impression that lions and elephants are the most hunted and exported species in much of Africa, perhaps notably South Africa and Namibia, but this only appears to be so because export data for non-CITES species is generally unavailable, and because this portrayal fits the antihunting agenda. USFWS import data, though they exclude hunters from Europe, provide a more accurate insight. Another reason that we do not hear much

---

[13] In addition to CITES not monitoring unappendixed species, CITES data is often deceptively difficult to interpret correctly and is frequently misrepresented even in peer-reviewed literature (see Foster et al. [2014] or Robinson and Sinovas [2018] for related discussions). Hunting trophy data for a lion, for example, may refer to pelts, skulls, claws, and possibly more, meaning that one cannot assume a 1:1 relationship between lions killed and lion trophies imported. A lion can make one trophy, so the trophy number is the same as the number of lions killed, or it can make one skin, one skull, 20 claws and numerous teeth, giving the impression that, rather than one lion being killed, there were perhaps 40 or more. Likewise, elephant tusks might be weighed, and that weight entered in the "number" column as kilograms. It is all too easy to mistake an entry of, say, "35" for 35 elephants instead of 35 kg of ivory from one elephant. Macdonald (2016) writes that from 1991 to 2013, 80 wild lion trophies were reported exported to the UK, while the UK importation records for the same period show that 4 lion trophies were imported. Also, CITES parties may collect and report data in different ways, have joined in different years, etc. Export data and import data often just do not match. 35,000 black bear trophies (hunted in Canada by Americans) were traded in the years 2013–2017 according to CITES data for mammals as reported by importers, which is a dominant first place with lion trophies coming in second at roughly 5000 trophies. Yet black bears do not figure at all in data reported by exporters, while lions remain the same, now seemingly the most traded.

about the trophy hunting of African antelopes is perhaps that these species are not considered as charismatic and iconic as the Big Five. They are much more similar to domestic ungulates in the US and Europe, which Americans and Europeans are more accustomed to as hunting quarry.

This all makes ungulates rather un-newsworthy, but it also makes them cheap. Di Minin et al. (2016) note that while most hunted individuals are among the less famous species, the majority of hunting revenue comes from the few famous ones. Many antelope species can be shot for a few hundred dollars apiece while wild lions in 2019 sold for an average of $33,747, elephants for $45,013, leopards for $23,062, cape buffalos for $12,136, and hippos for $20,375, though note that the African Leadership University *State of the Wildlife Economy in Africa* report suggests lower average prices, including $16,923 for elephant, $10,931 for lion and $4749 for buffalo (Sheikh 2019: 6; Snyman et al. 2021: 91). Black rhinos can go for up to $400,000 on auctions. The Big Five are available only in South Africa and Namibia, whereas the Big Five minus the rhinoceros may be hunted in Mozambique, Tanzania, Zimbabwe, and Zambia. In some cases, hunting tourists may come for one or more of the Big Five but then add some of the cheaper species to fill out the remaining days of a trip, which is usually between 7 and 28 days. Twenty-one days is a classic length of stay for an African hunting safari, but some hunters, with the pressures of the modern world, opt for far shorter trips, including 5–7 days hunting packages, as we saw in Chapter 1.4.

The African trophy hunting debate, then, mostly concerns a handful of charismatic species in a handful of core countries. Most notable of the African trophy hunting range countries are South Africa, Namibia, Tanzania, and Zimbabwe. Kenya and Botswana are quite useful for comparisons with countries that have implemented bans of trophy hunting, in the case of Kenya since 1977 and, in Botswana, for a five-year period recently before its reversal in 2019.

Central problems for wildlife conservation in Africa stem from population growth. Africa's population has increased almost fivefold since 1950 to a population of more than 1.3 billion people today. It is the continent with the highest percentual growth rate, and the population is expected to hit around 2.5 billion in 2050. Africa's cultivated land is expected to increase by 21% and the livestock population to increase by 73% by 2050 (Funston et al. 2016). UN estimates a population in Africa in 2100 of 4.4 billion people (Holechek et al. 2016). IUCN's position paper on trophy hunting from 2016, *Informing Decisions on Trophy Hunting* sums up the scientific consensus as such:

> The primary causes of population declines of the large mammals subject to trophy hunting, such as the African Elephant, African Buffalo, White Rhino, Black Rhino, African Wild Dog and Hartmann's Mountain Zebra are habitat loss and degradation, competition with livestock, illegal or uncontrolled poaching for meat and trade in animal products (ivory, horn, etc.), and retribution killing for human-wildlife conflict. (IUCN 2016: 7)

The idea behind trophy hunting in Africa, and indeed many other locations, is that by letting hunting tourists pay to shoot a limited number of carefully selected individuals in a population, funding is obtained that can be used to protect wildlife species and habitats and support local economies. This prevents land from being claimed for agriculture or other industries, it allows for the hiring of wardens to prevent poaching, it can provide compensations for human-wildlife conflicts, etc. In short, trophy hunting is a land use that relies on wildlife, can provide revenue for people, and can be operated in areas that may not be attractive to other types of tourists.

This idea has worked very well on a number of occasions. One of the success stories is the rhinoceros—both black (*Diceros bicornis*) and white (*Ceratotherium simum*). White rhino numbers in South Africa have increased from around 1800 in 1968 to over 18,400 thanks in large part to trophy hunting and the incentives it offered to landowners and reserves to take on rhinos. White rhinos from South Africa have also been reintroduced to other countries. Their numbers have increased by 67% in South Africa and Namibia since CITES approved rhino hunting quotas in 2004. IUCN provides a detailed example here of the relative importance of the trophy hunting of only seven specimens over eight years on one South African reserve with around 200 rhinos:

> Analysis of eight years' data revealed that only ~18% of the total reserve's operational expenditure was generated from tourism, while trophy hunting generated the bulk of income needed to fund operational expenditure (63%). Over the last eight years, only seven (or <1% of the population annually) White Rhino have been hunted on the reserve, generating (inflation adjusted) US$617,000; with live sales of another 47 White Rhino over the period bringing in an additional US$973,000. (IUCN 2016: 11)

Stories like this one have encouraged over 300 landowners in South Africa to build up rhino herds and protect habitats for trophy hunting purposes (IUCN 2016).

But while trophy hunting works in theory and often works in practice, it can also be fraught with problems that include corruption (at all levels), setting quotas without the necessary population data, overshooting quotas (or ignoring them completely), not distributing revenues fairly, not properly involving rural communities and, in cases, displacement of people and other human rights abuses.

Trophy hunting is an extraordinarily complex topic. It involves numerous species (and in many cases, different populations of species and subspecies) across multiple countries and different land ownerships and land use categories (from community lands to hunting concessions and private reserves). There is no aspect of trophy hunting that is straightforward, and it is not possible to state whether "trophy hunting" is good or bad for conservation. It is only possible to state that "this type of hunting, of this population of that species in this area" or some other complex triangulation of factors is beneficial or not, and only when sufficient data are available. With that in mind, we will illustrate the intricacies of trophy hunting in Africa with an extended discussion of trophy hunting and lion conservation. As an apex predator and keystone species, the conservation issues pertaining to lions work well as a case study of African conservation issues in general, and the amount of data on lion trophy hunting and conservation exceeds that of any other species. Today's opposition to trophy hunting began to ramp up with the hunting of Cecil, and lions are the main focus of African trophy hunting discourse in the media, with elephants, rhinos and giraffes following on behind. Canned lion hunting, as practiced mainly in South

Africa, is only marginally related to the conservation of lions in the wild, so we will save discussion of this issue for Chapter 6.4.

5.2.1 Lion Conservation

MacKenzie writes that the lion

> lies at the centre of hunting books, the most fearsome foe, a dragon substitute, a source of awe and fascination to most African hunters. All were proudest of their careful tally of lion kills. They had contributed to the successful annexation of the 'king of beasts' as a national and imperial symbol. The lion was everywhere in pioneering and hunting images, in a whole variety of moods, the epitome of empire itself. It was as though the virile imperialist and the lion – in India the tiger – were locked in deadly combat for control of the natural world. (MacKenzie 1988: 47)

Lions, it must be understood, hold a special place in the heart of many hunters. Its symbolism to western hunters, dating back to colonial Africa, is that of the African frontier, and the defeat of this apex predator was a symbol of the conquering of Africa by Europeans, just as it was a symbol of men dominating the most ferocious of animals in the Roman colosseum. The lion is the "king of beasts," the "king of the jungle," and the personal symbolic significance to trophy hunters of having killed this mightiest of animals is obvious. Grobbelaar describes the lion as

> the ultimate power animal—a beautiful majestic creature whose name, nature, and image we have used through the millennia as symbols of strength, divinity, royalty, dignity, wisdom, courage, justice, authority, and grace. (2020: 24)

The African lion is one of the species whose CITES Appendix listing has received most public, scientific, and political attention, and especially so since Cecil's death in 2015. Lions are currently listed as Appendix II (except for lions in India, which are Appendix I). This means that they can be hunted and exported if the producer country allows it, deems the hunting sustainable, and issues an export permit, and provided that the consumer country does not restrict its import. Some countries unilaterally do this for lions now, as discussed in Chapter 5.1. Animal rights and animal welfare advocates generally lobby to push lions to Appendix I to further protect them from hunting, while hunting organisations and most lion range countries lobby to keep it in Appendix II, arguing that lion hunting provides for rural livelihoods while sustaining and protecting lion habitat, and mediating human-wildlife conflict.

The lion's IUCN conservation status is "vulnerable," which means that it is likely to become endangered unless its circumstances improve, but its actual status varies greatly from population to population. The latest reliable data on lion populations across Africa comes from WildCRU's Dickman and Hinks (2019), who put the total number at 22,509, though, as they say, some of the country estimates are "guesstimates at best" (2019: 4). Trophy hunting of lions currently takes place in 10 countries, here followed by their estimated lion populations, based on Dickman and Hinks (2019): Tanzania (8176), South Africa (2070), Zimbabwe (1710), Mozambique (1295), Zambia (1095), Namibia (755), Ethiopia (648), Cameroon (274), Benin (200), and Burkina Faso (198). Botswana (2774) is likely to permit trophy hunting of lions again, as they have recently done for elephants (Amy Dickman, personal communication).

The total population has declined by 43% or more between 1993 and 2014, and lions have disappeared from 94% of their historic range. 56 wild lion populations remain in only 25 Africa countries. Only 6 populations have more than 1000 lions, and more than half have 100 or fewer. 40% of the current range of wild lions is in Protected Areas (PAs)[14] and 42% is in areas where trophy hunting of lions is legal. Bauer et al. (2015) note that the mostly unfenced lion populations in the East-, West-, and Central Africa are almost all declining while the commonly fenced populations in the south (Namibia and South Africa) tend to be stable or increasing.[15] Lion population trends seem to correlate with trends in prey species populations; between 1970 and 2005, herbivore numbers declined by 52% in East Africa and by 85% in West- and Central Africa and increased by 24% in southern Africa. This corresponds with Lindsey et al. (2017), who wrote that lions and their prey are only doing relatively well in PAs in Botswana, Namibia, South Africa, and Kenya. The declining lion populations are estimated to decrease by between a third and half of their 2015 population sizes within the next few decades (Amy Dickman, personal communication; Bauer et al. 2015; Lindsey et al. 2017; Macdonald et al. 2017; Dickman and Hinks 2019).

Bauer et al. further concluded that the most severely declining populations were the least monitored populations, as monitoring tends to correlate with conservation efforts. Populations not part of this study were thus expected to decline even further. Monitoring and human presence in general prevents poaching and in so doing can slow or prevent the decline in wildlife populations. This became clear in the wake of Covid-19. Not only does the African tourism industry—which includes both hunting- and photo tourism—provide billions of dollars and millions of jobs, but it also provides a constant tourist presence in the vicinity of wildlife, which by itself prevents poaching. The Covid-19 crisis ultimately and ironically resulted in

---

[14] "Protected Area" is a categorization used by IUCN. 7 kinds of PAs exist, ranging from category Ia, a strict nature reserve where human presence is hardly allowed, to VI in which sustainable use of natural resources—including hunting—may be allowed. Generally, areas managed for hunting are category IV or VI (Booth and Chardonnet 2015: 13–14). See also Dudley (2008) for more information.

[15] Fencing in PAs are commonly advocated as an effective conservation tool (see Packer et al. 2013). One of the leading causes of lion deaths is retaliatory killings, and fences keep lions away from livestock and pastoralists away from lions. It is, however, a very expensive solution because the areas that need to be fenced are large, and fences need maintenance. It is also a problem that fencing provides poachers with materials for snares. This endangers more wildlife and further increases maintenance costs. In the Niassa National Reserve in Mozambique, poachers' snares are responsible for more than half of all lion deaths (Funston et al. 2016: 1). According to Funston et al. (2016: 17), fences also break up ecosystems, obstruct migration routes, and isolate populations genetically. Fenced PAs are still considered "wild" and so is the enclosed wildlife, as the areas are so large. This fencing practice of large PAs should not be confused with the fencing of lions in much smaller areas for canned hunting or other commercial purposes in South Africa.

now unemployed game wardens, whose job it was to protect wildlife from poachers, sometimes being forced to resort to poaching for meat to stay alive.

The threats to lions are generally summarized as habitat loss and degradation, declining prey species, retaliatory and pre-emptive killings to protect humans and livestock, poorly managed trophy hunting, and poaching for traditional Chinese medicine (TCM). A recent study of threats to lions over the period 2005–2018 by Bauer et al. (2020) characterized and ranked eleven major threats to lions in Africa and concluded that human-lion conflicts and bushmeat poaching (prey depletion) are the greatest threats. Then come the small size of some populations (lacking genetic diversity and other associated problems), cropland expansion, livestock expansion, and then finally trophy hunting, right before resource extraction. In general, these conclusions are shared by IUCN's Cat Specialist Group (CITES 2019: 9), and by, for example, Bauer et al. (2015), Macdonald (2016), Lindsey et al. (2017), Macdonald et al. (2017), Loveridge (2018), and Funston et al. (2016). Overall, trophy hunting is not a major threat to lions, and well-managed trophy hunting is often considered part of the solution to those threats through its ability to provide value to habitat and to provide benefits that outweigh the costs of lion presence.

As poaching, habitat loss and degradation, human-wildlife conflicts, and declining prey species are often considered the most serious of the threats to lions, the best way to address these threats is by funding and supporting PAs (Funston et al. 2016). While PAs cover a large part of the African continent, most of them are "chronically underfunded, and only 31% of PAs with lions currently maintain the species at 50% or greater of the natural density they would reach if only suffering natural mortality" (Funston et al. 2016: 15).

Trophy hunting is one way to direct funding to PAs through a form of tourism. Where trophy hunting of lions is properly managed, the fees paid by trophy hunters contribute a significant proportion of state wildlife authorities' budgets, provide incentives to preserve PAs and protect lions from poaching and retaliatory killings, and provide employment in rural communities (Funston et al. 2016: 13). Not only is trophy hunting one way of channelling tourism income to rural communities and maintain PAs; it is often the only way. Photographic tourism relies on scenic beauty, a wide array of wildlife, advanced infrastructure- and service facilities, relatively close proximity to airports, and political stability. In many places in Africa, photographic tourism is simply not a viable alternative to trophy hunting, and in most places where it is, it is already employed. As Craig Packer writes:

> Over a quarter of Tanzania's surface area is devoted to trophy hunting—most of the land is dull, hot, and filled with tsetse flies. In short, it is no place for a photo safari. So hunting could well provide the best possible incentive for conserving vast tracks of land. (Packer 2015: 30)

Photo tourism is well-developed in many key areas in Africa that combine scenic beauty and splendour with easily-viewed well-known wildlife species. The Big 5 and other large and charismatic species are what tourists usually want to see, and they want to see them easily. After they have seen animals, tourists want to relax in high-end lodges, enjoying fine dining. Opportunities to do this abound throughout much of southern and eastern Africa but occupancy rates are usually far below 100%, and often below even 50%. If you want to travel to Africa and see African wildlife in wonderful landscapes with fine food and good wine then you most certainly can and the limitations will be your wallet, not the availability of accommodation.

Finding viable income alternatives to trophy hunting in many areas has proved to be exceptionally challenging. A comprehensive recent study by the Luc Hoffman Institute found no obvious alternatives to trophy hunting and other tourism:

> To date, trophy hunting and tourism have generated significant returns on a large scale to enable wildlife conservation on community-owned or community-managed land. With challenges facing both those models, there are no easy or obvious new business models that can generate the same returns on the same scale. A range of approaches has emerged. Some of them are geared towards increasing the value of nature to communities. Others aim to generate funds to diversify income for communities. However, all have constraints and all are likely to be practicable only in certain contexts. (Roe et al. 2020: 32)

Trophy hunting has often been a vital element to lion conservation. The Bubye Valley Conservancy (BVC) in Zimbabwe, which was converted from cattle ranching to wildlife reserve with trophy hunting is one such example. More than 3700 km^2, BVC has, through strict quotas, fences, and anti-poaching units (all funded by trophy hunting) ensured that the 13 lions introduced in 199 have increased to over 500 today. The BVC also has populations of around 700 elephants, 5000 buffalos, 82 white rhinos, and 211 black rhinos, which is the third-largest population in Africa. The BVC generated almost $1.4 million in 2015 from trophy hunting fees, employs 400 people, and invests $200,000 a year in community development (IUCN 2016: 15; Tensen et al. 2018). By any measure, BVC is a conservation success story. What was once cattle ranching land is now wildlife-rich land, supporting and developing communities. Without trophy hunting this would likely not have been possible.

The Savé Valley Conservancy (SVC), also in Zimbabwe and close to BVC, tells a similar success story. SVC had only five to ten lions when it was converted from a cattle ranch to a wildlife reserve in 1992. In 2005, ten more lions were introduced to the area, and by 2016, the SVC had a population of 200–280 lions, 1500 elephants, 42 white rhinos, and 121 black rhinos (IUCN 2016; Tensen et al. 2018).[16]

But as all lion conservation experts also recognize, trophy hunting can be problematic, and trophy hunters are often only grudgingly accepted by conservationists as part of a solution. Sometimes the trophy hunting of lions has been unsustainable and has led to declining lion populations. In Zimbabwe (notably around Hwange National Park) lion hunting has been, in the past, a cause of population decline, although improved management and regulation have largely solved these issues (see below). Sometimes, trophy hunting has been tolerated only because its absence would lead to lion populations declining even faster. Packer conveys the sentiment well:

> [W]e knew perfectly well that trophy hunters were no angels. But the simple truth of the matter was that they controlled four times as much lion habitat in Africa than was protected by the national parks. So 80 percent of the lions left in the world were in their hands—regardless of the blood on those hands. (Packer 2015: 31)

---

[16] The question remains for both the BVC and the SVC population, as Tensen et al. point out, whether they might be subject to inbreeding in the long run, given the small founding populations.

Similar ambivalence about trophy hunting is evident in a recent letter to *Science* by Dickman et al. and 128 signatories (including AH), who confirm that

> [s]ome people find trophy hunting repugnant (including many of us), but conservation policy that is not based on science threatens habitat and biodiversity and risks disempowering and impoverishing rural communities. (Dickman et al. 2019: 874)

Loveridge likewise makes no attempt in his 2018 book to hide his disdain for the trophy hunters that come to Africa to kill lions, and he provides many examples of trophy hunting's shortcomings. Yet he ultimately comes to the same conclusion:

> Banning lion trophy hunting will not halt the decline of Africa's lion populations. Loss of habitat and conflict with livestock owners are the most important causes of lion population decline. […] The threat of habitat loss may actually be exacerbated if hunting areas are no longer protected. (250–251)

One of the potential problems with trophy hunting as lion conservation occurs if too many lions are added to hunting quotas. This has been a problem in Zimbabwe, as documented by Loveridge (2018). The Gwaai Conservancy near Hwange National Park in Zimbabwe consisted, in 2003, of thirty private properties (one of which is Antoinette farm where Cecil was killed). Each of these small properties was granted a hunting quota for at least one male and one female lion, supposedly based on lion populations resident on the properties. This amounted to granted hunting quotas of 13 female and 65 male lions for 2003. However, as Loveridge writes:

> Given what we now knew about the ranging behaviour of lions, it seemed improbable that each and every relatively small property had its own resident lions. It was much more likely that a few lions were roaming across a wide area encompassing many small, unfenced tracts of land. If indeed quotas were based on any credible population estimate, it seemed that landowners were double, triple or quadruple counting the lions moving through the area. (Loveridge 2018: 90)

As he further notes, most lions getting killed in the Gwaai Conservancy were lions that normally lived in Hwange. Many of them wore tracking collars, so it was easy to identify where they spent most of their time. It seemed therefore that the population of lions in the park was roughly the same population on which the quota of 65 male and 13 female lions in the Gwaai Conservancy was based. The lion population in the park was at most 30 adult males, meaning that every adult male in the entire area would be killed more than twice-over before the hunting quotas in the Gwaai Conservancy were filled (Loveridge 2018: 89–91). Too many lions on quota and consequent overhunting led to serious declines of lion populations near Hwange until 2005, but evidence also showed that once proper management was implemented and enforced, lion populations recovered quickly and were very resilient to a limited and sustainable trophy hunting offtake (Loveridge et al. 2016).

Ideally, if trophy hunting is to serve lion conservation, the total offtake should be only a small percentage of the male lions in a population. With more than twice as many male quota-lions as actual male lions, if chance had it that all the lions should stray out of the National Park (where hunting is not permitted) then all the lions could end up shot by trophy hunters. While not guaranteed to seriously affect lion populations, this kind of quota setting very much leaves lion conservation up to the vagaries of lion movements between non-hunting and hunting areas.

What this particular issue in Zimbabwe laid bare is that monitoring of lion populations is a problem. Arriving at population numbers for lions in the whole continent is difficult; arriving at numbers that make sense in terms of borders between countries, conservancies, parks, farms, and other lines on human maps is almost impossible. As Craig Packer noted about Tanzania's lion populations, "quotas were essentially arbitrary since no one knew how many lions they had in their hunting blocks. Trying to count lions was impossible, especially in areas where people liked to shoot them" (Packer 2015: 36).

Another potential problem is if the male lions that are killed are too young. This problem ties in with the problem of infanticide in lions, which has received some attention in mainstream media. In short, what can happen when a dominant male is killed or otherwise replaced in a pride is that a new dominant male, or coalition of males, kills all the nine-months old or younger cubs of the previous male. Such infanticide means that the lionesses will quickly return to estrus and be ready to mate with the new male or coalition of males. Repeatedly killing males whose sired cubs are younger than nine months runs the risk of a pride's cubs never maturing into independence. This has been a problem in Tanzania (Packer 2000; Whitman et al. 2004; Packer et al. 2010). Whitman et al. made the important discovery that

> if offtake included males as young as three years of age, the female population would invariably collapse if quotas were too high. But as the minimum age of trophy males was raised, the chances of population persistence increased markedly—to the point where removing males ≥ 6 yr in age had no substantial effect, regardless of quota size. (Whitman et al. 2004: 176)

The argument is that allowing male lions to reach six years of age ensures that they will have reared a cohort of cubs to independence. Creel et al. (2016) confirmed the findings of Whitman et al. that restricting harvest based on age drastically improved the sustainability of trophy hunting. However, their analysis did not support the conclusion that the quota size became unimportant at ages ≥ 6. Creel et al. found that the risk of extirpation fell consistently below 50% only at ages ≥ 7 and that the risk was still 5–11% at age 8. Age-based restrictions on lion offtake has so far been introduced in Mozambique, Tanzania, South Africa, Zambia, and Zimbabwe (Booth et al. 2020).

Another potential problem with trophy hunting as a conservation instrument is that money from hunting clients, even when trophy hunting fees are high enough, does not necessarily go where it ought to go, discussed further in the next section.

5.2.2 Rural Livelihoods

As leading causes of declining wildlife in Africa are poaching and human-wildlife conflicts, one of the purposes of wildlife management has become to ensure that rural communities have sufficient incentives to help conserve wildlife and wildlife habitats. Putting rural communities in charge of local wildlife and giving them the rights to use it for economic purposes has proven an effective way of facilitating such incentives and giving them a sense of ownership of, and responsibility for, their natural resources. The 2012 *IUCN SSC Guiding Principles on Trophy Hunting as a Tool for Creating Conservation Incentives* thus recognize that trophy hunting can serve as a successful conservation tool, when it "[r]espects local cultural values and practices, [...] and is accepted by [...] most members of the local community on whose land it occurs," and when it "[i]nvolves and benefits local residents in an equitable manner, and in ways that meet their priorities" (IUCN 2012: 6).

Lindsey et al. (2006: 288) found that 86% of hunting clients would prefer to hunt in an area where considerable parts of the proceeds went to support local communities. Furthermore, 46% of hunting clients would pay the going rate, or more, for a poor-quality trophy animal if the animal was a problem animal and so its removal would assist rural communities (a livestock-killing lion or a crop-raiding elephant, for example). Only 16.5% of hunting operators thought this would be the case. Hunting clients, it seems, might be more concerned about rural livelihoods than hunting operators both realize and utilize. Though it is also possible that some hunters will talk the talk in surveys but not walk the walk when they book their hunting trip.

The devolution of control over natural resources such as wildlife to local communities is, in conservation- and wildlife management literature, often referred to as Community-Based Natural Resource Management (CBNRM). CBNRM has been implemented in a number of African countries up through the 1990s, replacing the predominantly state-run and top-down conservation-by-coercion systems, here criticized by Brian Child:

> CAMPFIRE grew out of the folly and failure of colonial legislation which banned all utilization of wildlife for commercial as well as traditional hunting. A few people were even moved to make way for National Parks. Communities were disenfranchised from their resources, which became valueless, but also a symbol of oppression, so that their destruction was encouraged and poachers were heroes. This phase of alienation of people from their wildlife can be categorized as the period of wildlife management against the people. (Child 1993: 285)

Zimbabwe's Communal Areas Management Programme for Indigenous Resources (CAMPFIRE), introduced in 1989, is one of the earliest CBNRM programmes. Several others in Africa and elsewhere have been based on this. CAMPFIRE transferred use-rights to wildlife from the central government to Rural District Councils (RDCs), who would in turn lease the use rights to private trophy hunting or safari tourism operators. CAMPFIRE was introduced in areas of Zimbabwe that have little potential for safari tourism but significant populations of large mammals for hunting tourism purposes, so hunting revenue has been the main source of revenue for the RDCs. Bond (2001: 230) writes that around $8.6 million out of a $9.4 million total (93%) of RDC income between 1989 and 1996 came from sport hunting leases (Child 1993; Bond 2001).

Judged on its ability to conserve wildlife through trophy hunting, CAMPFIRE has largely been a success. In its first ten years, wildlife populations in CAMPFIRE areas increased by about 50%, elephant populations doubled, and both poaching and land use conflicts were reduced. Revenues from trophy hunting have been reinvested in the building of schools, in boreholes for wells, in lodges and other tourism facilities that provide employment opportunities, and in cash dividends for rural households. It is also testament to CAMPFIRE's relative success and communal faith in the programme that it grew from covering 8880 households in the beginning to 777,000 households by 2002 (Frost and Bond 2008; Mbaiwa 2008; Jones 2009). In a 2019 assessment of CAMPFIRE, Tchakatumba et al. concluded that

> [d]espite all the shortcomings of the CAMPFIRE programme [including a reduction in household benefits from it and lack of transparency at times in income distribution], the vast majority of our household sample opted for its continuation, which is an indication of the programme resilience despite a severe national economic depression. (Tchakatumba et al. 2019: 131)

CBNRM management schemes similar to CAMPFIRE have been implemented in other African states and lessons have been learned from mistakes that were made with CAMPFIRE. In Namibia, for example, hunting revenues are being distributed at the basic level of community representation in what is called Conservancies to avoid revenues accumulating at the higher RDC-level, as has been the case with CAMPFIRE.

Namibia has since its inception in 1996 arguably become the best example of successful CBNRM in Africa with strong support among rural residents. In 2016, 82 communal conservancies covered about 20% of Namibia's total area and supported work for 184,000 people out of a population of less than 2.5 million. Twenty-nine species were hunted in 2013, which generated revenue of US$1,671,379 (IUCN 2016). The wildlife conservation successes of Namibia's CBNRM implementation are summarized by IUCN here:

> Wildlife populations have shown dramatic increases since the beginning of the communal conservancy programme in Namibia. On communal lands in northeast Namibia, from 1994–2011, the Sable Antelope population increased from 724 to 1474 and the Impala from 439 to 9374. In the conservancy region of northwest Namibia, from the early 1980s to 2011, the threatened Hartmann's Mountain Zebra population increased from less than 1000 to an estimated 27,000, and the number of Black Rhino more than tripled, making it the largest free-roaming population in Africa (conservancies are unfenced). The growth of communal conservancies and protection offered by national parks has enabled elephants to increase their population from around 7500 in 1995 to more than 20,000 today. The Kunene Conservancy's Lion population grew from roughly 25 in 1995 to 150 today, and Namibia now has a large free-roaming Lion population outside of national parks. (IUCN 2016: 16)

One potential problem with trophy hunting as conservation in Africa is that rural residents may lose their traditional access to the hunting of wildlife for meat and ritualistic purposes because the wildlife has more value if it is sold to trophy hunters. Since trophy hunters are overwhelmingly foreign, the loss of local rights to use wildlife can be interpreted through a neo-colonial lens, an interpretation of both trophy hunting and attempts to ban trophy hunting that we will return to shortly. Namibia has sought to remedy this by dividing wildlife into categories of "specially-protected game," "protected game" and "huntable game." Hunting of the first two categories—the rarest and most valuable species—requires a permit from the Ministry of Environment and Tourism, who will first evaluate the sustainability of the offtake. The right of ownership of the final category is devolved to the conservancies, who may use it themselves or sell it as they please (Weaver and Petersen 2008).

A study by Naidoo et al. (2016) compared the benefits to rural Namibian communities of trophy hunting tourism with those of photographic tourism. They found that trophy hunting began generating benefits twice as fast as photo tourism (3 vs. 6 years) after being implemented. They also found that the benefits from photo tourism was mainly in the form of employment opportunities, while the benefits from trophy hunting was cash for conservancy operations and (very importantly, but often underestimated) considerable amounts of meat for rural households. A simulation of a trophy hunting ban showed that the number of Namibia's conservancies with incomes greater than their operating costs in 2013 would drop from 74 to 16% if trophy hunting incomes were eliminated. Conversely, if photographic tourism disappeared, 59% of conservancies would still make a profit. As will be discussed further in Chapter 5.6, trophy hunting can generally complement photographic tourism in areas where the latter is a viable strategy, but some areas cannot support photographic tourism at all and must rely on trophy hunting. Naidoo et al. concluded that 54% of Namibia's conservancies that generated benefits to rural communities relied solely on hunting, while only 12% specialized in photo tourism (Naidoo et al. 2016).

Angula et al. (2018) surveyed 32 of Namibia's communal conservancies to find out what the rural populations thought of trophy hunting. They found that 91% of 160 respondents would not be in favour of a ban on trophy hunting, and only 11% thought that wildlife could still be supported on communal lands if trophy hunting were banned. In an open-ended question about what would happen if trophy hunting were banned, 65% answered that they would be deprived of income and employment opportunities critical to their livelihoods. Fourteen percent noted the importance of the meat that they would lose and 6% indicated that human-wildlife conflicts and poaching would increase. On the other hand, 18% thought that trophy hunting might be bad for wildlife, and 13% were not averse to a ban in principle. However, they could also not think of alternatives to replace the hunting benefits. One response to the authors' question "what would you like to say to people who want to ban trophy hunting" is perhaps worth highlighting here:

> Why do you want to take away my job? Why do you want me to suffer from elephants destroying my crops without any benefits, just so you can satisfy your needs of looking at these animals? Trophy hunting is well managed and beneficial for us, we cannot ban it. (Angula et al. 2018: 28)[17]

---

[17] AH visited Namibia in 2021 and had the opportunity to speak to some community representatives in several conservancies in north and west Namibia. All of them echoed the sentiment expressed here, with some being clearly angry that people overseas were interfering with a form of conservation that communities felt worked for both wildlife and people. The film *Beyond the Trigger* by podcaster Ryan Dalton (Into the Wild Podcast) records some of this testimony and is free on YouTube.

5.2 Hunting and Conservation in Africa

A country that actually did ban trophy hunting, and which is therefore suitable for comparison with nations that have it, is Botswana. In response to declining wildlife populations, Botswana banned all hunting in 2014, but it recently reinstated hunting, with hunting quotas for elephants being introduced due to unsustainably large population increases. The country received much negative publicity in the international popular media in 2021 for auctioning off 170 live wild elephants, ostensibly for trophy hunting.

Joseph Mbaiwa (2017) examined the community consequences of Botswana's hunting ban. He found that banning trophy hunting caused severe job- and income losses in rural communities, which had until then relied on trophy hunting for about two thirds of their revenue. Several areas that were unsuitable for photo tourism and relied solely on trophy hunting lost all their tourism industry. In addition to job- and income losses came the loss of important meat, which used to be partly distributed to the poorest in the communities and partly auctioned off. Several community projects such as providing houses for the needy, funeral insurances, scholarships, and cash dividends had to shut down as well. Poaching incidents began increasing shortly after the ban after having been on the decline ever since CBNRM was implemented. While poaching levels of largely unmonitored populations are extremely hard to estimate, they are believed to be causing drastic declines of certain large herbivore species in Botswana's Okavango Delta. According to Mbaiwa, the ban also re-introduced negative attitudes towards wildlife among rural residents, which had otherwise become positive while trophy hunting was allowed. One villager was credited with the remark that

> before, when there was hunting, we wanted to protect those animals because we knew we earned something out of them. Now we don't benefit at all from the animals. The elephants and buffaloes leave after destroying our ploughing fields during the day. Then, at night, the lions come into our kraals. (Mbaiwa 2017)

Generally speaking, trophy hunting implemented as part of CBNRM has the potential to benefit rural communities in many different ways, as noted by for example Weaver and Petersen (2008) and Jones (2009). Meat, employment opportunities, investments in rural communities and cash subsidies are the obvious benefits. A less obvious benefit is that giving wildlife a monetary value can change people's perception of wildlife from a negative to a positive. Lions, rhinos, and elephants do not have to just be pests that destroy your livelihood; if managed properly, they can *be* a livelihood. In situations where trophy hunting is not sustainably managed, overhunting can be a benefit to rural Africans simply because it is not sustainable and gets rid of troublesome wildlife (Dube 2019). Though of course not a welcome thought from an ecological perspective, if rural communities do not benefit from lions and elephants, then they are pests, and they would in many cases prefer them extinct, as noted here by Craig Packer:

> [A]sk most rural Tanzanians what they think about wildlife, and they'll probably tell you that if they can't eat it, they'd just as soon eradicate it.
>
> Elephants and Lions? The two most hated species in the country. (Packer 2015: 18)

These comments reflect a general, and important, disconnect between western views of wildlife and the views of those who live alongside wildlife, a theme we will return to shortly.

In terms of benefiting rural communities through trophy hunting, Namibia has done relatively well, but CBNRM is not a guaranteed success or without its problems. An issue in Namibia and elsewhere is that the upper echelons of the trophy hunting industry are heavily dominated by white Africans. 70% of freehold land (commercial, privately owned land) in Namibia is white-owned. In 2019, over 90% of the total revenue generated by trophy hunting in Namibia was attributed to these freehold lands (which contain over 80% of Namibia's wildlife) and not to communal conservancies, which covered 15.7% of the country in 2019 and where all the revenue goes to black rural communities (Namibia Land Statistics 2018; MacLaren et al. 2019; Becker 2022). Porters and trackers are generally black, while professional hunters are almost exclusively white. A project from 2013 to develop trophy hunting among non-white farmers reported that 554 out of 555 hunting operators at the time were white (Abbiati et al. 2013). Brian Jones similarly noted that in Zimbabwe, safari operators are predominantly European, tend to keep local people out of high-level employment opportunities, and treat even the essential local trackers as unskilled laborers (Weaver and Petersen 2008; Jones 2009: 168). Another problem is that the benefits do not always reach the poorest in the rural communities that they are meant to benefit. In Brian Jones' review of community benefits from safari hunting, he notes that revenues from hunting have often mostly benefited the people in the rural communities who are involved with the CBNRM administration. These tend to be the people with relatively more money, education, and connections. In the case of CAMPFIRE, as indicated earlier, this has meant that more revenue than intended stays at the RDC-level and does not benefit the ward-level, which is the sub-district level of employees and beneficiaries beneath the RDCs, more equivalent to Namibia's Conservancy level. The jobs that are created in the tourism sector by CBNRM generally do not go to the poorest, and the reliance of the poorest on some levels of subsistence hunting for meat is often compromised because wildlife is being reserved for hunting tourism purposes (Frost and Bond 2008; Jones 2009). Nqobizitha Dube found that Zimbabwean villagers today think CAMPFIRE has become a programme that benefits only a few individuals who do not even reside in the villages that CAMPFIRE was supposed to benefit (Dube 2019), and Kalvelage et al. (2021) found that tourism-driven development in the Zambesi region of Namibia only reaches a limited number of rural residents employed in low-wage jobs and conservancy management.

Corruption and mismanagement are among the reasons that trophy hunting may fail to benefit those who need its benefits the most. On its list of perceived corruption levels in 180 ranked countries (higher rank = less corruption), Transparency International listed Botswana at #34, Namibia #56, South Africa #70, Tanzania #96, and Zimbabwe #158 in 2020. This ranking might explain why the academic literature also seems to contain more occurrences of corrupt hunting management in Tanzania and Zimbabwe. The lion books by Loveridge (2018) and especially Packer (2015) contain thorough portrayals of the problems with corruption and general mismanagement in Zimbabwe and Tanzania, respectively. Both authors relate their experiences with having their research permits revoked when their research displeased the wrong people in governments or in the hunting industry.

Of course, not all mismanagement is corruption. As related here by Packer, problems with the trophy hunting business often come down to a tragedy of the commons:

> [T]here might only be four or five honest operators out of fifty-plus companies in Tanzania. Only a tiny minority was willing to take the longer view of conservation; the rest were fly-by-night operations that only cared about short-term profits. They had no incentive to ensure healthy populations in the distant future. If they overshot the lions in Tanzania, they could move operations down to Zambia or Zimbabwe the following year. (Packer 2015: 48)

Bill Wall and Brian Child (2009) sought to explain when hunting might contribute to conservation and rural development. They note, based on the work of Douglass North, that countries can have either personal or impersonal economies and that conservation through hunting is much more likely to be successful in impersonal systems. The difference is described here:

> At one end we have relatively effective economies characterised by persistent competition in the political and economic market-places, where a level competitive field is kept in place by strong judicial systems and well-defined property rights. These systems protect the conditions for impersonal transaction by providing judicial recourse where people are cheated, and generate considerable economic participation and wealth. They are called 'impersonal economies'. At the other extreme we find weak or inequitable economies characterised by highly interdependent political and economic elites, where individuals seldom trust political, judicial or economic institutions (including property rights) so that economic transactions depend heavily on personalised exchanges and reciprocity, i.e. who you know. [...] These systems are called 'personalised economies' and within them the potential for many ordinary people to generate value-added products or services is reduced by a lack of access to resources and a dependence on personalised exchange. (Wall and Child 2009: 256–257)

This dichotomy illuminates some of the difficulties with CBNRM in African countries, which can to a large extent be characterized as personalised economies. The importance of knowing the right people explains—at least partly—why disproportionate benefits from trophy hunting in, for example, the CAMPFIRE programme stay at the management level and may not reach the poorest and least connected residents of rural Zimbabwean communities.

In Tanzania, as summarized by Leader-Williams et al. (2009), corruption issues have ranged from the level of operators (baiting near national parks, hunting from vehicles, near watering holes, or at night with spotlights, overshooting quotas, and shooting females) to the ministerial level (nepotism, hunting lease prices kept artificially low, blocking fair competition, purposely delaying reforms, favouring non-resident hunting operators who pull revenues out of the country, etc.).

Widespread corruption is one of the most common arguments from antihunters as to why trophy hunting does not work for conservation, but this objection fails to address that corruption is not a unique problem for trophy hunting. Many other activities within trophy hunting countries operate within the reach of corruption at all levels, including photo tourism. Currently, in Tanzania for example, there are continuing reports of pastoralists being relocated, forcefully according to some, from the Loliondo region of the Ngorongoro District to make land available both for photo tourism and hunting. This is just the latest in a long history of land-based conflicts in the wider district of Ngorongoro, which includes the Ngorongoro crater, a famous tourist site. Corruption, where it exists, is a symptom of national, institutional flaws rather than an intrinsic feature of trophy hunting.

5.2.3 Eco-Imperialism and Living with Lions

African trophy hunting is easily characterized as neo-colonialist, as it commonly involves wealthy westerners killing wildlife in Africa that most rural Africans are financially excluded from killing. This lends weight to the claim that when white people shoot wild animals, it is called hunting, and when indigenous people do it, it is called poaching. As demonstrated mainly in Chapter 2.7 but also in connection with Roosevelt's African journey in Chapter 2.8, white people killing especially African and Indian wildlife in gross excess has a long and depressing history. Any hunting in Africa by westerners today, and even speaking on the subject of hunting in Africa as a white westerner, is surely tainted by these racist and exploitative connotations. As also discussed above, white land ownership and white people being heavily overrepresented at the top level of the trophy hunting industry are endemic issues.

However, western interventions to stop trophy hunting can be neo-colonialist as well. Resolution 17.9 of CITES acknowledges "that peoples and States are and should be the best protectors of their own wild fauna and flora." The issue at hand is that not everyone acknowledges the principle that CITES acknowledges. Animal rights activists and other trophy hunting opponents in Europe and USA routinely and vehemently demand that trophy hunting should be banned outright in Africa or that—at least—trophy imports should be banned in the importing countries, which tends to be the only thing said countries can realistically do to ban trophy hunting in Africa. What is rarely taken into consideration here are the attitudes toward wildlife and trophy hunting of African governments and African communities, or indeed how such bans might affect rural economies and livelihoods. Westerners, as Andrew Loveridge rightly points out,

> have largely forgotten what it is to be afraid of dangerous predators on a dark night. They no longer live on the edge of the wilderness. Rural Africans in lion country still live with this terror. For them, lions are perilous killers and sometimes even supernatural monsters. It is no wonder they are frequently less enamored of large, dangerous animals than Westerners who live in safe, urban environments. (Loveridge 2018: 143)

The attempts to put an end to trophy hunting regardless of African attitudes to wildlife or to trophy hunting's importance to African economies might constitute *eco-imperialism*, referring to the "forceful imposition of Northern environmental

5.2 Hunting and Conservation in Africa

views on the Global South" (Nygren 2013: 59). In recent years, eco-imperialism is evident in the discourse surrounding global anthropogenic climate change, as countries in the global south are essentially forced to develop in a greener and more climate-friendly way than the global north did. For trophy hunting, similarly, the global north's message is that it cannot let Africa's wildlife be pushed towards, or over, the edge of extinction in the same way that it has done, and is doing, with its own wildlife.

CBNRM ideally counteracts eco-imperialism by putting the management of wildlife in the hands of those who live amongst it and are most affected by it. This has been key to its success. Putting rural Africans in charge of wildlife management demonstrates that wildlife can have value, and—as shown earlier—can improve attitudes toward wildlife. As also covered already, human-wildlife conflicts and consequent retaliatory killings are one of the primary threats to much of Africa's wildlife, most prominently lions and elephants. Lions are of course a danger to people as well as to livestock. While elephants will trample people, lions may even come to treat people as food. Packer noted the following about lion attacks near the Selous in Tanzania:

> These were not cases of big cats defending themselves against angry pastoralists; these lions know what they are doing: they want to eat *you*, and your little dog, too.
>
> These cases included attacks in the middle of town and lions breaking into people's houses—pushing through the walls, digging down through thatched roofs, snatching babies out of the arms of nursing mothers, grabbing people as they went to the outhouse at night.
>
> Over a hundred people a year. (Packer 2015: 39, emphasis in original)

Yet, despite the serious threat to both lives and livelihoods from lions, elephants, and other wildlife, western eco-imperialism puts many rural Africans in a situation where they are unable to defend their livelihoods or even themselves. John Akama, for example, explains that in Kenya, where the animal rights lobby has had tremendous success in shaping wildlife policies such as the hunting ban in 1977,

> cultivators and pastoralists cannot protect themselves or their property from wildlife despite considerable injury and severe damage to farms and livestock (Akama 1998). State law prohibits any form of destruction and killing of wildlife. Consequently, peasants are reduced to guarding crops and livestock by making noise, beating drums and lighting night fires so that someone else may make a profit from tourists willing to view and photograph an animal that local opinion would wish dead. Hence, local people's attitude toward protected wildlife areas varies from that of indifference to intense hostility. (Akama 2008: 83)

Wildlife in Kenya has declined tremendously ever since the hunting ban, though the reasons for this are more complicated than just having banned hunting. Nonetheless, in 2016 Ogutu et al., in an extensive analysis of the cause of wildlife declines in Kenya (and concurrent increases in livestock), point towards the shift in wildlife ownership and value provoked by the ban as a factor, concluding that

> [t]he replacement of wildlife by livestock is exacerbated because restrictions on the use of wildlife are made easy because they are owned by the state and not by the landholders or users in the rangelands, and this underprices wildlife. The net result is that the competitive advantages of multi-species wildlife systems with multiple values (tourism, hunting,

biodiversity, ecological adaptability, ecosystem services) are replaced by individually owned commodity production systems in the form of cattle, shoats, donkeys and camels. Wildlife typically earns far less on the rangelands than it should, and most of what it earns is captured by the elites. (Ogutu et al. 2016: 28)

The perception of many rural Africans who live with problematic wildlife is that the global north cares more about Africa's animals than about Africa's people. A representation of this attitude is eloquently provided by Goodwell Nzou, who wrote the following in an op-ed for the *New York Times* in 2015, shortly after the death of Cecil:

My mind was absorbed by the biochemistry of gene editing when the text messages and Facebook posts distracted me.

So sorry about Cecil.

Did Cecil live near your place in Zimbabwe?

Cecil who? I wondered. When I turned on the news and discovered that the messages were about a lion killed by an American dentist, the village boy inside me instinctively cheered: One lion fewer to menace families like mine.

My excitement was doused when I realized that the lion killer was being painted as the villain. I faced the starkest cultural contradiction I'd experienced during my five years studying in the United States.

Did all those Americans signing petitions understand that lions actually kill people? That all the talk about Cecil being "beloved" or a "local favorite" was media hype? Did Jimmy Kimmel choke up because Cecil was murdered or because he confused him with Simba from "The Lion King"?

Nzou proceeds to recount his own childhood, growing up in a village terrorized by lions, and ends the article as follows:

We Zimbabweans are left shaking our heads, wondering why Americans care more about African animals than about African people.

Don't tell us what to do with our animals when you allowed your own mountain lions to be hunted to near extinction in the eastern United States. Don't bemoan the clear-cutting of our forests when you turned yours into concrete jungles.

And please, don't offer me condolences about Cecil unless you're also willing to offer me condolences for villagers killed or left hungry by his brethren, by political violence, or by hunger. (Nzou 2015)

Many more examples of such attitudes are found in e.g. Mbaiwa (2017), Angula et al. (2018) and Dube (2019).

Mkono (2019) provides additional examples but other perspectives as well. Mkono analysed African attitudes to trophy hunting in the comments sections of trophy hunting news items on three African news sites with a presence on Facebook (BBC News Africa, News24.com, and NewsDay-Zimbabwe). She concluded that Africans think westerners care more about Africa's wildlife than Africa's people (80% of posts represented this attitude) and found that trophy hunting is by many considered objectionable in principle because of the neo-colonialist connotations of rich white men being given power over Africa's wildlife (70% represented this attitude) and not because of any rights or welfare of animals.

It is important to note, however, that the attitudes represented in this study are the attitudes of Africans who read and write English and browse and comment on Facebook. This excludes many rural Africans who actually live among wildlife and suffer livestock- and human losses to lions. So, while the attitudes of Americans and Europeans represent attitudes from privileged positions compared to most African attitudes, so do the attitudes in this study presumably represent the privilege of not actually being terrorized by lions and other wildlife on a daily basis. This comparative privilege affords a set of ethical considerations—such as the view that trophy hunting represents neo-colonialism—that is largely not afforded to those rural Africans whose immediate survival depends on the livelihoods provided by trophy hunting. Many of those citing this study in online discussion of neo-colonialism also miss the point that Africa is not a single country, and a commentator from Morocco or Egypt has as much experience living with lions as someone in the UK or the USA. Simply living on the same continent does not give the automatic right to appropriate the lived experience of those thousands of miles away on the same continent, living a very different life.

Van Houdt et al. (2021) conducted a study of divergent views on trophy hunting in Africa and found that African and North American respondents as well as respondents with a conservation background showed significantly more support for trophy hunting than European respondents and respondents without a conservation background. Prioritizing above all else the lives of animals that are a danger to people or to livelihoods is a luxury that comes with not being one of those people at existential or economic risk. As Europe-based authors, we are in no danger from lions where we live, so our opinion about the trophy hunting of lions should matter very little. So should any opinion that comes from a position of privilege, which is also the position of most antihunters. The opinions of urban Africans on Facebook should matter more perhaps (though see above), but even they should not matter nearly as much as the opinions of the rural Africans who actually live with lions. Ironically and unfortunately, the rural Africans who live with lions are the people whose opinions are heard the least, while the loudest and most visible objections to trophy hunting come from the people whose opinions should matter the least. These objections to trophy hunting in Africa often come from people who accuse all of their opponents of lacking empathy with animals. One might wonder, though, what happened to their own empathy with rural Africans? Animal rights activists and antihunters tend to be the same white, western population segments that are staunchly anti-racist and anti-Trump, and have their social media profiles adorned with "Black Lives Matter" slogans. This makes it curious to witness how often they end up promoting the distinctly racist and imperialist argument that the lived experiences of rural Africans and the science of African and Africa-based biologists must be discredited or ignored.

Trophy hunting does have strong colonialist connotations and represents a commodification of wildlife that many find objectionable, but if African communities decide to embrace it anyway on their own terms, then this should be their choice and only theirs. A recent statement from the Community Leaders Network—representing over 50 leaders from African communities involved in CBNRM—puts this very well:

Peoples who were colonised across the world have adopted many interventions, technologies and approaches to life from the colonial period. These include medicines, transportation and communications systems and tools, and likewise, conservation hunting. It is our right to decide what adopted interventions we retain or discard based on their tangible and intrinsic value. For Westerners to determine what it is suitable for Africans to retain is the epitome of the very thing that is being criticised – a colonial mindset.

You must acknowledge that rural African people sustainably managed biodiversity, including charismatic wild animals, for millennia before the arrival of colonists and continue to do so today in many post-colonial contexts which are based on inclusionary policies. These policies not only recognise our rights and provide appropriate incentives for us to control our land, but also allow us to benefit economically from sustainably harvesting the biodiversity that surrounds us and of which we are custodians. (Resource Africa 2021)

5.3 Hunting and Conservation in Eurasia and the Americas

The African trophy hunting covered in this chapter so far fits the commonly held preconception of trophy hunters being wealthy, white Europeans and US-Americans travelling to relatively poorer countries (mainly in Africa) to hunt and bring home trophies. While African countries are the best-known locations for such hunting, there are many important examples of this kind of trophy hunting in non-African countries as well. Prominent examples include USA and Canada in North America, Iran, Kazakhstan, Kyrgyzstan, Mongolia, Pakistan, Russia, Tajikistan, and Uzbekistan in Asia and Central Asia and Argentina in South America. The most prestigious and expensive trophies in all of these countries tend to be the horns of sheep and goats (*Ovis* and *Capra*, together Caprinae) including but not limited to argali (*Ovis ammon*), bighorn sheep (*Ovis canadensis*), markhor (*Capra falconeri*), urial (*Ovis orienalis*), and many different species of ibex.

Common for the species above, and for much other wildlife in the countries just mentioned, is that populations have been hugely reduced in, and sometimes extirpated from, much of their historic ranges by unregulated market- and subsistence hunting, deforestation, and large-scale conversion of wildlife habitat mainly to agriculture. Poaching is an important detrimental factor as well and has, in Central Asia, unfortunately often been carried out by underpaid law enforcement officers and PA-staff. Other forms of corruption in the Central Asia trophy hunting industry have been common as well. In Kyrgyzstan for example, insufficient funding for conservation has led to park rangers being underpaid, in turn leading to rangers taking bribes from trophy hunting outfitters to allow hunting in PAs (Mallon 2013).

That said, trophy hunting and CBNRM have been rather successful and led to impressive recoveries of some species in several of the aforementioned countries. The hunting of markhor and urial in the Torghar District of Pakistan has been one of the greatest trophy hunting conservation success stories. As local tribal leaders worried about uncontrolled and illegal community hunting for markhor and urial in the mid-1980s, they developed a conservation project that would employ community members as game wardens, sponsored by limited trophy hunting offtake. Markhor and urial numbers were respectively down to fewer than 100 and 200 individuals

5.3 Hunting and Conservation in Eurasia and the Americas

in the valley at the outset of the project. Between 1986 and 2012, $US2.71 million was generated from trophy hunting for the local communities. This has covered the salaries of upwards of a hundred game guards and funded community projects such as schools and healthcare facilities. Markhor and urial populations have since increased to over 3500 and 2500, respectively (Bellon 2008; Cooney et al. 2017). In the Gilgit-Baltistan province of northern Pakistan (a bit less than a tenth of Pakistan's total area), 80% of trophy hunting income from the hunting of Himalayan ibex (*Capra sibirica*), blue sheep (*Pseudois nayaur*) and markhor goes directly to local communities. From 2000 to 2014, 262 ibex, 30 markhor, and 19 blue sheep were hunted in this province by 198 foreigners and 111 locals, generating $US1,387,832 for communities and $US346,958 for the government wildlife department while aiding in the recovery of these species:

> The program's biggest achievement is the discernible recovery of ibex, markhor, and blue sheep populations that had previously reached alarming levels due to illicit and unsustainable hunting by local communities. The program has instilled a sense of ownership of wild ungulates among local communities; it is now much more difficult for outsiders to openly hunt wild ungulates. (Nawaz et al. 2016: 226)

In Tajikistan, community conservancies have been established since 2008 for the hunting of markhor and ibex. From populations numbering in the low hundreds, they have increased to more than 2500 markhor and 2000 ibex. Argali, urial, and snow leopard (*Panthera uncia*) populations (which benefit from an increase in these prey species) are all protected by community conservancies that rely solely on trophy hunting:

> "I have many issues with trophy hunting, but we don't have the luxury not to do it," says Karimov. "And without this level of sustainable trophy hunting, our populations of ungulates, and subsequently snow leopards, would rapidly decrease." (Bolingbroke-Kent 2022)

Rural communities have few or no other possible source of income, so trophy hunting has provided livelihoods, and even supported the empowerment of women, as further explained here by Khalil Karimov:

> The communities involved in hunting and wildlife conservation are remote people living largely traditional lives, with tribal bonds still intact. They draw on local traditional knowledge of the habitats and ranging behavior of the ibex and Markhor, successfully blending it with modern systematic knowledge for good management.
>
> Community members are involved in all aspects of carrying out hunting. Community members are rangers, who monitor the populations and carry out anti-poaching activities; providers of homestays; suppliers of food for hunting tourists; drivers; makers of handicrafts to sell to hunters (traditional socks, gloves, musical instruments and other small souvenirs); and performers who present traditional songs, music and dance to hunting tourists.
>
> The employment and empowerment of women in wildlife conservation is increasing, and the number of female rangers is slowly growing. There are currently five female rangers, with two of them actively involved in hunting. Women are also involved in hunting tourism as homestay owners, cooks, handicraft makers and even tourist and hunting guides. (Karimov 2019: 2)

In Mongolia, the major threats to argali and ibex are subsistence poaching, competition with domestic sheep and goats, and natural resource extraction. Mongolian argali are some of the largest in the world, which should make trophy hunting very viable for conservation and CBNRM benefits. The implementation of CBNRM through trophy hunting of argali led to a population increase from 200 to 1500 between 2007 and 2014 (IUCN 2016), yet trophy hunting here has been fraught with more problems than has been the case for other sheep-hunting countries. Very high trophy fees and a lack of transparency in how the government issues licenses has led to widespread allegations of government corruption and of no money going towards conservation, a lack of argali surveys has led to management in the dark, and trophy hunting companies and their staff being based in Mongolia's capital of Ulaanbaatar has prevented trophy hunting benefits from reaching local communities in argali range country (Reading and Amgalanbaatar 2016).

On Seri Indian lands in Mexico, trophy hunting and CBNRM has led to several reintroduced bighorn sheep populations growing from 20 or so to several hundreds or even thousands, and livelihoods of the Seri Indians in charge of the management have been greatly improved (Cooney et al. 2019).

Trophy hunting has been banned in China since 2006 due to public pressure from urban residents in Beijing and other major cities. In Dulan county in Western China's Qinqhai provice, where blue sheep compete with domestic yak, sheep, and goats for grazing, rural communities wish to reinstate trophy hunting to control and reduce blue sheep population numbers and support rural development. The Chinese government supports this and even attempted to do so in 2011, but a strong public backlash from China's urban centres made tourism operators pull out, and the government dropped the plans. Survey results by Zhou et al. (2020) showed significant differences between rural (Dulan) and urban (Beijing) residents in the acceptance of trophy hunting as a wildlife management and rural development tool and also between poor and wealthy rural residents, i.e. those dependent on livestock grazing for their livelihoods and those not. These results also showed that while trophy hunting was still practiced until 2006, it was mostly porters and hunting guides who benefited economically because of end-of-hunt tips from the hunters of up to a year's annual salary ($US1000), as rural communities saw very little of the trophy fees that hunters paid to the government (Zhou et al. 2020).

In contrast to China, blue sheep are still hunted in Nepal's Dhorpatan Hunting Reserve and have been so since 1987. The population of blue sheep in the reserve increased from 400 in 1990 to over 1500 in 2011 and money collected by rural communities from hunters have been used to fund school- and health facilities and community infrastructure, and the presence of Reserve staff has reduced poaching. There have however been concerns about the male to female ratio of blue sheep due to trophy hunters only targeting males, and a payment system split between the government and local communities has been confusing for hunters, and led to some level of illegal activity (Aryal et al. 2015).

Caprinae, as illustrated by most of the examples above, have proven relatively suitable for CBNRM and conservation via trophy hunting. A position statement on trophy hunting of the IUCN Caprinae specialist group explains here why Caprinae may be more suitable for trophy hunting than for example lions:

> Because they seek adult males, trophy hunters are unlikely to have a negative short-term impact on most healthy populations. It is unlikely that trophy hunting of mature males will cause extinction. A population of Caprinae that produces mature males (aged 6–12 years, depending on the species) and therefore provides trophy hunting opportunities, is likely reasonably healthy. Because trophy hunting is incompatible with poaching or habitat destruction, several of the goals of trophy-hunting programs are shared with the goals of the Caprinae Specialist Group. Other specialist groups may not have such convergent goals, because for some taxa (for example large carnivores) trophy hunting is not as selective for mature males, or has negative consequences by disrupting the social structure. (IUCN Caprinae Specialist Group 2008: 23)

Community managers can demand hefty sums, even surpassing almost all African game, for some Caprinae trophies. An exceptional markhor, for example, can sell for over $US100,000. This puts a large amount of money towards conservation and rural development in exchange for a very low offtake.

Trophy hunting for polar bears (*Ursus maritimus*), banned in much of its range today because of historical overhunting, has been a sustainable and important source of indigenous employment and income in Canada since the 1970s. Freeman and Wenzel (2006: 26–27) describe the implementation of a trophy hunting regime for polar bears at this time as

> a rare example of science-based management overcoming emotion-based political campaigns, which at that time were proclaiming that total protection against consumptive use was the only way to conserve depleted wildlife resources.

This became even more important in 1982–1983, thanks to heavy-handed campaigning by animal rights activists against the indigenous trade in sealskin- and ivory products (walrus ivory). As these items represented the main source of income in many communities, Inuit hunters found themselves in an economic crisis when the campaigns caused prices to plummet. This created a market for young hunters to turn to the job of trophy hunting guide instead, and not only for the income:

> In addition to providing economic benefits, becoming a guide or an outfitter enables the individual to engage in an occupation that remains strongly associated with subsistence-related pursuits that many Inuit continue to value and enjoy in a profoundly personal way. [...]
>
> The presence of visiting sport hunters in these isolated Inuit communities is also appreciated as an affirmation that not all Southerners share animal protectionists' opposition to Inuit cultural values and subsistence practices, an opposition that has added demoralizing insult to economic and social injury. (Freeman and Wenzel 2006: 26–27)

Resistance to trophy hunting of polar bears has nevertheless not subsided. Despite 40 years of conservation successes and the insistence of scientists, indigenous communities, and officials that trade is not a current threat, hunting opponents have attempted (and failed) to add polar bears to Appendix I of CITES in both 2009 and 2013 (Weber et al. 2015).

All these examples aside, most trophy hunting that takes place in the world is not a tourism activity. As we are concerned with more than just the trophy hunting stereotype, it is time to discuss domestic trophy hunting in Europe and the US.

As we depart from the subject of international and intercontinental big game hunting tourism, we also deviate from what many—academics and otherwise—consider the very definition of trophy hunting. As briefly discussed in the introduction chapter, domestic North American and European trophy hunting is more commonly referred to as big game hunting, recreational hunting, meat hunting, sport hunting, or just plain hunting. The fact remains, however, that American or European hunters who would shoot an antlered buck of any kind and leave the antlers behind in the woods are extremely rare. Antlers—be they on their own, on a half-skull mount, or on a shoulder mount—are by far the most common kind of hunting trophy in the world, and the biggest of them are perhaps the most prestigious hunting trophies in the world.

This fact also addresses the question of why we choose to focus on deer hunting in this Euro-American conservation chapter; the prevalence of deer hunting simply overshadows all other domestic Euro-American hunting that might fall under the category of trophy hunting. Many other kinds of trophy hunting exist—bears, wolves, cougars, boars, walruses, sheep, goats, boar and more are all potentially hunting trophies[18]—but deer hunting is just far more common. As noted in Chapter 4.1, based on the most recent USFWS national survey, 8.1 million out of 9.2 million big game hunters in the US are deer hunters, and most of them hunt whitetail deer. Most of the remaining are turkey hunters.[19]

Another reason that we focus on deer hunting is that many of the other possible kinds of trophy hunting resemble the stereotypical kind of international trophy hunting tourism. These tend to involve expensive hunting for species that are far less abundant than deer, and it is often out-of-state hunting in USA or requires travelling to another European country. The hunting of, for example, big horn sheep in USA is not the kind of semi-subsistence meat-hunting for locally abundant species that characterizes much deer hunting.

Deer hunting in USA and Europe is fraught with issues that are altogether different from the issues with trophy hunting in Africa discussed above. The core question in an African context tends to be how trophy hunting can prevent vulnerable or endangered species from declining further or going extinct while also supporting poor rural communities. A core question in a Euro-American context, conversely, is often how wildlife agencies can use hunting offtake to keep cervid populations in check to avoid the problems associated with deer overabundance, while also funding conservation costs in general. Unsurprisingly, using hunting to keep wildlife populations

---

[18] Again, SCI's *World Hunting Awards Field Journal* is useful as an indicator of trophy hunted species.

[19] Turkeys are considered big game in the National Survey, and while it is rarer, some turkey hunters do keep trophies. SCI also has a "Turkeys of the World" award. The usual way to prepare a turkey trophy seems to be to make a fanned-out display of the tailfeathers on a wooden plaque.

from being unsustainably small involves different management problems than using hunting to keep populations from being unsustainable large.

5.3.1 An Overabundance of Deer

> Today, with habitat restoration and careful monitoring of deer herds, the Michigan herd has grown to nearly 2 million. A healthy doe gives birth to twins or triplets every year, so each spring some 900,000 fawns are born. By October 1, when archery season begins, 500,000 of those fawns will still be alive. During the three seasons—archery, rifle, and muzzleloader—that last from early October to late December, hunters in Michigan kill more than 400,000 deer a year, nearly a 50 percent success rate, the second largest harvest in the United States, and even that isn't enough to stabilize the herd population. Despite the size of the harvest, deer kills by car have jumped, crop predation has increased, and as many as 125,000 deer a year starve to death. Managing the Michigan deer herd these days is even more challenging; as 75 percent of the deer now are found on private land, an increasing amount of which prohibit hunting. (Swan 1995: 76)

When Roosevelt and his hunter-conservationist contemporaries put forests and wildlife conservation on the political agenda in the US, the prompt was to save dwindling populations of wildlife from extinction, including several species of deer. This was accomplished, and more. Côté et al. summed up this development in an important scholarly review article about the problems of deer overabundance, saying that

> within a century, deer management has reversed course from a preoccupation with augmenting population growth through habitat protection, hunting regulations, and predator control to serious concerns about how best to limit deer densities and the consequent impacts of these animals on other ecosystem constituents and functions. (Côté et al. 2004: 113)

The overabundance of deer has even spawned a scientific sub-discipline of its own, reflected in edited volumes with titles such as *The Science of Overabundance—Deer Ecology and Population Management* (McShea et al. 1997).

US whitetail populations were at a low of probably fewer than 300,000 in 1930. Populations have since recovered to an estimated high of over 35 million around the late 1990s, shrinking a little now to between 30 and 35 million[20] (Carpenter 1997; Hewitt 2015). Five whitetails introduced in Finland about 80 years ago have increased to a problematic population of over 100,000 today; the North American Elk, also known as wapiti, was estimated at around one million individuals in the late 90s and is considered overabundant in some of its range today; and mule deer were estimated at around three million in the late 90s, with current populations believed to be stable (Carpenter 1997; Sanchez-Rojas and Gallina-Tessaro 2016).

In Europe, roe deer number over 15 million and are increasing (Lovari et al. 2018). Red deer have doubled from around 1.25 million in 1985 to more than 2.5

---

[20] We would normally rely on IUCN Red List data for population trends, but the estimation of above 11 million individuals by Gallina and Arevalo (2016) is in this case far from other common estimates of somewhere close to 30 million.

million today (Lovari et al. 2016) and pose serious problems as an invasive species in Australia. European populations of the moose (*Alces alces*) have increased by three to five times since 1970 to at least 440,000 in recent decades despite a harvest rate of about 50% per year. The North American population is at about one million and is also still increasing (Côté et al. 2004; Hundertmark 2016). Sika deer in Japan have increased from an estimated fewer than two million in 2007 to over three million in 2014, despite doubling culling efforts to approximately 11.5% of the population per year (Noguchi 2017). Reeves muntjac deer in the UK is an invasive non-native species that has increased steadily over the past few decades and in parts of southern England they can be a serious pest in conservation woodland, forestry, and gardens (Ward et al. 2021).

There are several reasons for deer populations often increasing to the point of being considered overabundant. First, the natural predators of deer have all but disappeared. Wolves, bears, and mountain lions have been extirpated from most of their historical ranges and have limited impact on populations in many places where they still exist. Second, the unregulated market hunting of deer has been ended in the US for just over a century now and was never an issue in much of Europe, so the extent to which humans can act as the limiting factor is tightly controlled. Hunting is banned in many deer habitats and hunting ranges continue to shrink, which means that the only limiting factor for population growth in many cases is the availability of forage. Where hunting *is* allowed, the total numbers of hunters, as shown earlier, have generally decreased. Also, game laws still tend to emphasize the shooting of males, which has only a limited impact on population growth. Where the shooting of females and fawns is encouraged, the reluctance from hunters to do so has been considerable. More on that shortly. Third, available forage for deer has increased greatly over the years. Logging activities in forests and subsequent replanting provides ample sources of food for the species of deer such as whitetails and roe deer that browse the fresh shoots of newly planted trees. Furthermore, modern agricultural practices of planting winter seeds have greatly increased the winter survivability of many grazing species that would otherwise face starvation during harsh winters. Fourth, climate change decreases the negative impact of winters on deer populations, especially in the global north. More hunters also feed deer in the winter to ensure that more of them survive till hunting season (Putman and Staines 2004).

5.3.2 *Problems of Overabundance*

> If all hunting in America were to stop tomorrow, the consequences for many species would be devastating. In short order roadkill crashes would skyrocket; crop predation would rise; outbreaks of disease among animals would jump dramatically, in some cases causing large die-offs and massive suffering; negative encounters between people and animals would increase; the incidence of rabies would rise; Lyme disease would become an even more widespread public health menace; some habitats would be destroyed; and the populations of many species, such as deer, rabbits, and Canada geese, would rise dramatically. In time,

a new self-limiting population would be established, but there are no guarantees about what would go on between now and then. In the meantime the suffering of the animals would be far greater than it is now. (Swan 1995: 160)

Of course, as noted throughout the academic literature on deer management, overabundance is largely a value judgement. If deer populations, regardless of size, did not cause any problems, then we would have no grounds to call them overabundant. Judgements of overabundance therefore rely on human concerns over perceived problems caused by the abundances of deer.

In a famous book chapter "Overpopulation" from the book *Problems in Management of Locally Abundant Wild Animals* (1981), Graeme Caughley presented four classes of overpopulation problems, still widely used and referred to in the academic literature about deer management today. We shall elaborate in the following on how problems caused by increasing deer populations fit into Caughley's classes of problems.

1. **The animals threaten human life or livelihood.**

This first of Caughley's classes of overpopulation problems refers to the perceived negative impacts on humans or human livelihoods. Deer overabundance impacts humans negatively in a number of ways. We may refer to the limit of how much negative impact we are willing to tolerate as cultural carrying capacity, which is different from ecological carrying capacity, to be covered later.

A good summary of this class of problems for deer is provided by Conover (2011) who mentions deer-vehicle collisions, physical attacks on people, zoonotic diseases, losses to agricultural production, losses to the timber industry, and damage to households. Out of these, deer-vehicle collisions (DVCs), zoonotic diseases, and animal attacks can impact human health directly.

Conover estimates well over 1 million DVCs per year in the US, injuring around 27,000 people, killing around 50, and costing society $1.7 billion a year. DVCs represent a likely 4% of all motor vehicle accidents, and an estimated 3–4% of Americans every year are involved in a DVC. Bissonette et al. (2008) report an increase of deer killed in DVCs in the US from 200,000 in 1980 to 538,000 in 1992. To put it differently, the risk of DVCs, especially in unhunted areas, is high enough that unscrupulous poachers can even take advantage of it:

> There is no public land in Marin where deer hunting is legal, and as a result deer are everywhere. There are so many of them that firearms are not necessary for poaching here, because the deer are so numerous and have little fear of humans. A game warden recently caught a group of deer poachers in Marin who had been using high-speed cars to kill deer. In the trunk of a Cadillac with a special, reinforced front bumper, pulled over for speeding on country roads in the early morning hours, were three freshly killed deer. (Swan 1995: 11)

In Europe, Groot Bruinderink and Hazebroek (1996) showed a tenfold increase of moose killed by cars and trains in Norway over the preceding 30–40 years; an increase of roe deer DVCs in Sweden from 15,000 in 1989 to 50,000 in 1993 and from 1937 to 2338 in the same years in The Netherlands. 66,744 DVCs in Germany in 1976 increased to 145,636 in 1993. The authors further estimated a total annual number of DVCs in Europe of 507,000, injuring 30,000 people, killing 300, and costing $1 billion in damages. Outside Europe and the US, Japan also has problems with increasing DVCs involving sika deer, reported to number between 110,000 and 370,000 in 1998 (Saeki and Macdonald 2004).

Zoonotic diseases have received extra attention after Covid-19. This includes many calls to further restrict trade in and consumption of wildlife. Deer overabundances have become more relevant in this context as well, and there have even been calls to ban hunting on grounds of preventing further zoonotic disease outbreaks. Most zoonotic risks in relation to deer are not so much related directly to the deer themselves but rather the ticks that deer carry. Deer tick bites are known to transmit granulocytic anaplasmosis (caused by the bacterium *Anaplasma phagocytophilum*), babesiosis (caused by the apicomplexan parasite *Babesia*), and most importantly borreliosis (more commonly known as Lyme disease, caused by the bacterium *Borrelia*) to humans. Lyme disease is the most common zoonotic disease in Western Europe and the US. Lyme disease cases reported to health officials in the US increased from 8257 in 1993 to 27,444 in 2007 (Conover 2011). This number of reported cases should however be much lower than the total number of cases, which has also increased since 2007; the US Center for Disease Control suggests an estimate of 300,000 cases annually (CDC 2019). In Europe, incidences have been drastically increasing since 1990, and Sykes and Makiello (2017) estimate in excess of 200,000 European cases a year. While climate change seems to be a more significant cause of the increase in Lyme disease incidences than the increasing numbers of deer, Kugeler et al. (2015) have shown that reducing deer populations also substantially reduces the numbers of ticks and therefore, presumably, Lyme disease transmissions. A summary of studies with matching conclusions was conducted by Kilpatrick et al. (2014), who also found that after deer density in a suburban area that suffered from adverse effects of overabundant deer was reduced by 87% by allowing hunting, resident-reported cases of Lyme disease were reduced by 80%.

While deer attacks on humans are a minor problem compared to DVCs and zoonotic diseases, they do represent another type of human-wildlife conflict that happens increasingly often. Deer attacks occur when growing and unhunted deer populations become too habituated to humans and come to see people as rivals during the rut (when bucks attack) or as threats to their fawns (when does attack). Attacks are frequent in deer parks but see also for example Hubbard and Nielsen (2009) for descriptions of moose, elk, and whitetails attacking people on university campuses.

The other overabundance problems that affect humans do so financially. These are agricultural damages, timber damages, and household damages. Deer pose problems for agriculture because they eat, trample, and lie down in crops. Whitetails—and to a lesser extent elk and mule deer—in the US (Conover 1994), sika deer in Japan (Noguchi 2017), and red deer in Europe are the primary culprits (Bleier et al. 2012). European roe deer, while more than five times as numerous as red deer, have comparatively little impact in continental Europe, but cause more damage—along with fallow deer—in England and Wales (Putman and Moore 1998). Serious damages are incurred especially by fruit farmers, and farmers who grow other deer-favourites such as maize (sweet corn).

Richard K. Nelson conveys here a sense of the scale of agricultural damages in the US in relation to deer population growth and hunting efforts:

> In Wisconsin, I asked a farmer what would happen to him if hunting were completely eliminated. "These deer would multiply so fast that we'd be put out of the farm business in three to five years," he replied. After I heard similar opinions from other farmers, I put the question to a leading state deer biologist. "It's scary to think about the impacts that one wildlife species can potentially have on farming," he agreed. "Deer are already causing huge amounts of damage. But if that herd was allowed to grow statewide without hunting as a control means, I would agree with those farmers, and I'd say the farming economy in Wisconsin could be virtually eliminated, probably within three to five years". (Nelson 1996: 3)

Fifty-three percent of US agricultural producers experienced problems with deer in 1994 and spent a mean of 43.6 h and $1002 that year to prevent wildlife damages, which amounts to $2 billion countrywide (Conover 2011). In Japan, sika deer cause so much damage to agriculture that some farmers are forced to give up and emigrate from rural areas (Noguchi 2017). Other examples are in Hungary where compensation payments for crop damages increased from 2.8 million euros in 1997 to 6 million in 2008 (Bleier et al. 2012); in Italy, where a study showed that increasing red deer densities cause estimated losses to maize farmers in excess of 20,000 euros per farm/year (Corgatelli et al. 2019); and in Denmark, where 24% of surveyed farmers in 2017 reported damages up to DKK10,000 ($1600) per year, 11% reported DKK10,000–25,000 ($4000), and 11% reported in excess of DKK25,000 (Løbner 2017).

Timber industries also feel the impact of expanding deer populations. Deer impact forestry mainly by browsing, fraying, and bark-stripping. Browsing may refer to the selective eating of seeds, seedlings, saplings, and fresh shoots of trees and shrubs. When deer eat the terminal buds of a tree, the growth is stunted, and the tree develops (if it develops at all) multiple stems. Such a tree is less suitable for timber than a single stem tree. Fraying refers to the rubbing of antlers against a tree to remove antler velvet or to mark territory. Excessive fraying removes tree bark, breaks branches, and can kill affected trees. Bark stripping is when bark is stripped off a tree and eaten, which also often kills the tree. Conover estimated deer damage costs to the forestry industry of $1.6 billion a year in the Midwestern and Eastern US (Putman and Moore 1998; Conover 2011).

Household damages refers to damages to landscape plantings and gardens. Sixty-one percent of metropolitan US residents reported wildlife damages in 1993, but only 4% were related to deer (Conover 2011). This appears as one of the comparatively minor human-deer conflicts, as one might additionally assume that suburban residents are more likely to accept minor damages to their gardens if it means having more deer around to look at. Furthering this line of reasoning, evidence suggests that people are generally willing to accept increasing deer populations if the only inconvenience is to humans and human livelihoods. Respondents in a 2004 New Hampshire survey generally supported increases in deer populations if the negative consequences only impacted human lives (Duda et al. 2010: 37). Seventy-seven percent indicated that they would still support an increase in the deer population if it meant "An increased likelihood that deer would damage your garden or landscaping," 65% if it meant that "You would have an increased likelihood of having an automobile accident with a deer," and 64% if it meant "An increased likelihood of losses to farmers and timberland owners." Conversely, if overpopulation negatively impacted the deer themselves or wildlife habitats in general, the numbers were far lower. Only 29% would still support an increase in the deer population if it meant that "More deer would die from starvation during winters," and 27% if it meant "Less food or poorer quality of habitat for other wildlife" or "Poorer health overall for the deer herd." These answers suggest that ecological carrying capacity becomes an important issue at least for urban people long before cultural carrying capacity (Duda et al. 2010).

2. **The animals depress the densities of favoured species.**

Overabundances of deer commonly lead to forest overbrowsing. Overbrowsing refers to a situation where deer browse the fresh buds of certain species of trees and shrubs to the extent that none of them can outgrow deer-browsing height, resulting in the suppression and possible elimination of the deer's preferred forage species from the ecosystem.

Aldo Leopold was one of the first to address the problem of overbrowsing, writing here about the consequences of having extirpated the wolf from most of its historical range:

> I have watched the face of many a newly wolfless mountain, and seen the south-facing slopes wrinkle with a maze of new deer trails. I have seen every edible bush and seedling browsed, first to anaemic desuetude, and then to death. I have seen every edible tree defoliated to the height of a saddlehorn. Such a mountain looks as if someone had given God a new pruning shears, and forbidden Him all other exercise. In the end the starved bones of the hoped-for deer herd, dead of its own too-much, bleach with the bones of the dead sage, or molder under the high-lined junipers. (Leopold 1968: 130–132)

Overbrowsing affects more than just the deer's preferred species. The overbrowsing of some plants leads to advantages for other plants, which in turn may advantage some animals and disadvantage others. Deer overbrowsing generally leads to a reduced biodiversity and biomass. A high density of deer can depress the understory growth of a forest to an extent that little forest floor food and habitat remains for birds, small mammals, and invertebrates (Côté et al. 2004; Côté 2011). An illustrative example is provided by Goetsch et al. (2011), who compared regular unfenced forest in Pennsylvania to a plot of forest that had been a fenced deer exclosure for 60 years. They found that the unfenced forest contained no shrubs and 99% less forbs.[21] Total understory plant cover was seven times higher in the fenced plot, and deer browsing had decreased diversity of species by 70% and species richness by 80% when comparing the unfenced to the fenced plot. The unfenced understory was almost entirely dominated by one single species of fern that did not appeal to deer. See McShea et al. (1997) for many further examples of negative deer overabundance impacts on forest regeneration and biodiversity.

---

[21] Herbaceous non-grass flowering plants.

3. **The animals are too numerous for their own good.**

Deer overabundance does not just impact the plants that they consume and other animal species whose existence relies on those plants; when the overabundance is at a level where the food is scarce, it has clear negative impacts on the deer themselves. Zoonotic diseases (above) are a potential risk to humans, but diseases and parasites have a considerably larger impact on deer than they do on humans. Overabundance greatly exacerbates these problems, firstly because food scarcity reduces the general health of the herd, making them more susceptible to diseases, and secondly because a higher density of deer facilitates and accelerates the transmission of diseases from animal to animal (Davidson and Doster 1997; Gortázar et al. 2006; Campbell and VerCauteren 2011).

Aside from diseases, higher deer densities can, if they result in reduced forage, lead to less healthy deer due to malnutrition. Stewart et al. (2005) found that elk at densities approaching the ecological carrying capacity of their habitat were in much poorer physical condition and had lower pregnancy rates than did elk in areas with a lower density.

4. **The system of plants and animals is off its equilibrium.**

Although the population may be plagued by diseases and the deer are smaller and in poorer health than if the population was smaller, the Class 3 deer overpopulation may still be in equilibrium within the carrying capacity of its environment. "Overpopulation" in this case is, as Caughley notes, an expression of human preferences for healthier and bigger deer and not any imbalance in an ecological sense.

If, on the other hand, a population *exceeds* the carrying capacity of its environment, then the overpopulation is Class 4. Caughley writes that

> [t]he commonest form of Class 4 overpopulation that exercises conservationists is an eruption of terrestrial herbivores past the level to which they will eventually stabilize. Usually the eruption reflects the response of the population to abundant food, as when a new population is started by translocation of a nucleus to an area presently lacking that species. Sometimes it reflects the relaxation of hunting pressure, as when an area is dedicated as a national park. Over the next twenty years or so the population climbs to its equilibrium density, overshoots it, crashes, and then converges to relative stability. (1981: 10)

What Caughley refers to as eruption here was coined by Aldo Leopold as irruption, and both terms are used to describe a drastic increase in population that comes to exceed the ecological carrying capacity of its environment (Leopold et al. 1947; McCullough 1997).

Irruptions and the eventual mass-death from disease and starvation are, as has been covered earlier, one of the common arguments of hunters for keeping populations in check with hunting. If hunters did not kill a portion of the deer—the pro-hunting animal welfare argument goes—they would eventually face a more prolonged and agonizing death from disease and starvation. This argument is also used, albeit slightly modified, for those arguing that trophy hunting in southern Africa and other locations reduces poaching. By providing revenue and physical presence, hunters reduce the snaring, poisoning and other forms of illegal killing that cause the deaths of far more animals than they hunt, and in ways that are far less humane.

5.3.3 The Role of Hunting in Deer Management and the Problems with Fair Chase

As ensured by the sportsmen who saved big game species on the verge of extinction in the early twentieth century, the North American Model for Wildlife Conservation is synonymous with fair chase hunting and the view that wildlife is a renewable natural resource. The contribution of hunters to conservation is of course not just to keep deer populations in check; the primary contribution of hunters is the funding of conservation through revenue raised by hunting and related activities. As mentioned in Chapter 4.1, US hunters spend around $25 billion a year on hunting, more than half of which is spent on big game hunting trips and equipment, producing over $5 billion annually in federal taxes. The Pittman-Robertson Act of 1937, establishing an excise tax on all hunting- and fishing related equipment, has according to the USFWS transferred more than $20 billion to conservation in its lifetime—more than $1.1 billion in 2018 alone. These numbers may not mean much to most people, so suffice to repeat from earlier that funds generated by hunters comprise over 75% of state wildlife conservation budgets on average. A similar percentage of those funds are generated by white-tailed deer alone (Heberlein 1991; Mahoney 2009; Heffelfinger et al. 2013; Hewitt 2015; US Department of the Interior 2018; USFWS 2018).

This reliance of wildlife agencies on hunting and hunters gives hunters and the hunting lobby great power over how North American wildlife is managed and prioritized, and over what sort of management research is funded. The setup has given rise to common criticisms that the purpose of wildlife management agencies has become to produce an artificial overabundance of deer for hunting purposes, and that managers have little to no incentive to explore population management alternatives to hunting, or to support science that questions the status quo (Decker and Brown 1987; Simon 2016; Heister 2019).

It is hard to argue against the claim that funding priorities of wildlife management agencies are skewed towards game species. McCabe and McCabe, for example, write that

> [i]t seems untenable, from a professional standpoint, that game species in the United States, which represent less than 5% of all wildlife species, should receive nearly 97% of the available dollars for management; that the endangered species, or 11% of all wildlife species, receives approximately 2.8% of the available management dollars; and that the majority, 85% nongame and nonendangered species, receives less than 0.2% of the management funds. (McCabe and McCabe 1997: 21)

This distribution of funding might favour hunting even more than hunters themselves would favour hunting. Fix et al. (1998) found that Colorado hunters' desired mean distribution of hunting license fees were 44% to hunting recreation, 29% to fishing recreation, 15% to endangered/non-game species, and 11% to wildlife viewing. In contrast to this, nonconsumptive wildlife users have for example shown little interest in supporting fees and taxes for bird watching and bird watching equipment to support a wildlife management system that they think only benefits hunters

(Heberlein 1991). The result is a vicious circle where funding priorities are unlikely to change until nonhunting stakeholders begin contributing to conservation, which they will not do because of unfair, hunting-biased, funding priorities. Defenders of the status quo distribution tend to argue that game species—especially in the absence of apex predators—are also keystone species, the management of which equates to the general conservation of the biodiversity of their habitats (McShea and Rappole 1992; Rooney and Waller 2003).

While the wildlife management models of USA and Europe have proved their worth when it comes to saving endangered big game species, their reliance on hunting is not unproblematic when the problem is overabundance, and the management goals are to reduce deer density or the ratio of does to bucks. David G. Hewitt explains the matter nicely as it relates to the historical development in whitetail populations:

> During the early- to mid-1900s when whitetail populations were depleted and far below population objectives, state agencies used regulations and educational campaigns to encourage hunters to shoot only male deer. This approach worked well when the objective was to increase deer populations and soon became established as a standard of good hunting practice. In areas of good habitat where the high reproductive capacity of white-tailed deer could be fully expressed, deer populations increased rapidly. Deer became overabundant in portions of Pennsylvania as early as the 1930s and deer managers had considerable difficulty in encouraging hunters to harvest female deer also. This cultural reluctance by hunters to harvest females posed a challenge for the next 50 years of deer management. Hunters had been taught to shoot only male deer and once harvest of females became necessary, hunters were reluctant to participate. (Hewitt 2015: 844)

Wildlife managers work with three different types of management for deer: Traditional Deer Management, Quality Deer Management (QDM), and Trophy Deer Management (TDM). The transition in the 1900s that Hewitt describes here is the transition from Traditional Deer Management, where the objective is simply to increase population sizes, to QDM, where possible objectives can be to reduce or limit population size to fit the ecological carrying capacity of the habitat, increase the ratio of mature bucks to does, improve deer health, or remove unwanted genetics from populations (Adams and Hamilton 2011). We will get back to TDM in Chapters 5.4 and 6.4.

In practice, QDM usually involves an emphasis on the hunting of females and juveniles, which is referred to as "antlerless harvest" in management literature. While getting hunters to shoot females was difficult, state wildlife agencies in the US managed to achieve an antlerless harvest that surpassed the antlered harvest in 1999. The trend has reversed, however, and antlerless whitetail harvests in the US have declined by 20% since 2007; 2017 and 2018 were the first years in over a decade that more antlered whitetail bucks than antlerless whitetails were harvested nationwide. The Quality Deer Management Association notes that while population increases are needed in some places from time to time, few states ought to be harvesting more antlered bucks than antlerless deer on a regular basis, so current trends could be considered a problem for QDM practices (Adams and Hamilton 2011; Hewitt 2015; QDMA 2020).

Ironically, the fair chase ethic that enabled hunters to save wildlife from extinction a hundred years ago also cultivates the unwillingness to participate in antlerless harvest today, which limits the usefulness of hunters for QDM. Further reinforcing hunter reluctance to participate in antlerless harvest is hunting's historical connection with training for war, rite of passage, and proof of manhood; one demonstrates masculinity and skill by killing the biggest and most cunning and dangerous animals, which for game species is usually the male. Bronner, for example, writes that "[s]ome hunters find it offensive to shoot the feminine does [...] The "fair game" that provides the challenge of combat is in metaphorically masculine animals, including bears and bucks" (Bronner 2008: 161). Here is also Allen Morris Jones, emphasizing instead of the sex, the rarity of trophy bucks:

> What kind of hunt is involved in shooting a spike or a doe. There is a drastic overabundance of these animals as related to our ability to kill them. So again, we must consciously limit ourselves. We are hunting, but there should be one very scarce type of animal to attract our skills. And the trophies with the largest racks, the oldest, the wisest, are almost always the most difficult to kill, even with the best equipment. (Jones 2012: 41)

Jones writes as a fair chase hunter of the old school here, but from a QDM-perspective the overabundance of spikes and does is exactly the reason that hunters *must* shoot them. Hunters' desire to challenge themselves should come second to their responsibilities as participants in wildlife management if their activities are to be taken seriously as anything other than egotistical. In other words, if hunters want to use "management" as a justification for hunting then they should manage, and that means hunting what is necessary rather than what is desired.

Then, of course, there is the matter of the trophy. Only bucks have antlers, so asking hunters not to shoot bucks is to ask hunters to forego the trophy. While hunters only rarely claim in surveys that trophies are a primary motivator for going hunting, there is little doubt that bringing home a nice rack is important to many hunters once they are out there.

Jan E. Dizard provided a good example of the hunter-QDM problem: When the Metropolitan District Commission at the Quabbin reservoir wanted deer herds reduced as fast and efficiently as possible, hunters proved completely unsuited for the task, as their fair chase ideals prevented them from shooting does. To make matters worse, the hitherto unhunted does at the Quabbin did not consider humans a threat and were already half-tame, which made the fair-chase minded hunters even less willing to shoot them (Dizard 1994).

How many hunters are part of the problem and how many are part of the solution is hard to say, but Ward et al. (2008) might give us an indication with their categorization of hunters according to what might be interpreted as QDM-attitudes. They characterised about two thirds of the hunters they surveyed as "No-Damage Traditionalists," meaning hunters who generally do not support antlerless harvest or antler restrictions, do not believe that deer damage to forests is a problem, do not think that deer are overabundant or cause serious human-wildlife conflicts, and do not consider the management of deer populations a very important part of hunting. The remaining third were categorized as "Damage-Control Managers,", who oppose the No-Damage Traditionalists on each of these issues. If this is representative of hunters in general, then only a third of hunters can be relied on to participate in QDM willingly while two thirds need to be persuaded, tricked, or coerced into being useful for wildlife management purposes.

Interpreting this research in the light of Kellert's three basic hunting motivations (naturalistic, utilitarian, and dominionistic), it seems an obvious assumption that the third of hunters who support QDM are mostly naturalistic/utilitarian while the two thirds who are unsupportive are mostly dominionistic/utilitarian. Dominionistic hunters, as discussed already, hunt for sport, challenge, competition, and trophies and have considerably less knowledge about wildlife than naturalistic hunters. It follows that they would be less willing to shoot females. Ward et al. seem to support this conclusion:

> Because deer populations are controlled through antlerless deer harvest, hunters who rate *obtaining venison* and *to help manage the deer population* as important or very important would be potentially more useful for game management than those hunters who do not. (Ward et al. 2008: 219, emphasis in original)

Hunters' reluctance to shoot females and juveniles has led to management measures designed specifically to change the bucks-only attitudes of hunters or—failing that—to force or trick hunters into furthering management objectives by participating in antlerless harvest regardless of their attitudes. "Earn-a-buck" is one such trick, requiring hunters to shoot a predetermined number of does before they are allowed to shoot a buck. Enck and Brown (2001) found that hunters at Cornell University's Arnot Forest generally embraced such management measures if they thought it would result in bigger bucks in the future, showing perhaps that even the two thirds of hunters who are unwilling to manage responsibly can be made to participate if properly incentivized.

Other common QDM measures include quotas (issuing permissions to shoot many does and only few bucks); closed seasons (e.g. no bucks in the rut); antler restrictions (allowing only bucks with certain antler points or antler spread to be harvested); and area requirements (allotting buck harvests based on hunting ground size). Non-harvest QDM can be the management of habitat, predators, other herbivores, habitat, diseases, or supplemental feeding.

Here, we describe the situation in Denmark in some detail, as it represents a recent and interestingly multifaceted perspective on the potential problems of using hunters as deer managers.

Denmark is noteworthy from a wildlife management perspective because it is one of the only countries in the world that uses hunting as a wildlife management tool without having any bag limits. Four species are relevant in a trophy hunting discussion for Denmark: red deer, roe deer, fallow deer, and sika deer with roe being more and sika being less common than the others. The Environmental Ministry regulates the hunting seasons for these species based on recommendations from the National Wildlife Management Council, which receives input from localized councils.

Danish hunters need to qualify for a hunting license for either rifle, shotgun, or bow. With the proper hunting license in hand, hunters can then hunt either on their own private land or on private land of others, where they have been given express permission to hunt. Hunting may also take places on certain public lands, for which the hunting rights may be rented from the environmental ministry. None of these options come cheap. Hunting is an expensive hobby in Denmark, and if one owns a piece of land with hunting opportunities, then one has paid dearly for that land, and for that reason. Most hunting by far takes place on private land, but only about a third of hunters own the land on which they hunt (Gjerris et al. 2016: 18).

Landowners who own one or more hectares of land may hunt on their own land. Any landowner who owns five or more hectares of land may rent out the hunting rights.[22] Hunters who do not own one hectare of land must therefore rent hunting permissions from landowners who own five or more. A landowner may allow anyone to hunt on their land free of charge, but this is very rare, as they can generate a considerable income by renting out the exclusive hunting rights. For larger hunting grounds, hunters often band together in hunting consortia to share the rent and hunting between them.

Here is where it gets interesting. With no bag limits, hunters who own one hectare or more can shoot as many deer as they want on their own land, provided they adhere to the hunting seasons and other hunting regulations. Landowners who rent out their five hectares or more to hunters are free to stipulate any restrictions and bag limits they want as part of the rental agreement. They almost never do, however, as they can get considerably more money for the hunting rights if there are no rules.

With no bag limits, this sounds like a way of expediently exterminating Denmark's populations of deer, but that is not what happens, and Danish hunters' reluctance to participate in antlerless harvest seems to be the reason.

Danish hunters, whether hunting on their own land or on land that they have rented, shoot too few deer overall, but too many bucks. The problem appears to be a combination of trophy fixation and small hunting grounds. A rifle shot is loud, and since hunting grounds are small,[23] shooting a doe or fawn scares away any potential bucks nearby, likely sending them to the neighbours instead. Most hunters therefore leave the does and fawns alone and wait for a buck. Yet the hunters do not wait for the mature trophy buck. If hunters see a buck of maybe two to four years then they are likely to shoot it because bucks are rare, and they may be afraid that if they pass

---

[22] One km^2 is equal to 100 hectares, so the requirements of 1 and 5 hectares of land are very low. The Danish hunting management philosophy is to ensure that hunting opportunities remain broadly accessible and not just a privilege of wealthy landowners and royalty. Yet, as mentioned, it remains a very expensive hobby.

[23] Large hunting estates do not have the problem of bucks fleeing to neighbouring properties. They also often have wildlife managers employed to ensure that the offtake contributes to the healthiest possible population. Large trophy bucks are therefore often killed on such estates or on the small farmlands lucky enough to surround them.

it up it will wander away and get shot by the neighbour. Then, since many hunters shoot immature bucks, the mature bucks become fewer and fewer, which provides even more reason for hunters to be satisfied with immature bucks.

Here are some of the consequences of this scenario:

- About 1:10 may be considered a healthy mature buck (8 years or older) to doe ratio in a red deer population. Denmark's ratio is commonly about 1:100 because too many bucks are killed. There are too few mature bucks and too many does and fawns, which in most populations means too many deer overall.
- The general overpopulation of does and fawns leads to considerable agricultural and forest damages, for which farmers (who often also own land for hunting, which they rent out) seek compensation from the government. If more does and fawns were shot, agricultural damages would be far less severe, but if farmers specified in their rental agreements with hunters that only does and fawns were to be shot, they would receive only a fraction of the rent, because most hunters want bucks. It can therefore be in the farmers' best interests to have more deer, even if it means more damages to their own crops.
- Too few of the bucks with the strongest genes get to survive and pass on their genes for more generations because most of them are killed indiscriminately at a young age and before strong genes (big body and large antlers) can be identified. Weak and strong genes have equal chances of being carried on.
- Despite the trophy fixation of hunters, the impatience and fear of losing the buck to the hunter next door means that most noteworthy trophy animals are killed on large estates that employ professional wildlife managers.

Denmark has until recently relied on voluntary guidelines and codes of ethical conduct to get hunters to manage deer populations responsibly, but it seems this has not been enough.

Summarizing the current situation in Denmark and the historical development in the US, there are indications that hunters can indeed help managers with QDM through higher antlerless harvest rates, but the majority need to be forced or at the very least incentivised to participate. Left to their own devices and guided by their own fair chase ideals and trophy fixations, deer hunters shoot too many bucks for a healthy management regime.

5.4 Trophies and Genetics

> Man represents a new kind of predator and parasite in nature's world. The hawk, the weasel, the mountain lion weed out the diseased and unfit.[24] They play their part in evolution. The hunter seeks the finest victims. He represents evolution in reverse. (Teale 1990: 161)

What Edwin Way Teale here in his 1966 book *Wandering Through Winter* refers to as "evolution in reverse" is the common criticism levelled against trophy hunting that it leads to genetic deterioration by removing the males with the "best genes" (in itself an extremely difficult thing to define) from breeding populations. The pro-hunting counterargument is that trophy males are old and have thus had plenty of years in which to pass on their genes. This section examines these claims along with the issue of artificial production of unnaturally large trophies for trophy hunting purposes in intensively managed ungulate populations. Sticking with the deer management theme, we will discuss this latter issue first.

5.4.1 Trophy Deer Management and Deer Farming

TDM is defined by Adams and Hamilton as

> the approach in which only fully mature bucks, generally greater than 5.5 years of age, with high-scoring antlers are harvested and does are harvested to maintain low deer density and optimum nutrition for the remaining animals. (2011: 362)

TDM is the management of deer with the goal of maximizing B&C scores (Van Brackle and McDonald 2007). Top-class antlers—and top-class trophies in general—rely on the three factors of nutrition, age, and genetics, and each of these factors can be managed. Nutrition can be controlled simply by keeping the population size below the ecological carrying capacity of the habitat by harvesting females, or by supplementary feeding. Age and genetics are simultaneously controlled by culling the males that have undesirable genes as early as they can be identified while letting the males with desired genes (large bodies and large, symmetrical trophies, traits favoured by hunters but not necessarily those that lead to better survival and reproduction) live and contribute their genes to the population for as long as possible. This is TDM in its most basic form. A continual harvest of trophy males in a deer population can be ensured by way of "stacking," which refers to the protection of trophy bucks until they are as old as 7 or 8 years. There will be a greater natural mortality rate at this age than if bucks are harvested at 5–6 years, but otherwise the harvests are merely delayed for a few more years after initial implementation of the management (Jacobson et al. 2011).

While adequate nutrition can be ensured by keeping populations below ecological carrying capacity, additional antler growth can be stimulated in a deer population

---

[24] One might add winters and disease to Teale's examples. Unfit specimens are also at greater risk of dying of non-predator natural causes than the fittest specimens.

5.4 Trophies and Genetics

through habitat management measures such as manipulating the amounts and nutritional qualities of forage. Deer can also be provided with the energy, minerals (mainly calcium, phosphorous, and magnesium), and most of all proteins required to produce larger bodies and antlers through supplementary feed such as deer pellets or high-nutrient vegetables (Brown 1990; Demarais and Strickland 2011; Jacobson et al. 2011).

Genetic potential for large antlers can be controlled—in addition to culling bad genes and protecting good genes in a wild population—by introducing deer from other populations or even by raising and protecting deer in captivity and breeding them with females of equal genetic potential. The breeding of trophy bucks is comparable to the breeding of racehorses, and so are the prices involved. These bucks have lineages of named deer that go back many generations and are much too valuable to be trophy hunted; their use is to have their sperm harvested, frozen, and sold for artificial insemination purposes to trophy-buck breeders. Intensive artificial breeding is also a feature of some parts of the South African game ranching model. Exceedingly large or otherwise desirable animals (such as unusual colour variants) and their offspring can command enormous prices.

At this stage, we are of course long past the point of discussing wildlife management. These animals are usually kept in high-fenced enclosures and may be entirely tame. This is farming and the species so selected are livestock.

Deer farming is increasingly common in many parts of the world. Farmed deer are commonly white-tailed deer and wapiti in the US, fallow deer and red deer in Europe and New Zealand, and sika deer in Asia. Most of the farming is focused on venison production rather than trophy hunting. China has the largest deer farming industry in Asia, producing venison for consumption and antler velvet for TCM. Other minor products of the deer industry in general are shed antlers for furniture and decorations, and deer urine for scent-masking hunting products (Frosch et al. 2008). Many European countries have hundreds of deer farms with thousands of deer, chiefly focusing on providing venison for local markets. Though selling the most exceptional males with the biggest antlers to hunting estates is not uncommon, it is not the focus of most deer farms. European countries that do focus on the engineering of trophy bucks for estate hunting tourism include France, Italy, Spain, Hungary, Czech Republic, and Russia.

A focus on high-fenced trophy hunting is also characteristic of deer farming in the US. According to Frosch et al. (2008), there were 7828 deer farms in the US in 2008. As with most things hunting, Texas is one of the leading states, representing more than 1000 such farms. Half of these farms are operations that breed trophy bucks and host the hunting of them, about a third are breeding-only, and the rest are hunting-only.

Most importantly in a deer farming overview is New Zealand, which is estimated to hold in excess of half of the world's farmed deer. New Zealand produces venison and antler velvet for markets worldwide and trophy bucks for trophy hunting tourism on local hunting estates. Commonly the deer are a cross between the European red deer and the North American wapiti because they grow larger and faster than either of the purebred species (Kuba et al. 2015).

5.4.2 Evolution in Reverse

That human predation and human influence can heavily affect trait changes in flora and fauna is uncontroversial (Darimont et al. 2009). Sometimes termed "unnatural selection," human-induced evolution through selective harvesting, which defines trophy hunting, has been studied in the last few decades. Coltman et al. examined in 2003 how 30 years of trophy hunting had affected bighorn sheep at Ram Mountain in Alberta, Canada. They found that

> [u]nrestricted harvesting of trophy rams has thus contributed to a decline in the very traits that determine trophy quality. Hunters have selectively targeted rams of high genetic quality before their reproductive peak, depleting the genes that confer rapid early body and horn growth. (Coltman et al. 2003: 657)

This article, published in *Nature*, received a lot of attention and was cause for controversy in both popular media and scholarly literature following its publication. The criticism of the last two decades, that trophy hunting can cause genetically inferior individuals of hunted species to prosper, largely stems from our knowledge of bighorn sheep in Alberta. A conclusion of the paper was that the pro-trophy hunting argument—that trophy hunters only target old and "surplus" specimens that have had plenty of time to contribute their genes to the population—might not hold true for species that can reach peak body- and trophy size *before* they reach their reproductive peak. This "surplus animals" pro-hunting argument is also criticized by the Caprinae Specialist Group of the IUCN:

> The argument that trophy males are past their prime, and therefore are surplus animals, is invalid. On the contrary, trophy hunters selectively target those males that would be responsible for much of the reproduction in an unhunted population. (IUCN Caprinae Specialist Group 2008: 23)

Festa-Bianchet et al. (2014) likewise examined 7100 bighorn rams harvested over 37 years in Alberta, and their conclusions largely reinforced the findings of Coltman et al. that horn sizes decrease when trophy rams are harvested before their larger horns have had a chance to contribute to their increased mating success. They further concluded that the continued harvesting of rams once they reached the minimum legal horn size had the genetic consequence that rams become older and older before they reach the minimum horn size, and that more rams do not reach the minimum size at all. The rams that never reach minimum size at all, logically, have a whole natural lifetime in which to reproduce, whereas the rams that grow the largest horns very quickly have the least amount of time, given a hunting intensity where a large portion of the minimum-size rams are harvested.

Douhard et al. (2016) showed that the hunting of Stone's sheep (*Ovis dalli stonei*) over 37 years in British Columbia, Canada also led to hunting-induced decrease in horn sizes. The authors compared an area with low hunting pressure to an area with high hunting pressure and found that early-life horn growth decreased by 12% under high hunting pressure with no changes observed at low hunting pressure in the same time period. As only males with full-curl horns are hunted, the decrease

5.4 Trophies and Genetics

in horn growth led to a 45% decrease in harvested males under high hunting pressure. Harvests remained the same in the area with low hunting pressure. The study also found that since hunters target sheep based on horn length/curl and not thickness/mass, this selection caused horn length to decrease more than horn circumference in the study period. However, non-resident trophy hunters had less effect than resident hunters, the reason being that non-resident hunters are required to hire a guide who will go to great lengths to find exemplary trophy males. These are generally older and have had ample time to reproduce. Resident hunters would more often shoot any full-curl male, which results in a higher offtake of immature sheep.

Monteith et al. (2013) analysed 15,778 trophies of antlered game and 6526 of horned game submitted to *B&C Records*, looking for trends in trophy sizes. The authors found mean trophy size reductions of 1.87% for antlers and 0.68% for horns between 1950 and 2008. As noted by Pelletier et al. (2012), however, using record book entries can be an unreliable measure, as minimum thresholds for trophy records preclude estimations at the lower end.

Antlered species, just like horned species, may reach peak trophy size before they reach peak reproductivity, as noted here by Festa-Bianchet and Lee (2009: 100):

> [I]n red deer *Cervus elaphus* both age and antler size play a role in mating success: average antler mass peaks at about ten years of age and remains high thereafter, but mating success peaks at eight to eleven years of age, then declines (Kruuk et al., 2002). The genetic impacts of harvesting a large-antlered male when he is seven or when he is 12 are, therefore, very different.

While red deer antler mass may peak at around ten years of age, they are often close to their antler mass peak already at around age seven. Telling the difference between a 3-year-old and a 7-year-old based on the antlers is generally easy; telling the difference between a 7-year-old and an 11-year-old is often impossible in a hunting situation. To do so requires knowledge about other signs of age such as posture, facial expression, neck size, colouring, and the shape of the stomach and spine. Many hunters do not possess this knowledge, and among the ones that do, letting a 7-year-old red deer with a large set of antlers go because it might not have reached its reproductive peak yet is a lot to ask.

A similar dynamic may be relevant for roe deer. In Lithuania, low quality roe deer bucks (abnormal antler shapes and low body weights) are routinely culled, and trophy bucks have a minimum legal age of five years. This is contrasted with Latvia and Estonia where roe deer hunting is not strictly regulated. Balčiauskas et al. (2017) found that the biggest roe deer antlers in Lithuania are around 40% bigger than they are in Latvia and Estonia. Eliminating factors other than hunting strategy that influence antler size of roe deer, the authors attribute the difference in trophy sizes largely to the age minimum in Lithuania and the considerable pre-sexual reproductivity peak hunting offtake in Latvia and Estonia.

Overall, however, the literature seems to suggest that trophy hunting asserting a negative evolutionary pressure on trophy hunted species is a minor problem. The strong evidence for negative evolutionary pressure caused by hunting has mostly been in sheep. Yet hunting is just one of many environmental factors affecting evolutionary traits, and the problem has often been a too high hunting pressure more so than trophy hunting per se (Heffelfinger 2018; LaSharr et al. 2019). While some potential for artificial selection by trophy hunters is a concern in isolated cases, the overall threat level seems incomparable to the threats posed by habitat loss, poaching, and diseases, which money from trophy hunting can generally help to prevent.

According to what evidence we have today—again mostly sheep—selective hunting seems only to lead to negative evolution if the targeted trait has strong heritability, if the harvest pressure is high and lasts for many decades, if the selective pressure sets in before sexual reproduction peak, and if immigration is either prevented or immigrants are killed off (Festa-Bianchet and Lee 2009; Festa-Bianchet and Mysterud 2018).

For managers, it is important to ensure that the biggest trophy animals have lived long enough past the age at which they attain maximum body- and trophy size that they have also been at their reproductive peak for enough time to have passed on their genes to the next generation. Given that ensuring this is a very difficult task, it is also important to ensure that many of the trophy animals with preferred genetics are not harvested at all. However, each species, and indeed each population, is different and needs management that corresponds to biological, environmental, and geographical specifics. A more intensive and selective hunting pressure may for example be applied to hunted populations that are close to large PAs with unhunted populations, ensuring an exchange of genetic material. Adverse effects of selective hunting can potentially also be counterbalanced by early and targeted culling of males with undesired genes (Mysterud and Bischof 2010).

5.5 Population Management Alternatives

As noted in Chapter 5.3.3, the fact that hunters contribute most of the funding of state wildlife management agencies in the US means that management alternatives to hunting often go unexplored. That deer populations require management and that management equals hunting, critics say, is taken for granted. Allen Rutberg writes that wildlife managers rely on two dogmas to justify deer management:

> The first fundamental dogma is that deer management is always necessary to prevent deer overpopulation. [...] The second fundamental dogma is that hunting is essential for deer management. In wildlife management conferences, "hunting" is often used synonymously with "management," and "antimanagement" is used as a misleading codeword for "antihunting". (Rutberg 1997: 40)

Yet alternative ways to manage deer populations have been tried and are in use. The search for management alternatives to hunting is on some occasions based on public concerns about gunfire in suburban areas. Both stray bullets and the noise of gunshots are possible concerns. In these cases, hunting with bows and crossbows can be a safer and far quieter management option, albeit less efficient at reducing deer densities than hunting with firearms.

Prompted by the preconception that wounding rates of hunters are too high and that hunters are not safe enough for suburban areas, another option is to hire professional sharpshooters to kill deer. Sharpshooters are presumably safer in suburban environments and less noisy than hunters because, depending on local legislation, they may be more likely to use suppressors.[25] Sharpshooters are also uninhibited by ideals of fair chase because their purpose, as cullers, is to kill as quickly and painlessly as possible, and they do not seek to challenge themselves by making the killing "fair," as shown here by DeNicola et al.:

> The following methods are recommended for sharpshooting programs: (1) use baits to attract deer to designated areas prior to removal efforts, (2) shoot deer from portable tree stands, ground blinds, or from a vehicle during the day or night, (3) when possible, select head (brain) or neck (spine) shots to ensure quick and humane death, (4) process deer in a closed and sheltered facility, and (5) donate meat to food banks for distribution to needy people in the community. (DeNicola et al. 2000: 27)

The downside to sharpshooters is that they can cost $600 or more per deer, whereas hunting can be an income. This affects the available budgets of state wildlife management agencies for other conservation tasks (DeNicola et al. 2000; Kilpatrick et al. 2007).

A population management option that does not involve people killing deer directly is predator restoration, which has tended in recent times to fall under the complex and often controversial umbrella of "rewilding." This is rarely a feasible solution though; if the issue with deer overabundance is an unacceptable level of human-wildlife conflicts in a suburban or peri-urban area, then substituting deer for wolves or mountain lions and a small risk (but guaranteed worry) of fatal human-wildlife conflicts seems like a suboptimal solution at best. If the deer population is far from urban and suburban areas, and the problem is not human-wildlife conflict but over-browsing and reduced forest regeneration, then predator restoration can be a success; the reintroduction and recovery of wolves in Yellowstone National Park is a good example. Smaller predators like bobcats can also be (re-)introduced with success to suburban areas to reduce deer densities (Warren 2011).

Sometimes the opposition to public hunting as population management is based on a fundamental objection to killing, meaning that a nonlethal population management alternative must be found. Such alternatives can include relocation and immunocontraception.

Trapping and relocating deer is rarely considered a viable option. It is difficult to achieve and expensive, the deer often die post-release due to the stress involved with both the capture (capture myopathy) and the new and unfamiliar environment into which they are released, and there is significant risk of disease transmission caused by relocating animals. Finding suitable places for release is also increasingly

---

[25] A suppressor, or moderator, is a can-shaped device that can be attached to the muzzle of a firearm to reduce the intensity of the sound produced when a bullet is fired. Popularly called a "silencer," suppressors do not "silence" the calibres of rifle typically used in deer hunting. Rifle shots are very loud, and even a suppressed rifle shot in an urban area would not be "neighbour friendly." Small calibre rifles (such as .22LR rimfire) shooting subsonic ammunition through a suppressor are virtually silent, but are not legal for deer shooting in many jurisdictions.

challenging, as deer overabundance is the norm more so than the exception (Craven et al. 1998; Hewitt 2015).

This brings us to the final nonlethal set of options, which is fertility control, sometimes used in urban/suburban environments where lethal management is unfeasible or undesired. Two of the most relevant methods in this category are immunocontraception, which is when a vaccine is administered that stimulates immune response infertility, and surgical sterilization.

The latter option, similar to the option of trapping and relocating, has often turned out to be prohibitively expensive and impractical. It is only uncommonly considered a viable option for use in small, suburban populations with little to no immigration. Surgical sterilization requires expensive capture and veterinary involvement, and it is ineffective if there is any immigration of fertile deer. Some evidence suggests that it can be occasionally viable for suburban populations that tend to have small home-ranges and smaller population sizes, and it is cost-effective in the sense that each deer only needs to be treated once (Boulanger et al. 2012; Boulanger and Curtis 2016; Evans et al. 2016).

Immunocontraception vaccines can be delivered by hand or via a dart fired from a specially designed veterinary dart- or capture-gun. Although much additional effort is required for hand delivery, it does ensure a more consistent delivery dosage, and it allows for ear-tagging, making it possible to recognize which deer have already been treated. While surgical sterilization is permanent, deer treated with immunocontraceptives require retreatment with booster vaccinations every few years. This can make sterilization programs preferable in some cases, but immunocontraceptives are generally a more affordable option (Evans et al. 2016).

To achieve significant reductions of deer densities, both immunocontraception and sterilization require a long timeframe. Lethal population management options reduce current deer densities right away while also reducing future densities; fertility control only removes the potential for future offspring while doing nothing to resolve immediate problems caused by overabundance. Furthermore, the natural mortality of suburban deer populations is so low that a suggested 70–90% of a population needs to be treated/retreated annually before a desired density reduction can be expected, and then only after seven years or more (DeNicola et al. 2000; Boulanger et al. 2012).

Aside from the extended timeframe, the biggest downside to deer management methods that do not involve hunters is the cost; hunting can be an income or a cheap expense, depending on the micromanagement required, but the alternatives are all very expensive, exemplified here by Kilpatrick et al. (2007: 2095):

> In Georgia, USA, a community paid an average of $144 per deer to shoot 1127 deer over a 3-year period (Butfiloski et al. 1997). In Connecticut, USA, cost to a community to shoot 80 deer in 4 nights was $646 per deer (Kilpatrick 2005). In New Jersey, USA, a community paid $1000 per deer to treat about 145 deer with fertility control over a 3-year period. (S. Predl, New Jersey Division of Fish and Wildlife, personal communication)

Or here by Hewitt (2015: 845):

> The expense of removing deer using traditional deer hunting is borne by the hunters. Organized recreational hunts in suburban areas cost US$14 – 177/deer, depending on organizational and law enforcement costs whereas removing deer by trapping (greater than US

$1000/deer) or using sharp-shooters (US$90 – 260/deer) is expensive and finding places to release trapped deer is increasingly difficult. (Hewitt 2015: 845)

Advocates of animal rights might argue that the incurred costs of non-lethal management are irrelevant; deer have rights so lethal methods are unacceptable regardless of their efficiency and income generation.

5.6 Photo Safari and Photographic Hunting

Photo safari and photographic hunting are, on the surface, the same thing. Both are ways to generate income for conservation without killing animals, by using wildlife photography. But this is where the similarities end, and we need to explain the distinction here. The photo safari is an activity associated with wildlife tourism, and it is especially associated with Africa. The familiar image is a group of white tourists sitting in the back of a safari van on a savannah, taking pictures of nearby wildlife. The objective of the photo safari is to experience nature and wildlife from a distance (although close proximity to the subject is often demanded by photographic clients) and to create the best possible conditions for the wildlife viewing and photo-taking. Comfort is often prioritized, and the photographic equipment is nearly always expensive. The photo safari is aimed at nonhunters, and the people who join photo safaris are likely to be against hunting, wondering why hunters cannot enjoy wildlife in the same way as they do. The photo safari is often suggested by hunting critics as an alternative to trophy hunting for conserving African wildlife, but the target audience for photo safari tourism is not trophy hunters.

Photographic hunting is something else entirely. Photographic hunting is the same as hunting, except that the weapon is replaced with a camera, and the shot that kills is replaced with the nonlethal opening and closing of a shutter. The photographic hunt, often suggested to hunters by the antihunter, is meant specifically as a replacement for hunting, and it is as such meant to give the hunter the very experience of hunting in every possible way—finding the prey, stalking the prey, and getting close enough to deliver a killing shot—but without actually killing. The hunting trophy is the photo, which proves that the hunter who took it got close enough to have killed the animal, if he or she had so chosen.

While the Photo Safari is generally a guided tourist activity, the photographic hunt is more commonly meant as a replacement for domestic hunting—perhaps deer hunting most of all. Photographic hunting is not relevant in an African context; if the trophy hunter who hunts in Africa were satisfied with a photo, they would just join a photo safari instead, and pay far less for the experience.

Photographic hunting is an antihunting fantasy more so than a common practice. The meat is an essential objective of hunting for most deer hunters, and photographic hunting obviously yields no meat. Another objection to the practice is that the intention to kill is an essential part of hunting and that taking it away—replacing it with intention to photograph—would make the hunt meaningless:

> One can refuse to hunt, but if one hunts one has to accept certain ultimate requirements without which the reality "hunting" evaporates. The overpowering of the game, the tactile drama of its actual capture, and usually even more the tragedy of its death nurture the hunter's interest through anticipation and give liveliness and authenticity to all the previous work: the harsh confrontation with the animal's fierceness, the struggle with its energetic defense, the point of orgiastic intoxication aroused by the sight of blood, and even the hint of criminal suspicion which claws the hunter's conscience. Without these ingredients the spirit of the hunt disappears. The animal's behavior is wholly inspired by the conviction that his life is at stake, and if it turns out that this is a complete fiction, that it is only a matter of taking his picture, the hunt becomes a farce and its specific tension evaporates. All of hunting becomes spectral when a photographic image, which is an apparition, is substituted for the prey. (Gasset 1942: 94–95)

The essence of hunting is participation in nature; if the weapon is replaced with a camera, the participant *in* nature becomes an observer *of* nature instead. The hunter becomes a member of the audience rather than an actor in the play. Or as Forrest Wood Jr. wrote:

> A tourist who is sightseeing, a photographer who is taking pictures of wildlife, a hiker who is observing, a bird watcher who admires, a naturalist who studies objectively, all are not participating in the processes of nature. Their actions are those of an individual entity who is above, beyond outside, or distinct from that which he observes. (Wood Jr. 1997: 23)

Whereas photographic hunting is idealized by antihunters as a replacement for hunting at an individual and personal level for animal rights or animal welfare reasons, photo tourism is meant to compliment or replace hunting tourism at an industry level and serve as an alternative source of financing for conservation. Photo tourism is routinely brought up as the way to fund conservation if trophy hunting were banned, yet as argued already, this is mostly already being done to the extent that photo tourism is a feasible option. A photo safari, according to IUCN,

> requires political stability, proximity to good transport links, minimal disease risks, high density wildlife populations to guarantee viewing, scenic landscapes, high capital investment, infrastructure (hotels, food and water supply, waste management), and local skills and capacity. (IUCN 2016: 9)

Most areas that meet all these requirements already have photo safari tourism, but even ecologists who have spent most of their lives in Africa have to admit that much of the continent is "dull, hot, and filled with tsetse flies" (Packer 2015: 30) or "hot, dry, dusty, and worryingly insecure" (Loveridge 2018: 242) and simply not viable for photo tourism. As Loveridge continues,

> [p]hoto-tourism has great potential to protect prime wildlife areas in politically stable regions. But is likely not a universal substitute for trophy hunting, particularly in areas that your average vacationer does not want to visit. (Loveridge 2018: 242)

Also, as is often overlooked, photo tourism can have negative impacts on wildlife as well, as here noted by IUCN in relation to cheetahs:

> Large numbers of tourist vehicles or insensitive tourist behaviour can lead to a number of negative effects such as interference with Cheetahs hunting, scaring Cheetah away from kills to which they are unlikely to return, and separation of mothers from cubs. (Durant et al. 2015)

Another study, interviewing seven consumptive (hunting and fishing) and nonconsumptive tourism managers working at a large game reserve in Namibia that caters to both kinds of tourists, found that

> although both consumptive and nonconsumptive wildlife tourists impact the environment at the game reserve, the behaviour of nonconsumptive wildlife tourists seems to be more negative than that of consumptive wildlife tourists. The study further found that hunters behave in an eco-friendlier manner towards the environment and tend to be more concerned about their own impact on nature. (van der Merwe et al. 2021: 218)

While the scope of this study is rather narrow, it does bring into question the common claim among ecotourists who are critical of trophy hunting that their attitudes and activities are less harmful.

Beside financial viability, it is worth mentioning that photo tourism and hunting tourism may differ in the form of care for habitats, wildlife, and communities that they foster in tourists. Adrian Franklin argues that the potential for nonconsumptive tourism to match the effectiveness of consumptive tourism in fostering genuine care for, and connectedness with, particular natural areas is doubtful. Photo tourists are generally fleeting, and their interest is shallow. They may visit many different national parks or other wildlife destinations, but they mostly visit just once and, in Franklin's words,

> there is nothing about these experiences that galvanises a longer-term relation of care and, indeed, the proliferation of destinations clambering for attention may only serve to create a blasé indifference to any one location, even if, overall, environmental concerns are raised. (Franklin 2008: 41)

Hunters and anglers tend to build stronger connections with certain localities. They will often return to the same places and emphasize their intimate familiarity with particularities of these places—a familiarity that can be necessary for hunting and fishing success. This argument is weaker when it comes to stereotypical trophy hunting tourists than those who hunt locally. Granted, some Europeans or Americans may travel to the same place in Africa a handful of times, but they will not approach the level of intimate ecological familiarity with any natural area that Franklin refers to. Guided hunting tourism does not generally lend itself well to intimate familiarity with a location; trophy hunters in Africa are not able to roam freely and slowly familiarize themselves with the land and its flora and fauna; they are taken from place to place by a professional hunter who might seek to produce a kill and trophy for the client in the shortest time possible. Hunting tourists also might have a tight schedule and not much time to secure each desired trophy. That said, some hunters will return time and again to the same location, but overall, the quest for novelty is likely, in most, to exceed the desire for familiarity.

On the other hand, trophy hunters pay a lot more money than photo tourists do for their respective experiences. Granted, they may have had more money to spend to begin with, so the percentage is the same, but the much bigger individual contributions to wildlife conservation and possibly rural livelihoods are more likely to create a lasting impression. If, for example, a hunter witnesses the happiness of rural villagers when he rids them of a problem-elephant and supplies them with its meat, one could imagine a bond being created that no photo tourism experience could compete with. Lasting connections like this might benefit conservation efforts in the future. It would however be hopeful fantasy to think all hunters build up such meaningful connections.

5.7 Poaching

Antihunters and animal rights advocates often conflate trophy hunting with poaching. Sometimes no distinction is made between poachers and trophy hunters, and sometimes there is a firm conviction that trophy hunting leads to poaching. It seems prudent at this point to discuss what poaching is and expand on its relation—if any—to trophy hunting.

The Oxford English Dictionary defines poaching as "[t]he action of trespassing in pursuit of game, fish, etc." This definition seems inadequate; deliberately killing a deer out of season or without the required tag or quota or otherwise illegally on one's own land would still be considered poaching, even though no trespassing was involved. Also, poaching must be intentional; a hunter who misjudges the age or sex of legal and illegal prey, trespasses without knowing it, shoots too close to a public road out of eagerness, hunts with the wrong kind of license, or breaks the law in some other way by accident and then confesses to the crime, is not a poacher (though many poachers will of course claim to have hunted illegally by mistake if they are caught).

A more precise understanding of poaching is given by Muth and Bowe (1998: 11) as

> any act that intentionally contravenes the laws and regulations established to protect wild, renewable resources, such as plants, mammals, birds, insects, reptiles, amphibians, fish, and shellfish.

Poaching cannot be hunting, and hunting cannot be poaching. The actions by which wildlife is killed may be the same, but if those actions are legal, it is generally hunting (it can also be culling or regulation) and if they are illegal and premeditated then it is poaching. Trophy hunting cannot be poaching, as trophy hunting by definition is legal. If a trophy hunt is illegal, it would be trophy poaching and not trophy hunting. This should clear up the linguistic relationship between poaching and trophy hunting, but to understand whether trophy hunting can lead to poaching, we need to be more specific about different kinds of poaching in different parts of the world and more importantly their motivations.

Just as there is a stereotypical trophy hunter, so is there a stereotypical poacher. Today, most of what we hear about poaching in the media relates to Africa, and Europeans and Americans are largely unaware of the scale of poaching in their own countries. Most westerners would probably think of elephants and rhinos being killed for their tusks and horns in Africa before they would think of deer poaching locally. However, AH lives in a part of rural Gloucestershire, UK, where almost every farm gate has a sign warning against deer poaching and providing contact details to report incidents to local police. Illegal fishing, a problem in the same area on the River Severn and tributaries, is also poaching.

In an African context, it can be pretty safely said that trophy hunting does not lead to poaching. As covered in Chapter 5.2, trophy hunting largely funds anti-poaching efforts and habitat protection in African countries. Were trophy hunting to disappear from areas that rely on it, and in the absence of any viable replacement, wildlife populations would lose their economic value, their habitats would be converted to agriculture, locals would have no incentive to tolerate human-wildlife conflicts, and there would be no funding for rangers to protect wildlife from poaching. These have been the very real consequences of long-term (Kenya) and short-term (Botswana) bans of trophy hunting and have also occurred locally where hunting has ceased, such as in abandoned hunting blocks in Tanzania (Strampelli et al. 2022).

Antihunters may claim that it makes no difference to the animals if they are killed by a poacher or a trophy hunter, but there is a big difference in both the scale and the methods of killing. 1342 rhinos (black and white combined) were killed by poachers in all of Africa in 2015 while only 69 were hunted legally by trophy hunters (Cooney et al. 2017). For lions, the number of deaths from poaching can be up to 100 times larger than the offtake from regulated trophy hunting (Dickman and Semcer 2020). In India, where sport hunting has been illegal since 1972, poaching is a major threat to all kinds of wildlife, especially in the form of large-scale and organized ritualistic "hunts":

> Ritual communal hunting usually commences during the spring period or sometimes as a celebration of the post-harvest season. During such hunts, thousands of Adivasi[26] men go en masse into the forests to hunt. For instance, in places like the Singhbhum-western West Bengal landscape as many as a lakh (100,000) or even more hunters will fan across the forests spread over seven forest divisions between the two states. Such communal hunts are an integral part of social life of many forest communities. [...]
>
> While earlier such ritual hunts had rules that governed them, they were diluted over time. Now such mass hunts are almost always a 'kill-all hunt', i.e. indiscriminate killing of wild animals without any restriction on the numbers of kills or what is being hunted – a bird or a mammal, a common species or an endangered one, a male or a female, a young one or an adult/old animal. Such indiscriminate hunting has especially plagued the Bastar landscape. (Kazmi 2021)

What further exacerbates the impact of deaths by poaching, as indicated here, is that poachers often kill males, females, and infants indiscriminately. Animals may suffer for hours or days when poachers use snares or poison, and poisons can cause excessive secondary kills of scavengers who feed on poisoned animals, a scourge of Africa's vultures. Trophy hunters—when it is properly managed—usually kill old males and ensure quick deaths—within seconds if all goes well. Hunters or poachers with rifles control what they kill, but snares and poisons kill wildlife randomly and in potentially huge numbers. In some areas, dogs are used by poachers to put animals at bay or to remove animals (especially warthogs) from burrows. Without firearms,

---

[26] Indigenous or tribal.

the animals are killed with axes, sharpened rebar, clubs, rocks, or other tools. Such a death is very far from a humane dispatch.

In Europe and North America, the relationship between trophy hunting and poaching is different than in Africa, but to help us understand this difference, let us first take a closer look at some attempts to categorize motivations for poaching. Musgrave et al. (1993) give us a rough classification of poaching as either commercial, violating the law for profit, or non-commercial, violating the law for other reasons. Trophy poachers, "who will go to any length to take a trophy animal, even if it means violating the law" (Musgrave et al. 1993: 983) are included in the non-commercial category.

Montgomery (2020) outlines three categories specifically for poaching in the global south: *Trophy Poaching* "refers to instances in which wildlife is killed illegally to meet black market demands for animal trophies," which includes e.g. ivory for carvings; *Medicative Poaching* "involves the harvest of animal parts for specific physical properties that are of purported medicinal, aphrodisiac, or therapeutic value" (which includes rhino poaching for horn); and *Consumptive Poaching*, which is poaching for meat for personal consumption (Montgomery 2020: 475). While these definitions may work for Africa, they do not encompass the trophy poacher who wants the trophy for himself and not for resale. Furthermore, whether someone poaches a rhino to sell its head as a trophy or to sell its horn for TCM may not be an essential distinction, as both can be considered commercial poaching.

Muth and Bowe (1998) created a more thorough typology of ten different motivations for poaching:

> (1) commercial gain, (2) household consumption, (3) recreational satisfactions, (4) trophy poaching, (5) thrill killing, (6) protection of self and property, (7) poaching as rebellion, (8) poaching as a traditional right of use, (9) disagreement with specific regulations, and (10) gamesmanship. (Muth and Bowe 1998: 13)

In Africa, poaching is usually carried out by relatively poor rural men because they need the money that ivory or rhino horn or other animal parts can be exchanged for on the black market (category 1), or because they need the bushmeat for themselves or for resale (category 1–2), or because they need to protect themselves or their livelihoods from wildlife conflicts (category 6).

In Europe and the US, all of the above motivations and more may apply. Unbeknownst to many, the scale of poaching in the US especially, is staggering:

> Almost nowhere is poaching of wildlife more lucrative, more dangerous and more difficult to control than in the United States, yet the poaching problem in this country seldom receives international attention. The level of poaching has reached a crisis here at home both financially and in loss of numbers of game and nongame wildlife. (Musgrave et al. 1993: 977)

If we are to uncover whether trophy hunting leads to poaching, it is Muth and Bowe's *trophy poaching* category (4) that we should pay attention to. According to Muth and Bowe, more and more hunters turn to poaching in National Parks for trophies that they cannot legally get access to, and they are willing to pay unscrupulous outfitters more and more money for their assistance. Wildlife agents have reported that most of their caseload involves trophy poaching. One apprehended trophy poacher is quoted as saying "I felt that my hunting career would not be complete without a Boone and Crockett [record book] buck" (Muth and Bowe 1998: 16) and that he did it because of the notoriety and recognition of having shot a trophy animal.

The problem with trophy poaching is perhaps documented best by Eliason (2012) who interviewed game wardens in Montana about the extent of the problem. One of them told Eliason that the problem with trophy poaching is

> [n]ot serious in terms of wildlife populations. Very serious in terms of reducing quality of hunting opportunities to legal hunters in local areas. Damages the public perception of hunters. The corruption and lack of ethics involved in the pursuit of trophy animals as status symbols would shock the public if they knew the extent of it. Boone and Crockett, Safari Club International, and Pope and Young are a big part of the problem. (Eliason 2012: 75)

Other wardens called the problem with trophy poaching "extensive and getting worse," that it takes place "[m]ore than one would want to know," and that it constitutes "by far the most serious problem wardens face in today's world. Most of our laws were geared toward meat hunting. States are just now starting to tackle the trophy problem in their legislatures" (Eliason 2012: 75).

The wardens interviewed by Eliason agreed that the desire to gain status among peers by owning the biggest hunting trophies and having them recognized in record books is a driving motivator for the trophy poaching of deer. There is no interest in the meat and no financial incentive; only a desire to obtain trophies and be recognized among peers.

A recent Danish example is from September of 2020. Two men were caught by the police at night in their car with the severed head of a 12-point red deer buck in the trunk. They had poached the deer at night on another man's property, sawed off its head, and left the body behind in the forest (Jørgensen 2020).

Trophy poaching, as also supported by Eliason, is often facilitated by corrupt outfitters. In October 2020 in Nebraska, 30 trophy poachers pleaded guilty in the biggest poaching case in the history of the state. Nancy Gaarder from the *Omaha World-Herald* reported on this:

> The hunters came from at least 21 states to the deep-cut canyons and Sand Hills of Nebraska to bag trophies for their walls back home.
>
> More than 100 people from states like New York, Wisconsin, Virginia and Utah paid $2500 to $7000 to take aim at big bucks, pronghorn antelopes and turkeys, all with the help of Hidden Hills Outfitters near Broken Bow.
>
> But they weren't really big game hunters. They were poachers — using bait, spotlights at night and other illegal tactics to guarantee their success. (Gaarder 2020)

Hidden Hill Outfitters, investigators found, had facilitated trophy poaching for at least five years. Their clients, in addition to baiting and hunting at night, had been using supressed firearms during archery season and collected only the heads of deer, leaving the bodies behind to rot. They had also—seemingly just for the fun of it—been shooting nongame birds such as hawks and falcons when they were resting on fences or power lines. Co-owner of Hidden Hill Outfitters Jacob Hueftle had killed at least 100 nongame birds. He was not even allowed to hunt legally because he was on probation for earlier hunting violations (Gaarder 2020).

So, does trophy hunting lead to poaching in this context? It certainly looks plausible that record books and the trophy hunting mentality leads to a desire to possess trophies, and that the inability to obtain them legally can lead to poaching. Game wardens interviewed by Eliason

> claimed that the trophy hunting industry (magazines, television programs) has exacerbated the situation by creating unrealistic expectations for hunters, which serve to fuel the desire for trophy poaching. (Eliason 2012: 76)

Domestic trophy poaching, and the trophy records mentality that leads to it, is something that should perhaps be getting a lot more attention from celebrities and the popular media than legal and often beneficial trophy hunting in Africa. It is noteworthy, in the light of earlier discussions of neo-colonialism, that domestic hunting in general receives very little media and celebrity interest. Indeed, while giving evidence to the UK Parliament in 2021, the founder of the Campaign to Ban Trophy Hunting, Eduardo Gonçalves, stated that "As far as I am concerned the issue of trophy exports from Britain is to all intents and purposes a non-issue." He based that statement on the fact that there is only one export from the UK listed by CITES since 2019, a pre-CITES tiger skin. In fact, a great many red deer trophies (certainly in the order of thousands) are exported from estates in Scotland by overseas clients, along with fallow, roe, muntjac, sika and Chinese water deer trophies from the UK more generally. None of these species are in the CITES appendices and their trade is therefore not monitored by CITES, so naturally, it will look like the UK has no trophy hunting if one goes by the CITES trade database. Wild boar trophies are also exported from the UK. If British campaigners and politicians think that trophy hunting is fundamentally wrong and want to ban it on these grounds, then we think that the choice to target only imports (where livelihoods abroad will bear the consequences) and not exports (where British livelihoods and voters would bear the consequences), are ethically questionable.

There are clearly nuances to the moral reprehensibility of poaching. It is by definition wrong because it is illegal, but the motivations for illegal actions make a moral difference. Robin Hood and his merry men who poached the king's deer as an act of rebellion as well as to feed themselves were not the villains of their stories. We should equally be entertaining the idea that market- or meat poaching in Africa can be a moral grey area. Killing a rhino for the money that its horn fetches on the black market seems horrifyingly greedy on the surface, but if one considers that the poacher is likely to be poor and unemployed and may do it because he saw no other way to feed his family, then the black and white of the issue turns distinctly grey. On the other hand, a well-off European or American who poaches a trophy deer for the sole purpose of boosting his own ego by having the antlers registered in a record book and displayed on his wall under false pretences is perhaps the best example of morally bankrupt poaching.

On September 25, 2020, the British Top Gear celebrity Jeremy Clarkson wrote a typically bullish but uninformed column in UK tabloid *The Sun* about how the British

Army should be deployed in a kill-on-sight war against poachers in Africa (Clarkson 2020). The article picks up on a trend of increasing militarization of conservation efforts. Kenya is particularly famous or infamous in this respect, depending on who you ask. The government announced in 2020 that the death penalty would be enacted for poaching, and Kenyan rangers have often both killed and been killed in armed conflicts with poachers.

The Clarkson article is forgettable were it not for it being a good example of why celebrities should stop weighing in on complex ecopolitical matters that they know nothing about. But if we indulge his suggestion for a moment, why not deploy them in North America or Europe instead? Granted, trophy whitetails and red deer are not in the same precarious situation as much of Africa's wildlife, but the West's domestic trophy poachers are certainly a lot more deserving of this kind of capital punishment than Africans who poach to survive or out of desperation. Yet as we find eco-imperialism in Western nations trying to dictate how Africa should manage its wildlife, so do we find it in the difference between those who get to define hunting and poaching and those who do not—or as Loveridge writes:

> The majority of Africans do not hunt for pleasure or sport but rather to feed their families or to protect their livelihoods. When they do so in protected areas, they are called poachers. Wealthy Westerners who pay large sums to hunt the same animals, sometimes in dubious circumstances, are called "sportsmen." The former face legal penalties, including lengthy prison terms; the latter receive the accolades of their peers. (Loveridge 2018: 248)

References

Abbiati, George P., John Paul Croteau, Lindsay Marie Mitchel, and Melissa Ann Samaroo. 2013. Development of Trophy Hunting Among Previously Disadvantaged Farmers in Namibia. Bachelor's Thesis, Worcester Polytechnic Institute. https://1library.net/document/yenl351y-development-trophy-hunting-previously-disadvantaged-farmers-namibia.html (accessed 22 June 2022).

Adams, Kip P., and R. Joseph Hamilton. 2011. Management History. In *Biology and Management of White-Tailed Deer*, ed. David G. Hewitt, 355–378. Boca Raton: CRC Press.

Akama, John S. 2008. The Ban on Wildlife Hunting in Kenya. In *Tourism and the Consumption of Wildlife—Hunting, Shooting and Sport Fishing*, ed. Brent Lovelock, 73–86. London and New York: Routledge.

Angula, Hilma N., Greg Stuart-Hill, David Ward, Greenwell Matongo, Richard W. Diggle, and Robin Naidoo. 2018. Local Perceptions of Trophy Hunting on Communal Lands in Namibia. *Biological Conservation* 218: 26–31.

Ares, Elena. 2019. Briefing Paper 7908—Trophy Hunting. House of Commons Library. https://commonslibrary.parliament.uk/research-briefings/cbp-7908/ (accessed 51 May 2020).

Aryal, Achyut, Maheshwar Dhakal, Saroj Panth, Bhupendra Prasad Yadav, Uttam Babu Shrestha, Roberta Bencini, David Raubenheimer, and Weihong Ji. 2015. Is Trophy Hunting of Bharal (Blue Sheep) and Himalayan Tahr Contributing to Their Conservation in Nepal? *Hystrix, the Italian Journal of Mammalogy* 26 (2): 85–88.

Balčiauskas, Linas, Romualdas Varanauskas, and Egidijus Bukelskis. 2017. Impact of Selective Hunting on the Trophy Size of Roe Deer: Baltic Example. *North-Western Journal of Zoology* 13 (1): 118–127.

Bauer, Hans, Guillaume Chapron, Kristin Nowell, Philipp Henschel, Paul Funston, Luke T.B. Hunter, David W. Macdonald, and Craig Packer. 2015. Lion (*Panthera leo*) Populations Are Declining Rapidly Across Africa, Except in Intensively Managed Areas. *Proceedings of the National Academy of Sciences of the United States of America* (*PNAS*) 112: 14894–14899.

Bauer, Hans, Amy Dickman, Guillaume Chapron, Alayne Oriol-Cotterill, Samantha K. Nicholson, Claudio Sillero-Zubiri, Luke Hunter, Peter Lindsey, and David W. Macdonald. 2020. Threat Analysis for More Effective Lion Conservation. *Oryx* 56: 108–115.

Becker, Frowin. 2022. Reframing Trophy Hunting's Socio-Economic Benefits in Namibia (Commentary). Mongabay, April 28. https://news.mongabay.com/2022/04/reframing-trophy-huntings-socio-economic-benefits-in-namibia-commentary/ (accessed 22 June 2022).

Bellon, Luc. 2008. Sustainable Conservation and Grassroots Realities Lessons from the Conservation Programme in Torghar, Balochistan, Pakistan. In *Best Practices in Sustainable Hunting—A Guide to Best Practices from Around the World*, ed. Rolf D. Baldus, Gerhard R. Damm, and Kai-Uwe Wollscheid, 27–31. CIC Technical Series Publication No. 1. http://www.cic-wildlife.org/wp-content/uploads/2012/12/Technical_series_1.pdf (accessed 16 May 2019).

Bissonette, John A., Christine A. Kassar, and Lawrence J. Cook. 2008. Assessment of Costs Associated with Deer-Vehicle Collisions: Human Death and Injury, Vehicle Damage, and Deer Loss. *Human-Wildlife Conflicts* 2 (1): 17–27.

Bleier, Norbert, Róbert Lehoczki, Dóra Újváry, László Szemethy, and Sándor Csányi. 2012. Relationship Between Wild Ungulates Density and Crop Damage in Hungary. *Acta Theriologica* 57 (351): 351–359.

Bolingbroke-Kent, Antonia. 2022. Why Trophy Hunters in Tajikistan Are Unlikely Saviours of the Snow Leopard. *The Guardian*, April 19. https://www.theguardian.com/world/2022/apr/19/trophy-hunters-tajikistan-unlikely-saviours-snow-leopard-conservation-aoe (accessed 12 October 2022).

Bond, Ivan. 2001. CAMPFIRE & the Incentives for Institutional Change. In *African Wildlife & Livelihoods—The Promise and Performance of Community Conservation*, ed. David Hulme and Marshall Murphree, 227–243. Oxford: James Currey Ltd.

Booth, Vernon R., and Philippe Chardonnet, eds. 2015. *Guidelines for Improving the Administration of Sustainable Hunting in Sub-Saharan Africa*. FAO: Sub-Regional Office for Southern Africa. http://www.fao.org/3/a-bo583e.pdf (accessed 6 March 2020).

Booth, Vernon R., Jones Masonde, Chuma Simukonda, and David H.M. Cumming. 2020. Managing Hunting Quotas of African Lions (*Panthera leo*): A Case Study from Zambia. *Journal for Nature Conservation* 55: 1–8.

Boulanger, Jason R., and Paul D. Curtis. 2016. Efficacy of Surgical Sterilization for Managing Overabundant Suburban White-Tailed Deer. *Wildlife Society Bulletin* 40 (4): 727–735.

Boulanger, Jason R., Paul D. Curtis, Evan G. Cooch, and Anthony J. DeNicola. 2012. Sterilization as an Alternative Deer Control Technique: A Review. *Human-Wildlife Interactions* 6 (2): 273–282.

Bronner, Simon. 2008. *Killing Tradition—Inside Hunting and Animal Rights Controversies*. Lexington: The University Press of Kentucky.

Brown, Robert D. 1990. Nutrition and Antler Development. In *Horns, Pronghorns, and Antlers*, ed. George A. Bubenik and Anthony B. Bubenik. New York: Springer.

Campbell, Tyler A., and Kurt C. VerCauteren. 2011. Diseases and Parasites. In *Biology and Management of White-Tailed Deer*, ed. David G. Hewitt, 219–250. Boca Raton: CRC Press.

Carpenter, Len H. 1997. Deer in the West. Proceedings—1997 Deer/Elk Workshop—Arizona: 1–10.

Caughley, Graeme. 1981. Overpopulation. In *Problems in Management of Locally Abundant Wild Animals*, ed. Peter A. Jewell and Sidney Holt, 7–19. New York: Academic Press.

CDC. 2019. Lyme Disease—Data and Surveillance. Centers for Disease Control and Prevention, November 22. https://www.cdc.gov/lyme/datasurveillance/index.html (accessed 31 July 2020).

Child, Brian. 1993. Zimbabwe's CAMPFIRE Programme: Using the High Value of Wildlife Recreation to Revolutionize Natural Resource Management in Communal Areas. *The Commonwealth Forestry Review* 72 (4): 284–296.

References

CITES. 1983. Convention on International Trade in Endangered Species of Wild Fauna and Flora. cites.org. https://www.cites.org/sites/default/files/eng/disc/CITES-Convention-EN.pdf (accessed 5 August 2020).

CITES. 2002. Conf. 12.3 (Rev. CoP18) Permits and Certificates. cites.org. https://www.cites.org/sites/default/files/document/E-Res-12-03-R18.pdf (accessed 5 August 2020).

CITES. 2019. CoP18 Doc. 76.1 (Rev. 1)—Species Specific Matters, African Lion (*Panthera leo*). Report of the Secretariat. cites.org. https://cites.org/sites/default/files/eng/cop/18/doc/E-CoP18-076-01-R1.pdf (accessed 6 September 2020).

Clarkson, Jeremy. 2020. There Is a War Our Squaddies Could Fight... One to Save Rhinos from Extinction. *The Sun*, September 25. https://www.thesun.co.uk/news/12771367/war-squaddies-fight-save-rhinos-extinction/ (accessed 20 October 2020).

Coltman, David W., Paul O'Donoghue, Jon T. Jorgenson, John T. Hogg, Curtis Strobeck, and Marco Festa-Bianchet. 2003. Undesirable Evolutionary Consequences of Trophy Hunting. *Nature* 426 (6967): 655–658.

Conover, Michael R. 1994. Perceptions of Grass-Roots Leaders of the Agricultural Community About Wildlife Damage on Their Farms and Ranches. *Wildlife Society Bulletin* 22 (1): 94–100.

Conover, Michael R. 2011. Impacts of Deer on Society. In *Biology and Management of White-Tailed Deer*, ed. David G. Hewitt, 399–408. Boca Raton: CRC Press.

Cooney, Rosie, C. Freese, H. Dublin, D. Roe, D. Mallon, M. Knight, R. Emslie, M. Pani, V. Booth, S. Mahoney, and C. Buyanaa. 2017. The Baby and the Bathwater: Trophy Hunting, Conservation and Rural Livelihoods. *unasylva* 249 (68): 3–16.

Cooney, Rosie, Paola Mosig Reidl, and Luis Guillermo Munõz Lacy. 2019. Community-Based Trophy Hunting of Bighorn Sheep in Mexico. CITES & Livelihoods Case Study 2019. https://cites.org/sites/default/files/eng/prog/Livelihoods/case_studies/6.%20Mexico_bighornsheep_long_Aug2.pdf (accessed 14 July 2020).

Corgatelli, Gabrielle, Silvana Mattiello, Stefania Colombini, and Gianni Matteo Crovetto. 2019. Impact of Red Deer (*Cervus elaphus*) on Forage Crops in a Protected Are. *Agricultural Systems* 169: 41–48.

Côté, Steeve D. 2011. Impacts on Ecosystems. In *Biology and Management of White-Tailed Deer*, ed. David G. Hewitt, 379–398. Boca Raton: CRC Press.

Côté, Steeve D., Thomas P. Rooney, Jean-Pierre Tremblay, Christian Dussault, and Donald M. Waller. 2004. Ecological Impacts of Deer Overabundance. *Annual Review of Ecology, Evolution, and Systematics* 35: 113–147.

Craven, Scott, Thomas Barnes, and Gary Kania. 1998. Toward a Professional Position on the Translocation of Problem Wildlife. *Wildlife Society Bulletin* 26 (1): 171–177.

Creel, Scott, Jassiel M'Soka, Egil Dröge, Eli Rosenblatt, Matthew S. Becker, Wigganson Matandiko, and Twakundine Simpamba. 2016. Assessing the Sustainability of African Lion Trophy Hunting, with Recommendations for Policy. *Ecological Applications* 26 (7): 2347–2357.

Darimont, Chris T., Stephanie M. Carlson, Michael T. Kinnison, Paul C. Paquet, Thomas E. Reimchen, and Christopher C. Wilmers. 2009. Human Predators Outpace Other Agents of Trait Change in the Wild. *Proceedings of the National Academy of Science of the United States of America (PNAS)* 106 (3): 952–954.

Davidson, William R., and Gary L. Doster. 1997. Health Characteristics and White-Tailed Deer Population Density in the Southeastern United States. In *The Science of Overabundance—Deer Ecology and Population Management*, ed. William J. McShea, H. Brian Underwood, and John H. Rappole, 164–184. Washington and London: Smithsonian Books.

Decker, Daniel J., and Tommy L. Brown. 1987. How Animal Rightists View the "Wildlife Management-Hunting System". *Wildlife Society Bulletin* 15 (4): 599–602.

Demarais, Steve, and Bronson K. Strickland. 2011. Antlers. In *Biology and Management of White-Tailed Deer*, ed. David G. Hewitt, 107–145. Boca Raton: CRC Press.

DeNicola, Anthony J., Kurt C. VerCauteren, Paul D. Curtis, and Scott E. Hygnstrom. 2000. *Managing White-Tailed Deer in Suburban Environments—A Technical Guide*. Ithaca: Cornell

Cooperative Extension, the Wildlife Society Wildlife Damage Management Working Group, and the Northeast Wildlife Damage Research and Outreach Cooperative.

Di Minin, Enrico, Nigel Leader-Williams, and Corey J.A. Bradshaw. 2016. Banning Trophy Hunting Will Exacerbate Biodiversity Loss. *Trends in Ecology and Evolution* 31 (2): 99–102.

Dickman, Amy, and Amy Hinks. 2019. State of the Lion: Fragility of a Flagship Species—Report prepared for the Lion Footprint Forum, May 2019. Unpublished Report Based on a Paper in Preparation.

Dickman, Amy, and Catherine E. Semcer. 2020. Saving Africa's Lions Will Rely on Evidence Around Trophy Hunting, Not Emotion. *The Hill*, February 5. https://thehill.com/changing-america/opinion/481669-saving-africas-last-lions-will-rely-on-evidence-not-emotion?fbclid=IwAR38fJl16bIIft_t1PhkoDntgkiDC0utmJqOwz9mORn1oaf0AvIYiIuJon0 (accessed 18 February 2020).

Dickman, Amy, Rosie Cooney, Paul J. Johnson, Maxi Pia Louis, Dilys Roe, and 128 signatories. 2019. Trophy Hunting Bans Imperil Biodiversity. *Science* 365 (6456): 874.

Dizard, Jan E. 1994. *Going Wild—Hunting, Animal Rights, and the Contested Meaning of Nature*. Amherst: University of Massachusetts Press.

Douhard, Mathieu, Marco Festa-Bianchet, Fanie Pelletier, Jean-Michel Gaillard, and Christophe Bonenfant. 2016. Changes in Horn Size of Stone's Sheep over Four Decades Correlate with Trophy Hunting Pressure. *Ecological Applications* 26 (1): 309–321.

Dube, Nqobizitha. 2019. Voices from the Village on Trophy Hunting in Hwange District, Zimbabwe. *Ecological Economics* 159: 335–343.

Duda, Mark Damian, Martin Jones, and Andrea Criscione. 2010. *The Sportsman's Voice—Hunting and Fishing in America*. State College, PA: Responsive Management/Venture Publishing.

Dudley, Nigel, ed. 2008. Guidelines for Applying Protected Area Management Categories. IUCN. https://portals.iucn.org/library/sites/library/files/documents/PAG-021.pdf (accessed 6 January 2020).

Durant, S., N. Mitchell, A. Ipavec, and R. Groom. 2015. *Acinonyx jubatus*, Cheetah. The IUCN Red List of Threatened Species. https://www.iucnredlist.org/species/219/50649567 (accessed 11 February 2021).

EC. 2020. The European Union and Trade in Wild Fauna and Flora. European Commission, May 14. https://ec.europa.eu/environment/cites/index_en.htm (accessed 21 May 2020).

Eliason, Stephen L. 2012. Trophy Poaching: A Routine Activities Perspective. *Deviant Behavior* 33 (1): 72–87.

Enck, Jody W., and Tommy L. Brown. 2001. Attitudes and Beliefs of Deer Hunters Participating in an "Earn-a-Buck" Program at Cornell University's Arnot Forest. Human Dimensions Research Unit Publications Series No. 01-10.

Evans, Charles S., Anthony J. DeNicola, and Robert J. Warren. 2016. Comparison of Treatment Effort for Immunocontraceptive Vaccines and Surgical Sterilization in Deer. *Wildlife Society Bulletin* 40: 593–598.

Festa-Bianchet, Marco, and Ray Lee. 2009. Guns, Sheep, and Genes: When and Why Trophy Hunting May Be a Selective Pressure. In *Recreational Hunting, Conservation and Rural Livelihoods*, ed. Barney Dickson, Jon Hutton, and William M. Adams. Oxford: Wiley-Blackwell.

Festa-Bianchet, Marco, and Atle Mysterud. 2018. Hunting and Evolution: Theory, Evidence, and Unknowns. *Journal of Mammalogy* 99: 1281–1292.

Festa-Bianchet, Marco, Fanie Pelletier, Jon T. Jorgenson, Chiarastella Feder, and Anne Hubbs. 2014. Decrease in Horn Size and Increase in Age of Trophy Sheep in Alberta Over 37 Years. *The Journal of Wildlife Management* 78 (1): 133–141.

Fix, Peter, Cynthia Pierce, Michael Manfredo, and Linda Sikorowski. 1998. Evaluating Hunters' Preferences for Wildlife Program Funding (Findings Abstract). *Human Dimensions of Wildlife* 3 (3): 75–75.

Foster, Sarah, Stefan Wiswedel, and Amanda Vincent. 2014. Opportunities and Challenges for Analysis of Wildlife Trade Using CITES Data—Seahorses as a Case Study. *Aquatic Conservation* 26 (1): 154–172.

References

Franklin, Adrian. 2008. The 'Animal Question' and the 'Consumption' of Wildlife. In *Tourism and the Consumption of Wildlife—Hunting, Shooting and Sport Fishing*, ed. Brent Lovelock, 31–44. London and New York: Routledge.

Freeman, M.M.R., and G.W. Wenzel. 2006. The Nature and Significance of Polar Bear Conservation Hunting in the Canadian Arctic. *Arctic* 59 (1): 21–30.

Frosch, Brian J., David P. Anderson, and Joe L. Outlaw. 2008. Economic Impact of Deer Breeding Operations in Texas. Selected Paper Prepared for Presentation at the Southern Agricultural Economics Association Annual Meetings, Dallas, Texas, February 2–6. https://core.ac.uk/reader/7083040 (accessed 28 August 2020).

Frost, Peter G.H., and Ivan Bond. 2008. The CAMPFIRE Programme in Zimbabwe: Payments for Wildlife Services. *Ecological Economics* 65 (4): 776–787.

Funston, Paul, Philipp Henschel, Luke Hunter, Peter Lindsey, Katarzyna Nowak, Christina Vallianos, and Karen Wood. 2016. Beyond Cecil: Africa's Lions in Crisis. Panthera, WildAid and WildCRU. https://wildaid.org/resources/beyond-cecil-africas-lions-in-crisis-english/ (accessed 28 May 2020).

Gaarder, Nancy. 2020. 30 Plead Guilty so Far in Largest Poaching Case in Nebraska History. *Omaha World-Herald*, October 19. https://omaha.com/news/state-and-regional/crime-and-courts/30-plead-guilty-so-far-in-largest-poaching-case-in-nebraska-history/article_9137385c-ab8b-5c85-acb6-531b58de42d7.html (accessed 20 October 2020).

Gallina, S., and H. Lopez Arevalo. 2016. *Odocoileus virginianus*, White-Tailed Deer. The IUCN Red List of Threatened Species. https://doi.org/10.2305/IUCN.UK.2016-2.RLTS.T42394A22162580.en (accessed 23 July 2020).

Gasset, José Ortega y. 1942. *Meditations on Hunting*, trans. Howard B. Wescott, 1985. New York: Charles Scribner's Sons.

Gjerris, Mickey, Sune Borkfelt, Christian Gamborg, Jes Harfeld, and Sara Kondrup. 2016. *Jagt—Natur, mennesker, dyr og drab*. Aarhus: Klim.

Goetsch, Chandra, Jennifer Wigg, Alejandro A. Royo, Todd Ristau, and Walter P. Carson. 2011. Chronic Over Browsing and Biodiversity Collapse in a Forest Understory in Pennsylvania: Results from a 60 Year-Old Deer Exclusion Plot. *The Journal of the Torrey Botanical Society* 138 (2): 220–224.

Gortázar, Christian, Pelayo Acevedo, Francisco Ruiz-Fons, and Joaquín Vicente. 2006. Disease Risks and Overabundance of Game Species. *European Journal of Wildlife Research* 52 (2): 81–87.

Grobbelaar, Denise. 2020. The White Lion as Symbol of the Archetype of the Self and the Cannibalization of the Self in Canned Hunting. *Jung Journal* 14 (2): 11–29.

Groot Bruinderink, G.W.T.A., and E. Hazebroek. 1996. Ungulate Traffic Collisions in Europe. *Conservation Biology* 10 (4): 1059–1067.

Heberlein, Thomas A. 1991. Changing Attitudes and Funding for Wildlife: Preserving the Sport Hunter. *Wildlife Society Bulletin* 19 (4): 528–534.

Heffelfinger, James R. 2018. Inefficiency of Evolutionarily Relevant Selection in Ungulate Trophy Hunting. *The Journal of Wildlife Management* 82 (1): 57–66.

Heffelfinger, James R., Valerius Geist, and William Wishart. 2013. The Role of Hunting in North American Wildlife Conservation. *International Journal of Environmental Studies* 70 (3): 399–413.

Heister, Anja. 2019. Beyond the North American Wildlife Conservation Model and Towards Earth Rights. *The Ecological Citizen* 3 (A): 67–74.

Hewitt, David G. 2015. Hunters and the Conservation and Management of White-Tailed Deer (Odocoileus virgianus). *International Journal of Environmental Studies* 72 (5): 839–849.

Holechek, Jerry L., Andres F. Cibils, Konimba Bengaly, and Jenesio I. Kinyamario. 2016. Human Population Growth, African Pastoralism, and Rangelands: A Perspective. *Rangeland Ecology & Management* 70 (3): 273–280.

Hubbard, Ryan D., and Clayton K. Nielsen. 2009. White-Tailed Deer Attacking Humans During the Fawning Season: A Unique Human–Wildlife Conflict on a University Campus. *Human-Wildlife Conflicts* 3 (1): 129–135.

Humane Society. 2016. Trophy Hunting by the Numbers—The United States' Role in Global Trophy Hunting. https://www.hsi.org/wp-content/uploads/assets/pdfs/report_trophy_hunting_by_the.pdf (accessed 6 February 2020).

Hundertmark, K. 2016. *Alces alces*, Moose. The IUCN Red List of Threatened Species. https://doi.org/10.2305/IUCN.UK.2016-1.RLTS.T56003281A22157381.en (accessed 23 July 2020).

IUCN. 2009. Big Game Hunting in West Africa—What Is Its Contribution to Conservation? Papaco Studies 2. https://www.iucn.org/content/big-game-hunting-west-africa-what-its-contribution-conservation (accessed 18 June 2020).

IUCN. 2012. IUCN SSC Guiding Principles on Trophy Hunting as a Tool for Creating Conservation Incentives. https://portals.iucn.org/library/node/29197 (accessed 22 June 2020).

IUCN. 2016. IUCN Briefing Paper: Informing Decisions on Trophy Hunting. https://www.iucn.org/sites/dev/files/iucn_sept_briefing_paper_-_informingdecisionstrophyhunting.pdf (accessed 6 January 2020).

IUCN Caprinae Specialist Group. 2008. Position Statement of the IUCN Caprinae Specialist Group on Trophy Hunting. In *Best Practices in Sustainable Hunting—A Guide to Best Practices from Around the World*, ed. Rolf D. Baldus, Gerhard R. Damm, and Kai-Uwe Wollscheid, 23–25. CIC Technical Series Publication No. 1. http://www.cic-wildlife.org/wp-content/uploads/2012/12/Technical_series_1.pdf (accessed 16 May 2019).

Jacobson, Harry A., Charles A. DeYoung, Randy W. DeYoung, Timothy E. Fulbright, and David G. Hewitt. 2011. Management on Private Property. In *Biology and Management of White-Tailed Deer*, ed. David G. Hewitt, 453–479. Boca Raton: CRC Press.

Jones, Brian T.B. 2009. Community Benefits from Safari Hunting and Related Activities in Southern Africa. In *Recreational Hunting, Conservation and Rural Livelihoods*, ed. Barney Dickson, Jon Hutton, and William M. Adams, 157–177. Oxford: Wiley-Blackwell.

Jones, Allen Morris. 2012 (First 1997). *A Quiet Place of Violence—Hunting and Ethics in the Missouri River Breaks*. Montana: Bangtail Press.

Jørgensen, Anders. 2020. Krybskytter anholdt i klitplantage—havde skåret hovedet af et krondyr. *Jyllands-Posten*, September 20. https://jyllands-posten.dk/indland/politiretsvaesen/ECE12426294/krybskytter-anholdt-i-klitplantage-havde-skaaret-hovedet-af-et-krondyr/ (accessed 20 October 2020).

Kalvelage, Linus, Javier Revilla Diez, and Michael Bollig. 2021. Do Tar Roads Bring Tourism? Growth Corridor Policy and Tourism Development in the Zambezi Region, Namibia. *The European Journal of Development Research* 33: 1000–1021.

Karimov, Khalil. 2019. Community-Based Trophy Hunting of Ibex and Markhor in Tajikistan. CITES & Livelihoods Case Study 2019. https://cites.org/sites/default/files/eng/prog/Livelihoods/case_studies/Tajikistan_ibex%26markhor_long_revSept26.pdf (accessed 14 July 2020).

Kazmi, Raza. 2021. Conservation in East-central India. https://india-seminar.com/2021/744/744_raza_kazmi.htm (accessed 28 October 2022).

Kilpatrick, Howard J., Andrew M. Labonte, and John S. Barclay. 2007. Acceptance of Deer Management Strategies by Suburban Homeowners and Bowhunters. *Journal of Wildlife Management* 71 (6): 2095–2101.

Kilpatrick, Howard J., Andrew M. Labonte, and Kirby C. Stafford. 2014. The Relationship Between Deer Density, Tick Abundance, and Human Cases of Lyme Disease in a Residential Community. *Journal of Medical Entomology* 51 (4): 777–784.

Kuba, Jarosław, Tomas Landete-Castillejos, and Jan Udała. 2015. Red Deer Farming: Breeding Practice, Trends and Potential in Poland—A Review. *Annals of Animal Science* 15 (3): 591–599.

Kugeler, K.J., R.A. Jordan, T.L. Schulze, K.S. Griffith, and P.S. Mead. 2015. Will Culling White-Tailed Deer Prevent Lyme Disease? *Zoonoses and Public Health* 63 (5): 337–345.

LaSharr, Tayler N., Ryan A. Long, James R. Heffelfinger, Vernon C. Bleich, Paul R. Krausman, R. Terry Bowyer, Justin M. Shannon, Robert W. Klaver, Clay E. Brewer, A. Mike Cox, Andrew

References

Holland, Anne Hubbs, Chadwick P. Lehman, Jonathan D. Muir, Bruce Sterling, and Kevin L. Monteith. 2019. Hunting and Mountain Sheep: Do Current Harvest Practices Affect Horn Growth? *Evolutionary Applications* 12 (9): 1823–1836.

Leader-Williams, Nigel, Rolf D. Baldus, and R.J. Smith. 2009. The Influence of Corruption on the Conduct of Recreational Hunting. In *Recreational Hunting, Conservation and Rural Livelihoods*, ed. Barney Dickson, Jon Hutton, and William M. Adams, 296–316. Oxford: Wiley-Blackwell.

Leopold, Aldo. 1968 (First 1949). *A Sand County Almanac, and Sketches Here and There*. New York: Oxford University Press.

Leopold, Aldo, Lyle K. Sowls, and David L. Spencer. 1947. A Survey of Over-Populated Deer Ranges in the United States. *The Journal of Wildlife Management* 11 (2): 162–177.

Lindsey, Peter A., R. Alexander, L.G. Frank, A. Mathieson, and S.S. Romañach. 2006. Potential of Trophy Hunting to Create Incentives for Wildlife Conservation in Africa Where Alternative Wildlife-Based Land Uses May Not Be Viable. *Animal Conservation* 9 (3): 283–291.

Lindsey, Peter A., L.S. Petracca, P.J. Funston, H. Bauer, A. Dickman, K. Everatt, M. Flyman, P. Henschel, A.E. Hinks, S. Kasiki, A. Loveridge, D.W. Macdonald, R. Mandisodza, W. Mgoola, S.M. Miller, S. Nazerali, L. Siege, K. Uiseb, and L.T.B. Hunter. 2017. The Performance of African Protected Areas for Lions and Their Prey. *Biological Conservation* 209: 137–149.

Lindsey, Peter A., James Allan, Peadar Brehony, Amy Dickman, Ashley Robinson, Colleen Begg, Hasita Bhammar, Lisa Blanken, Thomas Breuer, Kathleen Fitzgerald, Michael Flyman, Patience Gandiwa, Nicia Giva, Dickson Kaelo, Simon Nampindo, Nyambe Nyambe, Kurt Steiner, Andrew Parker, Dilys Roe, Paul Thomson, Morgan Trimble, Alexandre Caron, and Peter Tyrrell. 2020. Conserving Africa's Wildlife and Wildlands Through the COVID-19 Crisis and Beyond. *Nature Ecology & Evolution* 4: 1300–1310.

Lovari, S., J. Herrero, M. Masseti, H. Ambarli, R. Lorenzini, and G. Giannatos. 2016. *Capreolus capreolus*, European Roe Deer. The IUCN Red List of Threatened Species. https://doi.org/10.2305/IUCN.UK.2016-1.RLTS.T42395A22161386.en (accessed 23 July 2020).

Lovari, S., R. Lorenzini, M. Masseti, O. Pereladova, R.F. Carden, S.M. Brook, and S. Mattioli. 2018. *Cervus elaphus*, Red Deer. The IUCN Red List of Threatened Species. https://doi.org/10.2305/IUCN.UK.2018-2.RLTS.T55997072A142404453.en (accessed 23 July 2020).

Loveridge, Andrew J. 2018. *Lion Hearted: The Life and Death of Cecil & the Future of Africa's Iconic Cats*. New York: Regan Arts.

Loveridge, Andrew J., M. Valeix, G. Chapron, Z. Davidson, G. Mtare, and D.W. Macdonald. 2016. Conservation of Large Predator Populations: Demographic and Spatial Responses of African Lions to the Intensity of Trophy Hunting. *Biological Conservation* 204: 247–254.

Løbner, Rasmus Filsø. 2017. Afgrødetab forårsaget af kron- og dåvildt i SAGRO's område. SAGRO – Viden og Vækst. https://www.sagro.dk/media/9492/vildtskaderapport.pdf (accessed 4 August 2020).

Macdonald, David W. 2016. Report on Lion Conservation with Particular Respect to the Issue of Trophy Hunting. Wildlife Conservation Research Unit (WildCRU), University of Oxford.

Macdonald, David W., Andrew J. Loveridge, Amy Dickman, Paul J. Johnson, Kim S. Jacobsen, and Byron du Preez. 2017. Lions, Trophy Hunting and Beyond: Knowledge Gaps and Why They Matter. *Mammal Review* 47 (4): 247–253 (1–7).

MacKenzie, John M. 1988. *The Empire of Nature—Hunting, Conservation and British Imperialism*. Manchester and New York: Manchester University Press.

MacLaren, C., J. Perche, and A. Middleton. 2019. The Value of Hunting for Conservation in the Context of the Biodiversity Economy. Ministry of Environment and Tourism. https://resmob.org/wp-content/uploads/2019/06/2019-06-Hunting_report_Draft.pdf (accessed 22 June 2022).

Mahoney, Shane. 2009. Recreational Hunting and Sustainable Wildlife Use in North America. In *Recreational Hunting, Conservation and Rural Livelihoods*, ed. Barney Dickson, Jon Hutton, and William M. Adams, 266–271. Oxford: Wiley-Blackwell.

Mallon, David. 2013. Trophy Hunting of CITES-Listed Species in Central Asia—A TRAFFIC Report. CITES Secretariat. https://www.traffic.org/publications/reports/trophy-hunting-of-cites-listed-species-in-central-asia/ (accessed 16 July 2020).

Mbaiwa, Joseph E. 2008. The Success and Sustainability of Consumptive Wildlife Tourism in Africa. In *Tourism and the Consumption of Wildlife—Hunting, Shooting and Sport Fishing*, ed. Brent Lovelock, 141–154. London and New York: Routledge.

Mbaiwa, Joseph E. 2017. Effects of the Safari Hunting Tourism Ban on Rural Livelihoods and Wildlife Conservation in Northern Botswana. *South African Geographic Journal* 100 (1): 41–61.

McCabe, Thomas R., and Richard E. McCabe. 1997. Recounting Whitetails Past. In *The Science of Overabundance—Deer Ecology and Population Management*, ed. William J. McShea, H. Brian Underwood, and John H. Rappole, 11–26. Washington and London: Smithsonian Books.

McCullough, Dale R. 1997. Irruptive Behaviour in Ungulates. In *The Science of Overabundance—Deer Ecology and Population Management*, ed. William J. McShea, H. Brian Underwood, and John H. Rappole, 69–78. Washington and London: Smithsonian Books.

McShea, William J., and John H. Rappole. 1992. White-Tailed Deer as Keystone Species Within Forest Habitats of Virginia. *Virginia Journal of Science* 43 (1B): 177–186.

McShea, William J., H. Brian Underwood, and John H. Rappole, eds. 1997. *The Science of Overabundance—Deer Ecology and Population Management*. Washington and London: Smithsonian Books.

Mkono, Mucha. 2019. Neo-colonialism and Greed: Africans' View on Trophy Hunting in Social Media. *Journal of Sustainable Tourism* 27 (5): 689–704.

Monteith, Kevin L., Ryan A. Long, Vernon C. Bleich, James R. Heffelfinger, Paul R. Krausman, and R. Terry Bowyer. 2013. Effects of Harvest, Culture, and Climate on Trends in Size of Horn-Like Structures in Trophy Ungulates. *The Journal of Wildlife Management* 78 (1): 133–141.

Montgomery, Robert A. 2020. Poaching Is Not One Big Thing. *Trends in Ecology and Evolution* 35 (6): 472–475.

Musgrave, Ruth S., Sara Parker, and Miriam Wolok. 1993. The Status of Poaching in the United States—Are We Protecting Our Wildlife. *Wildlife Law and Policy Issues* 33 (4): 977–1014.

Muth, Robert M., and John F. Bowe Jr. 1998. Illegal Harvest of Renewable Resources in North America: Toward a Typology of the Motivations for Poaching. *Society & Natural Resources: An International Journal* 11 (1): 9–24.

Mysterud, Atle, and Richard Bischof. 2010. Can Compensatory Culling Offset Undesirable Evolutionary Consequences of Trophy Hunting? *Methods in Ecology and Evolution* 79: 148–160.

Naidoo, Robin, L. Chris Weaver, Richard W. Diggle, Greenwell Matongo, Greg Stuart-Hill, and Chris Thouless. 2016. Complementary Benefits of Tourism and Hunting to Communal Conservancies in Namibia. *Conservation Biology* 30 (3): 628–638.

Namibia Land Statistics. 2018. Namibia Land Statistics Booklet 2018. Namibia Statistics Agency. https://d3rp5jatom3eyn.cloudfront.net/cms/assets/documents/Namibia_Land_Statistics_2018.pdf (accessed 22 June 2022).

Nawaz, Muhammad Ali, Jaffar Ud Din, Safdar Ali Shah, and Ashiq Ahmad Khan. 2016. The Trophy Hunting program: Enhancing Snow Leopard Prey Populations Through Community Participation. In *Snow Leopards: Biodiversity of the World: Conservation from Genes to Landscape*, ed. Thomas McCarthy and David Mallon, 220–229. Manchester: Manchester Metropolitan University.

Nelson, Richard K. 1996. Introduction: Finding Common Ground. In *A Hunter's Heart—Honest Essays on Blood Sport*, ed. David Petersen, 1997 (First 1996), 1–10. New York: Henry Holt and Company

Noguchi, Junko. 2017. Overabundance of Sika Deer and Immunocontraception. *Journal of Reproduction and Development* 63 (1): 13–16.

Nygren, Anja. 2013. Eco-Imperialism and Environmental Justice. In *Routledge International Handbook of Social and Environmental Change*, ed. Stewart Lockie, David A. Sonnenfeld, and Dana R. Fisher, 58–69. London: Routledge.

Nzou, Goodwell. 2015. In Zimbabwe, We Don't Cry for Lions. *The New York Times*, August 4. https://www.nytimes.com/2015/08/05/opinion/in-zimbabwe-we-dont-cry-for-lions.html (accessed 2 July 2020).

Ogutu, Joseph O., Hans-Peter Piepho, Mohamed Y. Said, Gordon O. Ojwang, Lucy W. Njino, Shem C. Kifugo, and Patrick W. Wargute. 2016. Extreme Wildlife Declines and Concurrent Increase in Livestock Numbers in Kenya: What Are the Causes? *PLoS ONE* 11 (9): 1–46.
Packer, Craig. 2000. Infanticide Is No Fallacy. *American Anthropologist* 102 (4): 829–831.
Packer, Craig. 2015. *Lions in the Balance—Man-Eaters, Manes, and Men with Guns*. Chicago and London: The University of Chicago Press.
Packer, Craig, H. Brink, B.M. Kissui, H. Maliti, H. Kushnir, and T. Caro. 2010. Effects of Trophy Hunting on Lion and Leopard Populations in Tanzania. *Conservation Biology* 25 (1): 142–153.
Packer, Craig et al. (58 authors). 2013. Conserving Large Carnivores: Dollars and Fence. *Ecology Letters* 16 (5): 635–641.
Pelletier, Fanie, Marco Festa-Bianchet, and Jon T. Jorgenson. 2012. Data from Selective Harvests Underestimate Temporal Trends in Quantitative Traits. *Biology Letters* 8: 878–881.
Putman, R.J., and N.P. Moore. 1998. Impact of Deer in Lowland Britain on Agriculture, Forestry and Conservation Habitats. *Mammal Review* 28 (4): 141–164.
Putman, R.J., and B.W. Staines. 2004. Supplementary Winter Feeding of Wild Red Deer *Cervus elaphus* in Europe and North America: Justifications, Feeding Practice and Effectiveness. *Mammal Review* 34 (4): 285–306.
QDMA. 2020. QDMA's Whitetail Report 2020—An Annual Report on the Status of White-Tailed Deer—The Foundation of the Hunting Industry in North America. Quality Deer Management Association. https://www.qdma.com/2020-whitetail-report/ (accessed 20 August 2020).
Reading, Richard P., and Sukh Amgalanbaatar. 2016. Argali Sheep (*Ovis ammon*) and Siberian Ibex (*Capra sibirica*) Trophy Hunting in Mongolia. In *Snow Leopards: Biodiversity of the World: Conservation from Genes to Landscape*, ed. Thomas McCarthy and David Mallon, 230–235. Manchester: Manchester Metropolitan University.
Resource Africa. 2021. Conservation and Colonialism—A Statement from the Southern Africa Community Leaders Network. Resource Africa, 20 January. https://resourceafrica.net/conservation-and-colonialism-a-statement-from-the-southern-africa-community-leaders-network/ (accessed 6 February 2021).
Robinson, Janine E., and Pablo Sinovas. 2018. Challenges of Analyzing the Global Trade in CITES-Listed Wildlife. *Conservation Biology* 32 (5): 1203–1206.
Roe, Dilys, Francesca Booker, Olivia Wilson-Holt, and Rosie Cooney. 2020. Diversifying Local Livelihoods while Sustaining Wildlife—Exploring Incentives for Community-Based Conservation. Luc Hoffman Institute. https://luchoffmanninstitute.org/wp-content/uploads/2020/02/Diversifying_Local_Livelihoods-2020_publication-FINAL_compressed.pdf (accessed 6 February 2021).
Rooney, Thomas P., and Donald M. Waller. 2003. Direct and Indirect Effects of White-Tailed Deer in Forest Ecosystems. *Forest Ecology and Management* 181 (1–2): 165–176.
Rutberg, Allen T. 1997. The Science of Deer Management—An Animal Welfare Perspective. In *The Science of Overabundance—Deer Ecology and Population Management*, ed. William J. McShea, H. Brian Underwood, and John H. Rappole, 37–54. Washington and London: Smithsonian Books.
Saeki, M., and D.W. Macdonald. 2004. The Effects of Traffic on the Raccoon Dog (*Nyctereutes procyonoides viverrinus*) and Other Mammals in Japan. *Biological Conservation* 118 (5): 559–571.
Sanchez-Rojas, G., and S. Gallina-Tessaro. 2016. *Odocoileus hemionus*, Mule Deer. The IUCN Red List of Threatened Species. https://doi.org/10.2305/IUCN.UK.2016-1.RLTS.T42393A22162113.en (accessed 23 July 2020).
Sheikh, Pervaze A. 2019. International Trophy Hunting. Congressional Research Service, R45615.
Sheikh, Pervaze A., and M. Lynne Corn. 2016. The Convention on International Trade in Endangered Species of Wild Fauna and Flora (CITES). Congressional Research Service, RL32751.
Simon, Alexander. 2016. Against Trophy Hunting—A Marxian-Leopoldian Critique. *Monthly Review* 68 (4).
Snyman, Sue, Daudi Sumba, Francis Vorhies, Elizabeth Gitari, Christina Ender, Albert Ahenkan, Aurélie Flore Koumba Pambo, and Ona Abiaga Natacha. 2021. State of the Wildlife Economy

in Africa. ALU School of Wildlife Conservation. https://sowc.alueducation.com/programs/research/ (accessed 22 June 2022).

Stewart, Kelley M., R. Terry Bowyer, Brian L. Dick, Bruce K. Johnson, and John G. Kie. 2005. Density-Dependent Effects on Physical Condition and Reproduction in North American Elk: An Experimental Test. *Population Ecology* 143 (1): 85–93.

Strampelli, Paolo, Philipp Henschel, Charlotte E. Searle, David W. Macdonald, and Amy Dickman. 2022. Habitat Use of and Threats to African Large Carnivores in a Mixed-Use Landscape. *Conservation Biology* 36: e13943.

Swan, James A. 1995. *In Defense of Hunting*. New York: HarperSanFrancisco.

Sykes, Robert A., and Phoebe Makiello. 2017. An Estimate of Lyme borreliosis Incidence in Western Europe. *Journal of Public Health* 39 (1): 74–81.

Tchakatumba, Pierre Kabeya, Edson Gandiwa, Emmanuel Mwakiwa, Bruce Clegg, and Simukayi Nyasha. 2019. Does the CAMPFIRE Programme Ensure Economic Benefits from Wildlife to Households in Zimbabwe? *Ecosystems and People* 15 (1): 119–135.

Teale, Edwin W. 1990 (First 1965). *Wandering Through Winter*. New York: St. Martin's Press.

Tensen, Laura, Rosemary J. Groom, Joy Khuzwayo, and Bettine Jansen van Vuuren. 2018. The Genetic Tale of a Recovering Lion Population (*Panthera leo*) in the Savé Valley Region (Zimbabwe): A Better Understanding of the History and Managing the Future. *PLoS ONE* 13 (2): 1–13.

Timbavati. 2022. Sustainability and the Funding of the Timbavati Private Nature Reserve. Timbavati Private Nature Reserve, June 29. https://timbavati.co.za/sustainability-and-the-funding-of-the-timbavati-private-nature-reserve/ (accessed 29 June 2022).

US Department of the Interior. 2018. Secretary Zinke Announces More Than $1.1 Billion for Sportsmen & Conservation. US Department of the Interior Press Releases. https://www.doi.gov/pressreleases/secretary-zinke-announces-more-that-11-billion-sportsmen-conservation (accessed 11 August 2020).

USFWS. 2018. 2016 National Survey of Fishing, Hunting, and Wildlife-Associated Recreation. U.S. Department of the Interior, U.S. Fish and Wildlife Service, and U.S. Department of Commerce, U.S. Census Bureau.

Van Brackle, Michael D., and J. Scott McDonald. 2007 (First 1995). Common Misconceptions. In *Quality Whitetails—The Why and How of Quality Deer Management*, ed. Karl V. Miller and R. Larry Marchinton, 58–68. Mechanicsburg: Stackpole Books.

van der Merwe, Peet, Elmarie Slabbert, and Machiel Koch. 2021. Environmental Sustainability of Consumptive and Non-consumptive Wildlife Tourists: The Case of a Game Reserve in Namibia. *International Journal of Biodiversity and Conservation* 13 (4): 218–227.

van Houdt, Shaya, Richard P. Brown, Thomas C. Wanger, Wayne Twine, Richard Fynn, Kenneth Uiseb, Rosie Cooney, and Lochran W. Traill. 2021. Divergent Views on Trophy Hunting in Africa, and What This May Mean for Research and Policy. *Conservation Letters* 14: 1–8.

Wall, Bill, and Brian Child. 2009. When Does Hunting Contribute to Conservation and Rural Development? In *Recreational Hunting, Conservation and Rural Livelihoods*, ed. Barney Dickson, Jon Hutton, and William M. Adams, 255–265. Oxford: Wiley-Blackwell.

Ward, Kelly J., Richard C. Stedman, A.E. Luloff, James S. Shortle, and James C. Finley. 2008. Categorizing Deer Hunters by Typologies Useful to Game Managers: A Latent-Class Model. *Society and Natural Resources* 21: 215–229.

Ward, Alastair I., Suzanne Richardson, and Joachim Mergeay. 2021. Reeves' Muntjac Populations Continue to Grow and Spread Across Great Britain and Are Invading Continental Europe. *European Journal of Wildlife Research* 67 (34): 1–10.

Warren, R.J. 2011. Deer Overabundance in the USA: Recent Advances in Population Control. *Animal Production Science* 51: 259–266.

Weaver, L. Chris, and Theunis Petersen. 2008. Namibia Communal Area Conservancies. In *Best Practices in Sustainable Hunting—A Guide to Best Practices from Around the World*, ed. Rolf D. Baldus, Gerhard R. Damm, and Kai-Uwe Wollscheid, 48–52. CIC Technical Series

Publication No. 1. http://www.cic-wildlife.org/wp-content/uploads/2012/12/Technical_series_1.pdf (accessed 16 May 2019).

Weber, Diana S., Tait Mandler, Markus Dyck, Peter J. Van Coeverden De Groot, David S. Lee, and Douglas A. Clark. 2015. Unexpected and Undesired Conservation Outcomes of Wildlife Trade Bans—An Emerging Problem for Stakeholders? *Global Ecology and Conservation* 3: 389–400.

Whitman, Karyl, Anthony M. Starfield, Henley S. Quadling, and Craig Packer. 2004. Sustainable Trophy Hunting of African Lions. *Nature* 428 (6979): 175–178.

Wijnstekers, Willem. 2011. *The Evolution of CITES*, 9th ed. International Council for Game and Wildlife Conservation.

Wood, Forrest, Jr. 1997. *The Delights and Dilemmas of Hunting—The Hunting Versus Anti-Hunting Debate*. Lanham, New York, and London: University Press of America, Inc.

Zhou, X., D.C. MacMillan, W. Zhang, Q. Wang, Y. Jin, and D. Verissimo. 2020. Understanding the Public Debate About Trophy Hunting in China as a Rural Development Mechanism. *Animal Conservation* 24 (3): 346–354.

Chapter 6
Fair Chase and Sport Hunting

In the modern world, "sport" is synonymous with competition and competition is about winning. When combined with "hunting," we have a term that implies ego, record books, and the killing of animals for fun and bragging rights. Sport hunting has become a toxic term in the modern world because of these negative connotations but the term derives from older meanings of the word sport, which lean into the notion of deriving pleasure or entertainment from an activity, and the idea of behaving in an honourable, or sporting, way. Sport hunting was a term that gained favour amongst those who hunted for recreation (although of course, often deriving meat and trophies through their activities) to distinguish their hunting from that of the market hunters, exploiting species for commercial gain. However, the negative interpretation of sport hunting is common and logical because this is how we normally use the word sport. It is as intuitive to assume that sport hunting is hunting for "sport" (meaning competition) as it is to assume that trophy hunting is a threat to wildlife. James Swan writes that:

> [t]he use of the term *sport* increases bias in surveys. It tends to portray hunting as a sort of lethal tag game, and conjures up memories of the 1800s when buffalo were shot for pleasure from trains or passenger pigeons were caught and released as live targets for shotgunners, or images in modern times of a bunch of drunks out in the woods shooting at animals for target practice. (Swan 1995: 162, emphasis in original)

This is a negative stereotype that is very hard to get rid of, because anyone who dislikes hunting has an interest in repeating and reinforcing the idea that sport hunting is hunting for sport.

While there are certainly hunters who do treat hunting as if it were a sport—note here Chapter 4.3.3 about dominionistic hunters in particular—this is not the original meaning. As noted throughout Chapter 2, the ideals of sport hunting and fair chase (the topics of this chapter) have roots dating back to Ancient Greece at least. This notion of proper and improper, honourable and dishonourable, ways to hunt persisted through the Middle Ages and until the late nineteenth century in USA where Roosevelt and other conservationists ushered in a renaissance of fair chase and sportsmanship ideals in response to the devastation of North America's wildlife by market hunters. The late nineteenth century sport hunters, not reliant on hunting for food like subsistence hunters or for income like market hunters, could afford themselves an approach to hunting where the manner in which they hunted was more important than hunting success. The essence of the sport hunter's ideal of fair chase is to give one's prey "a sporting chance." If there are elements of competition in hunting for the true sportsman, it is a competition with oneself. The goal is not to outdo oneself in size or quantity, mind, but to outdo oneself in sporting behaviour. Dizard emphasizes how the sportsman may also be interpreted as a Nature Hunter:

> The ideal of the sportsman is now thoroughly embedded in the hunting fraternity's sense of itself. In their publications and in hunters' public statements, hunters are described as the vanguard of conservation, true environmentalists, bound by a code of honor that respects property, the nobility of wild animals, and the safety of others, hunters and nonhunters alike. Hunters, in this self-styled view, hunt for sport in the transcendent sense of the word. It is not cheap thrills or base pleasure they are after. Rather, it is deep communion with nature stripped to its bare essentials. (Dizard 1994: 98–99)

But what is fair chase and sporting behaviour in the modern world and what role, if any, do these concepts play in trophy hunting? Chapter 3.4.2 provided an overview of the fair chase standards of the Boone and Crockett Club, which probably approximate the average hunter's understanding of the meaning of fair chase. Chapter 5.3.3 explored how fair chase notions of it being unethical to shoot females can hamper wildlife management efforts. But we can dig a little deeper still by expanding on the ideas of Chapter 4.3.1, specifically the relationship between project and process.

Fair chase, stripped down to the bone, is about ensuring that there is a very precise power dynamic between hunter and prey. To fully understand this, we may turn to *Meditations on Hunting* again and examine José Ortega y Gasset's suggestion that the understanding of hunting itself depends on a specific disparity of power. The first principle is that the hunter must be more powerful than the prey. Gasset wrote that

> hunting is an occurrence between two animals, one of which is the agent and the other the subject, one the hunter and the other the hunted. If the hunted is also, on the same occasion, a hunter, this is not hunting; it is combat. (Gasset 1942: 47)

If a man, alone and armed only with a knife, walks around on a savannah somewhere in Africa, looking for a lion to kill, he is not hunting, because he is much more likely to be killed by a lion than to kill a lion. His activity, as Gasset indicates, would not be hunting, because if he finds a lion or a lion finds him, it would either be combat, or he would die before it even qualified as combat; if there is a hunter in this scenario, it is the lion that finds him. For any quest to kill a wild animal to be hunting, the disparity in power levels between hunter and prey must be big enough that there is no confusion about who is the hunter and who is the quarry.

At the same time—and this is the second principle—the difference in power levels must not grow too large, lest the hunting turn into simple slaughter. In other words,

> it is necessary that the hunted animal have his *chance*, that he be able, in principle, to avoid capture; that is to say, that he possess means of some degree of effectiveness to escape from the pursuit, because hunting is precisely the series of efforts and skills which the hunter has

to exercise to dominate with sufficient frequency the countermeasures of the animal which is the object of the hunt. [...]

> The hunting species and the hunted species, therefore, have to find themselves a very specific distance apart on the zoological scale. The venatic relationship cannot be provoked at less than nor more than that distance between them. As I have already said, one does not hunt a superior or an almost equal, but neither does one hunt the excessively inferior, because then the latter cannot use his wiles. (Gasset 1942: 49–50, emphasis in original)

Fair chase, we may infer, is when we regulate the disparity in power between hunter and prey so that the prey has a very good chance of escaping. How good this chance needs to be for hunting to be fair chase is a matter of opinion. To some hunters, fair chase may be nothing more than obeying the law. As Forrest Wood, Jr. writes:

> if one wants to know the most efficient and effective methods and devices to kill wildlife, one need only to consult the hunting rules and regulations and note what is prohibited by law. (Wood, Jr. 1997: 6)

Others might say that fair chase is living up to the fair chase standards in B&C's fair chase affidavit for acceptance in the B&C record book. Bowhunters might argue that hunting with a scoped rifle, allowing shots to taken over distances of several hundred metres or more, precludes fair chase altogether. Using a bow rather than a rifle would be an example of technological handicapping, whereby a hunter deliberately follows a course of actions that regulate the power disparity between themselves and their quarry. Hunters may also handicap themselves behaviourally. Ways in which hunters can regulate the power disparity between them and their quarry for purposes of fair chase are discussed in the next section.

6.1 Technological Handicaps

Gasset says that man

> [c]ould annihilate quickly and easily most animal species, or at least precisely those that he delights in hunting. Far from doing that, he restrains his destructive power, limits and regulates it. (Gasset 1942: 50)

This is much truer in 2022 than it was in 1942. Technology is the primary reason that killing wild animals, if we desired it, would present no challenge at all. So, we limit our technology in order to turn killing into hunting, and to ensure that our quarry can still evade us. In other words, if hunting, and by extension hunters, were solely about killing then our technological advantages would mean that there be very little left to hunt. Hunters may be inclined towards technological restrictions, but studies indicate that nonhunters are more inclined: support for hunting with high-tech gear such as tripwires, cameras, and hearing devices in the general US population was at 20% in 2008 while opposition was at 70%. Among hunters, 50% were in opposition and 40% in support. The use of special scents to attract game had 36% support and

50% opposition in the general public and 68% support and 24% opposition among hunters (Duda et al. 2010: 203).

Limiting ourselves technologically for fair chase purposes is not a new idea. In ideas of proper hunting from ancient Greece, catching animals with nets was not seen as sporting. This was something only the lowly food hunters did. And when Rowland Ward wrote the passage below in 1906 in the 9th edition of his *Sportsman's Handbook*, he was referring to the effectiveness of then-modern weaponry, specifically the development and widespread adoption of more powerful rifle cartridges, more reliable actions (the part of the rifle that handles cartridges and fires them), and more accurately made rifle barrels, questioning whether their use was in fact necessary and sporting:

> The progress of invention has of late years strengthened the position of the sportsman in respect of his armament, especially as regards great and dangerous game. He has wide choice of excellent weapons. In recent times this abundant facility has somewhat complicated the question, and some good sportsmen thought they recognised the obtrusion not indeed of a fresh element, but more prominence for it—the endeavour to cloak deficiency of skill by increase of mere mechanical power in dealing with great game. In reality, the conditions have not changed for the true sportsman, who seeks rather an exercise of his skill and courage than mere butchery. (Ward 1906: 2)

Next, contemplate this passage by Aldo Leopold:

> [C]onsider the duck-hunter, sitting in a steel boat behind composition decoys. A put-put motor has brought him to the blind without exercise. Canned heat stands by to warm him in case of a chilling wind. He talks to the passing flocks on a factory caller, in what he hopes are seductive tones; home lessons from a phonograph record have taught him how. The decoys work, despite the caller; a flock circles in. It must be shot at before it circles twice, for the marsh bristles with other sportsmen, similarly accoutred, who might shoot first. He opens up at 70 yards, for his polychoke is set for infinity, and the advertisements have told him that Super-Z shells, and plenty of them, have a long reach. The flock flares. A couple of cripples scale off to die elsewhere. Is this sportsman absorbing cultural value? Or is he just feeding minks? The next blind opens up at 75 yards; how else is a fellow to get some shooting? This is duck-shooting, current model. It is typical of all public grounds, and of many clubs. Where is the go-light idea, the one-bullet tradition?

> The answer is not a simple one. Roosevelt did not disdain the modern rifle; White used freely the aluminum pot, the silk tent, dehydrated foods. Somehow they used mechanical aids, in moderation, without being used by them.

> I do not pretend to know what is moderation, or where the line is between legitimate and illegitimate gadgets. It seems clear, though, that the origin of gadgets has much to do with their cultural effects. Homemade aids to sport or outdoor life often enhance, rather than destroy, the man-earth drama; he who kills a trout with his own fly has scored two coups, not one. I use many factory-made gadgets myself. Yet there must be some limit beyond which money-bought aids to sport destroy the cultural value of sport. (Leopold 1968: 180–181)

When Leopold wrote this in 1949—forty-three years after Ward raised the same issue—he was dismayed by poor hunting ethics, but more so than that, Leopold was concerned about the negative influence that technology had on fair chase principles and the cultural value of hunting. The issue, according to Leopold, is not only technology that makes it too easy to kill, but that technology makes it too easy, too comfortable, and too convenient to be out in the wilderness in the first place. If hunting is supposed to be an escape from culture and modern conveniences, then that purpose is defeated when so much modern technology is brought along on the hunt.

Many more nature- and hunting writers have since debated the relationship between hunting, technology, and fair chase. Now, imagine the gadget-hungry hunter's selection of toys today. Hunters now have access to much more powerful and precise weapons and optics. In addition, they can buy such things as GPS (for navigation and for attaching to hunting dogs), night-vision goggles, thermal imaging equipment, motion-sensor wildlife cameras, all-terrain vehicles (ATVs), breathable waterproofs, thousands of different types of camouflaged clothing, rangefinders, drones, solar-powered smartphone chargers, and electronic scent-elimination devices. All of these are products that are promoted by outdoor magazines, which are so reliant on advertising revenues that they cannot say no or publish anything that is critical of this gadgetization of hunting. David Petersen writes on this subject, that

> commercial outdoor magazines love to praise what they hawk. Sticking with the example of ATVs, many hook-and-bullet publications run regular "off-road" columns, features and even special issues touting the latest in vehicles and accessories, thus endorsing and glorifying the bucket-butt slobbishness that motorized hunting promotes—all in hopes of attracting more ads for same. And it works. To hell with ethics. (Petersen 2010a: 28)

The standard or default weapon of any big game hunter today is a rifle with a scope. This, therefore, seems like a sensible baseline from which fair chase may either increase or decrease technologically. Fair chase decreases as the rifle and scope increase in potential range and with technological gadgets and gear that make it easier to find and track quarry or to get around in the field. Increasing fair chase from this baseline can be done in many ways. The optics can be dispensed with (shooting over "open sights" attached directly to the rifle and requiring considerable skill and practice to use in the field), or the rifle can be replaced with a muzzleloader (which requires fiddly reloading and is practically limited to a single shot), a handgun (requiring closer range and therefore better fieldcraft), or a bow (combining multiple challenges not faced by the rifle shooter). As Leopold writes, it is not easy to answer where the line should be drawn for hunters' use of technology. Nor is it even easy to distinguish anymore what a technological handicap is. The bow is supposed to be a simpler and more archaic weapon than the firearm. Yet compare a modern compound bow with all of its available accessories to a rifle without optics, and it is hard to see which of those weapons is supposed to represent fair chase. Muzzleloaders are also a puzzling way to increase the difficulty of hunting. If hunters were using antiquated black powder weapons, then there would be some sentimental appeal perhaps, but new muzzleloaders are essentially modern rifles and look like modern rifles, only loaded using an outdated principle.

When hunters choose weapons that are supposed to be inferior to the scoped rifle, it seems in some cases as if the purpose was only to be able to say that they had hunted with a novelty weapon. YouTube, for example, is rife with videos of hunters who demonstrate the killing of whitetails with various alternative armaments, and the purpose is rarely fair chase. A hunter, sitting in the comfort of a blind with a baiting station a few metres away, killing a feeding whitetail deer with a Glock pistol has not demonstrated fair chase principles; he has merely demonstrated that a whitetail deer can—of course—die by being shot with a bullet from a gun that was never meant for hunting. A YouTube video like this gets a lot of views and comments from "hunters" who also think it would be cool to execute a deer with a handgun. Do such "hunters" handicap themselves technologically? In terms of the weapon they use, sure, but they push their behaviour to the limit to compensate for this handicap. The result is not fair chase. Nor is it hunting.

6.2 Behavioural Handicaps

The other fair chase factor that hunters can adjust is their behaviour. Taking the B&C fair chase affidavit for B&C trophy record book entry as a guideline, it is not considered fair chase to herd animals with planes, ATVs, or other motorized equipment, which are examples of behavioural handicaps combined with technological handicaps. Nor is it acceptable to hunt animals that are confined by game-proof barriers, animals that are transplanted for the purpose of commercial shooting, animals swimming, or animals helpless in deep snow. It is a common fair chase perception that one does not shoot an animal that is lying down, and some hunters think that shooting a buck in the rut is not fair chase. It is usually illegal as well as highly unethical to hunt at night and especially so to jacklight and shoot animals at night.[1] For bird hunting, excluding wild turkeys, it is a fair chase ground rule—no pun intended—that one does not shoot at a bird unless it is in flight. Ironically, as Dan Crockett describes here, the fair chase ethics of bird hunters can almost be a defensive feature of some birds, as a bird that refuses to escape is a bird that hunters cannot shoot:

> It soon became apparent the only challenging part of hunting ptarmigan was finding them. Trusting their mottled rock-lichen-snow camouflage, they allowed themselves to be approached almost to petting range. Once airborne, they had an unsettling tendency to land again before you could shoot without feeling you were in cahoots with ground sluicers. (Crockett 1996: 242)

Some big game hunters, similarly, will not want to shoot an animal that does not act as an animal is "supposed" to act. On rare occasions, a deer might act very strangely and come right up to a hunter, and many hunters will feel unable to shoot such a deer, just as they will be unable to shoot a deer that is lying down. Some diseases can cause this type of behaviour, and deer in the rut will sometimes approach people fearlessly, but sometimes deer are simply habituated to people being around, as long as those people have not been hunting. Deer being unafraid of people is why implementing hunting in order to manage deer herds in a previously unhunted area can be problematic, as discussed in Chapter 5.3.3 in relation to the Quabbin Reservoir; many hunters will not shoot half-tame deer.

---

[1] Jacklighting is to shine a bright light at wildlife at night, which causes them to stand motionless and stare into the light. This happens frequently when deer are caught in the headlights of a car at night, hence the expression.

One of the most divisive fair chase topics is hunting over bait, which incidentally is also a concern because of the spread of diseases (see Chapter 5.3.2). Baiting may or may not be illegal, depending on the country or, in the US, state and depending on the species. It is a practice associated especially with the hunting of black bears and whitetails in the US, various deer in Europe, and lions, leopards, hyenas, and bushpigs in Africa. According to survey results from 2008, 59% of the general population in the United States are opposed to hunting over bait, while 27% support it. Among hunters there was a more even split with 45% opposed and 44% in support (Duda et al. 2010: 202). A 2014 report from the Virginia Department of Game and Inland Fisheries on possible effects of removing a ban on hunting over bait substantiated the conclusion that the US public opinion is generally against hunting over bait. It noted further that baiting may not be considered fair chase, because it "can be seen as a substitute for learning how to scout for game, identify animal sign, or understand animal behavior" (Brown et al. 2014: 8). By accustoming hunted species to show up for bait in a particular location, hunters essentially reverse the hunter/hunted roles and eliminate the notion of fair chase by making animals come to them instead of hunters having to track them down. It is seen, perhaps rightly, as being fundamentally unskilled, amounting to little more than sitting, waiting and shooting, often at close range. While this may be unfair in some cases (placing bait and blinds may be a skill in itself) hunting over bait joins hunting with high-tech gear and hunting in high-fenced preserves as the three hunting practices that have the lowest public support in the US (Brown et al. 2014).

Knox (2011) describes the combination of supplemental feeding (which may or may not be used as bait) of deer and high fences as Intensive Deer Management, another word for Trophy Deer Management (see Chapter 5.4.1). He writes that "from a fair chase perspective it is difficult to mount an ethical defence of intensive deer management to deer hunters, and it is impossible to mount such an ethical defence to the nonhunting public" (Knox 2011: 46). High-fence hunting, also known as canned hunting, is important in relation to trophy hunting, and we will return to it in Chapter 6.4. Before that, we should draw some attention to a possible and hitherto unmentioned downside of fair chase hunting.

6.3 Fair Chase and Animal Welfare

Human technology and human intellect are what allow hunters to kill quickly and painlessly, which are both viewed as essential parts of ethical hunting. Yet forgoing advanced technology and certain kinds of behaviour for the sake of fair chase is also considered an essential part of ethical hunting. Hunters may therefore find themselves in a position where making the chase fairer increases the risk of animals suffering, while decreasing the risk of suffering makes the hunt less fair.

A concrete example of the fair chase vs. animal welfare dichotomy could be spotlighting, mentioned earlier. By shining a bright light on a deer in the dark, one can completely immobilize it for long enough to line up a perfect shot on a stationary target with the least possible risk of a wounding shot. Animals caught in a spotlight at night will often keep their head completely still, thereby allowing a hunter to take an instantly fatal head shot. Head shots are usually considered unethical because the head is a small and near-constantly moving target, and most codes of conduct direct hunters towards taking a "vitals" shot, aiming at the chest just behind or on the line of the front leg around a third to halfway up the body when the animal is broadside on. The bullet will pass through the heart or the vessels leading to the heart and one or both lungs, causing rapid death with a far larger margin of shot placement error than a head shot allows. Headshots may be preferred for culling, where deer may be shot from stationary vehicles and other stable platforms, and where a head shot risks far less meat damage than other shots. Yet despite the potential for far more rapid death that spotlighting provides, everything about this practice is highly unethical from a fair chase recreational hunting standpoint, and may also be illegal, depending on the jurisdiction.

A less straightforward but much more important example, not least because it is widely practiced—is bowhunting. Bowhunters argue that bowhunting is fairer than rifle hunting, as it requires hunters to get much closer to their prey. The closer proximity requires stealthier behaviour, and the bow itself (compound, recurve, or longbow—sometimes crossbow) is to varying degrees a more skill-intensive weapon than the rifle.[2]

The extent of the animal welfare issue with bowhunting is hotly debated. Antihunting groups claim that bowhunting has very high wounding rates and leads to prolonged suffering. A photo of a living deer with an arrow sticking out of its face incidentally also makes for great antihunting propaganda to drive this point home. Bowhunters and bowhunting advocates, meanwhile, do their best to prove that there are no significant differences in wounding rates between bowhunters and rifle hunter or even that bowhunting is more humane because it is virtually silent, and one might kill the deer without it even noticing that it has been hit (Mitchell 1981: 48–49; Swan 1995: 205).

Wounding rates are a problematic area of study though. First, there tend—especially in older publications—to be disparities in the definitions or calculations of wounding or wounding rates. By "wounding" we refer to an animal that has been hit with an arrow or a bullet and then escaped and not been retrieved. It may have then died shortly after, died slowly over days, or recovered completely; the hunter would not know this. An animal that has been hit and run off, and then been retrieved, or shot a second time and then retrieved, does not count as a wounding. Animals do not always "drop to the shot" even with ideal shot placement. AH witnessed the shooting of an impala for example that, later dissection showed, had caused both lungs to turn more-or-less to jelly and the heart to be mangled beyond all recognition. Despite this perfect and devastating shot placement, the impala managed to run some thirty meters before dropping, dead. Many hunting videos on YouTube

---

[2] It should be noted here that many US hunters and some in Europe also bowhunt because bowhunting seasons are often longer than rifle seasons, or the seasons are staggered, so someone who does both can hunt more.

show similar behaviour, often accompanied by comments lambasting the hunter's skill and lamenting the suffering caused, but such "adrenaline runs" rarely last more than a few seconds and certainly do not constitute "wounding." A basic definition of a wounding rate is given by Kurzejeski et al. (1999: 7) as "Hits/(Tagged + Hits) × 100" where "Hits = number of deer reported hit by arrows [or bullets] and Tagged = number of deer reported killed."

Second, wounding rates often rely heavily or even solely on hunter testimony. Rifle hunters may not always know if they have wounded or completely missed their target unless they find a blood trail. Bowhunters will know if they find a blood trail or recover their arrow, which will show signs on the shaft and head of having passed through the animal. Also, all hunters are likely to underestimate how often they wound an animal when questioned, as wounding an animal is a personal embarrassment and reflects poorly on both hunting and hunters in general. In other words, self-reported wounding rates are likely to be (potentially gross) underestimates.

Another problem is the lack of recent studies of wounding rates. A number of studies do back up antihunting claims of bowhunting wounding rates close to or even exceeding 50%, but most of these studies are from the 1940s to the 1980s, and bowhunting technology has come a long way since then. The compound bow did not become mainstream available until the late 1970s, so most of the early studies will rely on wounding rates with traditional bows (mainly longbows and recurve bows), which are much higher than with modern compound bows (Ditchkoff et al. 1998; Kurzejeski et al. 1999; Koppedrayer 2010).

Some recent studies do manage to produce estimates of wounding rates, although they still rely on the honesty of hunters. A thorough 2014 study in the UK of 102 deer stalkers[3] firing 2281 shots at various species of deer with rifles found that 28 (1.2%) of the deer were wounded and never recovered (Aebischer et al. 2014). Another study reported a 2% wounding rate for red deer rifle hunters in southwestern England (Bradshaw and Bateson 2000).

In Denmark, hunters report 120–140,000 deer killed each year. Schweissdog (bloodhounds trained to track deer) handlers are required by law to be called in to assist in finding deer that hunters think they wounded. This happens around 12–13,000 times a year. Deer biologist and schweissdog handler in Denmark, Mads Flinterup, estimates that for every 100 searches for a wounded animal, 30 turn out to be clean misses, 25 are deer that were hit perfectly and lie dead within 200 m of where they were hit, 25 are further away and will in many cases need to be put down, and 20 are deer that were hit but which he is unable to find. On one hand, if Flinterup's estimate is representative, 30% of the schweissdog searches represent misses and therefore not wounded animals. On the other hand, though it is required by law, not all hunters who wound animals bother to call for a schweissdog handler.

---

[3] By "deer stalkers" the article refers to hunters who have passed the Deer Stalking Certificate 1, managed by the Deer Management Qualifications organisation. It may be assumed that since every hunter in this study has taken and passed the non-obligatory DSC1 exam, their shooting skills might be better than that of the average deer hunter.

A realistic wounding rate based on the above is probably around 10% (Naturstyrelsen 2020; Mads Flinterup, personal communication).

Bowhunting wounding rates are higher than rifle hunting wounding rates for a number of reasons. First, a rifle hunter can cover more ground with any potential shot and so has more chances to shoot an animal. It seems likely that the rifle hunter would therefore be less bothered by having to wait for the perfect shot. A bowhunter has fewer chances to get a kill, and might be more likely to take a shot under imperfect conditions. Second, bows are, as mentioned, more skill-intensive weapons than rifles, so they require more practice. Not all hunters live up to this requirement. Third, a modern hunting bullet, designed to expand rapidly on impact, causes a devastating trauma to animals, especially when it hits the chest. The shockwaves causes by the impact and expansion travel throughout the surrounding tissues, and this so-called hydrostatic shock causes immense and usually rapidly fatal damage. On the other hand, an arrow relies on cutting a channel through the animal, severing major blood vessels or vital organs. Without the immense shock associated with a rifle bullet, an arrow must be placed with a high degree of accuracy. It is far more important for the bowhunter to hit vital organs or cause massive haemorrhaging with their shot. Arrows can also be deflected, or outright blocked, by shoulder blades. A rifle hunter usually just needs to hit in the correct general vicinity, and the expanding bullet and hydrostatic shock will take care of the rest. Fourth, the sound of a snapping bowstring is relevant, especially for traditional bows, as this sound reaches the animal before the arrow does. Depending on the distance, this gives the animal time to react to the sound before the arrow hits, which bowhunters term "jumping the string." With the exception of sub-sonic ammunition, which is not usually used for hunting larger animals like deer, the bullet from a rifle hits the animal before the sound of the shot arrives.

In a study by Ditchkoff et al. (1998) where all the hunted whitetails were radio-collared, 22 out of 80 collared bucks were shot by bowhunters in the study period (1995–1997), using traditional archery equipment, i.e., not a compound bow. Eleven of these bucks were not recovered, giving a wounding rate of 50%. However, only three of the eleven wounded bucks died from their wounds while eight survived, resulting in a wounding loss of 14%. This percentage echoes other findings from the 1980s.

Another study by Kilpatrick and Walter (1999) found that four whitetails out of 23 hits by bowhunters in Connecticut were unrecovered, giving a wounding rate of 17.4%. A more recent and extensive long-term study of hunting with advanced archery equipment (compound bows and crossbows) in Maryland reported that 104 bowhunters hit 908 whitetails from 1989 to 2006 out of which 746 were recovered immediately or within 24 h, meaning that 18% of the deer that were hit were wounded and not recovered (Pedersen et al. 2008).

On the website "texasbowhunter.com," some of the forum members self-reported their wounding rates of whitetails in a forum thread from 2007 (TexasBowhunter 2007). One of the forum members calculated the wounding rates of texasbowhunter forum members based on these reports. The final score here was that 59 deer out of 652 were wounded and not recovered, which is 9.05% wounding. Being aware that this is in no way a scientific study, it seems reasonable to assume that in this kind of informal forum setting where (a) the forum users are dedicated bowhunting enthusiasts, (b) boasting is probably not an unlikely occurrence, and (c) a point of the thread was to disprove the 50% or above wounding rates claimed by antihunters, a self-reported and unverifiable 9.05% wounding rate is probably much lower than the actual value.

Overall, reported wounding rates for bowhunters vary widely. Studies of bowhunting in USA spanning a period from the 1950s until today report wounding rates ranging from 3 to 62%. Many, especially of the earlier studies, are close to the 50%, but as mentioned, archery equipment has come a long way since. Based on this body of research, it is hard to say exactly what wounding rates are with various weapons, but a very rough estimate that ignores extreme outliers might be 3–15% with rifles and 10–50% with bows. Compound bows would then probably be close to the 10–20% and traditional bows closer to 50% (Kurzejeski 1999; Ditchkoff 2019). Wounding rates for hunting need a lot more research.

Regardless of how big the difference is in woundings between bowhunting and rifle hunting, it is hard to dispute that there is one. It therefore falls on hunters to have a good reason to hunt with a bow, as it is subject to at least some measure of justified criticism from an animal welfare perspective. This is even more the case when bows are used to hunt animals that are bigger or more robust than deer. Walter Palmer shooting Cecil with a bow is a good example that clearly led to prolonged suffering, and even elephants are hunted with compound bows in some places. The often preferred shot placement with a rifle for an elephant is a brain shot, from close range (often inside 30 m), using a solid bullet rather than an expanding bullet (to ensure it passes through the skull bones) fired from a large calibre rifle. Heart and lung shots can be taken but are less immediately effective than a brain shot, and are considered by many to be more effective as "insurance" shots or follow-up shots. An arrow has no chance of penetrating an elephant's skull, which means a bowhunter is limited in terms of shot placement, although an arrow passing through the heart and severing the main vessels will rapidly cause death, but not as rapidly as a well-placed brain shot. Regardless of species or hunting method, all wounding rates for hunting tourism in Africa are probably considerably lower than for deer hunting in the US and Europe, as hunting clients are usually accompanied by both a professional hunter, ready to take a second shot, and a tracker. Colleen Begg, Managing Director of the Mariri and Niassa Carnivore Projects in Mozambique, documents that out of 60–70 successful lion hunts and around 250 successful leopard hunts conducted in Niassa project areas between 2004 and 2019, only one lion and four leopards were reported wounded and not recovered. Hunting clients are commonly taken out to "zero in rifles" before a hunt, which gives the professional hunter a chance to gauge the shooting proficiency of clients and know what to expect (Colleen Begg, personal communication).

Hunting in Niassa is strictly managed and probably better managed than most hunting in Africa, but if wounding rates for hunting tourism in Africa in general are anywhere close to wounding rates for Niassa, then domestic hunting in Europe and the US have much bigger animal welfare problems with wounding than trophy hunting in Africa.

Sometimes wildlife managers or government authorities may decide that animals need culling, and that it needs to be done more efficiently than could be done via hunting, and therefore dispense with the notions of fair chase. One such historical example concerns red deer in New Zealand. Red deer were introduced in New Zealand for the first time in 1854 and by the twentieth century, there was a red deer population of around 300,000, all descended from the less than 1000, originally introduced (Figgins 2008).

With no natural predators of deer and not enough hunters to keep the herds in check, the populations grew out of control and the damage to New Zealand's ecosystem from over-browsing soon outweighed the red deer's contribution to the economy in the form of hunting tourism. The red deer was designated a pest, and between 1930 and 1956, approximately 670,000 deer were culled by the government to keep populations in check. This was not hunting, so it was not done with fair chase in mind:

> The use of planes and later helicopters to retrieve deer carcasses from the ground shooters later gave way to the use of helicopters as shooting platforms. Helicopter 'gunship' teams comprising pilot and shooter armed with a high-powered semi-automatic rifle roamed riverbeds and the open 'tops' of mountains. The deer would be shot from the air; the helicopter would then land to pick up the carcass or hover while they were slung underneath and then transport them back to the processing plant. (Figgins 2008: 94)

This method of killing deer was very efficient, and the deer populations were soon under control again, but with the export incomes from venison and deer velvet (for medicinal purposes) came in the 1970s the realization that farming red deer would be a highly profitable industry. Some of the culling efforts thus turned to live-capture efforts, and by 2005, New Zealand had about 5000 deer farms with approximately 2.2 million red deer. Wild red deer numbered about 250,000 in 2005 and are kept in check by about 45,000 recreational hunters (Figgins 2008).

Hunters are in a precarious situation when it comes to the fair chase vs. animal welfare dichotomy. Von Essen and Hansen write that modern hunters possess such technological advantages that they must continuously navigate competing interests that "involve a tension between honouring the art and propriety of the hunt and achieving efficiency" (2014: 5):

> [They] must achieve this balance between fairness and utility because the public today may only accept hunting if it is seen to satisfy utilitarian ends of wildlife management and meat procurement on the one hand, and encompasses the use of propriety and skill. (von Essen and Hansen 2018: 5)

Finding a balance that satisfies the general public is difficult. Finding a balance that is immune to attacks from antihunters is impossible. If hunters use a rifle, antihunters will complain that their quarry have no natural defences against rifles and never stood a chance; if hunters use a bow, the antihunters will say that hunters run greater risks of wounding animals for no other purpose than their own selfish sense of accomplishment.

6.3 Fair Chase and Animal Welfare

Morris Jones argues that,

> [i]f you're lazy, start walking. If you're a very good rifle shot, move to the bow. If you can shoot in the kill zone five times out of six from sixty yards, think about picking up a spear and atlatl. (Jones 2012: 42)

A more ethical conclusion here, in our eyes, would perhaps be that if you miss the kill zone one time out of six from sixty yards with a bow, think about getting closer or going back to the rifle. But the conclusion depends on where one positions oneself on the animal welfare vs. fair chase spectrum and that position is very personal. How much animal suffering do you risk in order to give the animal a greater chance of escape and yourself a greater feeling of accomplishment if you succeed?

It seems reasonable to assert that the most dedicated hunters who hunt with a homemade longbow are probably more responsible and wound less animals than the least dedicated hunters who hunt with a rifle and a scope. It is also likely that it is among traditional bowhunters that we could find some of the most skilled, ethical, and knowledgeable naturalists of all hunters. Aldo Leopold and David Petersen come to mind. The problem is that not all hunters are that dedicated, and many bowhunters do not possess the self-awareness required to realize that they are not among the few hunters who dedicate enough time to archery practice to hunt responsibly with a bow. Hunting with a modern rifle is such that most hunters can use one to kill animals without wounding much; hunting with a bow on the other hand will likely lead to more wounded animals for hunters who are not as dedicated as they think they are.

Risking animal suffering for the sake of fair chase made a lot more sense in the late nineteenth century in the US when deer populations needed to recover and killing in moderation was important. Today, as deer populations in Europe and the US are often overabundant, there is rarely a need to chase fairly for the sake of deer populations. For trophy hunting tourism in Africa and elsewhere abroad, giving an animal a greater chance to escape can be rather meaningless if you have paid for one specimen of that animal anyway, and plan to hunt until you shoot one.

Our position is that fair chase today, if it comes with a greater risk of wounding, is mostly a selfish endeavour. It serves to make hunters feel better about their activity and accomplishments and can bring them closer to nature, but it does little for the animal. To the animal, the only thing that matters is that it dies as quickly and painlessly as possible. The ethical hunter today should not increase the risk of wounding for the sake of fair chase. Fair chase is something to strive for when it can be adhered to in all the possible ways that do *not* increase wounding rates.

If you want to hunt ethically, emphasize fair chase as much as possible in getting to the field, moving around in it, and finding your quarry. Do not disturb nature unnecessarily and leave no trace of your presence. Forego electronics, bring a phone only for emergencies, prioritize fair chase principles, simplicity, and rough living in everything not related to the kill. But when you take your shot, use the weapon that gives you the highest chance of a fast kill with one shot. Use every means at your disposal to recover your quarry fast if you only wound it, including advanced technology and calling for help.

Aebischer et al. found that the following ten practices are the most important for rifle hunters to reduce the chance of wounding shots:

(a) Choose a comfortable shooting position.
(b) Avoid shooting off elbows or freehand, use a support (e.g. sticks, bipod).
(c) Aim at the chest or lower neck.
(d) Use bullets heavier than 75 grains [note that different jurisdictions may impose different bullet weight and calibre minimums].
(e) Avoid shooting when there is insufficient time.
(f) Shoot a distant animal only if there is sufficient time.
(g) Shoot an animal when it is stationary.
(h) Avoid shooting an animal that is heavily obscured or in thicket.
(i) Take extra care when the ground is not well known.
(j) Carry out shooting practice at least once a month (Aebischer et al. 2014: 9).

Follow these recommendations if you hunt big game with a rifle. Be sure that you are up to the challenging task if you hunt with a bow. If you wound and lose considerably more animals with the bow than you do with a rifle, hunt with a rifle. Read and reread *Beyond Fair Chase* by Jim Posewitz.

Theodore Roosevelt is famous for the phrase "speak softly and carry a big stick." The phrase refers to his foreign policy approach, but it works for striking an ethical balance between fair chase and animal welfare in hunting as well.

6.4 Canned Hunting

> Shooting in a private game preserve is but a dismal parody; the manliest and healthiest features of the sport are lost with the change of conditions. We need, in the interest of the community at large, a rigid system of game laws rigidly enforced, and it is not only admissible, but one may almost say necessary, to establish, under the control of the State, great national forest reserves, which shall also be breeding grounds and nurseries for wild game; but I should much regret to see grow up in this country a system of large private game preserves, kept for the enjoyment of the very rich. (Roosevelt 1893: 270)

Canned hunting – even more so than trophy hunting—is not a neutral description. Few hunters who practice trophy hunting would call themselves trophy hunters, but even fewer hunters who practice canned hunting would call their hunting canned hunting. We use the term canned hunting in this chapter for the same reason that we use trophy hunting throughout this book. While it may be a loaded term, and is often ill-defined, it is also by far the most common term. If we called it ranch hunting, estate hunting, put-and-take hunting, high-fence hunting, or a term used in South Africa for lions – captive-bred hunting, one might infer what is meant, but canned hunting is the term that is best known to the general public. That said, canned hunting in sufficiently large, fenced areas can be more of a fair chase than hunting wild animals over bait, so while there is some amount of negative judgment in labelling any hunting of a fenced animal canned hunting, it should not be assumed that it is always less fair chase than the hunting of wild animals. It is also worth being mindful of the size of some fenced areas in which hunting can take place. Bubye Valley Conservancy is fenced, but it is also not far off being twice the area of Luxemburg. Even a moderately sized and fully fenced hunting property in South Africa might be more than 5000 hectares (20 square miles).

So, what is canned hunting? We have covered this to some extent already. A canned hunt can most basically be defined as a hunt in which the quarry's chance of evading the hunter has been limited by physical or behavioural barriers. As we showed in Chapter 2.2, this is by no means a new practice; Egyptian and Chinese rulers dating back to 5000 BC or earlier—some of them reportedly as overweight and immobile as the negative slob trophy hunter stereotype of today—slaughtered hundreds or more animals at a time in canned hunts to demonstrate their wealth and power over nature. Hunting parks and canned hunting have existed somewhere in the world ever since. In fact, the region now known as South Africa has a long precolonial history of canned hunting. Large, community-based hunts, where beaters would drive animals into fenced areas and towards waiting hunters were used by the Ndebele, Xhosa and Zulu. The Zulu king Shaka kaSenzangakhona, often known as Shaka Zulu, was, according to some reports, especially keen on such hunts, where he would

> kill the driven animals with spears and even axes. He liked to kill the most dangerous animals – elephants, rhino, buffalo, lions and leopards. One hunt, thought to have taken place in August 1819, involved the driving of hundreds of animals towards Shaka and his hunters. There is no way of verifying the accounts, but they say that three prides of lions were driven towards the king. He speared one male and then a group of lions were trapped in an enclosed area into which Shaka and other warriors entered and, according to Ritter, engaged in single combat with lions, armed just with spears. (Somerville 2019b: 60-61)

In canned hunting today, escape is usually prevented by a physical barrier in the form of a fence, but other elements that can be relevant are if the animal is drugged or otherwise incapacitated, or if it is habituated to humans and thus makes no attempt to escape. These practices, illegal but documented to have occurred in some cases of lion hunting in South Africa, often come into play if the animal is already also behind a fence, so the fence seems to be the most defining characteristic of the canned hunt.

What most people probably associate with canned hunting is the hunting of fenced-in lions in South Africa because of how much media attention this practice has received. While it may have been practiced in South Africa since the late 1970s on first a small but then increasing scale, international public interest in South Africa's lion breeding and canned hunting industry began with an episode of UK-based The Cook Report in 1997 called *Making a Killing*, where investigative journalist Roger Cook posed as a hunting tourist in South Africa. Cook documented how only the most cursory pretence of fair chase – or any chase at all—was required in some facilities, and his footage of a female lion being killed up against a fence and in front of her unweaned cubs was seen around the world. South Africa was believed to have about 2000 captive-bred lions at the time. The public outrage and attention to lion breeding and canned lion hunting died down in the following decade, but interest was renewed in 2015 when the feature-length documentary *Blood Lions* came out, featuring Ian Michler's investigation of the industry. What *Blood Lions* revealed was that the only thing that had changed from 1997 to 2015 seemed to be the scale of the lion breeding industry. At the end of 2016, South Africa reportedly had 8437 lions—three to four times as many as the country has wild lions—in 321 lion facilities (Williams and 't Sas-Rolfes 2019). More than half of these facilities raise lions that are habituated to humans and used for pet-a-lion-cub and walk-with-lions tourism. Another source of revenue is gullible but well-meaning young westerners who pay thousands of dollars a month to "volunteer" at lion farms, where they are falsely led to believe that they support lion conservation. The common fate of the cubs that outgrow their age of being cuddled by tourists is however not rewilding, but often canned hunting or culling, and subsequent export of skeletons to supply the Asian trade in lion bones, used for, among other purposes, lion bone wine (Peirce 2018; Schroeder 2018; Dickman and Hinks 2019; Ashcroft 2020).

To unravel the lion breeding industry in South Africa – especially as concerns the lion bone export to Asia – is a book by itself, and indeed that book has been written more than once. Richard Peirce's *Cuddle Me, Kill Me* (2018) and Lord Michael Ashcroft's *Unfair Game* (2020) are recent examples of non-academic investigations of the industry. *Blood Lions* is a thorough documentary about this business as well, and the 2015 *Bones of Contention* report by Williams et al. is an in-depth examination into the lion bone trade. Because these subjects are so extensively covered already, and because the canned hunting of lions is just a small part of canned trophy hunting (and trophy hunting in general) worldwide, we will be relatively brief here.

Lion hunting in South Africa is almost exclusively the hunting of captive bred lions and would be considered by most to fall under the umbrella of canned hunting; only about 1% of successful lion hunts in South Africa in recent years have been of wild lions.[4] According to CITES data, almost 4000 lion trophies were exported by South Africa between 2013 and 2017, making lions South Africa's top trophy export species of those listed on CITES Appendices.[5] However, following Cecil's killing in 2015—which, to reiterate, was an illegal hunt but not a canned hunt—several countries banned the import of lion trophies, drastically affecting trophy export numbers. Australia, France, and the Netherlands banned all imports in 2015 and 2016, but more importantly, as discussed in Chapter 5.1.2, USA listed lions as threatened in the ESA, effective from January 2016. US Lion trophy imports have since been allowed only if they can be proven to enhance the conservation status of the species. This effectively ended canned lion trophy exports to USA, heavily impacting the lion breeding industry, as evidenced in a survey by Williams and 't Sas-Rolfes:

---

[4] This is a strongly self-reinforcing trend, as most hunters who only want to hunt a wild lion will not go to South Africa, since everyone would assume that their lion was canned.

[5] We have noted this previously, but it is important to reiterate: CITES does not monitor trade of species that are not cause for conservation concern, so all we can tell from these CITES data is that South Africa seem to have exported more lions than other Appendix species, not more than any non-appendix species, which is most ungulates. Also, to reiterate, one animal can represent many trophies, so 4000 lion trophies exported does not mean 4000 dead lions.

6.4 Canned Hunting

The 2016 suspension reportedly had an impact on the businesses of 82% of the responding facilities [...] In response to how facilities, if affected by the suspension, were adapting to the impact, 86 responded in the following ways: (i) 63% had scaled down breeding production, (ii) 58% retrenched workers, (iii) 41% sold off lion stocks; (iv) 30% redirected the business to focus on the lion bone trade; (v) 26% euthanized lions; (vi) 10% continued 'business as usual', (vii) 5% redirected the business to focus on interactive tourism, and (viii) 17% gave 13 answers for 'other'. (2019: 14)

Roughly a third of South Africa's hunted lions are females, and hunters pay between around $US4000 for cheap females to $US40,000 or more for expensive males and even over US$100,000 for exemplary trophy males or custom-(in)bred white lions. Desired lions can commonly be browsed and ordered online, as part of the booking process.

Several factors may serve to explain the appeal of canned lion hunting, as described by Lindsey et al. (2012). While the hunting of wild lions usually takes up to three weeks and has no guarantee of success, a hunt for a canned lion can be done in 2–3 days and has a 99% chance of success. If you just want to shoot a lion then you can go to South Africa and virtually guarantee that you will shoot a lion. It can be done at a lower price than wild lion hunting, and with a higher trophy quality. Canned lions can also be hunted with little to no walking, which seems to be relevant to some canned hunting clients. Lindsey et al. found that 45.6% of clients who hunted wild lions had high mobility (not hindered by age or obesity) while this was the case for only 32% of captive-bred lion hunting clients. It should be noted though that not all canned hunts are sedentary, and clients may well cover considerable ground tracking and stalking a lion. Finding a lion in a thousand hectares or more, even if fenced, is not necessarily straightforward. A typical approach would be to drive tracks around a property, and look for where the lion (which will have been released a legally proscribed period of time before) has crossed. This allows the hunting party to pick up the lion's tracks and follow them into the bush.

Two of the central organisations concerning canned lion hunting in South Africa are the South African Predator Association (SAPA) and the Professional Hunters' Association of South Africa (PHASA). SAPA supports and represents "captive-bred" lion hunting, which means the breeding of animals for the sole purpose of hunting. To be accredited, a captive-bred lion hunting establishment must live up to the "SAPA Norms and standards for hunting managed ranch lions." This includes being kept as wild as possible with limited human contact, and the lions should be released on hunting grounds of a suggested 1000 hectares or more (SAPA 2021). The SAPA argument is that captive-bred hunting that lives up to their requirements is not canned hunting, which they claim not to support. PHASA has previously been against captive-bred hunting on fair chase grounds but have changed their stance a few times. They most recently supported captive-bred hunting under the same conditions approved by SAPA, which rules out most of South Africa's canned lion hunting establishments. "Custodians of Professional Hunting & Conservation—South Africa" was started by a group of professional hunters and outfitters in response to PHASA support of captive-bred hunting. This new group is expressly against the hunting of captive-bred predators and does not allow its members any connection with the industry.

Canned hunting, and canned hunting of lions in particular, represents one of the few points of agreement between antihunters and most wildlife conservationists, as it is generally perceived to not benefit conservation in any major way, despite the arguments of predator breeders and canned hunt operators. One of the most common conservation-based arguments in favour of canned hunting is that captive-bred animals can be reintroduced into the wild and therefore serve as a genetic backup if wild populations become threatened with extinction. Yet reintroduction has not been attempted or perhaps needed to any great extent, and when it has, the potential for success has seemed limited (Hunter et al. 2012; Packer 2015). Another conservation argument put forward by the lion breeding industry is that captive-bred hunting takes the hunting pressure off wild lion populations. Lindsey et al. (2012) noted that if just 20% of the captive-bred market moved to wild lions, this demand would increase by 42.9%. As this percentage should be higher today, the pressure would definitely increase, but as the authors also note, this might only be a problem in countries where lion hunting is poorly managed and quotas are too high or likely to be overshot. Well-managed lion hunting regimes could benefit from better utilisation of their quotas and possibly also higher prices, as canned lion hunting might currently be underbidding wild lion hunting (Lindsey et al. 2012). A final conservation argument with some merit for captive lion breeding for hunting could be that it provides economic incentives to keep these – sometimes quite large – hunting areas in a natural-but-fenced state instead of the alternative of reverting to conventional agriculture (Michael 't Sas-Rolfes, personal communication).

A *Colloquium on Captive Lion Breeding for Hunting in South Africa: Harming or Promoting the Conservation Image of the Country* was held in South Africa in 2018, which was followed up by a document that condemned the lion breeding industry and its negative impact on South Africa's image, heavily influenced by the Blood Lions documentary. One of the paragraphs in the document states the following about canned hunting:

> This most extreme type of trophy hunting serves the captive-bred lions to their hunters on a silver platter. […] The farms often advertise as wildlife sanctuaries to lure in foreign volunteers under the pretence of helping save the species. When the lions reach the trophy age of four to seven years they are then deemed appropriate to be sold for a trophy hunt. The laws require that the 'hunting' is not conducted on the same farm that the animal was bred at. Instead the lions are transported to other areas and shot there, some within days of being relocated. This practice guarantees a kill as the habituated lion has nowhere to go inside the 'can' or enclosure where it is shot. (SA Parliament 2018: 7)

This document, however, has since been criticized for its compromised objectivity and scientific foundation. The Environment Minister responded by appointing a panel to investigate and present their recommendations about what to do with the industry. The panel's remit includes the management, breeding, hunting, trade and handling of elephant, lion, leopard and rhinoceros. The report, published at the end of 2020, was extensive and covered far more ground than just captive lion breeding, but for this species it concluded that "[t]hree different approaches to captive lions are presented by the panel, with the majority view being that, in future, South Africa will not captive breed lions, keep lions in captivity, or use captive lions or their derivatives commercially" (High-Level Panel 2020: 10). To date, captive lion breeding continues in South Africa and it is unclear how, or when, the practice will be phased out.

One of the potential problems with ending the lion breeding industry would be what to do with the captive lions that have no prospects for rewilding. While the lion breeding and canned hunting industry damages South Africa's image, so would euthanizing several thousand lions. The general public will most likely not understand that they cannot just be set free as wild lions; all most people will hear from their popular- and social media sources is that South Africa "is going to slaughter thousands of endangered lions" or something similarly sensationalist. Irrespective of what happens to the lion breeding and lion bone export side of the industry, the canned hunting of lions will probably at the very least slow down even more than it already has because of the ESA-listing in the US. The terrible reputation of canned lion hunting means that more and more hunting organisations are disassociating themselves from it. SCI has not officially supported captive-bred lion hunting since 2018 (see Chapter 3.4.5), and CIC and the Dallas Safari Club published a joint statement in November 2020, stating that the practice – whether it is called canned hunting or captive-bred lion hunting – is unacceptable. Rowland Ward supported this statement in December of 2020. B&C does not accept canned hunting of any kind, not just for lions. With all the major hunting organisations and record books condemning the industry, it seems probable that canned lion hunting tourism will shrink further.

Canned lion hunting gets the brunt of public and media attention and it has been responsible for a large percentage of direct revenues from hunting in South Africa over the past decade or more (just over 11% of $US 93 million in 2013). Yet it is, as mentioned, merely one kind of canned hunting, and limited to one country. The scale of canned lion hunting is dwarfed by the game farming and fenced hunting industry in its entirety. South Africa has an estimated ten thousand fenced game ranches, averaging 8.2 to 49.2 km^2 in size, covering nearly 17% of the country, and housing 16–20 million large animals – three to four times as many as in South Africa's PAs. Most of these properties were formerly agricultural, converted because wildlife provides a better return. These ranches cater every year to around 200,000 local hunters (who mostly hunt antelopes for meat) and 7000 tourist hunters (Pasmans and Hebinck 2017; Schroeder 2018). According to Lindsey et al. (2006), 60% of the tourist hunters are willing to hunt exotic (non-native) species and 12.4% are willing to participate in canned hunting where the animal to be hunted is released into a larger enclosure immediately prior to a hunt. The animals in most of these properties are however entirely wild, and by managing habitat for antelope and other species it is possible for properties to become havens for many other species, providing ample conservation benefits. Arguably, the South African approach is a highly successful form of rewilding. However, some properties may be run more intensively and in those cases conservation benefits, as opposed to economic benefits, may be less clear-cut.

For some species, breeding intensification has diversified into artificial selection of traits that are perceived to be desirable to hunters. So-called colour-variant breeding has seen some game breeders select naturally occurring genetic variants of commonly hunted species, thereby producing differently coloured or marked individuals. Golden-coloured wildebeest, white impala and copper or black springbok are examples of such colour variants and their breeding, and subsequent sale on auction of breeding individuals resulted in a remarkable spike in price for these variants. In 2014, variants accounted for 16% of auction turnover, and unusually coloured individuals were being bought for breeding purposes at prices 400 times greater (or more) than normally coloured individuals. Described by many as an economic bubble (when the price of an asset is based on past performance rather than actual value), the colour variant market started to fall rapidly by 2016. The demand by hunters for novelty was simply not as great as breeders had imagined. The colour variant market was being artificially supported by breeders selling to each other at high prices, fuelled by the dreams of big rewards that never materialized (Hart 2017).

Canned hunting in the rest of the world receives even less public and scholarly attention. The industry is largely unmonitored in the US, so information about the industry is limited and not very recent. Lovelock (2008) wrote that over 1000 ranches for canned hunting of exotic species exist in the US and that the average size in Texas, which has the most, is only about 0.3 km^2. The following is a quote from an October 19, 2017 New York Times article by Manny Fernandez about a hog-shooting adventure on one such exotic game ranch in Texas, where hunters can pay to shoot various native and exotic animals:

> Pursuing the hogs, Ms. Schaan and her son go off-roading through the brush in near-total darkness, with a hunting guide behind the wheel. Aided by their night-vision goggles, they passed by the giraffes before rattling up and down the hilly terrain. Daniel fired at hogs from the passenger seat with a SIG Sauer 516 rifle, his spent shell casings flying into the back seat. (Fernandez 2017)

Fair chase indeed. This, however, probably has nothing to do with trophies. It is just for the fun of haphazardly shooting at animals with a semi-automatic rifle from a vehicle. Otherwise, convenience, guaranteed chance of success and the chance to collect exotic trophies without travelling are what brings in the clients to canned hunting. Frosch et al. (also in 2008) put the total number of cervid farms (most of them just for whitetail deer) in the US at 7828 with 1000 of them being in each of Texas and Pennsylvania. About two thirds of these were estimated to cater to hunters (canned hunting is illegal in some states) and were increasingly important to rural US economies.

Outdoor writer Craig Dougherty wrote an article for outdoorlife.com in 2011 about the trophy deer breeding industry that supplies the canned hunting industry, titled *Freak Show Bucks: A Hard Look at Breeding for Antlers* with the subtitle *Make no mistake—this isn't hunting* (Dougherty 2011). He wrote this about the bucks:

> One look at this pen-reared buck tells you there is something wrong, something terribly wrong. His obscenely disfigured antlers look more like something you would find growing on a coral reef or in a post nuclear war sci-fi thriller. They twist and turn and droop and bulge and fork and then fork again. Some of these deer seem to have multiple bases and all of the racks are preposterously large for the necks and shoulders that support them.

6.4 Canned Hunting

The whitetail bucks that bear these extreme antlers walk around with their heads bent from the unnatural weight, and they typically live little more than half as long as captive deer that are not bred for trophy size. They bear names like Sudden Impact, Ballistic, Hurricane, and Thundering Thor and sell as breeder animals for up to hundreds of thousands of dollars. Straws containing their semen for artificial insemination can sell for tens of thousands to other breeders. Lesser bucks—by breeding standards, but still freaks by wild deer standards—are sold to canned hunting enterprises where the fees for killing them and collecting their trophies are based on what they would score in the B&C record book—not that they would be allowed in the book of course. Outside of the US, Spanish hunters have a rather bleak reputation when it comes to canned hunting, being one of the biggest importers of canned lion trophies. According to Portuguese journalist Alberto Mancebo, on December 17 and 18, 2020, 16 Spanish hunters killed 540 deer and wild boar on a fenced hunting estate in Portugal. The Portuguese government is investigating the event and stated that this slaughter has nothing to do with responsible conservation hunting as otherwise practiced in Portugal (BBC 2020).

The non-African canned hunting industry is not just questionable from a hunting ethics and animal welfare perspective, but from a broader conservation perspective as well. Brown (2016: 20) adds the following about the industry and notes the potential risks to wild deer:

> The genetic manipulation of these animals for the benefit of wealthy hunters and the manner by which they are killed may sour the public's approval of hunting in general. The transportation of these animals within and across state lines risks the spread of disease, and escapes from confined facilities can spread disease to wild populations. Thus the opposition to the confinement of cervids raises both biological and social issues.

In 2008, 67% of the United States population opposed canned hunting while only 20% supported it. Among hunters, 61% opposed and 26% supported canned hunting for able-bodied hunters. 85% of hunters, however, thought that the practice should still be legal (Duda et al. 2010: 202–203).

This statistic is indicative of the unwritten rule among many hunters, but perhaps especially American hunters, that a hunter does not criticize other hunters. This attitude seems to only grow stronger as hunters face increasing external pressure; it has become very common to hear hunters say that hunters have to speak with one voice, despite internal disagreements about ethics. They say that hunters need to stop their internal squabbles and disagreements and focus on the real enemies: antihunters, and those who wish to take away hunting rights and, in case of the US, the right to bear arms. In our experience, these calls for unity among hunters usually come from the kind of hunters who are the reason that some of the ethical hunters stop hunting out of sheer embarrassment. Aldo Leopold wrote: "Let it by no chance be inferred that because I speak as a sportsman I defend the whole history of the sportsman's movement" (Leopold 1932: 168). On this, we have to agree. Slob hunters think of strength in numbers and seek to make themselves look better by association with ethical hunters. But hunting does not need more hunters who all speak with one voice; if hunting is to survive in the face of mounting public criticism, then it needs better hunters, and they need to call out the rotten apples.

New Zealand has about 2000 deer farms with about a million deer (see Chapter 5.4.1). Many of these do not breed trophy stags but are part of the venison and antler velvet industry.[6] Examples of trophy breeding for canned hunting are however still plentiful in New Zealand, and the *Deer Industry News* publication by Deer Industry New Zealand is full of trophy animal advertisements and articles about the trophy market. Trophy animals, according to Deer Industry "are stags that are bred out to carry a full hard antler head. The market for these animals are traditionally Game Estate Parks where they are released for hunting tourism ventures" (NZ DI 2021).

Wilkins Farming is one such venture. They write the following in the "trophy breeding" section of their website:

We strive to produce long, wide, heavy heads and stags with presence and impact.

Our ongoing commitment to our trophy herd is evident in the breeding stags we purchase (or retain), and our investment in world-class genetics to ensure production of animals that deliver upon these traits.

For information about the significant sires represented within our Trophy Herd:

Rock Me—Heinrich VI—Vladimir—Rio—Infinity. (Wilkins Farms 2021a)

Rock Me, Heinrich VI, Vladimir, Rio, and Infinity are of course the names of the superstar trophy stags that Wilkins Farms use for breeding. Each of them has a profile on the website where you can read about their lineages, antler characteristics, and more. Here is the description for Heinrich VI:

Heinrich VI is an unbelievable antlered Eastern sire. Originally from Stanfield Stud Heinrich VI contains some the of very best Eastern genetics available in Heinrich, Maximilian, Alexei and Nicholai. His strength is his incredible antler, at 5 years of age he was a remarkable 56 inches wide, 47 inches long, 12.5 kg HA and scored 469 SCI. His progeny continually show us the same wide heavy antler characteristics. We continue to breed with Heinrich VI with the limited semen stock we have available. (Wilkins Farms 2021b)

The final sentence refers to artificial insemination practices using frozen deer semen. Heinrich VI died in 2014 but is still producing trophy offspring.

Taking canned hunting to the next level, hunting by remote control was advertised by one Texas-based establishment for a short while in the early 2000s. A rifle, linked to a video camera and a remote control, was set up in a game ranch, so buyers could shoot animals from home with their mouse (Lovelock 2008). Calling this hunting—canned or otherwise—is of course a stretch of the imagination. Thankfully, this was stopped before it took off and banned in most states. Even SCI had to admit that this was not hunting.

Steven Rinella is a very famous hunter and outdoorsman, hunting author, television personality, and founder of the MeatEater company and brand. He wrote an article

---

[6] One might assume that antler velvet stags and trophy stags were the same, as both rely on large antlers, but this is not the case, as here described by the New Zealand Deer Industry organisation: "Velvet production is based on beam size, thickness, weight and blood component whereas a trophy head will be evaluated on the number and length of tynes, spread and style, none of which are critical for velvet" NZ DI 2021). However, velvet stags, when they reach peak trophy size still generally get sold to a hunting park.

about canned hunting in 2012, which reflects our perspective on the matter as well, so we conclude with this great passage by Rinella:

> Ultimately, my gripe with this kind of hunting isn't so much that it's immoral, because I don't believe that there's anything wrong with killing domestic animals for human purposes. Rather, my gripe with the activity is that it cheapens the meaning of hunting when guys insist on applying the term "hunting" to the practice. Just the other day, a good friend of mine named Doug had an emergency on his farm and had to dispatch one of his steers with a rifle. He described it as "putting it down," not hunting it, a distinction that this friend understands very well.
>
> In fact, I believe that all hunters—even those who do canned hunts—know the difference. I believe this because they intentionally fudge the lines when it comes to how they discuss their activities. How come you so infrequently meet a canned hunter who admits openly that the trophy on his wall was taken behind a fence? How come hunting shows that do canned hunts never bother to mention the fact that the animals they're supposedly stalking can't really get away? I know a guy who hunted New Zealand once, and he killed a huge stag there in a place that I know to be fenced. When he got home, he talked a lot about the experience. He told the history of New Zealand, how it was discovered by Captain Cook. He talked about the fact that it was once home to flightless birds that stood thirteen feet high, but that the birds were killed off by indigenous hunters; he talked about the weather where he hunted, and the kinds of plants there, and the sorts of terrain he walked on. He even talked about the quality of the light in the early morning. But through all that detail, he always failed to mention the fact that his stag was released into that enclosure from a truck, and that he knew before he arrived that he'd be killing it. I'm not sure why he doesn't mention these things, but perhaps the word "guilt" has something to do with it. (Rinella 2012)

References

Aebischer, Nicholas J., Christopher J. Wheatley, and Hugh R. Rose. 2014. Factors Associated with Shooting Accuracy and Wounding Rate of Four Managed Wild Deer Species in the UK, Based on Anonymous Field Records from Deer Stalkers. *PLoS ONE* 9 (10): 1–12.

Ashcroft, Lord M. 2020. *Unfair Game: An Exposé of South Africa's Captive-Bred Lion Industry* (Britback Publishing).

BBC. 2020. *Portugal Outrage after Spanish Hunters Massacre 500 Wild Animals.* https://www.bbc.co.uk/news/world-europe-55435940.

Bradshaw, E. L., and P. Bateson. 2000. Welfare Implications of Culling Red Deer (*Cervus Elaphus*). *Animal Welfare* 9 (1): 3–24.

Brown, Robert D. 2016. The Politics of Deer Farming in North Carolina—Lessons Learned. *Wildlife Society Bulletin* 40 (1): 20–24.

Brown, Ryan J., R. Cox, M. Dye, B. Ellis, M. Kirchgessner, M. Knox, N. Lafon, B. Lewis, B. Moyer, G. Norman, Jaime L. Sajecki, R. Shuler, and D. Steffen. 2014. *Report of Senate Joint Resolution 79 (2014) – A Study Report on the Effects of Removing the Prohibition Against Hunting Over Bait in Virginia* (Commonwealth of Virginia, Richmond: Bureau of Wildlife Resources, Virginia Department of Game and Inland Fisheries).

Crockett, Dan. 1996. All Birds Flying. In David Petersen (ed.), 1997 (First 1996). *A Hunter's Heart – Honest Essays on Blood Sport* (New York: Henry Holt and Company): 237–245.

Dickman, Amy and A. Hinks. 2019. *State of the Lion: Fragility of a Flagship Species – Report prepared for the Lion Footprint Forum, May 2019.* (Unpublished report based on a paper in preparation).

Ditchkoff, Stephen S., Edgar R. Welch, and Jr., Robert L. Lochmiller. 1998. Wounding Rates of White-tailed Deer with Traditional Archery Equipment. *Proceedings of the Southeastern Association of Fish and Wildlife Agencies* 52: 244–248.

Ditchkoff, Stephen S. 2019. The Truth about Bowhunting Wounding Rates and The Fate of Deer. *Deer & Deer Hunting*, September 4. https://www.deeranddeerhunting.com/content/articles/fate-of-deer-truth-bowhunting-wounding-rates (accessed 22–12–2020).

Dizard, Jan E. 1994. *Going Wild – Hunting, Animal Rights, and the Contested Meaning of Nature*. Amherst: University of Massachusetts Press.

Dougherty, Craig. 2011. Freak Show Bucks: A Hard Look at Breeding For Antlers. *OutdoorLife* https://www.outdoorlife.com/photos/gallery/hunting/2011/09/freak-show-bucks-look-genetically-altered-deer/ (accessed 8–1–2021).

Duda, Mark D., M. Jones, and A. Criscione. 2010. *The Sportsman's Voice – Hunting and Fishing in America* (Responsive Management/Venture Publishing).

Fernandez, Manny. 2017. Blood and Beauty on a Texas Exotic-Game Ranch. *The New York Times*, October 19. https://www.nytimes.com/2017/10/19/us/exotic-hunting-texas-ranch.html (accessed 26–3–2019).

Figgins, Guil. 2008. Game Estates and Guided Hunts. In Brent Lovelock (ed.), *Tourism and the Consumption of Wildlife – Hunting, Shooting and Sport Fishing* (London and New York: Routledge): 87–98.

Frosch, Brian J., David P. Anderson, and Joe L. Outlaw. 2008. Economic Impact of Deer Breeding Operations in Texas. *Selected Paper prepared for presentation at the Southern Agricultural Economics Association Annual Meetings, Dallas, Texas, February 2–6, 2008* https://core.ac.uk/reader/7083040 (accessed 28–8–2020).

Gasset, José Ortega Y. 1942. *Meditations on Hunting in Howard B. Wescott (trans.) 1985*. New York: Charles Scribner's Sons.

Hart, Adam. 2017. Conservation versus profit: South Africa's 'unique' game offer a sobering lesson. *The Conversation*, August 14. https://theconversation.com/conservation-versus-profit-south-africas-unique-game-offer-a-sobering-lesson-82029 (accessed 17–6–2022).

High-Level Panel. 2020. *The High-Level Panel of Experts for the Review of Policies, Legislation and Practices on Matters of Elephant, Lion, Leopard and Rhinoceros Management, Breeding, Hunting, Trade and Handling—High-Level Panel Report—For Submission to the Minister Of Environment, Forestry and Fisheries* https://www.dffe.gov.za/sites/default/files/reports/2020-12-22_high-levelpanel_report.pdf (accessed 17–6–2022).

Hunter, Luke T. B., P. White, P. Henschel, L. Frank, C. Burton, A. Loveridge, G. Balme, C. Breitenmoser, and U. Breitenmoser. 2012. Walking with lions: Why there is no role for captive-origin lions *Panthera leo* in species restoration. *Oryx* 47 (1): 19–24.

Jones, Allen M. 2012 (First 1997). *A Quiet Place of Violence – Hunting and Ethics in the Missouri River Breaks*. Montana: Bangtail Press.

Kilpatrick, H. J., and W. D. Walter. 1999. A controlled archery deer hunt in a residential community: cost, effectiveness and deer recovery rates. *Wildlife Society Bulletin* 27: 115–123.

Knox, Matt W. 2011. The Antler Religion. *Wildlife Society Bulletin* 35 (1): 45–48.

Koppedrayer, Kay. 2010. Big Game and Little Sticks—Bowmaking and Bowhunting. In Nathan Kowalsky (ed.) *Hunting Philosophy for Everyone: In Search of the Wild Life* (Wiley-Blackwell): 198–209.

Kurzejeski, E. W., J. L. Byford, K. Causey, J. W. Enck, L. P. Hansen, W. Krueger, K. Mayer, and K. McCaffery. 1999. The Role of Bowhunting in Wildlife Management. *The Wildlife Society Technical Review* 99 (1): 1–15.

Leopold, Aldo. 1968. (First 1949). *A Sand County Almanac, And Sketches Here and There*. New York: Oxford University Press.

Leopold, Aldo. 1932. Game and Wildlife Conservation. In Susan L. Flader and J. Baird Callicott (eds.), 1991 *The River of the Mother of God – and other essays by Aldo Leopold* (Wisconsin and London: The University of Wisconsin Press): 164–168.

References

Lindsey, Peter A., R. Alexander, L. G. Frank, A. Mathieson, and S. S. Romañach. 2006. Potential of Trophy Hunting to Create Incentives for Wildlife Conservation in Africa where Alternative Wildlife-Based Land Uses may not be Viable. *Animal Conservation* 9/3: 283–291.

Lindsey, P., R. Alexander, G. Balme, N. Midland, and J. Craig. 2012. Possible Relationships between the South African Captive-Bred Lion Hunting Industry and the Hunting and Conservation of Lions Elsewhere in Africa. *African Journal of Wildlife Research* 42 (1): 11–22.

Lovelock, Brent. 2008. An Introduction to Consumptive Wildlife Tourism. In Brent Lovelock (ed.), *Tourism and the Consumption of Wildlife—Hunting, Shooting and Sport Fishing* (London and New York: Routledge): 3–30.

Mitchell, John G., 1981 (First 1980). *The Hunt—A Provocative Exploration of Hunting in North America* (Penguin Books).

Naturstyrelsen. 2020. Schweiss-registret – Eftersøgninger i jagtåret 2019–2020. *Miljø- og Fødevareministeriet, Naturstyrelsen*, December 22. https://schweiss.dk/sites/default/files/jagteftersoegninger_2019-20.pdf (accessed 22–12–2020).

NZ DI. 2021. Deer Information – Antlers – Trophy. *Deer Industry New Zealand*, January 10. https://www.deernz.org/deerhub/deer-information/antlers/trophy (accessed 10–1–2021).

Packer, Craig. 2015. *Lions in the Balance – Man-Eaters, Manes, and Men with Guns*. Chicago and London: The University of Chicago Press.

SA Parliament. 2018. Announcements, Tablings and Committee reports – No 167–2018. *Parliament of the Republic of South Africa* https://www.parliament.gov.za/storage/app/media/Docs/atc/71d13135-15bc-44fc-89f0-0c92b2dd4bb9.pdf (accessed 4–1–2021).

Pasmans, Thijs, and Paul Hebinck. 2017. Rural Development and the Role of Game Farming in the Eastern Cape, South Africa. *Land Use Policy* 64: 440–450.

Pedersen, M. Andy., Seth M. Berry, and Jeffrey C. Bossart. 2008. Wounding Rates of White-tailed Deer with Modern Archery Equipment. *The Proceedings of the Southeastern Association of Fish and Wildlife Agencies* 62: 31–34.

Peirce, Richard. 2018. *Cuddle Me, Kill Me: A True Account of South Africa's Captive Lion Breeding and Canned Hunting Industry* (Struik Nature).

Petersen, David. 2010a (First 2000). *Heartsblood: Hunting, Spirituality, and Wildness in America*. Washington and Covelo: Island Press/Shearwater Books.

Rinella, Steven. 2012. "Canned Hunts," *Meateater*, October 14. https://www.themeateater.com/hunt/big-game/canned-hunts (accessed 10–1–2021).

Roosevelt, Theodore. 1893. *The Wilderness Hunter*. New York: G.P. Putnam's Sons.

SAPA, 2021. SAPA Norms and Standards for Hunting Managed Ranch lions. *South African Predator Association*, January 5. https://www.sapredators.co.za/p40.html (accessed 5–1–2021).

Schroeder, Richard A. 2018. Moving Targets: The 'Canned' Hunting of Captive-Bred Lions in South Africa. *African Studies Review* 61 (1): 1–25.

Swan, James A., 1995. *In Defense of Hunting* (HarperSanFrancisco).

TexasBowhunter. 2007. "Bowhunting wounding rate?" *TexasBowhunter.com*, March 12. http://discussions.texasbowhunter.com/showthread.php?t=24917 (accessed 25–2–2021).

von Essen and Hansen. 2018. Policing Peers and Selves between Law and Morality: A Socio-Legal Perspective on Managing Misconduct in Hunting. *International Journal of Rural Criminology* 4 (1): 1–26. https://doi.org/10.18061/1811/86117, https://doi.org/10.18061/1811/86155.

Ward, Rowland. 1906 (First 1880). *The Sportsman's Handbook to Collecting, Preserving, and Setting-Up Trophies & Specimens, Together with a Guide to the Hunting Grounds of the World* 9th ed (London: "The Jungle," 166, Piccadilly).

Wilkins Farms. 2021a. "Trophy Breeding," *Wilkins Farming Co*, January 10. http://wilkinsfarming.co.nz/deer-venison-velvet/trophy-breeding/ (accessed 10–1–2021a).

Wilkins Farms. 2021b. "Heinrich VI," *Wilkins Farming Co*, January 10. http://wilkinsfarming.co.nz/wp-content/uploads/2019/11/WFC_Heinrich-VI.pdf (accessed 10–1–2021b).

Williams, Vivienne L., Michael J. 't Sas-Rolfes. 2019. Born Captive: A Survey of the Lion Breeding, Keeping and Hunting Industries in South Africa. *PLoS ONE* 14 (5): 1–31.

Wood, Forrest, Jr. 1997. *The Delights and Dilemmas of Hunting —The Hunting versus Anti-Hunting Debate* (Lanham, New York, and London: University Press of America, Inc.).

Chapter 7
Anthropomorphism and "Viralability"

The definition of anthropomorphism in the Oxford English Dictionary is "The attribution of human personality or characteristics to something non-human, as an animal, object, etc." Other definitions refer particularly to the attribution of human characteristics to God or gods, which historically precedes animal anthropomorphism. Then, much later again than animal anthropomorphism, came machine/AI anthropomorphism, which will undoubtedly become an important concept over the coming decades. Anthropomorphism shapes many of our thoughts, and more so than we immediately might think. Anthropomorphism may also determine our understanding of more intangible concepts, such as our general perceptions of the universe and of our existence. We might think of nature as having a purpose for example, because we see ourselves as purposeful (Guthrie 1997: 51). And it is commonplace to describe our planet as "Mother Earth," and the natural world as "Mother Nature," entertaining the idea that our planet and its systems nurture us and care about our existence.

The principal purpose of this chapter is to establish the argument that animal anthropomorphism and the charisma of many trophy hunted animal species have been, and continue to be, important factors in fostering antihunting sentiments. Once again, Cecil the lion will provide an important case study. Cecil's life and death contained many key elements that contributed to his death going viral, and to the widespread and mainstream antihunting sentiment that developed as a consequence. We begin elsewhere, however, because the case of the captive Danish giraffe Marius is useful in illustrating how we come to care about certain individual animals but not others. We can then transition into a general discussion of which animal attributes elicit our sympathy, and ultimately evaluate how this relates to trophy hunting.

7.1 #SaveMarius

In early 2014, Copenhagen Zoo in Denmark housed a two-year old giraffe, unofficially nicknamed Marius. The zoo had determined that Marius was unfit for the

European Association of Zoos and Aquaria breeding programs. His genes were too common and would have led to inbreeding, and Marius was therefore taking up highly specialised space that could be used for a genetically more valuable giraffe. Selling him to zoos or individuals outside the European Association of Zoos and Aquaria was prohibited, and the zoo assessed that separating him from the other giraffes (preventing him from breeding) would greatly reduce his quality of life. Contraceptives or castration are also both options that would prevent him from breeding, but they would also reduce his quality of life, and they come with significant health risks and expense. Marius was also unfit for a wild release. Thus, a decision was made to kill Marius, use him for educational purposes by dissecting him in front of an audience in the zoo, and feed him to the zoo's lions. While shocking to many, zoos face such difficult decisions frequently. What was proposed for Marius was standard procedure for Copenhagen Zoo as well as most other European zoos, as noted by Cohen and Fennel, and, like the killing of Cecil the lion, it all seemed to be nothing to take particular notice of:

> Approximately 5000 'surplus' animals are annually eliminated from European zoos, among them approximately 100 larger animals, information which few zoos readily disclose. The Copenhagen Zoo puts down 20 – 30 animals annually; it recently killed four lions in connection with another of its breeding programmes. (Cohen and Fennel 2016: 2)

Except this, apparently, *was* something to take notice of. The news about Marius' planned euthanasia broke in the Danish tabloid newspaper *Ekstra Bladet* on February 5, and while it took almost a month for the death of Cecil to go viral, it only took the imminent death of Marius three days. In total, 232,323 social media posts about Copenhagen Zoo and Marius appeared between February 5 and 12 (Zimmerman et al. 2014). An online petition was created on February 8 to save Marius, gathering 27,065 signatures in just a day, and various individuals offered to buy Marius. Yet the zoo stood firm. Marius was shot on February 9 as planned, dissected in front of the zoo visitors who wished to see it (a large crowd showed up), and then fed to the lions. The Scientific Director of Copenhagen Zoo, Bengt Holst, stated that

> [w]e have been very steadfast because we know we've made this decision on a factual and proper basis. We can't all of a sudden change to something we know is worse because of some emotional events happening around us. It's important that we try to explain why we do it and then hope people understand it. If we are serious about our breeding activities, including participation in breeding programmes, then we have to follow what we know is right. And this is right. (Eriksen and Kennedy 2014)

Only then, and as the petition to save Marius was replaced by a petition to have Bengt Holst fired (124,303 supporters in 2–3 days), came the real storm. The reactions, as examined by Zimmerman et al., "ranged from rational justifications and emotional condemnations to nationalistic stereotyping and reported death threats to the Zoo employees" (2014: 131).

Zimmerman et al. analysed 315,000 social media posts (roughly 75% from Twitter and 20% from Facebook) about the Marius affair. They discovered that the post authors, judging from their social media profile descriptions, frequently used the words *vegan*, *animal rights*, *liberal*, and *activist* to describe themselves. Unsurprisingly, the negative comments about Marius' killing used mostly animal ethics and animal rights arguments, while positive comments typically appealed to science and conservation arguments. Cohen and Fennel argued that the zoo management acted according to utilitarian considerations and that members of the public were driven by their emotions rather than any concrete philosophical objections (Cohen and Fennel 2016).

What was so special about Marius? Why did his death cause such an uproar when so many other zoo animals are killed both in Copenhagen and in other European zoos without anyone taking notice? We may include Cecil in these questions as well. Loveridge writes that Cecil was "the forty-second collared male study animal to be trophy-hunted since the lion project started in 1999. He was the sixteenth collared lion to be shot on that particular piece of land (Loveridge 2018: 202)," but none of these previous kills brought the world to a halt. What was so special about Cecil? Two answers that stand out clearly are *individuality* and *empathy* and both Cecil and Marius serve as excellent examples of how these qualities influence the way that many view animals and key issues around their management and conservation, including zoo captivity and trophy hunting.

7.1.1 Individuality

Abigail Levin's 2015 article *Zoo Animals as Specimens, Zoo Animals as Friends: The Life and Death of Marius the Giraffe* is a good foundation for this discussion. Levin's chief objective in this article is to demonstrate "a tension between modern zoological display practices and modern zoological objectives" (Levin 2015: 3). Her argument is that one of the display practices of zoos is to deliberately anthropomorphize zoo animals (especially new-born and young animals) and portray them as "special" because this is what engages the public and attracts visitors to the zoos. Yet the zoological objectives of education, science, and conservation may then later require the zoos to treat the animals as mere specimens of species whose individual lives are expendable.[1]

Foremost of the instruments of anthropomorphism in zoos is the naming of animals:

> Marius, like many other zoo animals, was a named animal, and the practice of naming conveys quasi-personhood. Indeed, many philosophers claim that the speech act of naming

---

[1] A parallel can be drawn here to the tension between photo tourism and hunting tourism. Photo tourism operators have a financial interest in anthropomorphizing animals and creating personal narratives about individuals. Stories about the famous lion Cecil are more engaging to the average safari tourist than stories about nameless populations of lions. Anthropomorphism is how many people relate to animals and try to understand them. Hunting tourism operators, on the other hand, have an interest in deemphasizing individuals and avoiding names. Trophy hunters want to shoot a nice lion specimen; they do not want to shoot one that they know has a name. Doing so would take away the wildness (or the pretence of wildness) of the hunt.

an animal performatively contributes to its actual elevation to the level of quasi-personhood. [...] Most people think of their pets as special, and most people name their pets. Conversely, farm animals raised for meat are not typically named, nor are lab animals the experimenter intends to kill—again, I would argue, because in these cases we do not see them as special. Indeed, we often refrain from naming them in these cases in order that we do not run the risk of coming to see them as such. (Levin 2015: 11)

As zoos bestow names and thereby quasi-personhood on their animals, the animals become indispensable. Adoption programmes, whereby people can pay to "adopt" these named animals and contribute to their upkeep are commonplace in zoos all over the world. By giving animals names and allowing for their adoption, zoos nurture an emotional engagement in their lives as individuals and make a promise to the public that these named animals will be treated like individuals. Thus, interestingly, the moral wrong committed by zoos that give their animals names, according to Levin, is not towards the animals themselves but towards the zoo-going public. It is the public to whom the zoos fail to uphold their promises when they suddenly – based on scientific reasoning and to meet largely ecocentric zoo objectives – kill their named animals.

While this analysis fits the fate of many named celebrity zoo animals, it does not quite fit the story of Marius. Marius, contrary to popular belief, was never officially named Marius. Zoo director Bengt Holst stated that Copenhagen Zoo generally[2] does not name their animals because they want to avoid anthropomorphizing the animals. However, zookeepers occasionally and informally identify animals by nicknames among themselves. In this case, a journalist had overheard zookeepers referring to this giraffe as Marius, and so, unintentionally, the name spread[3] (Maach 2014). Once the name Marius had taken hold more widely, it became irrelevant whether the zoo had named him or not. The giraffe that was to be shot, dissected in public view, and fed to lion was "Marius the Giraffe."

Etienne Benson's historical account of animal naming practices in science further substantiates the claim that naming animals generates attention, public interest, and thereby funds, asserting that animals may be named because it increases their value as business assets:

A celebrity animal such as the elephant Jumbo, who was displayed in London's Regent's Park until he was purchased by the American circus entrepreneur P.T. Barnum in 1882, or the celebrity pandas loaned from the Chinese government to zoos around the world since the 1970s, could attract crowds not only because they were enormous, exotic, or endangered but also because they had been given names that allowed them to be developed as distinctive characters in the mass media. Behind the scenes, such animals might also be referred to by numbers, codes, or alternative names, a doubling that reflected the zoo's dual roles of

---

[2] Exceptions are occasionally made in Copenhagen Zoo for animals with very long lifespans such as elephants. These are never given human names, however, but names related to their geographical background (Maach 2014).

[3] The lesson learned, we suppose, should be that zookeepers even internally ought to only refer to the animals by numbers. It is however common for people who work with animals professionally or in science to name them informally, while referring to them by numbers officially. See e.g. Benson 2016 for historical examples. Cecil was also generally referred to as Cecil and not as MAGM1.

preservation and display. In public, their names provided a powerful marketing tool. (Benson 2016: 113–114)

An example that is central to Benson's article is Jane Goodall's Gombe chimpanzee research. The Gombe chimpanzees became famous outside academia because National Geographic's documentaries allowed viewers to develop personal relationships with Mike, David Greybeard, Flo, Frodo (who became infamous after snatching, killing and partially eating a 14-month old girl), and the rest. This meant funding for both National Geographic and for Goodall's continued research. Goodall has been criticized for naming her research subjects instead of numbering them, but she has always insisted that chimpanzees are individuals and should be studied as such. Marc Bekoff expresses similar sentiments, writing as follows in the book, *Animals Matter*:

> In doing research on or about animals, we were taught to number animals instead of giving them names, in order to discourage us from bonding with them. However, naming and bonding with the animals whom I study is one way for me to respect them. (Bekoff 2007: 3)

Returning to Levin's article, she compares the media coverage of the killing of Marius with that of the killing of two adult lions and their two cubs only seven weeks later, also by Copenhagen Zoo. These lions bore no names, and the public had no perception of these lions as *special* or as individuals like Marius. The deaths of these nameless lions – killed to make room for a new lion – never attracted any attention (Levin 2015: 12).

This brings us to another point about the naming of animals. It is possible that these lions' deaths never went viral because, in a very practical sense, news about animals with names goes viral very easily, and news about nameless animals does so only with great difficulty. If headlines for Marius had said only "giraffe to be killed in Copenhagen Zoo" or something similar, there would be no name to search for in Google, and activists – lacking any uniquely identifying word for the news – would have difficulty coordinating social media efforts. On Twitter, it was the #Marius and #SaveMarius identifiers that allowed everyone to find and spread the news. The same was the case with #Cecilthelion. Nameless animals are simply animals. Named animals have the practical and symbolic power to become causes.

While naming is certainly central to the viral potential of individual animals, there are still other ways to tell personal stories of animals and thereby establish their individuality. Levin expands:

> Other display practices such as celebrating the births and subsequent birthdays of zoo animals, naming them through public naming contests, having 24/7 webcams for popular animals, Facebook pages, twitter feeds, and blogs for and indeed "by" the animals (Knut, the celebrity German polar bear, "authored" a first person blog about his daily life), and zoo merchandise of particular celebrity animals are all commonplace social practices of contemporary zoos which encourage friendships to be formed, by inducing the feeling that zoo-goers are participating in some meaningful way (to the zoo-goer, at least) in the animal's life. (Levin 2015: 13)

Giving animals an online presence allows further opportunities for anthropomorphizing. When looking at an animal in a zoo, very few people would assert that the animal was speaking to them. Online, however, a zoo can maintain the illusion that the animals are conversing with us:

> The San Francisco Zoo posts a caption of two otters close enough so that one appears to be whispering in the other's ear. The caption reads: "Dude, be cool. I know where they hide the fish." Of course, no one believes that the otters are actually talking, but the idea is certainly to invest them with personality, which is to say, to individuate them. (Levin 2015: 14–15)

This seems to be a popular way of individuating animals, and amassing followers and fans that can convert into online and real-world visitors and profits. Aside from zoo animals, lots of pets have become famous on social media. Instagram and Pinterest are social media platforms tailored towards pictures accompanied by short texts, so they offer a perfect ecosystem for uploading pictures of an animal accompanied by what it "says." Letting animals – especially dogs – speak through subtitles on YouTube videos is another version of this popular trend, and they are often made to speak in intentionally cute-broken English, presumably because that is endearing and feels more realistic. Animals may even be presented as possessing individual styles and senses of fashion. The website InStyle gives access to articles such as "22 Insta-Famous Pets You Desperately Need to Follow," "29 Adorable Animals with Better Fashion Than Us," "The Most Popular Pets on Instagram in 2017 Are Seriously Adorable," and "13 Dogs You Need to Follow on Social Media for National Dog Day."

As important as individuality is for attaining animal celebrity status, it is usually not enough. It needs to go hand-in-hand with an ability to elicit empathy.

7.1.2 Empathy

In 2008 the octopus Paul was hatched in England and transferred to the Sea Life Centre in Oberhausen, Germany. He died in 2010 (a normal octopus lifespan), but in his short life he achieved considerable fame for what appeared to be an ability to predict the outcomes of football matches in the UEFA Euro 2010 championship. Paul managed this feat simply by selecting food from containers of identical foods presented to him complete with country-flag marks. Paul is interesting in the present context because he is a rarity among animal celebrities in not being a mammal. Animal celebrities are very rarely octopuses, and nor are they commonly cockroaches, spiders, sardines, snakes, toads, mosquitoes, or jellyfish. Names are rarely bestowed on octopuses in the first place, and even if they were, this would usually not lead to the same celebrity status as can be attained by named specimens of charismatic species like dogs, cats, apes, bears, giraffes, or lions. A cockroach called Mike is just a cockroach that someone calls Mike. Despite having a name, and being therefore identified as an individual, few people (if any) would develop any real emotional attachment to Mike. Individuality is usually not enough to elicit our full sense of caring; an animal must also elicit empathy. So which characteristics of animals elicit empathy in humans?

7.1 #SaveMarius

Individuality in the form of a name is a step towards eliciting empathy, but human suspension of disbelief only goes so far. The further an animal is from being something that we might perceive of as having the characteristics of an individual – of being *special* – the further it is from eliciting empathy based on its individuality. We can easily accept that chimpanzees have different personalities, and we can, but to a lesser extent, accept that giraffes or lions do. The popular 2020 film *My Octopus Teacher*, directed by Pippa Ehrlich and James Reed, documented filmmaker Craig Foster's relationship with a wild octopus. This added to well-demonstrated collection of evidence for octopus intelligence and thus our perception that they may have individual personalities as well. But what about an ant, or a sea cucumber? Ants seem to lack much individual purpose most of the time, performing whatever tasks will help ensure the survival and reproduction of their colony. Barring missing legs or some other glaring defect, it is impossible to distinguish one ant from another of the same species and giving ants names would not change that. A sea cucumber looks even less like us. It is often difficult to tell if a sea cucumber is alive or dead, or if it is even an animal at all. It certainly has no way to indicate individuality, even if it possesses it. There is little if anything to elicit any genuine empathy, and it would make little difference if someone had given it a name. Indeed, sea cucumbers and most other invertebrates, have few or none of the qualities that tend to elicit empathy in humans, as we shall see. This can be highly problematic from a conservation standpoint. As noted by Stephen R. Kellert in a study of perceptions of invertebrates, they constitute more than 90 percent of all animal species, perform most of the basic and essential ecological tasks, and can become endangered and extinct just as well as pandas (*Ailuropoda melanoleuca*) or tigers – possibly with much greater ecological consequences. Kellert further found that most people fear or dislike most invertebrates like insects and spiders. A large majority indicated dislike of ants, bugs, beetles, ticks, cockroaches, and crabs and any other invertebrate that stings or bites or carries diseases. The general public also largely disapproves of major expenditures on behalf of endangered invertebrates (Kellert 1993: 846–849). In *Knowledge, Affection and Basic Attitudes toward Animals in American Society* (1980), Kellert and Berry concluded the following about general perceptions of all animal species:

> A qualitative assessment of the animal preference results suggested 12, non-mutually exclusive factors as important determinants of public perceptions of varying animal species. These included:

1. Aesthetics
2. Intelligence of the animal
3. Phylogenetic relatedness to human beings
4. Size of the animal
5. Economic value of the animal
6. Perceived dangerousness to human beings
7. Likelihood of inflicting property damage
8. Cultural and historical importance
9. Animal relationship to human society (e.g., pet, farm animal, game, pest, native vs. exotic animal)

10. Predatory tendencies
11. Skin texture and morphological structure
12. Locomotion characteristics (e.g., aquatic, aerial, crawling, walking, etc.) (Kellert and Berry 1980: 35, 41).

Gene Myers in 2007 reduced these twelve characteristics to just four: Agency, affectivity, coherence, and continuity. Agency represents an animal's ability to take on social roles that are similar to the social roles of humans. Affectivity is the ability to display emotions. Coherence means having a body that is human-relatable and comprehensible in terms of arms, legs, face, and especially eyes. And continuity refers to a gradual increase in the understanding of the animal after spending time with it (Myers 2007; Young et al. 2018).

Plainly, sea cucumbers and the invertebrates described by Kellert possess few to none of these characteristics, while celebrity zoo animals like Marius and lions like Cecil possess all of them to a great extent. Primates, being the closest to humans, possess them more than any other animals.

Snakes are not invertebrates, but they tend to suffer from the same image-problems as the invertebrates mentioned by Kellert. Ballouard et al. (2013: 95) note the influence of western media in shaping and reinforcing stereotypical perceptions of certain species, arguing that the media propagates false narratives about snakes as deadly animals even in countries where there are no venomous snakes. Kaa in The Jungle Book, Voldemort's snake Nagini in JK Rowling's Harry Potter books and films, and Sir Hiss in Disney's Robin Hood are a few examples of snake-individual antagonists in fiction. Spiders fare no better, with Aragog in Harry Potter and JRR Tolkien's Shelob in Lord of the Rings. A whole host of horror movies rely on ophidiophobia (fear of snakes) and arachnophobia (fear of spiders) for their effects. Most snakes and spiders are harmless to humans, but the general public does not know how to tell the majority from the minority. We can clearly tell from their appearance that a crocodile can eat us and a rhino can trample us, but if you visit a country where you are unfamiliar with the native species, you cannot tell if any given spider or snake is harmless or potentially deadly.

Sometimes, as snakes and spiders show, our empathy with species will depend partly on how they are embedded in our culture, but our empathy may also be determined by the perceived aesthetics of a species. Moths, which Kellert's research shows the American public would like to eliminate, are very similar (on every level) to butterflies. Neither butterflies nor moths are meaningfully closer to resembling a human than the other, or possessing more of Myers' characteristics than the other, yet most people have much more positive feelings towards butterflies. Kellert and Berry placed butterflies at number five out of the most liked animals, and the only invertebrate in the top 12 (Kellert and Berry 1980: 31). Kellert found that "[a] more positive view of invertebrates generally occurred when taxa possessed aesthetic value (such as butterflies) or practical value (such as bees)" (Kellert 1993: 849). Jay Swanson, an author of fantasy novels, writes the following in a humorous blog post called *10 Reasons to Hate Moths*:

> People often ignorantly associate moths with their better-known cousin, the butterfly. The thing is, butterflies are gentle, beautiful creatures that flutter gently in the breeze and brighten your day. Moths are dirty, heavy, dusty creatures that would take your wallet as soon as spit on you. Both of which they're attempting to do every time they flop by. They're evil and annoying.

While not the most reliable as a scientific observation of animal behaviour, Swanson does describe the intuitively different feelings about butterflies and moths that Kellert refers to. We could draw a similar comparison between peacocks and turkeys. Few people in the west who eat meat would probably object to eating turkey, but many would object to eating a peacock, because it has a very high aesthetic value, despite being little more than a flamboyant turkey (which themselves can be pretty flamboyant when in full breeding plumage).

Hal Herzog (2010) and Melanie Joy (2010) also cover how our empathy with animals is closely related to our culture and customs. In western culture we eat pigs and turkeys and keep dogs as pets. In our minds, to cope psychologically with our meat-eating habits (what Joy calls *psychic numbing*), we have learned to associate dogs with playful and positive images (as well as names and individuality) and pigs with dirty and negative images. We think of pigs as food, dogs as pets, and rats as pests. Animal rights and vegan activists commonly to use this on social media to further the cause of veganism and to protest against factory farming, showing photos and videos of how pigs and lambs are no different from the animals that we call pets and how they should therefore be treated the same.

That we do not think of giraffes as food can explain some of the reactions to the feeding of Marius to Copenhagen Zoo's lions. While not all protesters may have reasoned this far, many of the people who objected must have been aware that the alternative to the zoo's lions eating giraffe that day would be eating something else that had to be killed, which is usually a cow. But a cow – unlike a giraffe or a lion – is perceived as food in western culture. However, it is hard to find a cogent argument as to why the life of a giraffe is worth more than the life of a cow when, as in this case, neither of them are wild animals nor of any conservation importance.

These perceptions may partly explain why trophy hunting in Africa is considered less acceptable than trophy hunting in Europe and North America; Europeans and Americans are more accustomed to the idea that deer (by far the most hunted type of animal) are food. And, indeed, hunting for food is widely supported, as documented at the end of Chapter 2. African megafauna, such as elephants, rhinos, giraffes and lions, are not considered food by the general public in Europe and the US, and their hunting is considered less acceptable.[4]

A butterfly's aesthetic value is one empathy-eliciting characteristic beyond agency, affectivity, coherence, and continuity. Another characteristic related to aesthetics is size. Elephants have the characteristics of agency, affectivity, and continuity, yet people's empathy for elephants seems to outweigh the extent to which they possess

---

[4] The writings of the hunter Frederick Courtney Selous provide ample evidence as to why this attitude is a western attitude. The elephants Selous shot for ivory were avidly consumed by his hunting party (especially the fat, heart, and trunk meat) and his writings contain several references to shooting rhino primarily to feed his camp.

these characteristics. Size, as also indicated by Kellert and Berry, must account for a lot. A very large animal is simply impressive and awe-inspiring due to its size alone, and most people's intuitive valuation will be that it is a greater moral wrong, all else being equal, to kill a large animal than to kill a small animal.

Notably though, small animals – when small is linked to human- and especially baby-like features – can also elicit empathy, and much more so than adult specimens of the same species, or other species without such characteristics. Puppies and kittens are the obvious examples of animals that evoke our desire to protect them from harm. We perceive of them as defenceless, and more so the smaller, clumsier, and more "infant" they are. Puppies and kittens have, like new-born babies, large eyes, and soft, rounded features. They also have another feature that is generally endearing to humans, which is soft fur, partly a visual quality and partly pleasing to our sense of touch. Most species that we can pick up and cuddle that have soft and warm fur are very effective at eliciting empathy, and even very large and dangerous animals like bears may elicit additional empathy because they have thick, soft fur. For the lion, the mane adds to the lion's aesthetic value through sheer magnificence. Partly relevant here is also our cultural association with lions as "kings of the jungle" with their manes representing their royal crowns. Large birds of prey like eagles often have impressive aesthetic value along with symbolic and cultural value like that of the lion.

The cuteness of puppies and kittens is one form of empathy-eliciting aesthetic value, appealing to our desire to protect and nurture. Another is the colourful beauty of butterflies. A third, more fitting to lions, eagles, bears, and elephants, is what we might call grandeur: an aesthetic value that evokes feelings of admiration and respect because of size and power. There is a wildness that stands in contrast to nature more tamed by humans, and we admire these animals because of it. Wolves have a special place in this context. Their "grandeur" is less than a bear or a lion, but perhaps more than any other species, they encapsulate, and symbolise, for many people this sense of wildness.

Jamie Lorimer (2007) worked on these questions and identified four components that determine which species we have any attitudes towards, and what the attitudes might be:

Detectability and distinctiveness refers notably to the "size, colours, shape, sound and speed of a species" (Lorimer 2007: 20–21) and has much to do with how easy or hard it is for human senses to detect and perceive of a species. Our most important senses are sight and hearing, so an animal that we cannot hear and hardly, or not at all, see is not one we would have many feelings about. *Socio-economic biases* refer to the costs and benefits of animals to society. Lorimer notes that useful species are rarely threatened, which of course is a very relevant point in relation to trophy hunting. Lions are considered a pest and danger by local communities in many African countries, so their value to foreign hunters can be what keeps the populations alive, at the cost of some individuals. Under *aesthetics*, Lorimer notes that these are "the characteristics most often referred to by nature conservationists when they speak of the ferocious tiger, the magnificent eagle or the wizened oak tree" (Lorimer 2007: 22). Lorimer uses a scale with cuddly charisma at one end and feral charisma at the other, which of course is comparable to the distinction made earlier between the aesthetic value of puppies and kittens on one hand and that of lions, eagles and wolves on the other. Lorimer's fourth component is *epiphanies and intellectual satisfaction*. Intellectual satisfaction closely resembles what Myers refers to as *continuity*: a gradual increase over time in (especially professional) understanding and therefore appreciation of a species. An epiphany may instead refer to a single monumental experience with a lasting impact. It, "relates to the epiphanies people experience in unexpected encounters with species and places. An epiphany can be understood as a moment of enchantment. It is often short-lived but has lasting effects" (Lorimer 2007: 23). Unexpected and short-lived encounters with species can, of course, work both ways. While experiencing otters playing in a stream, for example, may found a long-lasting fascination of otters and desire to protect them, seeing your child get eaten by a lion may equally but conversely create a life-long antipathy towards lions.

Agency, affectivity, coherence, and continuity, according to Myers, and additionally aesthetic values of different kinds, including size, and cultural or historical importance as well as Lorimer's four factors above seem to cover and support the empathy-eliciting characteristics first mentioned by Kellert and Berry in 1980. Species with these empathy-eliciting characteristics are the same species that we call charismatic, and they are often those that become what is called "flagship species" for conservationists.

> The initial use of 'flagship species' [...] recognised that charismatic megavertebrates might be the best vehicles for conveying the entire issue of conservation to the public. Primates, in particular, were considered the best flagships for tropical forest conservation, and have been subjects of major public awareness campaigns. Thus, primates became the flagships for entire regions, and the campaigns using them as symbols are excellent examples of the way in which key groups of animals can be used to sell the whole issue of conservation, both in tropical countries and in the developed world. Thus, there is a strategic, rather than an ecological or biological, sense to the original use of flagship species. (Leader-Williams and Dublin 2000: 55)

Leader-Williams and Dublin also discuss overlaps and differences between the concept of "flagship species" and those of "keystone species," "indicator species," and "umbrella species." Going into much detail here is beside the point of this chapter, so suffice to summarize that flagship species can, through their charisma, be used to draw public attention to conservation causes; keystone species are of key importance to the stability and integrity of their ecosystems; the health of indicator species can be good indicators of the health of their ecosystems; and umbrella species occupy such a large geographical area in such an ecologically integrated way that protecting them generally protects the whole ecosystem (Leader-Williams and Dublin 2000).

Macdonald et al. (2022) conducted a survey in which 3181 respondents from nine countries (Argentina, Australia, Brazil, India, Portugal, South Africa, Spain, UK, USA) were shown twenty-five sets of two side-by-side photographs (fifty photos in total) of different species, colour variations, or photo quality variations of carnivores. For each set of two images of carnivores, participants were then asked to prioritize one of them for conservation efforts. Eighty-seven terrestrial mammal species from five families in the order Carnivora were distributed across 122 photographs. The results showed that big cat species such as lions, tigers and leopards were prioritized much higher for conservation by all respondents than species like wolves and wild dogs (*Lycaon pictus*), and especially hyenas (notably the spotted hyena *Crocuta crocuta*). Tigers were by far the highest rated species and white tigers more so than tigers of normal colouration, just as white lions also scored highly and higher than regular lions. One photo of otters (Lutrinae) also came in the top-ten, which the authors attribute to "a 'cuteness' factor that is hard to quantify, perhaps associated with their almost anthropomorphic manual dexterity and maternal care" (Macdonald et al. 2022: 10). The respectively positive and negative portrayals in entertainment media of bears and hyenas may explain why bears (represented in the study by giant pandas and polar bears) were highly favoured for conservation, while hyenas had one of the lowest scores of all. In line with the findings of Kellert and Berry (1980) as well as Lorimer (2007), body size was found to be an important determining factor for conservation priority. Rarity, or perceived rarity, was another key factor, which the authors suggest might partly explain the higher scores for white lions and tigers than regular lions and tigers (Macdonald et al. 2022). For white lions in particular, and especially from a conservation perspective, this is a concerning result, as white lions – rarely occurring in the wild – have no particular conservation value over normal lions but are often bred (and indeed inbred) in captivity for the novelty of their colour and sometimes specifically for trophy hunting purposes (see Chapter 6.4). White and black-and-white animals scored highly in general, as evidenced by giant pandas, polar bears, and snow leopards all appearing in the top-ten (Macdonald et al. 2022).

7.1.3 Innocence and Defencelessness

In reading social media comments about incidents like the one with Marius or the one with Cecil, two words appear often when describing the animals in question: "beautiful" and "innocent." What hides behind the word "beautiful" has already been discussed under aesthetics in the previous section, but "innocent" appears central enough to the gut reaction of many to warrant further investigation.

So, what does it mean to be an innocent animal? One interpretation is innocent as in outside of moral judgment, which certainly makes sense. Another is innocent as opposed to guilty. Innocent is what you are when you are not guilty, so if an animal is not guilty it must be innocent.

Being guilty implies having done something that is wrong, but as animals arguably have no concept of right/wrong or good/bad, you might say that nothing any animal can do is subject to moral valuation, as in the first interpretation of innocence. If animals cannot be moral agents, only moral patients; then right and wrong can be done to animals, but animals cannot themselves do right or wrong. We may scold a dog for eating food off the dinner table and say that it is a bad dog, but this only implies a failure to live up to human expectations. It is not a wrong in any moral sense. If a dog attacks a child, we put the dog down not because it is an evil dog that *deserves* the punishment of death, but simply because there is a risk that it might attack humans again. The decision to kill the dog is based on the utilitarian reasoning that the risk of another person being harmed by the dog outweighs any interest that the dog might have in not dying, or that its owner might have in keeping their dog. This is not to say that we humans do not often act irrationally and hurt animals because we *feel* like they deserve punishment, but these actions, and their motivations, do not imply that they do. If you were to hit your head on a cupboard and punch the cupboard in anger, your action says nothing about the moral accountability or agency of cupboards. Morris Jones writes the following on this subject, emphasizing the importance of having a choice:

> The crux of it is that the deer is not innocent. Innocence implies the possibility for guilt. Innocence implies choice. The coyote is not guilty for eating a deer, and neither is the deer guilty for eating grass. They have no choice about their actions. In the same way, neither of them are innocent. That which is without freedom is without innocence or guilt. It simply *is*. (Jones 2012: 140, emphasis in original)

Despite all the above, when social media users describe animals as innocent, it is probably not to be interpreted as meaningful philosophical judgements about the moral responsibility of some animals as opposed to others. A comparison with human infants may be useful here to explain why the word innocent is so commonly used. Hurting infants is generally perceived as worse than hurting adults, and there can be several good reasons for this.

First, the younger a person is, the more life they have left to live. We think, logically and all else being equal, that if someone must die, better that it is someone old, because they are most likely robbed of less remaining life. Second, infants are, like animals, popularly considered "innocent." We cannot say with much confidence that an adult is innocent because they may not be. They are moral agents, and they may have acted wrongly in a moral sense. Small children, like animals, are moral patients, not moral agents, so when small children or animals are harmed, we *know* that they can have done nothing to deserve it.

Third, infants are defenceless. Infants, like puppies and kittens, to most people score very highly on cuddly charisma, as Lorimer would call it. They evoke our parental instincts and desire to protect them from harm, and they elicit our empathy because they are unable to protect themselves.

The defencelessness factor is one that carries a lot of force. George Bernard Shaw stated that

> just as the murder of a child is more shocking than the murder of an adult (because, I suppose, the child is so helpless and the breach of social faith therefore so unconscionable), the murder of an animal is an abuse of man's advantage over animals: the proof being that when an animal is powerful and dangerous, and the man unarmed, the repulsion vanishes and is replaced with congratulation. (Shaw 1914: xiii)

If we return to Marius, then, this comparison with infants can further help explain why people empathised so much: He was young, he was "innocent," and he was defenceless. Now, there is of course a difference between young giraffes and adult lions that must be noted when it comes to defencelessness. Cecil would seem to not be defenceless, but in reality, an adult lion in the wild might as well have been a young giraffe in a zoo for all the chances it had of defending itself and surviving against a hunter. The mercy for whatever reason of humans – empathy, conservation, utility, indifference – is for most animals the only defence against humans, so being physically defenceless and relying on human empathy may be the best defence there is. The lion is empathetically disadvantaged here because it is virtually defenceless against the hunter while appearing not to be.

7.2 Anthropomorphism, Empathy, Conservation, and Trophy Hunting

Evidence shows that anthropomorphizing animals leads to increased empathy with them, and that both empathy and anthropomorphism encourage better treatment of individual animals. Some studies indicate that the more capacity for reason people ascribe to animals, the less inclined they are to harm or eat them (Gray et al. 2007; Bastian et al. 2011). The respondents of another study were "significantly more willing to help dogs when they were described with anthropomorphic language compared with non-anthropomorphic language" (Butterfield et al. 2012: 958). But anthropomorphism and empathy also seem to encourage conservation behaviour, i.e. attitudes and actions for the benefit not just of individual animals but for species and ecosystems. This might explain the frequent use of anthropomorphism in environmental discourse. Tam et al. (2013), for example, showed that anthropomorphism generally encourages connectedness with nature, which in turn leads to conservation behaviour. Chawla (2009) and Tam (2013) indicated much the same about empathy.

While anthropomorphism and appeals to empathy can be useful to conservationists and zoo managers, they can easily backfire. Zoos may have interests in anthropomorphizing their young animals to attract visitors, but these individuals quickly become nameless "specimens" if the time comes when they must be killed. Many zoos do what Copenhagen Zoo did with Marius, except instead of making a public educational event of it, they keep it as secret and hidden as possible. Powell and Ardaiolo (2016) revealed that the culling of zoo animals, including large mammals, is supported by zoo managers and practiced all over the United States (where the public is much less accepting of it), but it is never advertised or publicly discussed. Ian Parker writes the following in the thoroughly researched article, *Killing Animals at the Zoo*:

> The United States experienced its own Marius affair three decades ago. In 1982, the Detroit Zoo appointed a new director, Steve Graham. Arriving from Baltimore, he chose to stop giving animals names, to stop selling surplus animals to dealers, and to publicly acknowledge the need to cull. Graham soon announced the impending euthanizations of four elderly tigers. He and the zoo were sued by the Fund for Animals, and there were well-publicized court hearings. "I wish I could put you in a small, slippery-floored cage," an anonymous correspondent wrote to Graham. "Then I would torture you each day . . . until you died." The zoo did kill three of the tigers. Graham, who is now retired, said in a phone conversation

that, during the worst of the tiger crisis, Coleman Young, Detroit's mayor, "lent me his Uzi guy"—a bodyguard who carried an automatic weapon in a gym bag.

American zoos did not follow Graham's lead. They chose instead to celebrate animals' birthdays, send surplus animals to roadside zoos,[5] and never talk about death. In current industry guidelines on population control, the phrase "non-living" is used eight times, and there's no use of a shorter, more common alternative. (Parker 2017)

Similarly, conservationists sometimes anthropomorphize and encourage empathy with charismatic species in the hopes that it will lead to increased support for saving these species. However, when they elsewhere defend angling, trapping, sport hunting or trophy hunting as means of conservation, they want as little anthropomorphism and appeal to empathy, and as much science and utilitarian rationality, in the equation as possible. Even when conservationists do employ anthropomorphism and appeals to empathy, there are no guarantees that the strategy will work. It may even be counterproductive.

It is a common perception that charismatic species receive a disproportionate amount of attention from conservationists, simply because they are charismatic and easy to get the public to empathize with and therefore raise funds for. However, Courchamp et al. (2018), defining the ten most charismatic animals as the tiger, lion, elephant, giraffe, leopard, panda, cheetah, polar bear, wolf, and gorilla (aware that some of these are more than one species), conclude that all of them are at risk of extinction in the wild, but that the general public, except in the case of the panda, tiger, and polar bear, is largely unaware of their plight. The authors reason that,

> these animals may be assumed to be abundant because of their omnipresence in our culture, as they are seen everywhere—in zoos and toys, on small and large screens, on advertisements and books alike. (Courchamp et al. 2018: 6)

Loveridge notes the same tendency:

> I'd hazard a guess that there are more images of lions in sculptures, carvings, gargoyles, and coats of arms in central London than there are living lions in the whole of West Africa. (2018: 5)

The constant public exposure of these animals creates "virtual populations," which greatly outnumber the real populations of these animals and – through their constant presence – insert in the minds of the public the misconception that the real populations are also abundant.

A further point is that some of the conservation practices that have great appeal to the empathetic public because they focus on individuals may be far less effective than other practices, which carry no appeal at all because they focus on ecosystems or populations. Here is Craig Packer, putting this very well in the magnificent book, *Lions in the Balance*:

---

[5] What the American public might feel even less comfortable knowing is that apart from roadside zoos, other common recipients of large surplus mammals from zoos are hunting ranches (Scully 2002: 64). So instead of being used for educational purposes and fed to lions as Marius did, an American zoo giraffe may end up getting shot behind a fence in Texas.

> The idea of releasing captive-born lions into the wild has resonated with the public ever since the 1950s, when George and Joy Adamson first released hand-reared Elsa back into the wild.
>
> Elsa quickly got sick and died.
>
> The Adamsons then released Elsa's orphaned cubs, Jespa, Gopa, and little Elsa, into the Serengeti.
>
> The orphans faltered, sickened, and disappeared.
>
> Joy hand-reared a cheetah and a leopard. George rehabilitated lions in the desert.
>
> None of their cats survived as truly wild animals. The *Born Free* strategy has been endlessly repeated in Zimbabwe, Zambia, and South Africa over the last fifty years but to no lasting effect. Individuals come and go; they blink out in a flash. It is the future of a population that really matters for the survival of a species.
>
> But once we start talking about lion populations, everything gets too vague, too diffuse. Eyes start to glaze.
>
> How I wish someone could pick up and cuddle the Selous or the Serengeti. (Packer 2015: 209)

But anthropomorphism and empathy with animals are not only double-edged swords for conservationists. They may be so for those who focus on care for individual animals as well. While anthropomorphism elicits empathy and thus the desire to protect animals, empathy without knowledge of animals can lead to misguided actions on behalf of individuals, as discussed by Young et al. (2018), who provide the example that if people

> project their personal experiences without trying to cognitively understand the animal's, it can lead to incorrect empathy that can negatively impact animals and people. [...] People will come across baby deer curled up and alone, assume that it is abandoned and bring it to an animal rescue not understanding that the mother has left the baby there purposefully and will be back to collect it. Also, there are the accounts of people believing they can communicate with wolves or bears and end up dying or becoming seriously injured due to their incorrect empathy. (Young et al. 2018: 9)

There are many such examples of people causing harm in the belief that they are doing something good. Many cat owners think that they are doing something overall good by letting cats roam freely outside, while really, they are causing the death of much local wildlife and often compromising the safety and welfare of their cats as well; people also feed wild animals out of empathy, which can lead to malnutrition or to human-wildlife conflicts because the animals become habituated to humans; or people may release animals into environments that they did not come from, which in the worst-case scenarios leads to the spread of new invasive species; and sometimes, people will buy wild-caught birds and set them free, which only creates perverse incentives to catch more wild birds.

Conservationists and occasionally even animal welfare organisations, in conclusion, may not always know how to use anthropomorphism and empathy with animals as a political instrument. It can be a great benefit to their cause, or it can backfire.

Trophy hunters, whether they consider themselves conservationists or not, rarely stand to benefit from anthropomorphism. What hunting trophies generally have in common, regardless of what the hunter values about them, is that they come from charismatic species. This poses a problem for hunters. If they killed cockroaches, mosquitoes, spiders, rats, and snakes, they probably would not have an image problem or antihunters to contend with. But they do not. They kill Bambi, Dumbo, Mufasa, Marius, and that teddy bear we loved as a child. They kill the kinds of species that are heavily anthropomorphized in popular culture and have the characteristics that elicit our empathy, so every trophy photo has the potential to go viral. And considerably more so if there is a name and fame attached to animal, hunter, or both. This was the case with Cecil.

What further intensifies the image problem of trophy hunters is the topic discussed in Chapter 3.6, namely the significance of the trophy itself and how opponents of hunting perceive it. Whatever empathy the antihunting public was already likely to have with a given charismatic animal is funnelled into anger, directed at the hunter, when that animal is transformed from a beautiful wild creature into what they commonly interpret as a glorification of the hunter and a reprehensive mockery and humiliation of everything the animal used to be.

Hunters are generally forced to argue against empathy and anthropomorphism with facts about conservation, because this is all they can appeal to. This is not to say that facts are always on the hunters' side, but only that hunters mostly have no choice but to rely on something other than empathy.

References

Ballouard, Jean-Marie., (18 Authors), et al. 2013. Schoolchildren and One of the Most Unpopular Animals: Are They Ready to Protect Snakes? *Anthrozoös* 26 (1): 93–109.
Bastian, Brock, S. Loughnan, N. Haslam, and Helena R. M. Radke. 2011. Don't Mind Meat? The Denial of Mind to Animals Used for Human Consumption. *Personality and Social Psychology Bulletin* 38 (2): 247–256.
Bekoff, Marc. 2007. (First 2000). *As Strolling with Our Kin: Speaking for and Respecting Voiceless Animals: Animals Matter – A Biologist Explains Why We Should Treat Animals with Compassion and Respect.* Boston and London: Shambhala.
Benson, Etienne S. 2016. Naming the Ethological Subject. *Science in Context* 29 (1): 107–128.
Butterfield, Max E., Sarah E. Hill, and Charles G. Lord. 2012. Mangy Mutt or Furry Friend? Anthropomorphism Promotes Animal Welfare. *Journal of Experimental Social Psychology* 48: 957–960.
Chawla, Louise. 2009. Growing up Green: Becoming an Agent of Care for the Natural World. *The Journal of Developmental Processes* 4 (1): 6–23.
Cohen, Erik, and D. Fennell. 2016. The Elimination of Marius, the Giraffe: Humanitarian act or Callous Management uecision? *Tourism Recreation Research* 2016: 1–9.
Courchamp, Franck, I. Jaric, C. Albert, Y. Meinard, William J. Ripple, and G. Chapron. 2018. The Paradoxical Extinction of the Most Charismatic Animals. *PLoS Biology* 16 (4): 1–13.
Eriksen, Lars, and M. Kennedy. 2014. Marius the Giraffe Killed at Copenhagen Zoo Despite Worldwide Protests. *The Guardian, International Edition*, February 9. https://www.theguardian.com/world/2014/feb/09/marius-giraffe-killed-copenhagen-zoo-protests (accessed 18–9–2018).
Gray, Heather M., K. Gray, and Daniel M. Wegner. 2007. Dimensions of Mind Perception. *Science* 315: 619.

Guthrie, Stewart E. 1997. Anthropomorphism: A Definition and a Theory. In Robert W. Mitchell, Nicholas S. Thompson, and H. Lyn Miles (eds.), *Anthropomorphism, Anecdotes, and Animals* (Albany: State University of New York Press): 50–58.

Herzog, Hal. 2010. *Some We Love, Some We Hate, Some We Eat: Why It's So Hard to Think Straight About Animals* (HarperCollins e-books; Reprint Edition).

Jones, Allen M. 2012. (First 1997). *A Quiet Place of Violence – Hunting and Ethics in the Missouri River Breaks*. Montana: Bangtail Press.

Joy, Melanie. 2010. *Why We Love Dogs, Eat Pigs, and Wear Cows—An Introduction to Carnism.* San Francisco: Conari Press.

Kellert, Stephen R. 1993. Values and Perceptions of Invertebrates. *Conservation Biology* 7 (4): 845–855.

Kellert, Stephen R. and Joyce K. Berry. 1980. Knowledge, Affection and Basic Attitudes toward Animals in American Society – Phase 3. *United States Department of the Interior Fish and Wildlife Service* (Washington: U.S. Government Printing Office 20402).

Leader-Williams, N. and Holly T. Dublin. 2000. Charismatic megafauna as 'flagship species'. In Abigail Entwistle and Nigel Dunstone (eds.), *Priorities for the Conservation of Mammalian Diversity – Has the Panda had its day?* (Cambridge: Cambridge University Press): 53–82.

Levin, Abigail. 2015. Zoo Animals as Specimens, Zoo Animals as Friends: The Life and Death of Marius the Giraffe. *Environmental Philosophy* 12 (1): 1–23.

Lorimer, Jamie. 2007. Nonhuman Charisma: Which Species Trigger Our Emotions and Why? *Economic and Social Research Council (ECOS)* 27 (1): 20–27.

Loveridge, Andrew J. 2018. *Lion Hearted: The Life and Death of Cecil & the Future of Africa's Iconic Cats*. New York: Regan Arts.

Lutts, Ralph H. 1990. *The Nature Fakers*. Golden: Fulcrum Publishing.

Maach, Maja L. 2014. Dyrenavne som Marius giver kun Zoo ballade. *DR Nyheder*, February 10. https://www.dr.dk/nyheder/indland/dyrenavne-som-marius-giver-kun-zoo-ballade (accessed 24-10-2018).

Macdonald, David W., P. J. Johnson, D. Burnham, A. Dickman, A. Hinks, C. Sillero-Zubiri, E. A. Macdonald. 2022. Understanding Nuanced Preferences for Carnivore Conservation: To Know them is not Always to Love them. *Global Ecology and Conservation* 37.

Myers, Gene. 2007. (First 1998). *The Significance of Children and Animals—Social Development and Our Connections to Other Species*, 2nd ed. Purdue University Press: West Lafayette.

Packer, Craig. 2015. *Lions in the Balance—Man-Eaters, Manes, and Men with Guns*. Chicago and London: The University of Chicago Press.

Parker, Ian. 2017. Killing Animals at the Zoo. *The New Yorker*, January 16. https://www.newyorker.com/magazine/2017/01/16/killing-animals-at-the-zoo (accessed 28-11-2018).

Powell, David M., and M. Ardaiolo. 2016. Survey of U.S. Zoo and Aquarium Animal Care Staff Attitudes Regarding Humane Euthanasia for Population Management. *Zoo Biology* 35: 187–200.

Scully, Matthew. 2002. *Dominion—The Power of Man, The Suffering of Animals, and the Call to Mercy* (New York: St. Martin's Press).

Shaw, George B. 1914. "Preface" in Henry Stephens Salt (ed.), 1915 *Killing for Sport—Essays by Various Writers* (London: G. Bell and Sons, Ltd.): xi–xxxiv.

Tam, Kim-Pong. 2013. Dispositional Empathy with Nature. *Journal of Environmental Psychology* 35: 92–104.

Tam, Kim-Pong., Sau-Lai Lee, and Melody M. Chao. 2013. Saving Mr. Nature: Anthropomorphism Enhances Connectedness to and Protectiveness Toward Nature. *Journal of Experimental Social Psychology* 49: 514–521.

Young, A., Kathayoon A. Khalil, and J. Wharton. 2018. Empathy for Animals: A Review of the Existing Literature. *Curator the Museum Journal* 61 (2): 327–343.

Zimmerman, C., Y. Chen, D. Hardt, and R. Vatrapu. 2014. Marius, the Giraffe: A Comparative Informatics Case Study of Linguistic Features of the Social Media Discourse. *5th ACM International Conference on Collaboration Across Boundaries: Culture, Distance & Technology Kyoto, Japan, August 20–22, 2014*: 131–140.

Chapter 8
The Morality of Trophy Hunting

The moral arguments for and against hunting have always interested philosophers and naturalists and some of the most common arguments and relevant thinkers were detailed in Chapter 2. Additional pertinent and frequently cited literature on the subject includes Loftin 1984; Rolston III 1988; Causey 1989 and 1996; Vitali 1990; King 1991; List 1997; Luke 1997; Cahoone 2009 and Dickson 2009.

While recognizing that it is often challenging and sometimes impossible to discuss trophy hunting without involving sport hunting in general, it is our hope to avoid rehashing the basic debates about animal philosophy and the morality of hunting that others have already thoroughly covered. Thus, we will try not to engage any further with fundamental questions about whether humans, animals, species, or ecosystems have rights, interests, inherent value, or instrumental value; about human or animal sentience, conscience, consciousness, suffering, or personhood; about how animals experience being hunted and killed; about hunting as human nature or culture; about whether hunting can be categorized as a sport; about the ethics of eating meat versus being a vegetarian or vegan; or about hunting compared to factory farming as a source of meat. While all of these are very important questions concerning hunting, they are not unique to trophy hunting, and all have received much attention already. Instead, we want to focus on the aspects of trophy hunting that make for unique moral dilemmas and questions that are not normally explored in the general environmental ethics or sport hunting discourses.

Some other moral issues that are quite relevant for trophy hunting have been sufficiently discussed elsewhere in this book already. These include matters related to record keeping and hunting organisations and the interpretations of hunting trophies in Chapter 3, to eco-imperialism in Chapter 5, and to fair chase and canned hunting in Chapter 6.

The vast majority of literature about trophy hunting is in the field of conservation science, usually written by ecologists and wildlife biologists. These authors sometimes footnote that although trophy hunting can be distasteful, it can often be justified by conservation benefits. The morality question is, however, usually an add-on to the main conservation subject and not front and centre. That said, others have written about the morality of trophy hunting specifically. Notable contributions include Gunn 2001; Dobson 2012; Lovelock 2015; Aejaz 2016; Macdonald 2016; Macdonald et al. 2016b; Nelson et al. 2016; Batavia et al. 2018; Hsiao 2018; Batavia et al. 2020; Ghasemi 2020 and Morris 2020. We will be referring to some of these authors in this chapter.

The most basic moral dilemma in sport hunting is the dispute between deontologists, arguing that animals have inalienable rights to life, and consequentialists, arguing that hunting can lead to less total suffering and the conservation of species and habitats. There is to a lesser extent also a dispute between pro—and antihunting consequentialists, disagreeing on whether conservation benefits outweigh the stress, pain, and suffering experienced by hunted animals.

Macdonald et al. of WildCRU concluded in *Conservation or the Moral High Ground: Siding with Bentham or Kant* that

> [t]hose in the Kantian camp on trophy hunting might reassure themselves with the knowledge that they occupy the moral high ground, but if they hold sway, there may be rather less African wildlife for them to see from that lofty position (2016b: 308).

The authors restate in this article the premise of the conflict over trophy hunting as deontologists (Kant) versus consequentialists/utilitarianists (Bentham). Macdonald et al.—being conservationists and mainly concerned with populations—argue in favour of the utilitarian view.

However, deontology vs. consequentialism is not the only important—or perhaps even the most important—conflict between paradigms of normative ethics in trophy hunting discourse. What seems to be unique about the conflict over trophy hunting compared to the conflict over sport hunting is that there is less emphasis on the death of animals per se and more emphasis on the persons who *cause* said deaths. The emphasis is on the character, motivations, behaviour, and attributes of the *hunter* rather than the act and consequences of *hunting*. What follows is that in order to understand the moral conflicts underlying the trophy hunting discourse it is no longer enough to understand the obvious incompatibilities between deontology and consequentialism. We must also be open to the possibility of incompatibilities between *virtue ethics* and consequentialism. This chapter largely picks up from Chapter 3.7 on how trophy photos can fuel online hatred towards hunters.

8.1 The Vices of Trophy Hunters

As virtue ethics have not traditionally played a major role in debates about hunting, unlike deontology and consequentialism, they were not covered in Chapter 2. A brief introduction is therefore in order.

Whereas deontology emphasizes duties and rights, and consequentialism emphasizes consequences of actions, virtue ethics emphasize moral *character* and *character traits*. Virtue ethics are concerned with questions of what constitutes a good life and how to be a good person. The assumption, put forth first by Aristotle, is that if one is a virtuous person, moral actions will tend to follow without needing the specific guidelines for moral behaviour offered by deontology and consequentialism. Conversely, if one is a vicious person, immoral actions are likely to follow. Virtue ethics provide a guide for how to live a moral life and be a moral person rather than a guide for how to act morally in isolated cases (Athanassoulis 2020). That said, a simple explanation of moral actions is provided by Rosalind Hursthouse as follows: "An action is right [if and only if] it is what a virtuous agent would characteristically (i.e. acting in character) do in the circumstances" (Hursthouse 1999: 28). Virtue ethics in the context of trophy hunting thus serve to question the morality of the trophy hunter, and only through that, vis-à-vis Hursthouse, the morality of the trophy hunting activity.

Environmental Virtue Ethics (EVE) have been especially developed from 1983 onwards, when Thomas Hill Jr. published the article *Ideals of Human Excellence and Preserving Natural Environments*. Authors such as Ronald Sandler, Philip Cafaro, and Louke van Wensveen have emerged as central figures for EVE since then. EVE seeks to identify environmental virtues, which can be characterized as virtues that, if people have them, lead to caring for and appreciation of the natural environment and our place in it:

> When a character trait is justified as a virtue at least in part by environmental goods and values—be they instrumental goods (e.g., natural resources, ecosystem services, or recreational opportunities), intrinsic values (e.g., aesthetic or spiritual values), or inherent worth (i.e., the value of environmental entities in and of themselves)—it is an environmentally justified virtue (Sandler 2013: 3).

Thomas Hill Jr. introduced his article as such:

> The moral significance of preserving natural environments is not entirely an issue of rights and social utility, for a person's attitude toward nature may be importantly connected with virtues or human excellences. The question is, "What sort of person would destroy the natural environment—or even see its value solely in cost/benefit terms?" (Hill Jr. 1983: 1).

"What sort of person…" is a question that comes up often on social media in relation to trophy hunters; "what sort of person wants to kill such a beautiful animal?" or "what sort of person smiles while sitting with an animal they just murdered?" What differentiates these questions from the questions asked by deontologists and consequentialists is that they are essentially unrelated to the practical implications of trophy hunting: A dead animal is equally dead no matter the particulars of the person who killed it or their motivations. If the money paid for killing the animal contributes to conservation, then it does not matter to the habitat and species protected if the hunter was a fat, lazy, racist, misogynistic, slob looking for a wall hanging or a nature-loving, recycling, outdoors enthusiast who finds a long-lost connection with his or her primitive self through hunting. It is, in a consequentialist sense, not important at all what *kind* of person kills an animal.

Imagine a scenario in which a Namibian conservancy can choose to (a) hunt animals themselves for their meat or (b) sell the permission to hunt the animals and retain the trophies to foreign trophy hunters and still get the meat, plus the money and business opportunities. The latter scenario seems in a practical and consequentialist sense to benefit everyone. The hunters get to purchase the products and experiences that they desire, and the local community gets meat, income, and employment instead of just meat. The same number of animals will die either way.

To a consequentialist, (b) is preferable to (a).

Deontologically, (a) and (b) are equally wrong.

If people therefore object specifically to trophy hunters killing animals whilst accepting local communities killing the same animals, then the objections must be based either on ignorance of the situation, or on the perceived character and motivations of the hunter. In short, if you object to trophy hunting, but not to other forms of hunting in the same scenario, then your objection is based on virtue ethics.

A perspective that serves to illuminate the role of virtue ethics in a trophy hunting context is provided by Batavia et al. (2020) and the concept of *moral residue*. According to them, trophy hunting can be seen as a moral dilemma in which there are two incompatible moral responsibilities: (a) We ought not to infringe on vital interests of sentient animals, and (b) we ought to protect and preserve species, biodiversity, and rural livelihoods. This, to reiterate, is the basic deontologist v. consequentialist conflict of trophy hunting. If we uphold (a), then trophy hunting is out of the question, which in some cases compromises (b).[1] But if trophy hunting is part of (b), we cannot uphold (a).

What the authors then argue is that when the choice falls on the consequentialist (b), we accept that we are in breach of the deontological (a), but this does not invalidate the moral requirement (a). What it does instead is create *moral residue*, which they explain as such:

> This experience of moral residue is described as an uncomfortable emotional response, which philosophers characterize variably, using terms such as guilt, sadness, anger, shame, remorse, regret, or moral distress (). Each of these emotions may have a place in conservation ethics. However, for a generic term we suggest grief, which captures the intractability and acute sense of loss often associated with moral residue in conservation (Batavia et al. 2020: 3).

Grief, the authors conclude, is a feeling that should be felt—and which conservationists should seek to make themselves feel—when they must decide to support or suggest actions or policies that violate (a) by fulfilling (b)—or vice versa presumably, although that scenario is not part of the article. We may identify a connection here with the virtue ethics criticism of trophy hunting. Conservationists feeling grief does nothing for the animals being killed. However, when a conservationist is expected to feel grief, it translates to an expectation that the conservationist is a virtuous person, because a virtuous person ought to feel some measure of grief over having condemned a number of individual animals to death for the greater good of conservation. If conservationists displayed no grief at all when they allow (or at least condone) animals to be trophy hunted for conservation purposes, then they would seem to not be virtuous, which would in turn make trophy hunting immoral.

The trophy hunters, if we apply the analysis to them, are in a different situation than conservation scientists, as they are not being asked to weigh the moral requirements (a) and (b) against each other and choose the lesser evil. Trophy hunters occupy the

---

[1] The implicit assumption here is a scenario in which there are no viable conservation alternatives to trophy hunting.

curious position of being the reason that there is a conflict between (a) and (b) to begin with, as well as the reason that (b) can be achieved, but at the expense of (a). Trophy hunters enter the stage once the decision to go with (b) has already been made. This changes the scenario somewhat. Trophy hunters are not being asked to take responsibility for a hard decision that they had to make; trophy hunters actually want to carry out the very same killing that conservationists incurred (or should incur) moral residue—grief—for justifying. That trophy hunters *want* to kill the animals and even pay a lot of money for it, is precisely what leaves them open to criticism on the basis of virtue ethics, because the questions starting with "what kind of person…" become so obvious.

Also, what the public generally sees of these trophy hunters are the trophy photos, which show hunters grinning from ear to ear and sometimes kissing, high-fiving, or demeaning the animal in some way that seems antithetical to the display of grief. As well an absence of grief, there appears to be no remorse, regret, or any other sign of moral residue. We are then immediately, and again, brought back to the question that according to Hill Jr. is central to EVE: "What sort of person would do a thing like that?" If trophy hunters are not perceived as virtuous persons, then the morality of their activity must also be questioned by the virtue ethicist.

Jon Jensen attempted in his article *The Virtues of Hunters* (2001) to justify certain kinds of hunting with EVE. Jensen mentions four common and closely related environmental virtues that can, and ideally should, apply to hunters: humility, connectedness, gratitude, and respect. Humility is the realization that we are a part of, and participants in, the biotic community rather than being above it; connectedness refers to the visceral experience of procuring our own food; gratitude is the heightened appreciation of food that has been acquired in such a way and at the expense of an animal's life; and respect—like gratitude—is a quality that is nurtured in the ways a hunter handles his quarry before and after its death (Jensen 2001).

However, these virtues seem to be relevant mostly in the context of naturalistic and, to a lesser extent, utilitarian hunters, as covered in Chapter 4.3 . Jensen makes clear that the environmental virtues that may characterize some hunters and be developed in a person by the activity of hunting do not apply to hunters whose prime motivation is a trophy. To this point, the virtues of humility and respect, at the very least, do appear strikingly absent in many of the trophy photos that can be found on social media. Wanting to get a hunting trophy scored in a record book or collecting a trophy in order to complete a collection likewise do not seem like humble motivations, and some trophy photos show hunters engaging in behaviour that is clearly disrespectful of the animal (see Chapter 3.7).

Other examples of virtue ethics-based perspectives on trophy hunting in academic literature can be found in *An elephant (head) in the room: A critical look at trophy hunting* in which Batavia et al. (2018) make the argument that most conservationists neglect the ethical significance of the trophy itself:

> [T]he consequentialist argument misses the core concern we seek to raise, namely, that collecting bodies or body parts as "trophies" is an ethically inappropriate way to interact with individual animals, regardless of the beneficial outcomes that do or do not follow (Batavia et al. 2018: 3).

This is again a critique that we can characterize as a virtue ethics position since it problematizes the disrespectful behaviour of the hunter more so than the death of animals. Hunting trophies collected by Western hunters, the authors continue, are "ethically inappropriate" because they can be situated within a narrative in which they are "fetishized as emblems of conquest, symbolizing the prowess of the (typically male) conqueror" (2018: 3).

As discussed in Chapter 1, this article by Batavia et al. considers as trophy hunting only that type of hunting conforming to the negative stereotype of the Great White Hunter in Africa, paying a fortune to shoot lions, elephants, and rhinos. Among this clientele, the authors are probably often correct in their assessment that trophies are prestigious show-off items, sometimes scored and compared to the trophies of others.

But to the many more hunters who collect trophies from their kills but do not fit the above stereotype—i.e. many domestic deer hunters all over Europe and the US—hunting trophies might only rarely be characterized as emblems of conquest; they can be reminders of the hunt in its entirety and objectified memories of—perhaps even tributes to—the hunted animals, or they can be items of decoration that add a certain rustic feeling to a home (see Chapter 3.6). The problem with dismissing trophies as ethically inappropriate is that it ignores all other motivations for keeping a hunting trophy than the most stereotypical and unsympathetic one. If hunters keep trophies as memories of—and perhaps even as a kind of atonement for—the animals that they have killed and eaten, does this not make them more sympathetic and virtuous rather than less? By working with a stereotypical definition of trophy hunting, one runs the risk of oversimplifying interpretations of both hunting motivations and hunting trophies.

It is worth considering that Batavia et al., as well as critics of trophy hunters in general, might assume worse motivations for hunting than is realistically the case. Walmsley and O'Madagain (2020) found that people,

> expect that agents are motivated primarily by the worst of the reasons that they have for a given action and that people expect others to be motivated by worse reasons than they are motivated by themselves (Walmsley and O'Madagain 2020: 7).

This might explain some of the hostility towards trophy hunters. The "worst-motive fallacy" suggests that people are negatively biased in their moral evaluations of the motivations of others and therefore more likely to assume, for example, that hunters are motivated by bloodlust or wanting to show off than by the hunting experience more broadly.

Nelson et al. (2016) argue that the general trophy hunting discourse "tends to skirt the broader ethical question sitting at the heart of the controversy. This question, put simply, is—what constitutes a good reason to kill an animal?" (Nelson et al. 2016: 303). In underscoring the ethical importance of hunters' motivations, Nelson et al. also open a virtue ethics avenue of attack on the consequentialist conservation paradigm. The main point of the article, however, is that conservationists often dismiss objections to consequentialist ethics as emotional. This, the authors argue, can be a mistake, as a history of psychological research has shown that emotional reactions can be a part of good decision making.

8.1 The Vices of Trophy Hunters

Ghasemi (2020) examined trophy hunting from utilitarian, deontological, and virtue ethics perspectives and argued that one ought to be in opposition to trophy hunting based on all three of these moral frameworks—with some provisions that it could be justified by utilitarianism if certain rules were followed.[2] His conclusion about trophy hunting from the virtue ethics perspective was as follows:

> As reviewed here, through a virtue ethics framework, there seems not to be an association between trophy hunting and trophy hunters' virtues and human flourishing. In contrast, studies done on the motivations and intentions of trophy hunters suggest that some vices are involved in it. Therefore, trophy hunting appears to be ethically problematic under a virtue ethics framework (Ghasemi 2020: 6).

This does seem to be the consensus. The environmental virtues that were personified in hunters such as Aldo Leopold seem all but absent in the stereotype of the trophy hunter, largely because the more virtuous of motivations for hunting seem far harder than vicious motivations to apply to trophy hunters in Africa. However, the virtue ethics case against trophy hunting is strong only because most common definitions of trophy hunting, as well as the mainstream understanding, are negatively stereotypical, and do not (generally) include domestic food hunters who also keep a trophy as memento. The virtue ethics criticism of trophy hunting largely comes to down to the presumption that we can consistently, and correctly, infer motivations based on the behaviour and demographics (Chapters 4.1 and 4.2) of trophy hunters.

8.1.1 Trophy Hunter Antipathy on Twitter

What the scholarly literature says is one thing. To gauge for ourselves the extent to which opposition to trophy hunting on social media might be influenced by virtue ethics, NB studied five different trophy photo tweets by a large antihunting account on Twitter, selected to represent different species of hunted animals and both male and female hunters.[3] For each tweet, the first 50 comments were examined for commentary relating to the hunter's character and of references to the death of the animal. Tweets with just one word (e.g. "horrible") that could apply to either the person or the action were not counted as either. Comments responding to other comments were excluded from the 50, meaning that only comments responding directly to the original photo tweet were considered.

---

[2] Ghasemi noted that a utilitarian framework can justify trophy hunting if it "(a) does not jeopardize wildlife populations, alter natural selection and ecosystem functioning or diminish native biodiversity; (b) creates equitable net conservation benefits to the local community that are not available through alternative ethical practices; (c) does not contribute to social inequality and injustice; (d) considers animal welfare and minimizes sentient animals' suffering and (e) does not cause public outrage and undermine public trust in conservation (Ghasemi 2020: 6)." In Ghasemi's view, these requirements are too often not met.

[3] This user tweets news, photos, and video clips about animals and various kinds of animal abuse, but is largely focused on criticizing trophy hunting, which it accomplishes by publishing the hunters' trophy photos, as discussed in Chapter 3.7.

The hypothesis was that the number of references to the hunter's character compared with the number of references to the animal's death might give an indication of whether virtue ethics or deontology best describe the current opposition to trophy hunting on Twitter. Below are observations about each of the five analysed tweets.

Tweet #1 from October 29, 2020 features a woman in a camouflage outfit lying behind a Kirk's dik-dik (*Madoqua kirkii*), which is an antelope little larger than a big housecat, native to Eastern Africa. No weapon is featured in the photo and the hunter has a relaxed smile, lying on her elbows behind the animal. The tweet features a question about what sort of person would do this, which indicates right away that there is a virtue ethics aspect to this criticism. The implication is that someone whose motivation for killing is to collect, and who would boast about it by appearing in a trophy photo is a vicious person.

Of the first 50 comments, 39 of them involve deriding the character of the hunter. Words used to describe her are predator, evil, cunt, big brave huntress (meant ironically), piece of crap, creature (in a negative rather than biological sense), ghastly creature, poor girl, sick, vile, psychopath, monster, monstrous, disgusting, heartless, sociopath, psychopath, killer, serial killer, murderer, selfish, a being with no soul, criminally insane, demon in female form, a waste of space, horrible, ropey old skank, hideous inside and out, disgusting excuse for a human being, sadistic, and a scumbag. Eight comments indicate that she deserves being harmed or killed or that karma will find her. One commenter hopes that she will die in protracted agony, one wants to strand her on an ice floe with a polar bear, and one blesses her infected heart.

Twelve of the 50 comments concern the death of the Kirk's dik-dik or the wrongness of killing it. Words used here are cruel, senseless, senselessly murdered, horrible killing, slaughtered, and "waste of a little life whose only crime was to be born." The animal was described only in a few different ways: Little, beautiful, innocent, and a wonder of nature.

Tweet #2 from October 10, 2020 features a man, standing with a very large spotted hyena held upright in front of him. Its head is much bigger than his. He has a handgun in his belt and a slight smile.

Of the first 50 comments to this tweet, 27 remark upon the character of the hunter: A beast, cowardly, murderer, sadist, bastard, truly evil, not a real man, inept, son of a whore, cruel, callous, vile, worthless human, disgusting, low-life, monster, sick, psychopath, a savage, disgusting motherfucker, demonic thug, a demon, don't know how he sleeps at night, an animal abuser, human waste, fucking cunt, revolting, and sad. Six commenters wish death or harm on the hunter. An additional one states that he should be in jail, and one other encourages finding out the identity of the hunter and ruining his business.

Eight to ten comments could be characterized as referencing the killing of the hyena, which is described as a beautiful creature, amazing animal, innocent, and priceless. Professor Amy Dickman was briefly involved in this thread. She made a comment that spotted hyenas should probably not have trophy hunting quotas, as hyena females are larger than males and therefore more desired as trophies, but hunting females has too much of an impact on the populations. A response to this

8.1 The Vices of Trophy Hunters

comment was "Know you support TH but it needs ending, killing for fun is not on." Here "for fun" can be interpreted as meaning that killing is unacceptable if the trophy hunter is the sort of person who would have fun doing it.

Tweet #3 from October 23, 2020, features a scrubland setting and a grey-haired man with a grey beard in a bright blue camouflage-pattern outfit, standing and smiling behind the body of a large giraffe, lying stretched out on its side. There is no weapon in the photo.

This tweet had more comments of other kinds, so only 12 of the first 50 comments reference the hunter. Words used are despicable, cruel, terrible devil, son of a whore, motherfucker, heartless, needs his hard drive checked (presumably a reference to him being a paedophile), human rot, psychopath, big piece of shit, ridiculous, sick, murderer, pervert, and a piece of shit. Two commenters again ask the central virtue ethics question: "What kind of human would take pride in shooting these beautiful creatures?" and "How can a person possibly take pleasure from that?" Another says that "few things enrage me as much as trophy-killing animals," implying it is the motivation of the trophy that makes the kill rage-inducing. Only four comments reference the killing of the giraffe, described as beautiful, friendly, and an incredible animal. Eleven comments wish for or threaten with death or harm to the hunter in the photo or hunters in general.

Tweet #4 from October 19, 2020 features a woman, sitting behind a bear that has been placed with its head and arms on a log. She grabs the fur of the bear's back with her hand. Blood is running out of its nose and staining the rocks below.

Twenty-six of the first 50 comments disparage the hunter. She is described as a disturbed and disturbing individual, someone who should not reproduce, whose parents should have used a condom, who should be behind bars, who should rot in hell, a piece of shit, murderer, the face of evil, ghastly woman, murderous cow, monster, demon, psychopath, sociopath, the type of scumbag who should not exist, does not deserve to breathe oxygen, soulless, an ugly soul, rotten to the core, demented, sick fuck, serial killer, an awesome cretin, vile, so young and already so evil, a symbol of a failed society with something fundamental missing from education and family life, disrespectful, unkind, a distorted mind, and abhorrently cruel. Only four comments specifically wished harm or death on the hunter in this tweet.

One person comments on tweet #4 that when they were a child their family hunted for food and that it was not done for pleasure. The comment got four likes and no disparaging responses, indicating again that the motivation for killing, the visual representation in the form of a trophy photo, and the character, behaviour, and emotions of the hunter could be deciding factors for the amount of hatred directed at trophy hunters more so than the death of animals per se. Only four of the first 50 comments to this tweet reference the death of the bear directly. The bear was described as beautiful and a magnificent creation of mother nature (harking back to the anthropomorphism discussions of the previous chapter).

Finally, tweet #5 features an overweight and smiling man sitting behind an aardwolf (*Proteles cristata*). The aardwolf is a hyena, but it is far smaller and more timid species than the better-known spotted hyena of tweet #2. In this case, it had been lain across a rock with its head propped up on a smaller rock.

About 32 of the first 50 comments allude to this hunter's person. Descriptions used are fat, small dick, impotent cock sucker, emasculated, oxygen thief, repulsive, disgusting blob of shit, cunt, not a man, giant fat loser, dumb vicious idiot, never touched a woman in his life, fat, useless prick, bloated face, horrible piece of wirj (?), scum, fat slob-sloth, creep, does not look like a sportsman, dollop of human scum, sick fuck, unerringly ugly, useless arsehole, piece of shit, monstrous fat bastard, horrible fat bastard, fat ugly cunt, nasty evil man with no heart, his BMI and lack of human brain will cause his heart to give out soon, sick and soulless, human turd, loser, burden on earth and deplorable piece of shit. The hunter's weight is referenced in a negative way in as many as 14–16 comments, and seven comments wish him harmed or dead.

The death of the aardwolf—described as noble, harmless, gorgeous, beautiful, and innocent—is referenced in 5–7 comments.

What is clear after having done this rudimentary analysis of the comments to five trophy photos on Twitter is that most of the comments that can be categorized as relating to either the character of the hunter or the death of an animal fall into the former category. One-hundred and thirty-six of the 250 comments (54%) could be characterized as highly negative comments about the hunters in the photos, and there was not a single positive comment. The two objectively rather attractive women were consistently disparaged for their looks, and only the overweight man in the fifth tweet received more negative comments about his appearance than the two women. The clear sense of this is that misogyny is a factor in motivating such comments. Thirty-eight comments in total (15%) wished or advocated for harm or death to befall the hunters, and the word "karma" was used very frequently. About 35 of the 250 comments (14%) referenced the deaths of the animals in the photos. Overall, the comments that could be associated with a virtue ethics objection to trophy hunting outnumbered the comments that could be associated with a deontological objection by almost four to one.

There could be several reasons why this is the case.

First, it is much easier to be mad at some*one* than at some*thing*. The anger, frustration, sadness and other emotions that people feel when they see a trophy photo need an outlet, and that outlet naturally becomes the smugly smiling hunter in the photo.

Second, there is the absence of grief, as described in Batavia et al. 2020—a moral residue not resulting in the grief that a virtuous person ought to feel. The emotion that people expect to see when someone has killed an animal is grief; when what they see instead is a smile, they question the hunter's character, and assume the worst. The smiling hunters simply do not deserve whatever pleasure they take from their trophy hunting activity. Antihunters may come to prefer an outcome (i.e. a ban on trophy hunting or trophy imports), which is worse for both individual animals and populations, simply because it also means that trophy hunters do not get to do what they enjoy. We can see evidence of a mentality similar to this in scenarios where sharpshooters are employed to keep deer populations in check. Sure, there are some potential deontological/utilitarian reasons for using sharpshooters instead of hunters such as cleaner kills/less suffering, safety concerns, and noise pollution in suburban areas, but a big reason for using sharpshooters may well be that people simply do not want someone who derives pleasure from it to carry out the killing. If such a thing as killing animals must be done, then it must be done clinically, professionally, and joylessly. Preferably, it should be done by an environmentally virtuous person, but failing that, it must at least not be done by an environmentally vicious person, which is how some people view sport hunters.

Third, antihunters are predisposed to dislike trophy hunters even before hunting enters the equation. Most of the world's trophy hunting tourists are Americans, and the conflict over trophy hunting is in many cases a conflict between American hunters and American animal rights activists. This means that the tensions between them are often about much more than just hunting. As covered in Chapter 4.1, trophy hunters tend to be straight, rich, white, conservative males, and antihunters tend to be the opposite in every way. Thus, when antihunters see photos of trophy hunters, smiling over dead animals, it is possible, indeed likely, that the anger directed at trophy hunters is about more than the killing of animals; it may also be about gender, race, wealth, gun laws, climate change, abortion, and many other issues where typically conservative hunters and typically liberal antihunters are very far apart.

8.2 Weighing Character Against Consequences

Macdonald et al. (2017: 2) note that "[s]ome hold trophy hunting in such moral repugnance that any benefit to conservation is insufficient to justify it." A virtue ethics version of this sentence could be this: some hold trophy hunters in such moral repugnance that any benefit to conservation is insufficient to justify trophy hunting. In the classical deontology versus consequentialism conflict, the question that needs answering is whether or not conservation goals justify the killing of individuals. In the virtue ethics versus consequentialism conflict the question could be formulated as whether conservation goals justify the accommodation of vicious people. To nuance this question further, we might ask how vicious trophy hunters need to be before we can no longer justify trophy hunting on the grounds of conservation benefits?

Examples of shameful hunting photos—rare, but they do exist—feature trophy hunters who dress up the animals they kill in hats or sunglasses or other accessories, pretend to ride them, pose them in humiliating or supposedly funny ways, drape them with American flags, kiss their partners near them, or even pretend to be performing sexual acts on or with them. Do these examples cross the line of how much human depravity we are willing to accept from horrible people in the name of conservation, bearing in mind that the animal is no better or worse off than if it had been treated respectfully?

The two extreme positions of either virtue ethics and consequentialism would be (a) that trophy hunting is unacceptable no matter how respectfully one goes about it because a virtuous person would not engage in such hunting, and (b) that the character and motivations of the hunter and all possible behaviour are irrelevant, as all that matters is the animal's death and the conservation good that may come from it. Most people are likely either at the extreme (a) (which perhaps makes it not that extreme) or somewhere in between (a) and (b).

The problem with position (a) or close to (a) is that it once again requires us to be right about many of our assumptions about the motivations of hunters, and as discussed previously (Walmsley and O'Madagain's worst-motive fallacy), we tend to attribute the worst motivations to others. The positive motivation of wanting to experience African nature and wildlife as a whole over an extended period is, for example, rarely assigned to trophy hunters that go to Africa. Sometimes there is a lack of respect for the whole hunting process and for animals in particular, which is evident from trophy photos, and in these cases, we possibly can rightfully infer vicious motivations for hunting as well. But these photos are so few compared to the total numbers of hunted animals, and they are heavily overrepresented online because outrageous photos go viral and show up again and again while mundane ones do not. We must not forget that they are shared with the purpose of encouraging antihunting sentiments, and the worst ones serve this purpose best.

Without interviewing hunters and assuming that they can articulate what motivates them and that they are truthful about it, as with the research discussed in Chapter 4.3, we can often only guess at motivations. Of course, sometimes guessing is easy; in the case from Denmark described in Chapter 5.7 where two men had poached a large buck on someone else's land, sawed off and collected the head, and left the body behind to rot, it is hard to imagine any motivations for the poaching other than the trophy.

In other instances, motivations can be inferred based on the method of hunting:

> More recently, heli-hunting has been introduced in Texas, USA, as a means by which hunter-tourists can hunt wild hogs (pigs). Hogs are considered by farmers to be pests and, in a similar way to New Zealand, tourism operators have taken advantage of this status to offer hunting as a tourist experience. High celebrity hunter-tourists such as rock star Ted Nugent have participated in highly inflammatory heli-hunting expeditions where they claim to use machine guns to kill hundreds of pigs. Not surprisingly, this has gained the attention of the animal rights movement, with the activity becoming ethically and politically contested. Meanwhile, the tourist-hunters defend their activity on the basis of providing help to Texan ranchers, and by contributing some wild pork to charity (Lovelock 2015: 92).

This example is clearly not trophy hunting, but it is a very good illustration of a situation where it is hard to imagine motivations for the hunter other than "shooting pigs from a helicopter is really fun!" A conservation justification often presented as motivation here will of course be that it removes invasive species.

It is also easy to guess that hunting tourists in Africa and elsewhere are not motivated mainly by the meat, as they might not eat any of that at all.[4] Meat is, on the other hand, very often a motivator for domestic trophy hunting. However, if a hunter kills a deer, brings all the meat home for the freezer, and then hangs the trophy on his wall, there is still really no way for us to tell if the trophy or the meat was the prime motivation. Usually, in trying to guess the motivations of hunters, we will

---

[4] While most trophy hunters in Africa do eat wild hunted game, it may not always come from the actual animal that they themselves have killed, as that animal is not always prepared in time.

8.2 Weighing Character Against Consequences

have nothing to go on but our imagination, and this poses a problem for any criticism of trophy hunting that is based on a virtue ethics framework.

Even if we can accurately guess the motivations of hunters, and we can conclude that these motivations make someone an environmentally vicious—or just plain vicious—person, we still have to determine whether any conservation benefits justify accommodating viciousness, and to what extent. If the heli-hunting fad, for example, were to become a thing in Africa and hunters could pay to fly around and shoot wildlife with machineguns from a helicopter, many conservationists (us included) might agree that the accommodation of viciousness had gone too far.

It seems that to many antihunters, trophy hunting is an activity that is morally wrong not because it involves the killing of animals per se, but because it involves the killing of animals by people who take pleasure in it. This brings us back to the notion of moral residue proposed by Batavia et al. To understand this position better, it might be useful to compare trophy hunting to certain actions or activities that we generally only find acceptable if the person carrying them out do not take pleasure in them. The most obvious example of this is someone who is in the military or in law enforcement. Individuals in these lines of work may end up in situations where they have to kill someone. Yet we do not expect or want them to like it. On the contrary, a soldier or a police officer who admitted that they liked to kill would be someone that we would consider highly unfit for the job (although undoubtedly such individuals do exist).

Jan E. Dizard brings up gynaecologists in a similar line of arguments:

> Just as gynaecologists and their patients define the pelvic exam in completely nonerotic terms, thus permitting intimate contact that in any other context would be utterly wrong, so hunters define their killing in ways that remove any hint of bloodlust or primal violent impulse (Dizard 2003: 129).

Dizard's argument here is that hunters seek to provide motivations for their activity that are acceptable to others. A gynaecologist—and probably a male gynaecologist more so than a female one—has to be very careful about how he takes pleasure in his work and how he talks about it. If there is even a hint of something that could be interpreted as sexual pleasure, he would be considered, at the very least, highly unfit for his job. Much more likely, he would be considered a dangerous sexual predator. Hunters, in the same way, need to be very careful about how they frame the pleasure they take in hunting; if they are suspected of taking pleasure in the act of killing, then the hunting becomes unacceptable to a lot of people.

These examples may help to explain why some antihunters—in situations where they have been convinced that culling is the only viable means of population control—would rather have the relevant authorities pay to have deer killed by professional sharpshooters than have them receive payment from hunters with a desire to do it. The same reasoning can explain why antihunters object even to the trophy hunting of problem animals that have to be killed no matter what. Although letting a trophy hunter kill the animal is an income while hiring someone to do it is an expense, the trophy hunt is unacceptable because it involves the animal being killed by someone who enjoys it. Anyone who enjoys killing is, under a virtue ethics position, de facto a bad person. But as both of us have found out all too commonly, the virtue ethics lens is not just focussed on trophy hunters. Those who publicly attempt to explain the complexities of trophy hunting, and the issues that will result if it is banned without replacement, are, through a "dirty" halo effect, often also deemed bad people. Thus, the insults directed at trophy hunters have also been heaped upon both of us and many other conservationists on social media. By focussing on actors rather than actions, the virtue ethics position has had a primary role in amplifying the toxicity of public debate over trophy hunting. It is however still trophy photos that fuel the fire that the heat of media coverage and the oxygen of virtue ethics keep burning.

References

Athanassoulis, Nafsika, 2020. "Virtue Ethics," *Internet Encyclopedia of Philosophy—A Peer-Reviewed Academic Resource* (https://iep.utm.edu/virtue/) [accessed 20 October 2020].
Aejaz, Ahmad, 2016. "The Trophy Hunting Debate: A Case for Ethics," *Economic and Political Weekly* 51/26–27, 29–31.
Batavia, Chelsea, Michael Paul Nelson, Chris T. Darimont, Paul C. Paquet, William J. Ripple, and Arian D. Wallach. 2018. The Elephant (Head) in the Room: A Critical Look at Trophy Hunting. *Conservation Letters* 12 (1): 1–6.
Batavia, Chelsea, Michael Paul Nelson, and Arian D. Wallach, 2020: "The Moral Residue of Conservation," *Conservation Biology*, 1–8.
Causey, Ann S. 1989. On the Morality of Hunting. *Environmental Ethics* 11 (4): 327–343.
Causey, Ann S. 1996. "Is Hunting Ethical" in David Petersen (ed.), 1997 (first 1996) *A Hunter's Heart —Honest Essays on Blood Sport*, 80–89, (New York: Henry Holt and Company).
Cahoone, Lawrence. 2009. Hunting as a Moral Good. *Environmental Values* 18: 67–89.
Dobson, John, 2012. "Ethical Issues in Trophy Hunting" In *Controversies in Tourism* eds. Omar Moufakkir and Peter M. Burns, 86–98, (Oxfordshire and Cambridge: CAB International).
Dickson, Barney, 2009. "The Ethics of Recreational Hunting" In *Recreational Hunting, Conservation and Rural Livelihoods*eds. Barney Dickson, Jon Hutton, and William M. Adams, 59–72, (Oxford: Wiley-Blackwell).
Dizard, Jan E. 2003. *Mortal Stakes – Hunting and Hunters in Contemporary America*. Amherst and Boston: University of Massachusetts Press.
Gunn, Alastair S. 2001. Environmental Ethics and Trophy Hunting. *Ethics and the Environment* 6 (1): 68–95.
Ghasemi, Benjamin, 2020. "Trophy Hunting and Conservation: Do the Major Ethical Theories Converge in Opposition to Trophy Hunting?" *People and Nature*, 1–11.
Hsiao, Timothy, 2018. "A Moral Defense of Trophy Hunting," *Sport, Ethics and Philosophy*, 1–9.
Hursthouse, Rosalind, 1999. *On Virtue Ethics* (Oxford University Press).
Jensen, Jon. 2001. The Virtues of Hunting. *Philosophy in the Contemporary World* 8 (2): 113–124.
King, Roger J. H. 1991. Environmental Ethics and the Case for Hunting. *Environmental Ethics* 13 (1): 59–85.
List, Charles. 1997. Is Hunting a Right Thing? *Environmental Ethics* 19 (4): 405–416.
Luke, Brian. 1997. A Critical Analysis of Hunters' Ethics. *Environmental Ethics* 19 (1): 25–44.
Loftin, Robert W. 1984. The Morality of Hunting. *Environmental Ethics* 6 (3): 241–250.
Lovelock, Brent, 2015. "Troubled-Shooting: The Ethics of Helicopter-assisted Guided Trophy Hunting by Tourists for Tahr" In *Animals and Tourism —Understanding Diverse Relationships*, ed. Kevin Markwell, 91–105, (Channel View Publications).
Macdonald, David W., University of Oxford, Wildlife Conservation Research Unit (WildCRU), 2016. *Report on Lion Conservation with Particular Respect to the Issue of Trophy Hunting*.

Macdonald, David W., Paul J. Johnson, Andrew J. Loveridge, Dawn Burnham, and Amy J. Dickman. 2016b. Conservation or the Moral High Ground: Siding with Bentham or Kant. *Conservation Letters* 9 (4): 307–308.

Macdonald, David W., Andrew J. Loveridge, Amy Dickman, Paul J. Johnson, Kim S. Jacobsen, Byron du Preez, 2017. "Lions, Trophy Hunting and Beyond: knowledge Gaps and Why They Matter," *Mammal Review* 47/4: 247–253 (1–7).

Morris, S. P., 2020: "A Moral Defense of Trophy Hunting and Why It Fails," *Sport, Ethics and Philosophy*, 1–14.

Nelson, Michael Paul, Jeremy T. Bruskotter, John A. Vucetich, and Guillaume Chapron. 2016. Emotions and the Ethics of Consequence in Conservation Decisions: Lessons from Cecil the Lion. *Conservation Letters* 9 (4): 302–306.

Rolston, I.I.I., and Holmes,. 1988. *Environmental Ethics—Duties to and Values in The Natural World*. Philadelphia: Temple University Press.

Sandler, Ronald L., 2013: "Environmental Virtue Ethics" In*The International Encyclopedia of Ethics*, ed. Hugh LaFolette, 1665–1674. (Blackwell Publishing).

Vitali, Theodore. 1990. Sport Hunting: Moral or Immoral? *Environmental Ethics* 12: 69–82.

Walmsley, Joel, and Cathal O'Madagain. 2020. The Worst-Motive Fallacy: A Negativity Bias in Motive Attribution. *Psychological Science* 31 (11): 1–9.

Chapter 9
Trophy Hunting Now

It is a tumultuous time in the world right now, but then perhaps it always has been. Nonetheless, at the time of writing (2022), the world is still in the grips of the Covid-19 pandemic, the negative effects of anthropogenic climate change are, daily, becoming apparent, and Russia has invaded Ukraine, challenging NATO and threatening to destabilise eastern Europe. Conservation is rarely a governmental priority in relatively stable times, and it drops further down the list when times get tough. In terms of trophy hunting though, the major external factor over the past two years, and a factor that highlights a key problem with funding that relies on tourism, has been the Covid-19 pandemic.

9.1 Covid-19

The Covid-19 pandemic has taught us a lot about conservation. Hunting and wildlife have been, and at the point of writing continue to be, affected differently depending on the wildlife management methods of different countries. Domestic hunting in the US increased, at least during the early part of the pandemic. After more than two decades of decreasing numbers of license holders, they were up more than 12% in 2020 according to an article in *Wall Street Journal*. With few restrictions on domestic hunting—mostly on hunting outside one's home state—and plenty of people with newly-found free time, many once-hunters are getting back out there, and new hunters are trying it out for the first time (Kesling and Bachman 2020). One might also speculate that American hunters who used to travel abroad to hunt are now hunting more domestically. Covid-19 has shown people how fragile our societies are, and that being able to feed oneself with venison might not be a bad idea. An idea like this resonates with a lot of rural Americans, and indeed people elsewhere.

The situation in the UK is the opposite. They do not have the same public deer hunting culture as the US or as many European countries. In the UK, deer stalking is connected with a venison industry that sells around 80% of harvested wild deer

© The Author(s), under exclusive license to Springer Nature Singapore Pte Ltd. 2023
N. Bichel and A. Hart, *Trophy Hunting*,
https://doi.org/10.1007/978-981-19-9976-5_9

to restaurants. When restaurants closed as part of the public health measures ("lockdown") introduced at the start of the pandemic, deer herds were not getting culled in anywhere near the numbers that they were. Here is an example from an article in *The Guardian*:

> One of the country's most prominent game chefs, Mike Robinson, who runs three restaurants, including the Michelin-starred Harwood Arms in Fulham, said deer had already been through two breeding cycles since the first national lockdown in March, with numbers set to grow rapidly. "People are not shooting deer because they can't sell them to restaurants at the moment," he said. "Perhaps 20% of the normal cull is being done ... we are at the beginning of a very large hike in the deer population" (Wall 2021).

Private consumers were being encouraged to eat more venison at home, but the infrastructure to get the venison to retailers is still largely missing, and the venison industry's freezers are full. Eventually, if the situation does not reverse (and there are signs in 2022 that it is reversing) the UK could face having either to cull deer without eating them or to allow deer to overpopulate and cause serious environmental problems.

As Covid-19 caused the worldwide tourism industry to grind to a halt, African countries began experiencing one of the worst wildlife conservation crises in many years. Lindsey et al. (2020) describe Covid-19 as wildlife conservation's most serious challenge in decades in Africa. They argue that "the crisis creates a 'perfect storm' of reduced funding, lower conservation capacity, and increased threats to wildlife and ecosystems" (2020: 1302). The crisis is a result of governments cutting non-essential spending, which includes wildlife conservation, and of the tourism industry having almost entirely collapsed. The tourism industry is in many countries the largest contributor to Africa's Protected Areas, and 90% of African tour operators have experienced booking declines in excess of 75% (Lindsey et al. 2020). The economic consequences of restrictions or bans on trade in wildlife because of Covid-19 and other zoonotic diseases also pose a threat to African conservation, as noted by Roe et al. (2020a) and Roe and Lee (2021).

Writing this in 2022, it is still too early to tell what the overall impact has been, and will be, on wildlife populations and habitats in Africa, but Lindsey et al. write that the anticipation is

> increased poaching, tree cutting for timber and charcoal, artisanal mining, PA encroachment by people and livestock, and conversion of natural habitat. We expect the threat posed by the increase in consumption of bushmeat to be particularly severe (2020: 1304).

As Covid-19 has halted all kinds of tourism in Africa and not just hunting tourism, it will not be easy to tell what the consequences would be of removing only trophy hunting. All we can tell from what is happening currently is that tourism in general is indeed extremely important for many African economies and that conservation is going to need international aid in the coming years if further deterioration of Africa's vulnerable habitats and wildlife is to be avoided. Less studied and discussed regions where trophy hunting also plays a valuable role in conservation funding have similarly suffered. Relying on overseas tourists to fund conservation is a good model as long as those tourists come. What the pandemic has shown, dramatically, is just how fragile

9.2 Post-Truth

On January 6, 2021, a horde of Donald Trump supporters stormed and vandalized the United States Capitol. They had been convinced by the sitting president that the democrats had committed voter fraud and were trying to steal the presidential election. Five people died. Trump became the first president in history to be impeached twice, and the second impeachment came with less than two weeks remaining of his presidency.

At the beginning of January 2021, the UK's Brexit transition period ended, and new rules concerning the EU-UK relationship took effect. As the UK enters into a period of economic decline, rising inflation and reduced global influence, many Britons who voted to leave the EU are now realising that the "broad sunlit uplands" that they had been promised by some of their politicians were a fantasy.

What Brexit and the storming of the US Capitol have in common is that they happened—at least in part—because a lot of people were unable to filter and analyse the information being given to them. They believed in media and politicians that lied to them. Political leaders lying—and getting caught lying—has become a trivial, every-day event. Lies can now be called "alternative facts" and unwanted news can be dismissed as "fake." It has become all too easy to find confirmation of biases in the media, and social media easily connects anyone with echo chambers where opinions and beliefs—no matter how ludicrous—are confirmed and never challenged. Conspiracy theories thrive in these echo chambers, and Covid-19 has proven fertile ground for theories about how those in power—be they politicians or the big tech companies—conspire against those without power. Voting machines were rigged, asking people to wear a mask is oppression, climate change is a hoax, and Covid-19 vaccines are a cover for Bill Gates' nanotech mind control robots.

Post-truth was the Oxford Dictionary word of the year in 2016. They defined it as "relating to or denoting circumstances in which objective facts are less influential in shaping public opinion than appeals to emotion and personal belief (Oxford 2016)." This seems even more fitting today than it did in 2016.

AH has been on Twitter since 2012, and NB made a Twitter account in early 2020 explicitly to follow trophy hunting debates and conservation scientists with knowledge about trophy hunting issues. Twitter has provided both of us with personal experiences of post-truth debating patterns in trophy hunting discussions online. We find there are a few common features of these debates that fit the post-truth characteristics and are worth exploring in this chapter. Two of these features have to do with characteristics of the people that we have engaged with, and they are *anonymity* and *reach*. The third feature is a phenomenon that we classify as *moral dumbfounding*. Needless to say, the anecdotes here are based on our own personal experiences on Twitter and should be treated as such.

Most conservationists who engage in trophy hunting discussions online (including both of us), use their full names as Twitter handles and elaborate on affiliations and credentials in the profile text. Individuals are generally easy to google, and employers—often universities—are easy to look up and contact. NB had only participated for a few months before someone contacted his university to get him in trouble, and AH has on several occasions experienced similar harassment. Other conservationists from the debate, notably Amy Dickman, have experienced worse, as it seems that being a woman in conservation exacerbates the harassment. This is one possible consequence of choosing not to be anonymous. Another is that we are personally and professionally responsible for everything that we write. If we lie or misinform, it damages our professional reputation and possibly even future careers. Our stance on trophy hunting has almost assuredly closed some doors for us professionally, while perhaps opening others. All things considered, we and others still choose to use our real identities because for many of us, Twitter is a professional networking, debating, promotional, and outreach tool, and because we think our accountability, education, publications, and experiences add weight to our arguments.

Our online discussions about trophy hunting have often been with people who refer to themselves as animal rights activists. What most of these people have in common is that their accounts—unlike ours—are anonymous. There is valid reason for such anonymity, as it is evident that there are also hunters and hunting advocates who will harass animal rights activists or others who speak out against hunting, when it is easy to do so. We are aware of many such stories. However, anonymity also makes these accounts personally unaccountable. Their Twitter profiles may be held accountable for spreading misinformation, but it does not implicate whoever is behind those profiles, and new profiles can always be created if a profile gets caught in too many lies or sanctioned for whatever reason. The result is a lopsided debate in terms of misinformation: peer-reviewed literature and the researchers behind it meet with unfounded accusations of corruption and of being "shills" for the trophy hunting industry; campaign websites, blogs, and antihunting opinion pieces are presented as if they were scientific research and evidence; and stories about what takes place in countries that allow for trophy hunting are made up on the spot.

The damage of this misinformation in the discourse is compounded by the reach of the profiles that spread it. While a few individuals (including AH) on the science side of our Twitter discussions have more than 10,000 followers, most have a few hundred to a few thousand. Some of the profiles of animal rights activists that we have debated with have from a few thousand and, in the case of antihunting celebrities, up to several million. Many are in the 5000–20,000 range. These accounts with more followers reach similar accounts that also have more followers and they retweet and therefore amplify the materials of animal rights activists very one-sidedly. Our conservative guess is that emotionally fuelled antihunting propaganda and misinformation probably reaches ten to fifty times as many people on Twitter as does information about trophy hunting that is based on peer-reviewed evidence and practical professional experiences from countries that use trophy hunting. AH once experimented by posting a trophy photo involving a man standing on an elephant drinking a beer. Calling this out as unacceptable behaviour in a tweet garnered hundreds of new followers within a few hours, far more than any post pointing out problems with habitat loss and poaching.

To provide another example, an anonymous Twitter account with around 25,000 followers made a post on December 17, 2020 about a newspaper headline that said, "Kenya is having an elephant baby boom with its population more than doubling." Referencing this headline, the person tweeted: "Amazing what happens when a certain group of people can't leave America to trophy hunt." The problem with this statement is that hunting has been banned in Kenya since 1977. In spite of this, by the end of January 2021, that tweet had been liked 247,400 times and retweeted 57,600 times. Comments by the few who noticed the error drowned in a sea of angry outbursts and emojis, and the spread of misinformation continued unabated. This is a very common occurrence.

According to Haidt et al. (2000), emotional reactions can be construed as intuitive moral judgments that are only rationalized afterwards. In other words, moral judgement often comes first, and then moral reasoning follows. This is especially likely to be the case when something brings about immediate and strong emotional reactions such as anger or disgust, which is exactly what trophy hunting does in many. Haidt et al. conducted a psychological experiment in which they confronted subjects with situations that were likely to elicit strong intuitive moral judgements, but the expected judgements were designed to be extremely difficult to justify rationally.[1] The authors found that subjects would generally react as expected according to their intuitions of right and wrong. They would, for example, say that a brother and sister having sex was wrong, even if they used several kinds of protection. They would also refuse to drink orange juice if a cockroach had been dipped in it, even if they knew the cockroach was sterile. Haidt et al. also found that subjects tended to stick with their intuitive responses even after failing to justify them rationally. Subjects would often explicitly state that they could not explain why something was wrong, but they just knew that it was (Haidt et al. 2000).

---

[1] The subjects in the study were presented with five situations/tasks: The first was a control task, where the moral reasoning was relatively straightforward. The second was a cannibalism story, where the subjects were asked if it was wrong for a medical student to bring home, cook, and eat a part of a human body that would be cremated the next day, only so it would not go to waste. The third was an incest story, where the subjects were asked if it was wrong for a pair of siblings to have sex just one time and in secret if they used both birth control pills and a condom. In the fourth task, the subject was asked if they would drink a glass of orange juice if a guaranteed to be sterile cockroach had been dipped in the water. The fifth task was that the subject could choose to sign a piece of paper that sold the subject's soul post mortem for two dollars. The paper however also said that it was not legal or binding in any way, and the subject was also free to tear up the contract and keep the pieces after signing it (Haidt et al. 2000: 15–16).

Seemingly identical to what Peter Singer called a 'yuk' response" (Chapter 2.10), Haidt et al. called this phenomenon *Moral Dumbfounding*:

> Dumbfounding seems to occur when a strong intuition is left unsupported by articulable reasons. The clearest evidence of dumbfounding is that participants will often directly state that they know or believe something, but cannot find reasons to support their belief (Haidt et al. 2000: 10).

In our Twitter experience, moral dumbfounding is a common reaction online among the same group of animal rights activists discussed above. This group, tending to be very passionate about their resistance to trophy hunting but often not very knowledgeable about its many aspects, and despite calling themselves animal rights activists, often focus on distinctly utilitarian reasoning as to why trophy hunting should be stopped. This includes overly simplistic and false claims that hunting is causing the extinction of the hunted animals, or that banning hunting would prevent animal suffering. As these claims are countered with evidence that hunting can, on the contrary, prevent extinction, that poaching deaths are worse than hunting deaths, or as other claims are proven false in a similar fashion, the activists eventually double down and resort to categorical statements clearly indicative of moral dumbfounding. Such statements include "trophy hunting is just sick and not acceptable for any reason," "killing can never be conservation" or "trophy hunters are psychopaths, and I will never accept trophy hunting." Of course, this is not to say that there are no valid moral arguments against trophy hunting. It is to say—curiously—that animal rights activists in online discussions about trophy hunting are often surprised and morally dumbfounded by the revelation that simplistic utilitarian arguments alone may not be enough. And their familiarity with animal rights arguments seems often to be insufficient to guide them through the debate on those terms.

Misinformation thrives in the way media and social media work, on the intuitive dislike of trophy hunting, and especially on the hatred of trophy hunters. One example of this is the common claim that only 3% of trophy hunting benefits reach local communities. This claim is based on a report by Economists at Large from 2013, which states the following in its summary:

> Research published by the pro-hunting International Council for Game and Wildlife Conservation and the UN Food and Agriculture Organisation, supported by other authors, finds that hunting companies contribute only 3% of their revenue to communities living in hunting areas (Campbell 2013: 3).

The research that is being referred to here is, as the quote correctly states, a report published by CIC and FAO, written by Vernon R. Booth in 2010. The problem is that the real number of 3.1% referred only to Tanzania, only in the year of 2008, only to a hypothetical case and, most damningly, only to a specific budget line relating to community payments rather than the large percentages that provide conservation taxes, local salaries and so on (Booth 2010: 23).

Yet because the summary section of the report leaves out the detail about the figure being for a fictional company in Tanzania in 2008, the 3% is now being consistently used by antihunters to say that only 3% of trophy hunting revenues filter down to communities today and across all of Africa. The report itself actually goes on to detail far better percentages in other nations, yet these are ignored. The African Leadership University published a report in 2021 that details some economics of trophy hunting in Africa and shows that revenues going to local communities from hunting on community land are over half in some of the most important African countries for trophy hunting (Snyman et al. 2021: 88). Despite this, "3%" is endlessly repeated as a fact and used to discredit trophy hunting as a conservation and community development tool, even by the UK Government.

The most damaging examples of powerful misinformation campaigning about trophy hunting probably come from Eduardo Gonçalves. Gonçalves is the founder of the "Campaign to Ban Trophy Hunting," (now called Ban Trophy Hunting) backed by several British celebrities. He is also author of several books, including *KILLING GAME: The Extinction Industry* (2020a), *Trophy Hunters Exposed: Inside the big game industry* (2020b), and *TROPHY LEAKS: Trophy Hunters and Industry Secrets Revealed* (2020c). As may be evident from the titles, these books are not science; they are antihunting campaign literature, self-published by Gonçalves through his own company, Green Future Books Ltd. While Gonçalves does reference much of the relevant scientific literature in these books, he cherry-picks information that makes trophy hunting look awful and ignores anything and everything that is positive. The science that does not fit the antihunting agenda is routinely ignored. Most sources, picked because they are critical of trophy hunting, are grey literature and not subjected to peer review.

Worst of all, Gonçalves misunderstands, misinterprets, and misrepresents research and trade data, which then gets picked up by journalists who think they are reporting either on science or on competent investigative journalism, or simply do not care because the story is good. An example of this is from Gonçalves' book, *Trophy Leaks* (2020c), in which he writes the following:

An estimated 1.7 million animals were shot by trophy hunters over the past decade—the equivalent of almost 500 animals a day, or one every 3 minutes (Gonçalves 2020c: 18).

Gonçalves' source material here is a research report from the International Fund for Animal Welfare, *Killing for Trophies—An Analysis of Global Trophy Hunting Trade* (2016). This report, based on CITES trade data from 2004 to 2014, states that "[b]ased on the CITES data reviewed for this report, we estimate that 1.7 million hunting trophies were traded between nations between 2004 and 2014" (Casamitjana et al. 2016: 21).

The first potential problem here is that Gonçalves is using an estimation—the calculations behind which are not explained—by an animal welfare organisation with a clear antihunting agenda.

The second problem is that he calls 2004 to 2014 "the past decade," leading newspapers to, understandably, interpret this as 2010 to 2020.

The third problem is that Gonçalves equates the estimated 1.7 million trophies to 1.7 million animals killed, despite the research report itself specifically stating that "it cannot be assumed that the number of trophies equates to the number of animals killed, as one hunted animal may have been reported to CITES as several trophies" (Casamitjana et al. 2016: 20).

The fourth problem is that CITES only monitors international trade of parties to CITES, and only of species that are listed in the CITES Appendices. This means that all domestic trophy hunting all over the world is excluded, and all trade in the most abundant trophy species is excluded.

Our main criticism here is not that Gonçalves' claim of one animal every three minutes is too high. In fact, it is probably too low, considering the fourth problem above. Our criticism is that there are so many errors and factors unaccounted for in this methodology that the number ends up being a complete fabrication. And this is generally the Campaign to Ban Trophy Hunting method.

Yet all the same, after Gonçalves' book was published, many newspapers used the faulty statistic in their headings. The *Daily Star* wrote, "Sick trophy hunters killed an animal every 3 min over last 10 years," The *Daily Mail* wrote, "Trophy hunters killed one animal every three minutes over the last decade, damning study reveals, with creatures gunned down from helicopters,[2]" *The South African* wrote, "Shocking scale of the slaughter: Trophy hunters kill an animal 'every 3 min'," and *News18* wrote, "Hunting for Sport Continued Unabated, One Animal Killed Every 3 min in Last Decade: Report." *Geo News* took the misinformation even further and reported that it was British hunters alone who were responsible: "British trophy hunters kill animals every three minutes, claim it helps conservation in Africa."

But it does not matter that the one animal every three minutes is both a fictional and outdated number. It is still reported as "breaking news" that "exposes" the trophy hunting industry by tabloids, which is then presented as "scientific fact" by animal rights activists on social media. The tabloids know that sensationalism and outrage generate clicks. Conversely, speaking out against a ban from a position based on science, professional experience, and community engagement takes a lot of work, generates fewer clicks, and may come with both hate mail and death threats. The official "Campaign to Ban Trophy Hunting" profile on Twitter has also persisted with the "one animal killed every three minutes" slogan, despite being informed that the number is completely fictional.

A list of Gonçalves' supporters can be found on the bantrophyhunting.org website. Many are TV- and movie celebrities, some are singers, and a great many are British politicians. This is unfortunate, as these people have a large audience. Ricky Gervais is the most influential of the British trophy hunting critics. In the US, host of *The Daily Show* Trevor Noah made a segment in 2019 for The Daily Show, which painted in an extremely negative light, and ridiculed, all the valid conservation arguments. The contribution to anti-science in the trophy hunting discourse of such well-meaning but often uninformed or misinformed celebrities is immense. As unfortunate as this is, however, it is even more unfortunate that Gonçalves also has the support of a few celebrity heavyweights that one would think represent the scientific community. Most notable of these is Dr Jane Goodall. Goodall wrote the foreword to the first book in the Gonçalves trilogy, in which she states the following:

---

[2] This headline refers to the shooting of feral hogs from helicopters, mentioned in Chapter 6. This trend has nothing to do with trophy hunting, so this is just another example of how the media often conflates any kind of hunting that seems unethical with trophy hunting.

9.2 Post-Truth

> 'Killing Game: the Extinction Industry' is a meticulously researched account of the effects of this type of hunting on the populations of the target species. Rather than generating funds to benefit the conservation of wildlife—as claimed by the hunters, the outfitters and the governments selling the licenses to kill—trophy hunting is leading to the extinction of a number of species (Goodall 2020: 10).

First, this book is not—and nor are any of his other books— "meticulously researched." Gonçalves' books are self-published antihunting campaign literature, and they are littered with misinformation. Second, as has been covered in Chapter 5, some populations of some species have in limited periods declined partly because of poorly managed trophy hunting but this has largely been managed through better regulation (notably lions around Hwange National Park in Zimbabwe and leopards in southern Africa): there is no peer-reviewed evidence whatsoever that modern trophy hunting is leading to the extinction of any trophy hunted species anywhere in the world. Goodall makes several other claims in this foreword that are either unfounded, false, or omit important information. She criticizes the fact that an American paid $350,000 to shoot a black rhino, despite the high prices and the low offtakes being the reason that trophy hunting has been instrumental in rhino conservation ('t Sas-Rolfes et al. 2022); she claims that giraffe populations are plummeting in part because of hunting, despite the hunted populations of giraffes in Namibia, South Africa and Zimbabwe increasing (Trouwborst et al. 2020) and findings that "the trophy hunting industry in South Africa has led to an increase in *G. giraffa* populations, which has been partly driven by conservation efforts within the trophy hunting industry" (Dunn et al. 2021: 12); and she makes various mental health diagnoses of hunters, providing further backing to the virtue ethics hypothesis that opposition to trophy hunting is dominated by an opposition to the perceived or imagined character of trophy hunters.

It is particularly worrying that Goodall appears to believe and endorse some of the most ridiculous of Gonçalves' claims, such as the one about monkey ranching: "And now I discover that there are ranches which breed baboons and monkeys (as they also breed lions) in order to provide hunters with easy to get trophies for their collections" (Goodall 2020: 12). Her comment refers to the following claim by Gonçalves:

> Over the last 10 years, over 3700 of these monkeys [vervet monkeys] have been shot by trophy hunters. Some were bred in captivity, including on monkey 'ranches' in Zambia (Gonçalves 2020: 101).

Searching through the CITES trade database in April of 2022, we find that 4121 vervet monkey items as reported by exporting countries with the purpose designation of H for Hunting Trophy have been registered by CITES since 2008, and none before.[3] The number reported by Gonçalves as "over 3700" must be referring to this number but at his time of writing. This brings us to the first glaring instance of misinformation in the quote—and an issue that we have raised before already—which is that 3700 vervet monkey trophies *does not* equate to 3700 vervet monkeys. Any number of

---

[3] The number of vervet monkey items as reported by importing countries is 2468. We refer here to the number of items as reported by exporting countries (4121) because that seems to be what Gonçalves used to arrive at 3700 items. Gonçalves probably used this figure because it is much higher, but it is not possible to say which of these numbers is more accurate.

body parts can be registered for one animal, so there is no way of knowing how many monkeys 3700 registered items represent. 3836 of the 4121 items were exported from South Africa (CITES country code ZA), 225 from Zimbabwe (ZW), 43 from Zambia (ZM), and 15 from 13 other countries combined. The second instance of misinformation is with the claim that some of these were bred on monkey ranches in Zambia. Out of the 4121 vervet monkey items, only 14 do not have source code W for "Wild." Eight of these have code R for "Ranched animal" and six have code "F" for "Born in captivity." One item (as registered by importer, not exporter) has code C for "Bred in captivity." The CITES guide to the application of source codes makes it clear that "ranched" does not involve any breeding of animals, so only one item with source code C for "Bred in captivity," shipped from Zambia to the US in 2014, fits Gonçalves' claim. Gonçalves' and Goodall's monkey ranching case does therefore not even rest on "some" monkeys that were bred in captivity and hunted—it rests on *one* monkey. And this is if we accept this CITES trade data outlier as fact. With strange outliers and CITES trade data being recorded by people—and people make mistakes—it is most likely that this one monkey is nothing but a clerical error. If not, then monkey breeding for trophy hunters was certainly a short-lived and small-scale industry.

Another example of Gonçalves uncritically portraying strange CITES database irregularities (and probable errors) as if they were real trends is his comment about chimpanzee- and gorilla hunting:

> In one year alone, no fewer than 79 chimpanzee hunting trophies came from Cameroon into the US. Gorilla trophies have come from the Central African Republic, from where some chimpanzee trophies were also acquired. Cameroon and Tanzania are among other countries of origin listed in CITES records (Gonçalves 2020: 101).

Clearly Gonçalves wants us to believe that chimpanzee and gorilla trophy hunting is a real thing, so let us examine this claim.

For starters, there is no legal hunting of chimpanzees or gorillas anywhere, so any such killing for trophies would be poaching and not hunting. This alone should be enough to dispel this nonsense, but we will look at the details of the claim anyway.

There was a total of 84 chimpanzee items traded between 1982 and 2014 (and none before or since) with purpose code H for Hunting Trophy. 79 of these items were shipped from Cameroon to the US in 2007. However, all of these 79 items, and three of the remaining five items, were coded with the trade term of "specimen." We reached out to Dr. Jonathan Kolby, former CITES Policy Specialist and wildlife trade law enforcement officer with the USFWS, to get his interpretation of these data. He responded that it appears most likely to be an error in the purpose code data (H for "Hunting Trophy"), since the description code of "specimens" is most often associated with scientific use. Also, the source code W for wild was recorded for these items, suggesting that they were lawfully traded specimens and not illegally traded trophies, which would have more appropriately received the source code I. These items, he argued, "probably represent a series of blood samples from a conservation project that should have been coded with purpose code S [for Scientific] rather than H [for Hunting Trophy]." Lastly, Kolby brought attention to the fact that these 79 chimpanzee specimens were reported as exported by Cameroon, but no corresponding importation records were reported by the United States, raising the possibility that either these chimpanzee items were not actually imported at all, or that they were imported without declaration. We invite you to consider whether this explanation seems more likely than the occurrence of a sudden one-year surge to poach chimpanzees for trophies in Cameroon in 2007, after which all these illegally killed apes were legally exported.

Subtracting all the "specimens" from these shipments, we are left with only two chimpanzee items, but one of these was not exported out of Africa but shipped from the US to Canada in 1982. The only item left was shipped from Tanzania to the US in 2014—so at most one chimpanzee. Taking into consideration again that chimpanzees or gorillas cannot legally be hunted anywhere, a clerical error or inaccuracy seems most likely. The total of five gorilla items that Gonçalves refers to are, for many of the same reasons, also most likely clerical errors.

Brandolini's Law states that "the amount of energy needed to refute bullshit is an order of magnitude larger than is needed to produce it." It took us several days and —as seen above—several pages to refute just a few short sentences by Goodall and Gonçalves. Correcting all the misinformation in Gonçalves' books would require us to write several books of our own.

Environmental icons like Jane Goodall letting their scientific integrity get compromised by their personal feelings about hunters and hunting is worrying, because the general public lacks awareness of the differences in expertise areas of conservation scientists. Goodall putting her stamp of approval on misinformation in the form of campaign materials like Gonçalves' books lends them a wholly undeserved air of scientific credibility. It compromises public trust in scientific research and undermines valuable conservation efforts.

Many other scientists have begun speaking out against antihunting misinformation in the past year. Hart et al. pointed out the problems with media disinformation in the trophy hunting debate in a recent publication:

> The dominant media narrative is beguilingly simple: well-loved animals are on the brink of extinction, people pay to kill them, this is horrible, and stopping it will save species. This narrative provides a perfect media concoction: peril (implied extinction risk), emotion (repugnance at killing for fun), a villain, a victim, and an action (donate and petition).[4] The clear message in most coverage is that stopping TH will have immediate positive effects on conservation. However, the reality is not so simple (Hart et al. 2020: 1).

African communities have also begun raising their voices. In July of 2020, more than 50 community leaders from Botswana, Malawi, Mozambique, Namibia, South Africa, Zambia, and Zimbabwe wrote an open letter addressed to some of the most outspoken British antihunting celebrities about how their antihunting campaigns are undermining indigenous rights to sustainably utilize African wildlife to incentivize coexistence and prevent conversion to agriculture. That letter began as such:

> Dear Ricky Gervais, Joanna Lumley, Peter Egan, Ed Sheeran, Dame Judi Dench and Piers Morgan

---

[4] See Chapters 7 and 8 for further discussions of the villains, victims, and emotions mentioned here.

> As representatives of millions of rural Africans, the majority of whom live below the poverty line, we are urgently appealing to you, as celebrities with status, to stop undermining our globally recognised conservation efforts and our basic human right to sustainably use the natural resources on which our communities' livelihoods depend (Resource Africa 2020a).

A narrated version of this letter is available on YouTube (Resource Africa 2020b).

More recently, on January 15, 2021, journalist Alex Morss published an article in *The Guardian* with the title "Celebrity power undermining global conservation efforts, scientists warn." The article explains how standing up for science leads to abuse and threats from animal rights activists and how the public is being misinformed and lied to about trophy hunting by celebrities (Morss 2021). Later that day, George Monbiot tweeted a link to Morss' article and added that

> I find this article by @morss_alex convincing. It persuades me that I got this issue wrong. I find trophy hunting repugnant, but the campaign against it appears to have damaged conservation efforts. Important insights (Monbiot 2021).

George Monbiot had previously been vocally against trophy hunting and was listed on the (as it was then) Campaign to Ban Trophy Hunting website as a supporter, although he has since been removed. He has almost half a million followers on Twitter (but lost thousands for changing his stance on trophy hunting) and his perspectives on environmental issues are taken more seriously by more academics and laypeople than this number reflects, though he is a very divisive figure. The days following Monbiot's tweet led to a measure of support in posts and articles for the scientific and Africa-centred approach to trophy hunting that had not been seen before, yet the effect gradually dissipated.

It must be emphasized here that the deliberate spread of misinformation about trophy hunting is by no means limited to antihunters. It is, however, in our view a bigger problem when it comes from the antihunting side because of their much bigger reach on social media. Simplistic and emotional antihunting arguments are what most people are predisposed to believe, and celebrities can benefit from the virtue signalling value of speaking out against trophy hunting. Any actor or singer or other mainstream celebrity has nothing to gain by arguing against a ban on trophy hunting. An example of someone who tried is Prince William. In a 2016 interview with ITV News, Prince William stated the following:

> There is a place for commercial hunting in Africa as there is round the world. It's not everyone's cup of tea, but the arguments for regulated, properly controlled commercial hunting is that the money that goes from shooting a very old infirm animal goes back into the protection of the other species. So when one is infertile—he's at the end of his life. If somebody out there wants to pay that money and it wouldn't be me, but if somebody did, then as long as that money goes back into the protection of the species then it is a justifiable means of conserving species that are under serious threat. And that isn't just me talking, there's a lot of eminent conservationists out there who truly believe that there is a balance to be had here (Prince William 2016).

Even being as measured and carefully worded as this is, Prince William still received much public criticism for this stance on trophy hunting, just as he had been criticized for deer- and boar hunting a few years earlier. Supporting hunting in any way, shape, or form is just not a mainstream crowd pleaser. Being against trophy hunting is the intuitive default position. No one who hears about trophy hunting for the first time would think that it could possibly play a positive role in conservation. It is too unintuitive. Antihunting misinformation simply confirms what people are already predisposed to believe, and it is much harder to convince someone they were wrong than it is to convince them they were right.

For this reason, as mentioned, pro-hunting misinformation is less of a concern than antihunting misinformation, but it is still a concern. Slob hunters in echo chambers can, and do, convince each other that any kind of hunting is ethical and serves a conservation purpose. But the bigger problem is with the large and organised pro-hunting groups. According to Facebook's monthly report on coordinated inauthentic behaviour from October of 2020, Facebook banned the company Rally Forge from Facebook and Instagram. Rally Forge is a US marketing firm that operated on Facebook and Instagram on behalf of Turning Point USA and Inclusive Conservation Group. Inclusive Conservation Group, in turn, had received funding from SCI to create a pro-hunting presence on social media. Two-hundred and two Facebook accounts, 54 Facebook pages, and 76 Instagram accounts were banned for coordinated inauthentic behaviour in the process. This included the banning of two prominent pro-hunting groups "Let Africa Live" and "Proud American Hunter," which had been working fake profiles for pro-hunting propaganda purposes (Gleicher 2020).

Ultimately, antihunters who have the view that trophy hunting is—or trophy hunters are—so morally repugnant that the activity cannot be justified by conservation benefits have a valid point. We do not agree with it, but we completely understand and accept it. We make no claims that a consequentialist stance like ours is inherently better or more "moral" than stances based on deontology or virtue ethics. However, we cannot accept dishonesty about conservation benefits. The public has to know that if trophy hunting is banned—either entirely in the range countries or if just imports are banned—it will most likely lead to more wildlife dying in more painful ways than under management that allows for trophy hunting where it has been shown to work. The public needs to know that banning all trophy hunting will in some locations lead to habitat loss, community impoverishment, and increased human-wildlife conflict. If people know all this and still want to ban trophy hunting, then so be it. But the evidence suggests this is not the case; an opinion poll commissioned by IUCN's Sustainable Use and Livelihoods Specialist Group undertaken in 2021 provided some background and framing to the issue and showed that fewer than 40% of UK people would support a ban on trophy hunting if that ban harmed conservation or local communities. A poll commissioned by the Campaign to Ban Trophy Hunting, which provided no framing and posed a leading question, found that more than 80% of UK people supported a total ban. It is at least heartening that most people, when provided with the nuances, adopt a nuanced position. Of course, the Campaign to Ban Trophy Hunting poll generated headlines and celebrity support, while the Sustainable Use and Livelihoods Specialist Group poll did not. Facts may matter, but they do not always sell (Survation 2019a, 2019b; Vinogradova 2021).

It is a widely held fear among conservationists, including us, that as a result of pervasive and deliberate misinformation campaigns, and despite our best efforts to combat them, neither the general public nor our politicians are properly and fully informed about the consequences of taking actions against trophy hunting. The fear is that the public may one day wake up to a world without lions outside of National Parks and be deeply confused, because they had been promised by virtue signalling or misguided celebrities and animal rights activists that banning trophy hunting was going to save lions. And then the blame will fall on the people who warned against this.

We have equal reason to fear that if trophy hunting is not managed properly, the world's well-meaning hunters could one day wake up to a lionless world and be perplexed, because SCI and other organisations had promised them that more hunting equates to more conservation. This seems much less likely, however, because hunters and hunting organisations do not have nearly the same influence on public and political opinion as do celebrities and animal rights campaigners.

As we seek to solve the problems of misinformation about trophy hunting, we also need to stop pretending that we can have meaningful discussions about it without reaching a shared understanding of what we are even discussing. Is trophy hunting defined by what we assume are the motivations of hunters or by the keeping of an animal part as a trophy? The first definition relies on assumptions that will not always be accurate. The second includes forms of hunting that many people have no problem with, and which are not normally considered trophy hunting. Or is trophy hunting perhaps defined by the species of animals that are hunted or where it takes place? It certainly seems like the trophy hunting of just a few charismatic species in Africa gets most of the attention from western media and celebrities while domestic trophy hunting remains largely ignored.

Perhaps now is the time to take a step back and return to fundamental discussions about hunting's role in modern societies and what makes it ethical or unethical? We may yet discover that the killing of animals is not even central to the controversy.

References

Booth, Vernon, 2010. "Contribution of Wildlife to National Economies—CIC Technical Series Publication No. 8," *Joint Publication of FAO and CIC* (http://www.cic-wildlife.org/wp-content/uploads/2012/12/Technical_series_8.pdf) (accessed 6 February 2020).

Campbell, Roderick, 2013. "The $200 million question—How much does trophy hunting really contribute to African communities?" *Economists at Large* (https://conservationaction.co.za//wp-content/uploads/2013/09/The-200-Million-Question-How-Much-Does-Trophy-Hunting-Really-Contribute-to-African-Communities.pdf) (accessed 16 February 2021).

Casamitjana, Jordi, Jennifer Tsang, Jeff Flocken, Scott Lajoie, and Shawna Moos, 2016. "Killing for Trophies—An Analysis of Global Trophy Hunting Trade" *International Fund for Animal Welfare - IFAW* (https://d1jyxxz9imt9yb.cloudfront.net/resource/36/attachment/original/Killing_For_Trophies.pdf) (accessed 14 January 2021).

Dunn, Matilda E, Kirstie Ruppert, Jenny Anne Glikman, David O'Connor, Stephanie Fennessy, Julian Fennessy, and Diogo Veríssimo. 2021. Investigating the International and Pan-African Trade in Giraffe Parts and Derivatives. *Conservation Science and Practice* 3 (5): 1–18.

References

Gleicher, Nathaniel, 8–10–2020. "Removing Coordinated inauthentic behavior," *Facebook* (https://about.fb.com/news/2020/10/removing-coordinated-inauthentic-behavior-september-report/) (accessed 10 February 2021).

Gonçalves, Eduardo, 2020a. *KILLING GAME: The Extinction Industry*, eBook version for Kindle (Green Future Books).

Gonçalves, Eduardo, 2020b. *Trophy Hunters Exposed: Inside the Big Game Industry*, eBook version for Kindle (Green Future Books).

Gonçalves, Eduardo, 2020c. *Trophy Leaks: Trophy Hunters and Industry Secrets Revealed*, eBook version for Kindle (Green Future Books).

Goodall, Jane, 2020. "Foreword" in Eduardo Gonçalves 2020a, *KILLING GAME: The extinction industry*, eBook version for Kindle (Green Future Books).

Hart, Adam, Rosie Cooney, Amy Dickman, Darragh Hare, Charles Jonga, Paul K. Johnson, Maxi Pia Louis, Rodgers Lubilo, Dilys Roe, Catherine Semcer, and Keith Somerville, 2020. "Threats Posed to Conservation by Media Misinformation," *Conservation Biology* 0/0: 1–2.

Haidt, Jonathan, Fredrik Björklund, and Scott Murphy, 2000. *Moral Dumbfounding: When Intuition Finds No Reason* (Unpublished Report).

Kesling, Ben and Rachel Bachman, 13–12–2020. "One of the Pandemic's Big Winners: Hunting," *The Wall Street Journal* (https://www.wsj.com/articles/one-of-the-pandemics-big-winners-hunting-11607871411) (accessed 26 February 2021).

Lindsey, Peter, James Allan, Peadar Brehony, Amy Dickman, Ashley Robinson, Colleen Begg, Hasita Bhammar, Lisa Blanken, Thomas Breuer, Kathleen Fitzgerald, Michael Flyman, Patience Gandiwa, Nicia Giva, Dickson Kaelo, Simon Nampindo, Nyambe Nyambe, Kurt Steiner, Andrew Parker, Dilys Roe, Paul Thomson, Morgan Trimble, Alexandre Caron, and Peter Tyrrell. 2020. Conserving Africa's Wildlife and Wildlands through the COVID-19 Crisis and Beyond. *Nature Ecology & Evolution* 4: 1300–1310.

Morss, Alex, 2021. "Celebrity power undermining global conservation efforts, scientists warn," *The Guardian* (https://www.theguardian.com/environment/2021/jan/15/celebrity-power-undermining-global-conservation-efforts-scientists-warn-trophy-hunting-dispute) (accessed 18 January 2021).

Monbiot, George, 2021. https://twitter.com/GeorgeMonbiot/status/1350073445975994374 (accessed 14 February 2021).

Oxford 2016. "*Word of the Year 2016*," Oxford University Press. (https://languages.oup.com/word-of-the-year/2016/) (accessed 26 February 2021).

Resource Africa, 14–7–2020a. "Open letter: Celebrity campaigns undermine successful conservation and human rights," *Resource Africa* (https://resourceafrica.net/open-letter-celebrity-campaigns-undermine-successful-conservation-and-human-rights/) (accessed 18 January 2021).

Resource Africa 28–9–2020b. "Let Africans decide open letter," *Resource Africa* (https://www.youtube.com/watch?v=JcZ1tQsFQx0&t=1s) (accessed 26 February 2021).

Roe, Dilys, Amy Dickman, E.J. Richard Kock, Elizabeth Rihoy Milner-Gulland, and Michael 't Sas-Rolfes. 2020. Beyond banning wildlife trade: COVID-19, conservation and development. *World Development* 136: 1–4.

Roe, Dilys, and Tien Ming Lee. 2021. Possible Negative Consequences of a Wildlife Trade Ban. *Nature Sustainability* 4: 5–6.

Survation 2019a. "17-Jul-19," *Survation* (https://www.survation.com/wp-content/uploads/2019a/07/Ban-Trophy-Hunting-Final-Tables.xlsx) (accessed 29 June 2022).

Survation 2019b:."13-Sep-19," *Survation* (https://www.survation.com/wp-content/uploads/2019b/10/Ban-Trophy-Hunting-Tables.xlsx) (accessed 29 June 2022).

Snyman, Sue, Daudi Sumba, Francis Vorhies, Elizabeth Gitari, Christina Ender, Albert Ahenkan, Aurélie Flore Koumba Pambo, and Ona Abiaga Natacha, 2021. "State of the Wildlife Economy in Africa," *ALU School of Wildlife Conservation* (https://sowc.alueducation.com/programs/research/) (accessed 22 June 2022].

't Sas-Rolfes, Michael, Richard Emslie, Keryn Adcock, & Michael Knight. 2022. Legal hunting for conservation of highly threatened species: The case of African rhinos. *Conservation Letters* 15 (3). https://doi.org/10.1111/conl.v15.3; https://doi.org/10.1111/conl.12877.

Trouwborst, Arie, Andrew J. Loveridge, and David W. Macdonald. 2020. Spotty Data: Managing International Leopard (Panthera pardus) Trophy Hunting Quotas Amidst Uncertainty. *Journal of Environmental Law* 32 (2): 253–278.

Vinogradova, Katie, 19–10–2021. "Conflicting attitudes around the trophy hunting ban," *Survation* (https://www.survation.com/conflicting-attitudes-around-the-trophy-hunting-ban/) (accessed 29 June 2022).

Wall, Tom, 10–1–2021. "Wild deer set to wreak havoc in UK woodlands as venison demand plunges," *The Guardian* (https://www.theguardian.com/environment/2021/jan/10/wild-deer-set-to-wreak-havoc-in-uk-woodlands-as-venison-demand-plunges) (accessed 14 February 2021).

William, Prince, 16–3–2016. "EXCLUSIVE: Prince William on Trophy Hunting and Poaching," *ITV News* (https://www.youtube.com/watch?v=_kWGnuyCmGo) (accessed 26 February 2021).

Correction to: Trophy Hunting and Conservation

Correction to:
Chapter 5 in: N. Bichel and A. Hart, *Trophy Hunting*,
International Political Economy Series,
https://doi.org/10.1007/978-981-19-9976-5_5

On pages 267 and 269, it was written that an area requirement law for hunting of red deer was implemented in Denmark in 2020. However, this law was, at this time, only at the proposal stage. This proposal was postponed until a later session and eventually, following an election, scrapped by the new sitting government. It has not become law.

The updated version of this chapter can be found at
https://doi.org/10.1007/978-981-19-9976-5_5

© The Author(s), under exclusive license to Springer Nature Singapore Pte Ltd. 2023
N. Bichel and A. Hart, *Trophy Hunting*,
https://doi.org/10.1007/978-981-19-9976-5_10

The manufacturer's authorised representative in the EU is Springer Nature Customer Service Centre GmbH, Europaplatz 3, 69115 Heidelberg, Germany. If you have any concerns regarding our products, please contact ProductSafety@springernature.com

Printed and bound by CPI Group (UK) Ltd, Croydon, CR0 4YY

25/03/2026

02078171-0013